A Guide to

Organic Chemistry Mechanisms

A Guided Inquiry Workbook
Conventional Curved Arrows

Peter Wepplo

Curved Arrow Press
Princeton, New Jersey

Address inquiries to P. Wepplo,
info@CurvedArrowPress.com
45 Wilton St., Princeton, NJ 08540
Printed in the United States of America
10 9 8 7 6 5 4 3 2

Acknowledgements

First, I would like to thank Donna Wepplo for encouraging me to write this book, and for her continued support to bring it to the marketplace. I would also like to thank the members of the Department of Chemistry, Medical Technology, and Physics of Monmouth University for giving me the opportunity to learn about teaching and to learn from their students.

I would like to thank Michael Dean, Evelyn Hampton, and Robert Topper of Monmouth University, and Joseph Cusick of Rutgers University for reviewing and proofreading portions of the book.

I would like to acknowledge Andre Straumanis and the "guided inquiry" community for having created a guided inquiry organic chemistry workbook. I would also like to thank Daniel Weeks for his book "Pushing Electrons" and as an unacknowledged contributor to the guided inquiry community. These provided invaluable models for this book and teaching organic chemistry.

I would like to invite anyone interested in this book, to submit suggestions or ask questions about it. This book is in its youth in teaching and learning organic chemistry.

To the Instructor

The examples contained in *A Guide to Organic Chemistry Mechanisms* are designed to supplement a standard organic chemistry textbook. They were also designed by how I perceived how our brains work and how teaching can make the best use of that model.

A quotation that guided my teaching is credited to Confucius, "I hear and I forget. I see and I remember. I do and I understand." I believed the greater the intellectual contribution a student made to their learning, the more they would understand. Therefore I sought ways to ask students to make the intellectual connections required for learning organic chemistry. Herein lies a maxim, "One cannot imagine the unimaginable." If you wish to lead a donkey with a carrot, the carrot must be within sight but still out of reach of the donkey. I think students lose sight of the connection between a mechanism they are given and the steps necessary to solve a problem from an assignment. This book connects the missing steps by changing the complexity of each mechanism. By simplifying the mechanisms, the solutions remain within intellectual reach of students. As students master the steps, they can repeat the problems in which they must add more information.

When used in a class, *A Guide to Organic Chemistry Mechanisms* can be used in large classes with transparencies of a worksheet or in smaller classes in a guided inquiry style with students working in small groups. This works well with examples from Part A. Because students can solve the problems on their own, you may also use Parts B or C if previously assigned.

To the Student

If *A Guide to Organic Chemistry Mechanisms* is being used in your class, your instructor should assign the examples that correspond with your textbook. If you are using *A Guide to Organic Chemistry Mechanisms* on your own, then you should find relevant examples by examining the index or table of contents. *A Guide to Organic Chemistry Mechanisms* will make it easy to learn the steps in a reaction mechanism. I recommend you work your way through Parts A, B, and C. Commonly, students need to repeat most mechanisms several times before they can apply that mechanism to a new problem. After you have succeeded in writing a mechanism from this book, you should test yourself with a new example from your class assignment or textbook. If you have difficulty, you may return to an example from *A Guide to Organic Chemistry Mechanisms*.

Table of Contents

Index

P-Q

R

S

T-Z

Functional Group Index
By Preparation

Functional Group Index
By Preparation

Preparation of	Starting Material	Chapter
Phenol		
amine		14.14
aryl ketone		6.11-12
ether		3.15
α-Substituted carbonyl compound		
aldehyde		9.1
ester		3.11, 9.5, 9.7-8, 9.10
ketone		3.19, 9.2-4, 9.6, 9.8, 9.15-17
nitrile		9.12

Preparation of	Starting Material	Chapter
Sulfur containing		
halide		3.3, 3.6
alcohol		10.3

[1] Requires an additional reduction reaction

[2] Step one of a two step Gabriel amine synthesis. It requires an additional hydrolysis reaction.

[3] A diol will result if methanol is replaced with water.

[4] See *Notes.*, p. 22.

Functional Group Index
By Reaction

Reaction of	Preparation of	Chapter
Alcohol		
alcohol		3.18
aldehyde		12.3, 12.7
alkene		4.13-14, 4.16
carboxylic acid		12.2
ester		8.20, 10.8
ether		3.2[5],3.14[5], 3.16
halide		3.5, 3.22, 3.25, 3.27, 10.2, 10.6-7
ketal		8.7
ketone		6.12[6],12.1, 12.5, 12.8
sulfonate		10.3
Aldehyde		
alcohol		8.2-3, 8.12, 9.1-2, 11.1
alkene		8.5-6
amine		11.9
carboxylic acid		6.3, 6.7, 12.2, 12.4
enone		9.1, 9.3
ester		8.12
ketone		9.4
phenol		6.11
Alkane		
halide		16.1, 16.4
Alkene		
alcohol		5.2, 5.16-19
aldehyde		12.9
alkane		15.1-3
alkene		13.1-2
alkyne		4.20-21
carboxylic acid		12.9
diol		12.10-11
epoxide		12.6
ester		5.3
halide		5.1, 5.4-9, 5.12-15, 16.2, 16.3
ketone		12.9

Reaction of	Preparation of	Chapter
Alkyne		5.20-24, 11.11
alcohol		3.17
aldehyde		5.24
alkyne		3.10, 3.17
halide		5.20-21
ketone		5.22-23
Amine		
alkene		4.17-18
amine		9.4, 9.15, 11.9
carboxylic acid amide		8.27
enamine		9.17
halide		14.14
ketone		9.4, 9.15
Aromatic compound, Substituted		
alkane		14.2, 14.9-10
amine		14.10[7], 14.11, 14.13
ester		14.6, 14.8
ether		14.12
halide		14.3, 14.5-6, 14.14
ketone		14.1, 14.5, 14.7, 14.9
nitro		14.4, 14.6, 14.11-12
phenol		14.14
Carboxylic acid amide		
amine		8.24-25, 11.6-7, 15.5
carboxylic acid		8.24-25
carboxylic acid amide		3.12
nitrile		10.1
Carboxylic acid		
alcohol		11.8
amine		15.4[8]
carboxylic acid amide		8.17-19[8]
carboxylic acid chloride		10.4-5
ester		3.4, 8.13-16[8], 8.20, 8.31, 10.8
ketone		13.6[8], 14.1[8]

Functional Group Index
By Reaction

Reaction of	Preparation of	Chapter
Diene, Diels Alder Rxn		
cyclohexene		7.1-15
halide		5.10-11
Diol		
aldehyde		12.12
ketone		6.8-9
Enone and unsat'd ester		
amine		9.15
cyclohexene		7.1-2, 7.4, 7.6-15
ester		9.16
ketone		9.15-16, 13.5
Ester and lactone		
alcohol		8.12, 8.26, 11.5
alcohol		8.21, 8.23, 11.10
aldehyde		11.10
alkene		8.22
carboxylic acid		8.21-23
carboxylic acid amide		8.27
ester		3.11, 8.12, 9.5-12
ketone		9.5-8, 9.10
Ether and epoxide		
alcohol		3.17-19
alkyne		3.17
halide		3.15
ketone		3.19
phenol		3.15
Halide		
alcohol		3.20-21, 3.23-24, 3.26, 3.28
alkene		4.1-12, 4.15, 8.5-6
alkylbenzene		14.2
alkyne		3.10, 4.20-23
amine		3.7, 3.12
enone		4.10
epoxide		3.16
ester		3.4, 3.11, 9.8-10
ether		3.2, 3.13-14
halide		3.1
ketone		9.8-10
nitrile		3.8-9, 9.12
Haloalkene or arene		
alkene		13.1-2, 13.4
alkyne		4.20-23
amine		14.11, 14.13
ether		14.12
halide		(3.6), 5.20-21
Imine, oxime, hydrazone		
alkylbenzene		11.12
Ketal		
ketone		8.8
Ketone		

Reaction of	Preparation of	Chapter
alcohol	6.10, 8.1, 8.4, 9.2, 11.2-4	
alkane		11.12
alkene		9.3
alkyne		8.4
amine		9.4
carboxylic acid		6.10, 9.13
enamine		9.17
enone		9.3
ester		6.1-2, 6.4-6
halide		9.14
ketal		8.7
ketone		9.2-4, 9.6, 9.14, 9.17
nitrile		8.10
oxime		8.9
Nitrile		
amine		8.28-29
carboxylic acid		8.28-29
carboxylic acid amide		8.29
ketone		8.30
nitrile		9.12
Organometallic		
alcohol		8.1-4, 8.26
alkane		13.3
alkene		13.1-2, 13.4
ketone		13.5-6
Phenol		
ether		3.13
α-Substituted carbonyl compound		
alcohol		3.19, 8.12, 9.1-2
aldehyde		9.1
alkane		9.8-10, 9.17
amine		9.4, 9.15
carboxylic acid		9.13
enone		9.1, 9.3
ester		3.11, 8.12, 9.5, 9.7-11, 9.16
halide		9.14
ketone		3.19, 9.2-10, 9.14-17, 13.5
nitrile		9.12
Sulfonate		
nitrile		3.8
sulfide		3.6

[5] An alcohol must first be converted to an alkoxide, see Chapter 9.7, ethoxide propagation. Also prepared with sodium metal.

[6] An alcohol must first be converted to a peroxide with hydrogen peroxide and acid.

[7] A benzamide conversion to an aniline.

[8] A carboxylic acid must first be converted to an acid chloride or anhydride.

Preface

About the Book

This book is a guide for learning organic chemistry reaction mechanisms. How should you use this book? Since I believed that I could always adapt a reaction to a new problem, I just need to know at least one example. This book is designed to teach that example. This book will help you to learn reactions in small portions making it easier for you to understand and remember.

However, before you start with the reactions, there are some things you should review. Therefore, Chapters 1 and 2 go over some fundamentals. In Chapter 1, I discuss some chemical principles that you can use to predict electron reactivity. This is followed by resonance structures. The problems are designed to be easy and illustrate patterns. This should appeal to our brains ability to find patterns that we can repeat. Therefore, you should succeed in completing all of these problems.

Resonance structures show how electrons move without forming any bonds to new atoms. In Chapter 2, you will do acid-base reactions. You will learn how to predict the equilibrium of a reaction and you will learn how to use the curved arrow. I have added an exercise to write English sentences with the curved arrows. I want to connect the logic of a curved arrow to a sentence as well as the graphical representation.

Now, start with any of the reactions or chapters and start with the Part A problems. Fill in the missing curved arrows. Everything you need to know is present in Part A. This shows the logic of a reaction. **You must write the correct curved arrows for each step before you go to Part B**. Go to that same reaction in Part B and repeat that reaction. Now you must add any missing curved arrows and structures. While I have removed some information from the reaction, the basic logic of the reaction remains. Finally, complete Part C. This is similar to problems in an organic chemistry textbook. I have retained the same number of reaction arrows as the original mechanism as additional hints. In addition, the reagents are written with their complete formulae and solvents are added but noted in italic type. You should strive to write out the complete mechanism. You may wish to photocopy a reaction or write it on a blank sheet of paper so you can go back and repeat an exercise. Once you have succeeded in writing a mechanism, you can move on to problems from your text.

The reactions are grouped by reaction type because it is easier to learn a series of related reactions. If your book is not organized in that manner, then select similar examples from the table of contents or from the index.

If you are able to write a mechanism for a problem in this book, you should be able to write a mechanism for other problems of the same mechanism. You should use this approach to solve the problems in your textbook. Clearly, you must be able to solve at least one problem if you are to solve another problem using that same mechanism.

You objective in studying should be to learn the *patterns* of the reaction mechanisms. The first problem always takes the greatest amount of time because you must learn the most to solve it. The more problems you solve, the less time it will take to solve them.

Writing Style (and Meaning)

I also wrote this book in the first person. Why write in the first person? First of all, it is a format that I am comfortable with. I like how it sounds. I also want to write in the first person because science can be gray. A scientific proof may not be as strong as we would like it to be. I think we are too frequently willing to accept *something as true simply because it is written in a book*. By writing in the first person, you will have a natural sense that an idea is my idea and other scientists may not accept it. As some of the mechanisms, topics and models contained in this text are different; I will leave it to you to determine whether they are useful or true.* I hope by doing so, you may go back to your regular textbook and measure the thoughts of that author in the same manner. †

The Curved Arrow

The 'curved arrow' is the symbol that represents how electrons move. They indicate which bonds are being made and broken AND which atoms are being joined. They are the language of organic chemistry and their use is also referred to as 'pushing electrons'.

Curved arrows are fundamental to understanding chemistry. The curved arrows are a **required element** in describing what is happening in a reaction. They describe the only electronic changes that can be made in any step in a reaction. If there is a curved arrow, that change **must** be made, and unless there is another curved arrow, **no other changes can be made**. Here is where students make a very common error. A student may write a curved arrow believed to lead to the formation of a product. However, the product they write will not be consistent with the curved arrows. *Any difference between a curved arrow and the predicted result is an error of **critical importance***.

Another error is to fail to start the curved arrow with a pair of electrons. Some students may start a curved arrow at a proton to show its movement. However, the curved arrow represents a movement of electrons, not protons. A useful device I often use and encourage students to use, especially initially, is to circle the electrons being moved.

Alternate Curved Arrow Usage

In this book, a conventional use of the curved arrows was employed. However, the conventional curved arrows are ambiguous if the curved arrow starts with a pair of electrons shared by two atoms AND indicate indicate a new bond being formed. To avoid this ambiguity, I have written another version of *A Guide to Organic Chemistry Mechanisms©*, ISBN 978-0-9779313-0-9. In this version, I have avoided the use of a curved arrow to indicate the formation of a new bond. To compensate for the formation of new bond, a "pre-bond" or dashed line has been added to indicate where new bonds will be formed. It is used with the curved arrow to show which electrons move to make or break bonds and makes pushing electrons consistent and unequivocal in meaning.‡ This book was written to make the mechanisms easier to understand. However, because 'pre-bonds' are not standard, I have offered the current version also.

A Guide to Organic Chemistry Mechanisms© Peter Wepplo, 2008

---

*   What is the difference? Models are never true, but they can be useful. If a model were true, then it wouldn't be a model for something.

†   Ideas presented in peer-reviewed journals will contain a reference to their source. Therefore, a reader understands that the idea belongs to the source. If authors accept the principles first laid out, then they may become commonly accepted. However, in the strictest use of logic, it does not become more true. It will remain only as true as the original proof or proofs.

‡   Many books use this convention, especially with Diels-Alder reactions. I have added the term 'pre-bond.

1 — Getting Ready for Reactions

About the Atom

Consider the following disassociation reaction. A large pK_a corresponds with a weak acid and a small pK_a with a strong acid.

$$X : H \longrightarrow \overset{\ominus}{:X} \qquad \overset{\oplus}{H}$$

Compound	X-H Bond Length	pK_a	Compound	X-H Bond Length	Halogen Radius	pK_a
$H-\overset{\underset{\mid}{H}}{\underset{\mid}{C}}-H$	1.10Å	50	H—Ï:	1.61Å	1.33Å	-10
C=C (H₂C=CH₂)	1.08Å	44	H—Br:	1.41Å	1.14Å	-8
HC≡C—H	1.06Å	26	H—Cl:	1.27Å	1.00Å	-7
H—F:	0.92Å	3.2	H—F:	0.92Å	0.71Å	3.2

1. If you compare the carbon compounds and HF, how does the bond length correlate with the acidity? The longer the bond, the …

2. Comparing HI, HBr, HCl, and HF, how does the bond length correlate with the acidity? The longer …

3. Why is there a difference? I suggest acidity can be predicted by a simplistic model of Coulombic attraction between the positively charged nuclei and the electron pair in a bond. Those forces can be determined by Coulomb's Law. However, without calculating the forces, can you tell which will be weaker? Since HF dissociates into H⁺ and F⁻, then which force must be weaker, that of the fluorine-electron pair or the proton–electron pair? (See *Notes*.)

    $$F = \frac{kq_1q_2}{r^2}$$

    Coulomb's Law: k is a constant, q_1 and q_2 are the charges, and r is the distance.

4. Considering the same principle, in which acid must the **proton**–electron pair distance be greater, hydrogen iodide or hydrogen fluoride? (Hint, compare bond length minus radius.)

5. Considering the same principle, in which compound must the **proton**–electron pair distance be the largest, methane (CH_4), ethene (CH_2CH_2), or ethyne (HCCH)?

6. How must the **carbon**-electron pair distances of methane (CH_4), ethene (CH_2CH_2), and ethyne (HCCH) differ?

7. The atomic radii for C, N, O, and F are 70, 65, 60, and 50 pm and HF is the most acidic. Since the charge for a pair of electrons is always the same, then suggest a property that you could you use to predict a compound's acidity? (Hint, use Coulomb's Law.)

8. Can you explain the bond length paradox noted in questions one and two?

Guide to Drawing Resonance Structures

Resonance Structures

If you are unfamiliar with the use of the curved arrow, refer to the discussion in the *Notes* section. A good place to start pushing electrons is in drawing resonance structures. They have the elements of electron movement, but the problems will be more limited in scope.

Resonance Structures of Anions

The principle for understanding resonance structures is to understand that electrons will operate by a push-pull mode or model. If there's a net negative charge, it will be the electrons of the atom with the negative charge that will push toward the pi bond. We will start our curved arrow with those electrons. Continue to move them toward any neighboring pi bonds (push) to create and break new bonds. You should note that two curved arrows are required to avoid structures with more than eight valence electrons.

For the following examples, add curved arrows, where needed, to show how the electrons move to form the next structure. For 1-6, the first and last structures are the same. In that case, you are converting it back to the starting structure.

1.

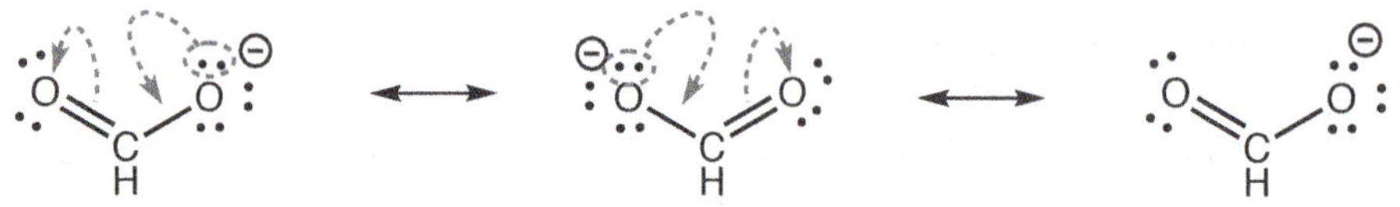

2.

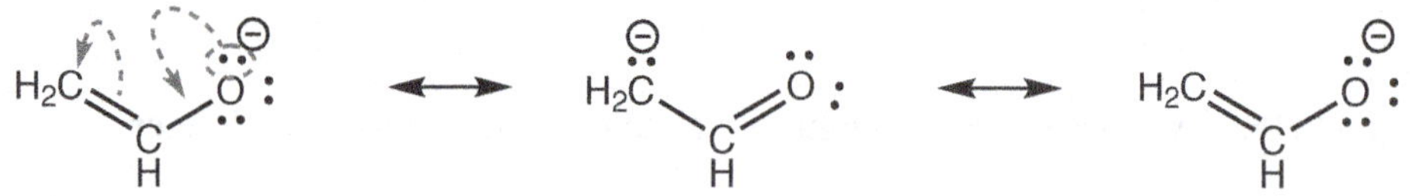

3.

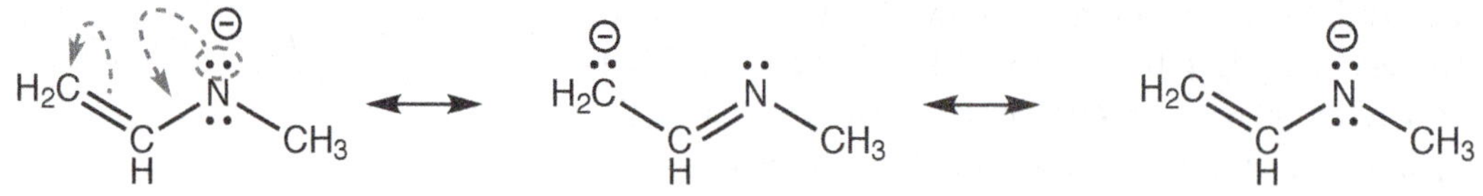

4.

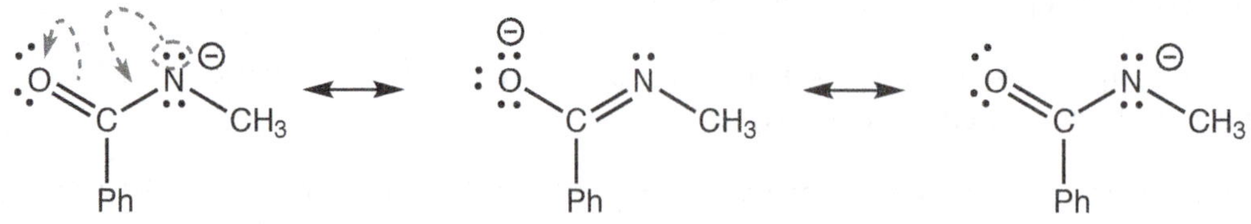

5.

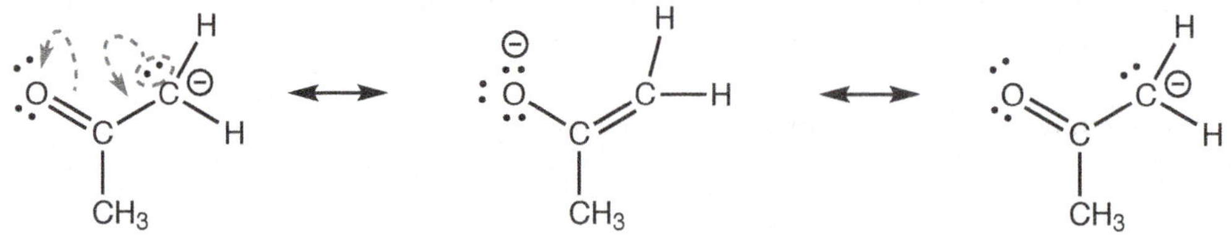

6.

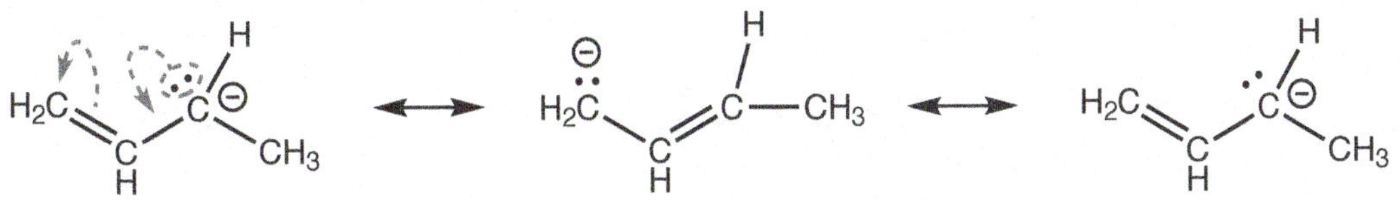

7.

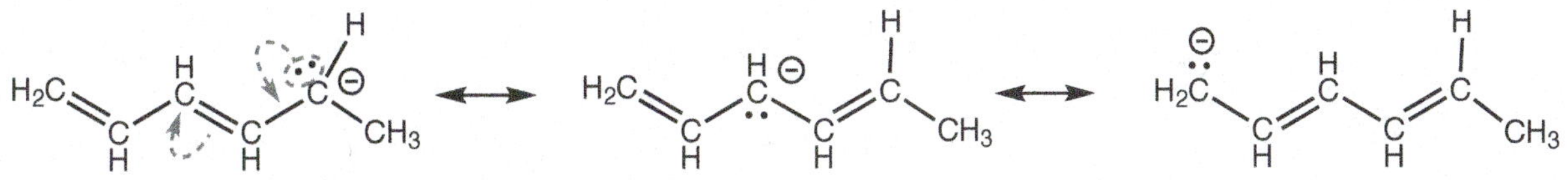

8.

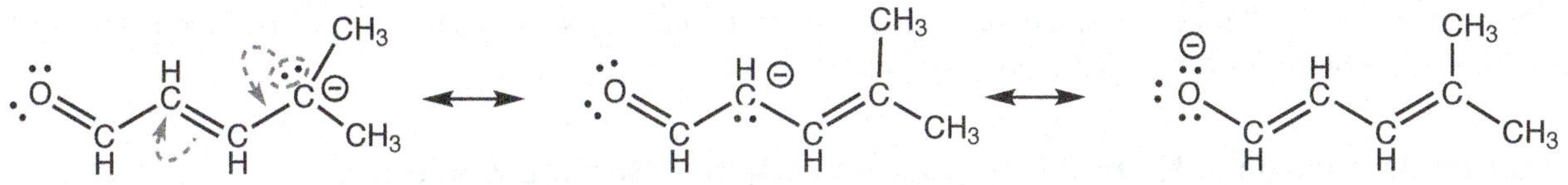

These examples show how the electrons can move, where the resulting charge will form, and how the charge can be distributed. It does not tell you on which atoms the greater charge density might exist nor upon which atom a reaction might next occur. You will note that since the original structure had a negative charge, the only charge that exists on any of the resulting resonance structures is a negative charge.

Resonance Structures of Cations

What if there is a positive charge? It is the positive charge that will attract (pull) electrons. Start a curved arrow from a neighboring pi bond or pair of non-bonded electrons and bring it toward the positive charge. You will note that only one curved arrow is necessary to create a new resonance structure for most of these examples. Because the original structure is a cation, completing its octet should not require further electron movement.

Add curved arrows to the following structures. For 9 and 10, the first and last structures are the same. In that case, you are converting it back to the starting structure.

9.

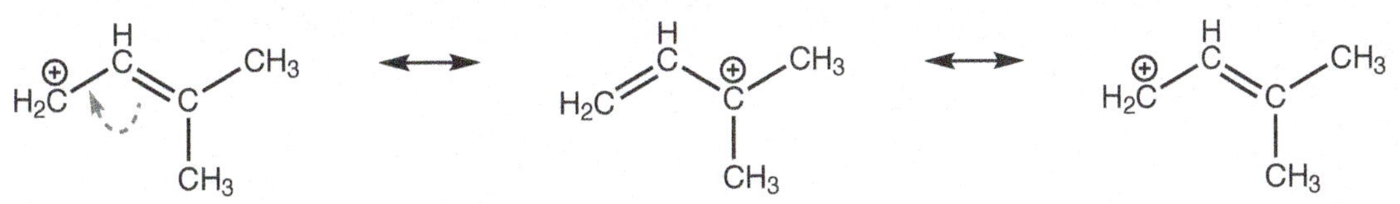

10.

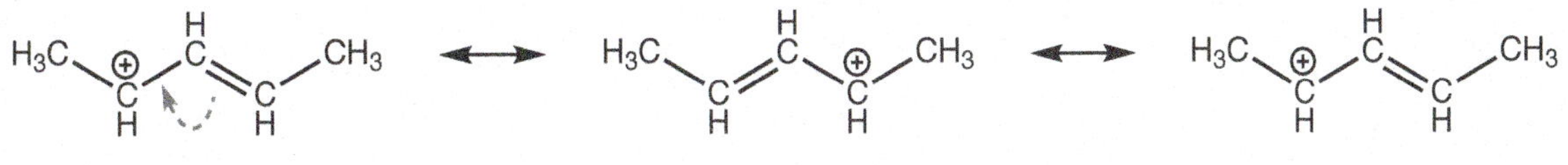

11.

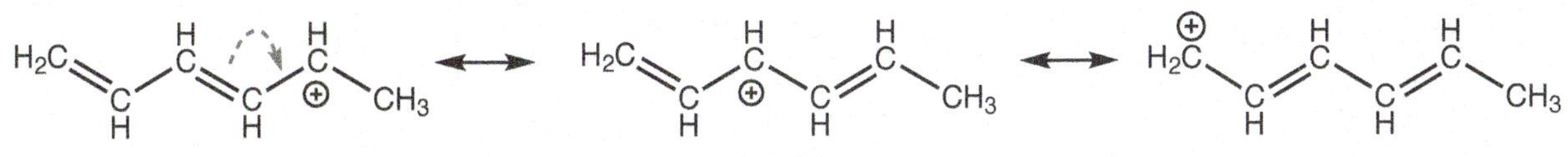

12.

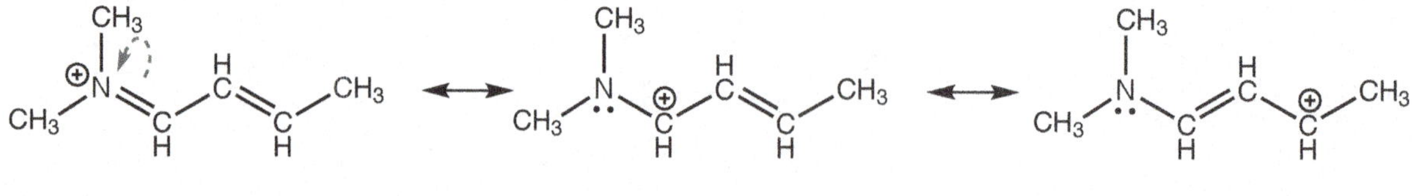

13.

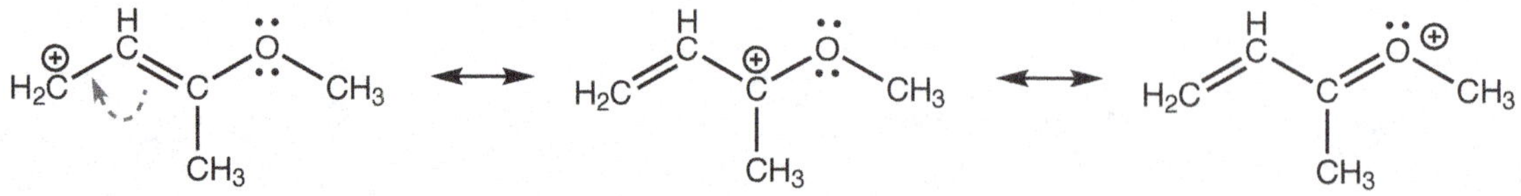

Again, these examples show how the electrons can move, where the resulting charge will form, and how the charge can be distributed. It does not tell you on which atoms the greater charge density might exist nor upon which atom a reaction might next occur. You will note that since the original structure had a positive charge, the only charges that exist on any of the resulting resonance structures are positive charges.

Resonance Structures of Neutral Compounds with Non-Bonded Electrons

What if there isn't a charge and there are adjacent non-bonded electrons? Then it will be the non-bonded electrons that will move (push) toward a neighboring pi bond. Start a curved arrow with the non-bonded electrons and direct them to the neighboring double bond.

Add curved arrows to the following structures. For 14-17, the first and last structures are the same. In that case, you are converting it back to the starting structure.

14.

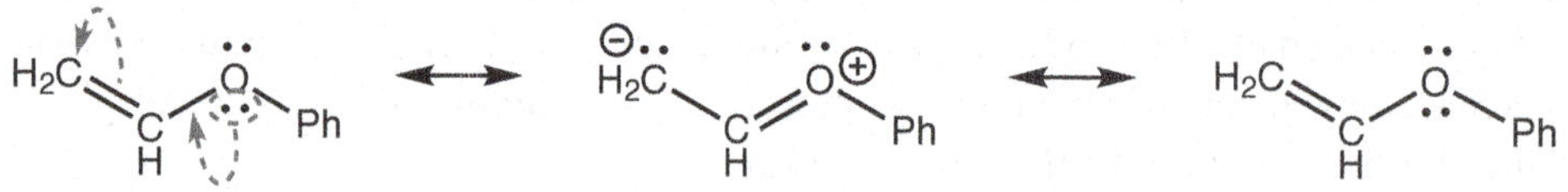

15.

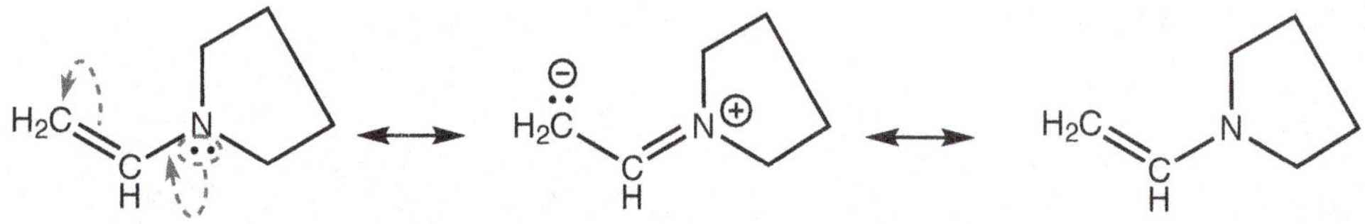

16.

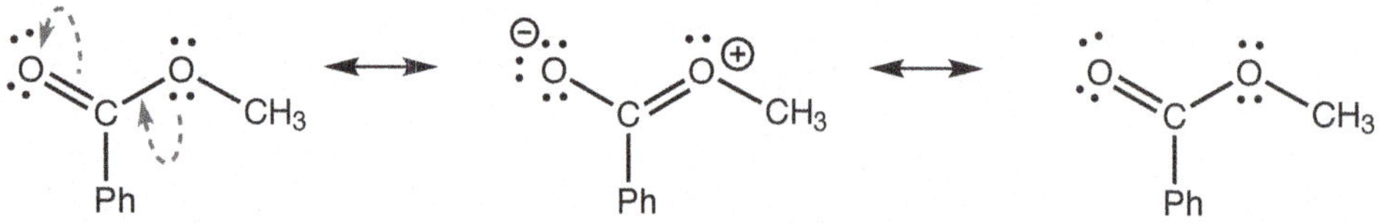

17.

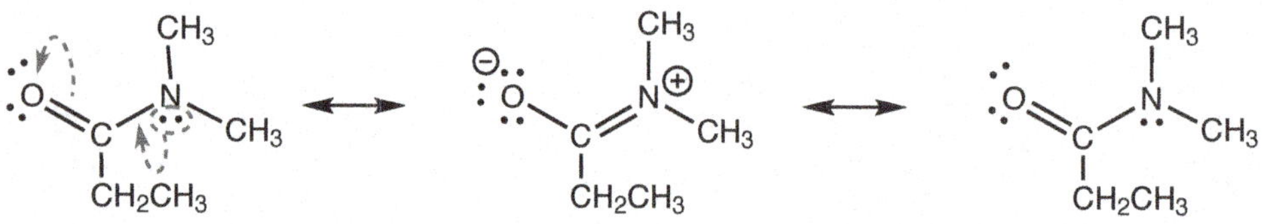

18.

19.

Again, these examples show how the electrons can move, where the resulting charges will form, and how the charges can be distributed. It does not tell you on which atoms the greater charge densities might exist nor upon which atoms a reaction might next occur. You will note that since the original structure was neutral, the net charges that exist on any of the resulting resonance structures are also neutral and only two atoms have a charge.

Resonance Structures of Neutral Compounds without Non-Bonded Electrons

What if there is no charge and there are no neighboring non-bonded electrons? Then we will push the pi electrons of a double bond toward the least substituted carbon or the most electron withdrawing atom, see Example 26. Start a curved arrow with the pi electrons of a double bond and direct them from the most to the least substituted carbon or for a C=O bond, toward the oxygen atom.

20.

21.

This is the same example as above, however the arrows are pointing in the opposite direction. If you are uncertain in which direction the electrons might move, a good strategy is to draw an arrow in the opposite direction and then to compare the results of the two possibilities. Compare the result below with the one above. Which is the more stable? If you do not recognize the lower arrangement as a lesser contributor, you may need to refer to your textbook for the rules of carbanion, carbocation, and resonance stability. Examples 23 and 24 are similar, which is more stable?

22.

23.

24. Compare Examples 23 and 24, which is preferred?

25.

26.

In these examples, we have incorporated chemical principles. The carbocations that are the most substituted are the most stable. Coinciding with this principle is that carbanions with the least substitution or a heteroatom (e.g., N or O) are the most stable.

Resonance Structures of Radicals

Radicals, compounds with unpaired electrons, are less stable than those with paired electrons. The fate of radical reactions is to form a paired-electron bond. However, sharing unpaired electrons with neighboring non-bonded electrons or pi-bonds can attain added stability. Also note the curved arrow has a single barb indicating the movement of a single electron. Two arrows are required for a pair of electrons.

27.

28.

2 -Acid-Base Chemistry

Bronsted-Lowrey Acids and Bases

Acid-base reactions are often the first intermolecular reaction you will encounter. A proton will be exchanged from the strongest acid to the strongest base.

In these examples, you must note the conjugate acid and conjugate base that result in each reaction. You should note the use and meaning of the curved arrows. If the example does not contain a curved arrow, you must supply one. The meaning of the curved arrow is important for you to understand. (See *Notes*.)

- The rule for predicting the product of an acid-base reaction is simple. A reaction will generally give the product that is the weakest base (or conjugate base). The base strength of a compound is related to the acidity of the acid, the stronger the acid, the weaker the base, or the corollary, the weaker the acid, the stronger the base. In order to compare the base strengths, the acidities of the acid and conjugate acid must be determined first.

- Look at each example and write the pK_a under each acid (on the left) and conjugate acid (on the right). Be careful that you correctly identify the acid and the corresponding pK_a. Strong acids have a small pK_a and weak acids have a large pK_a. You may need to use a table to find some values.

- For each acid (or conjugate acid), write the corresponding pK_a beneath it. Write the label "B(ase)" (or "CB") under each base (or conjugate base). Look at the pK_a of the acid or conjugate acid. The strongest acid corresponds with the weakest base. Label it, "weakest base". In example 1, the pK_a of HF is 3.2. It is placed under HF. The pK_a of acetic acid is 4.75 and 4.75 is placed under it. Because HF is the stronger acid (lowest pK_a), its conjugate base will be the weakest base. The equilibrium will shift to the right.

- Label the equilibrium of each reaction, L(eft) or R(ight). Example 1, R.

1. For this example, the acids and bases are labeled. HF is the strongest acid as it has the lower pK_a. Therefore, F⁻, its conjugate base, is the weakest base.

| **Base** | **Acid** | **Conjugate Acid** | **Conjugate Base** |

L/**R** pK_a 3.2 pK_a 4.75   weakest base

Notice the curved arrows. They describe the reaction that is taking place. We could write the following sentences to describe the curved arrows.

A bond is being made between the oxygen and hydrogen atom with the electrons from the oxygen atom.
A bond is being broken between the hydrogen and the fluorine atom with the electrons remaining attached to the fluorine atom.

2. Label acids, bases, and conjugate acids and bases.

L/**R**   base pK_a 3.2 pK_a 15.7   weakest base

Notice the curved arrows. We could write the following sentences to describe the curved arrows.

A bond is being made between the oxygen and hydrogen atom with the electrons from the oxygen atom.
A bond is being broken between the hydrogen and the fluorine atom with the electrons remaining attached to the fluorine atom.

Continue by completing the equation, adding curved arrows, pK$_a$ values, indicate the weakest base, the direction of the equilibrium, and write a sentence(s) describing any bonds being made or broken.

3.

H—O—H   +   H—F: ⇌   H—O⊕—H   +   :F:⊖

L/R   weakest base   pK$_a$ 3.2   pK$_a$ -1.7

4.

H—S—H   +   H—N—H   ⇌   H—S:⊖   +   H—N⊕—H
                    | |
                    H   H

L/**R**   7.0   weakest base   9.2

5.

H—F: +   :Cl:⊖   ⇌   :F:⊖   +   H—Cl:

L/R   3.2   -8

6.

⬡—O—H   +   ⊖:O—H   ⇌   ⬡—O:⊖   +   H—O—H

L/R   10.0   15.7

7.

        O   O
        || ||
H₃C—C—O—H   +   H₃C—N—H   ⇌   H₃C—C—O:⊖   +   H₃C—N⊕—H
                        | |
                        H   H

L/R   4.75   10.5

8.

H₃C·CH₂·O—H   +   H₃C—N—H   ⇌   H₃C·CH₂·O:⊖   +   H₃C—N⊕—H
                          | |
                          H   H

L/R   16.0   10.5

Continue by completing the equation, adding curved arrows, pK$_a$ values, indicate the weakest base, and the direction of the equilibrium.

9.

$$H_3C-\overset{\overset{\displaystyle O}{\|}}{C}\overset{\overset{\displaystyle H}{|}}{\underset{\underset{\displaystyle H}{|}}{C}}-H \quad + \quad \overset{\ominus}{:}\overset{..}{O}-\overset{\overset{\displaystyle CH_3}{|}}{\underset{\underset{\displaystyle CH_3}{|}}{C}}-CH_3 \quad \rightleftharpoons \quad H_3C-\overset{\overset{\displaystyle O}{\|}}{C}\overset{\overset{\displaystyle \ominus}{\underset{..}{C}}}{\underset{\underset{\displaystyle H}{|}}{}}-H \quad + \quad HO-\overset{\overset{\displaystyle CH_3}{|}}{\underset{\underset{\displaystyle CH_3}{|}}{C}}-CH_3$$

L/R   22   19

10.

$$H-\overset{\overset{\displaystyle H}{|}}{\underset{\underset{\displaystyle H}{|}}{C}}\overset{\ominus}{:}\overset{..}{} \quad + \quad :\overset{\overset{\displaystyle H}{|}}{\underset{\underset{\displaystyle H}{|}}{N}}-H \quad \rightleftharpoons \quad H-\overset{\overset{\displaystyle H}{|}}{\underset{\underset{\displaystyle H}{|}}{C}}-H \quad + \quad \overset{\ominus}{:}\overset{..}{\underset{\underset{\displaystyle H}{|}}{N}}-H$$

L/R   38   50

11.

$$H-\overset{..}{\underset{..}{O}}-H \quad + \quad H-\overset{\overset{\displaystyle H}{\underset{\displaystyle |\oplus}{}}}{\underset{\underset{\displaystyle H}{|}}{N}}-H \quad \rightleftharpoons \quad H-\overset{\oplus}{\underset{\underset{\displaystyle H}{|}}{O}}-H \quad + \quad H-\overset{..}{\underset{\underset{\displaystyle H}{|}}{N}}-H$$

L/R   9.2   -1.7

12.

$$H-\overset{..}{\underset{\underset{\displaystyle H}{|}}{N}}-H \quad + \quad Ph-\overset{\overset{\displaystyle H}{\underset{\displaystyle |\oplus}{}}}{\underset{\underset{\displaystyle H}{|}}{N}}-H \quad \rightleftharpoons \quad H-\overset{\overset{\displaystyle H}{\underset{\displaystyle |\oplus}{}}}{\underset{\underset{\displaystyle H}{|}}{N}}-H \quad + \quad Ph-\overset{..}{\underset{\underset{\displaystyle H}{|}}{N}}-H$$

L/R   4.6   9.2

13.

$$H_3C-\overset{\overset{\displaystyle O}{\|}}{C}-\overset{..}{\underset{..}{O}}-H \quad + \quad \overset{\ominus}{:}\overset{..}{O}-H \quad \rightleftharpoons \quad H_3C-\overset{\overset{\displaystyle O}{\|}}{C}-\overset{..}{\underset{..}{O}}:^{\ominus} \quad + \quad H-\overset{..}{\underset{..}{O}}-H$$

L/R   4.75   15.7

14.

$$H_3C-C\equiv C-H \quad + \quad :\overset{\ominus}{C}H_3 \quad \rightleftharpoons \quad H_3C-C\equiv C:^{\ominus} \quad + \quad H-CH_3$$

L/R   24   50

Continue by completing the equation, adding curved arrows, pK_a values, indicate the weakest base, and the direction of the equilibrium.

15.

L/R   24   16

16.

L/R   11.7   16

17. Sometimes we may be unsure how a reaction might proceed. Will H_2S and $(CH_3)_2NH$ react together? What will the products be if they do? In that case, draw ALL of the possible products and analyze the results. Complete this problem as before.

From the individual equilibria, can you predict the overall result?

38   <-5

7.0   10.7

18. What is the equilibrium between $HOCH_3$ and CH_3NHCH_3? First, determine the individual equilibria and then predict the overall result.

38   -2.2

15.5   10.7

Lewis Acids and Bases

The prior exercise showed a hydrogen atom accepting electrons. With Lewis acids, other atoms can also accept electrons. See your text for further discussion.

Add structures, non-bonded electrons, curved arrows, and formal charges as needed to complete the following.

19.

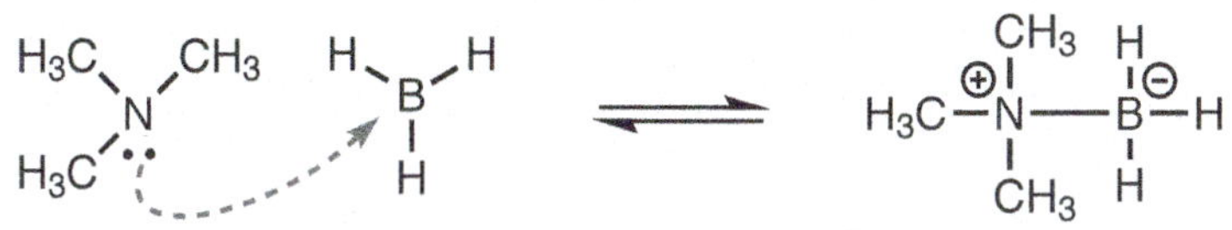

20.

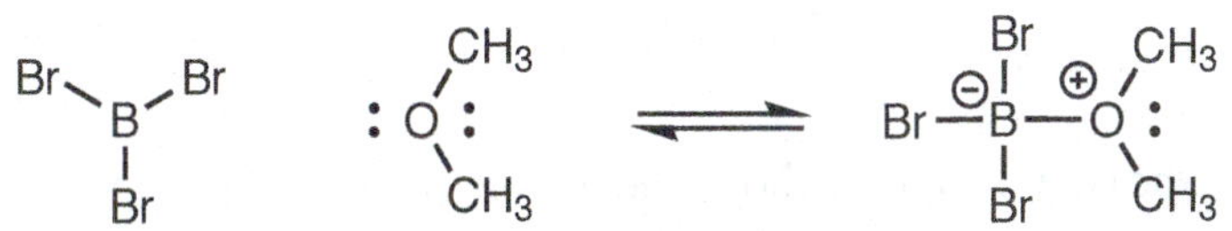

21.

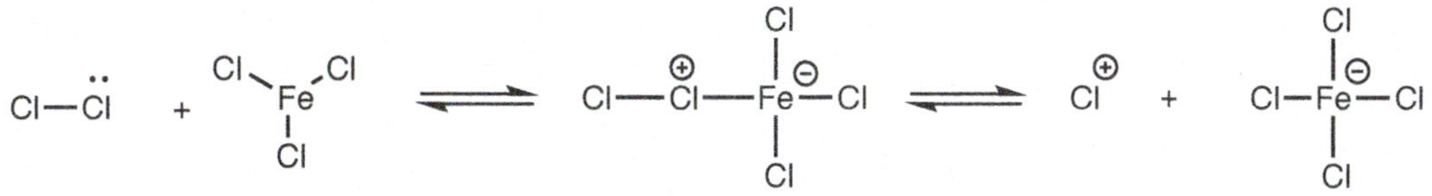

22.

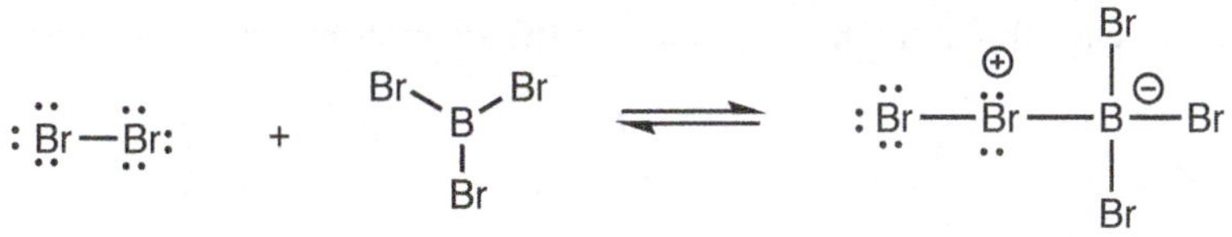

23.

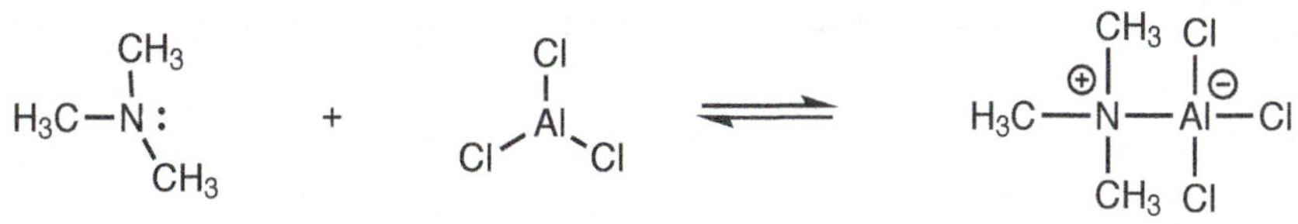

24.

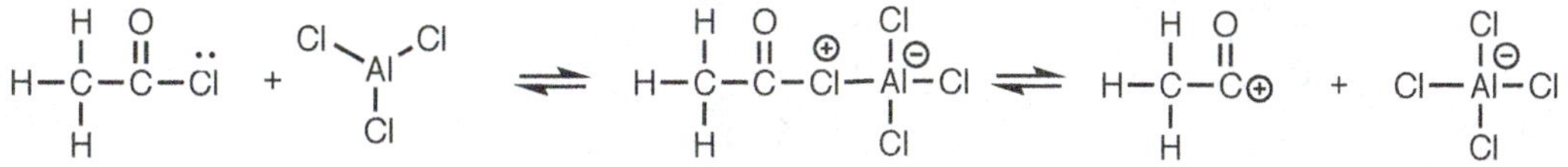

25.

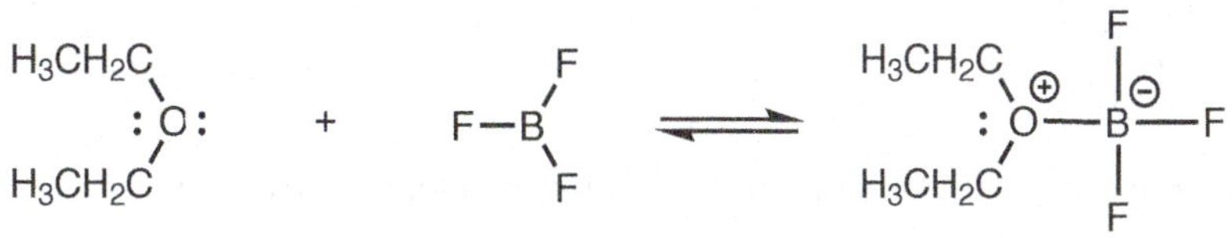

26.

3 - Substitution Reactions

S$_N$2 Substitution Reactions

Add curved arrows to the following reactions.

1. An S$_N$2 reaction of 1-chlorobutane with sodium iodide to give 1-iodobutane. (See *Notes*.)

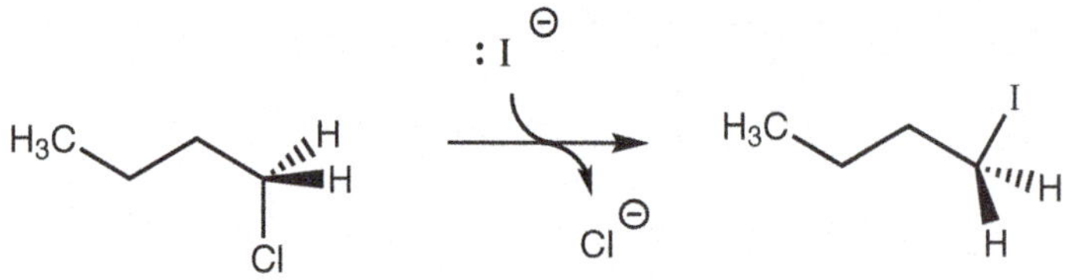

2. An S$_N$2 reaction of 1-bromobutane with ethoxide to give 1-ethoxybutane (butyl ethyl ether). (See *Notes*.)

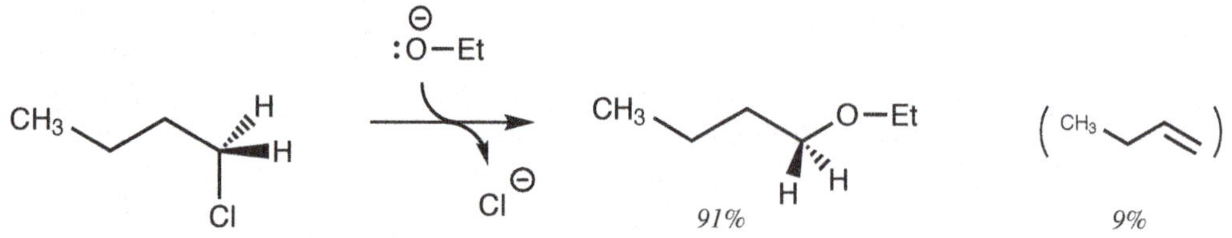

91%   9%

3. An S$_N$2 reaction of *(R)*-2-bromobutane with thiocyanate to give *(S)*-2-thiocyanatobutane. (See *Notes*.)

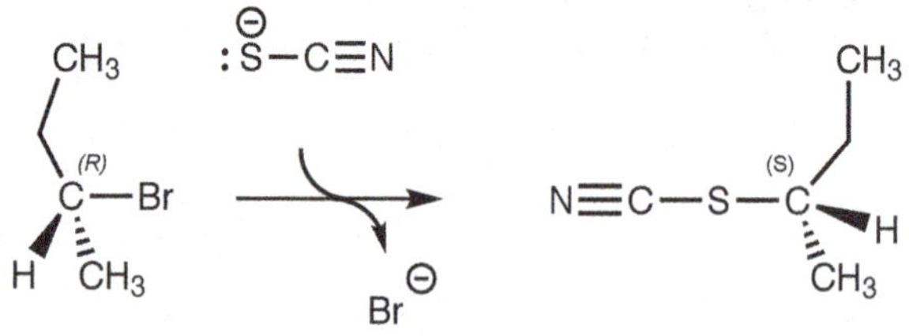

4. An S$_N$2 reaction of *(S)*-2-bromobutane with acetate to give *(R)-sec*-butyl acetate. (See *Notes*.)

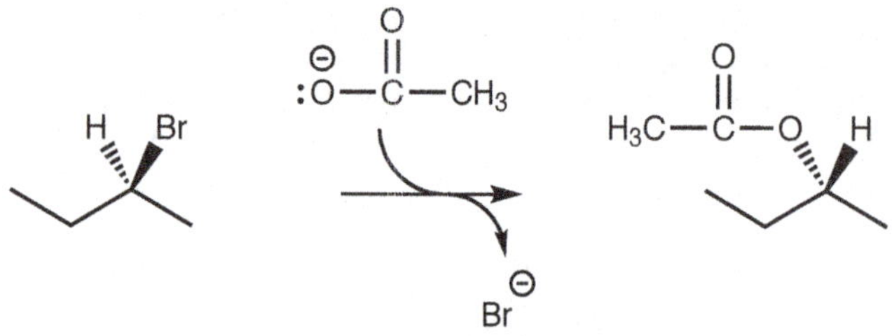

5. An S$_N$2 reaction of 1-butanol with hydrogen bromide to give 1-bromobutane. (See *Notes*.)

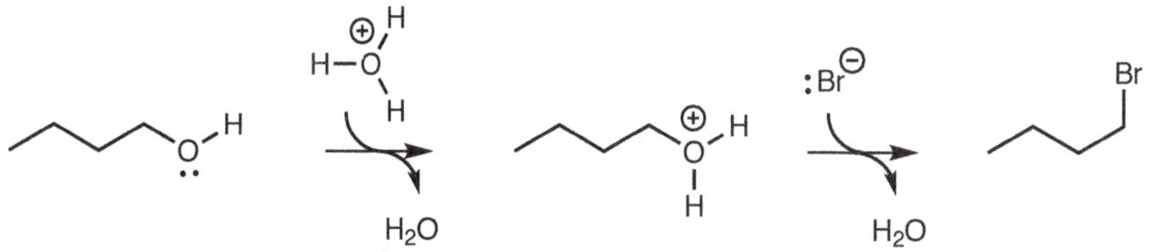

6. An S$_N$2 reaction of a triflate with sodium methanethiolate to give a thioether with inversion. (See *Notes.*)

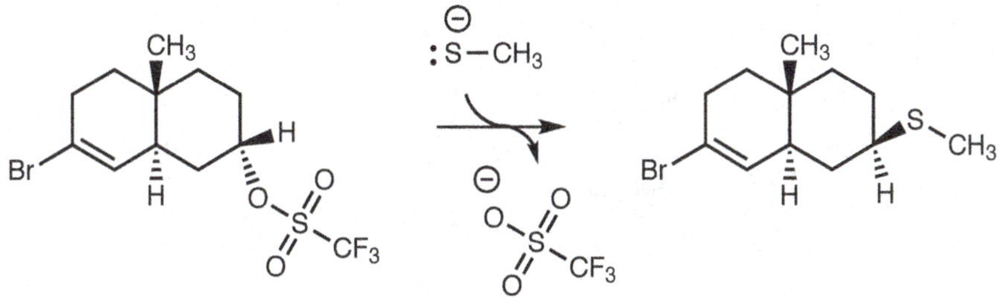

7. An S$_N$2 reaction of *cis*-1-chloro-4-methylcyclohexane with azide to give *trans*-1-azido-4-methylcyclohexane.

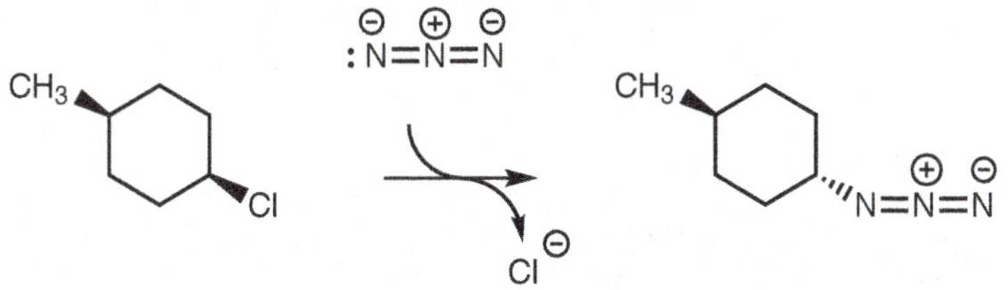

8. An S$_N$2 reaction of a ditosylate with one equivalent of cyanide to give a mono-nitrile. (See *Notes.*)

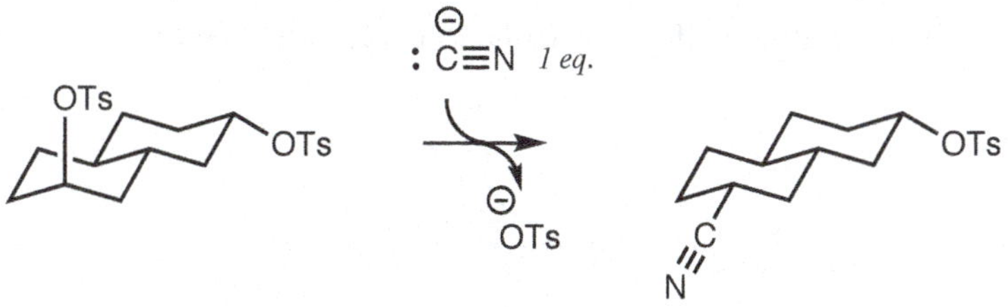

9. An S$_N$2 reaction of 1-bromo-3-chloropropane with cyanide to give 4-chlorobutanenitrile. (See Notes.)

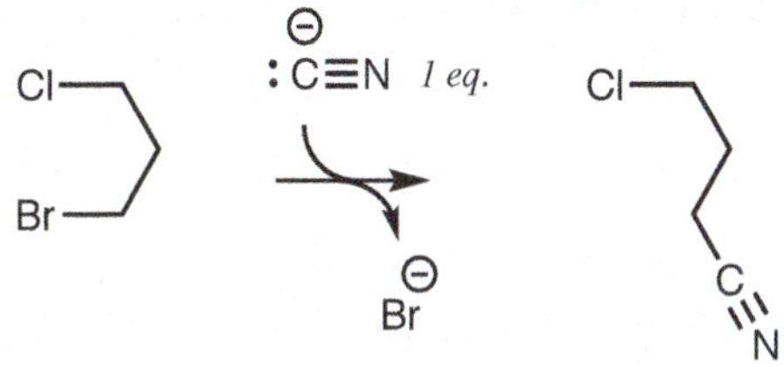

10. An S$_N$2 reaction of 1-bromobutane with acetylide anion to give 1-hexyne. (See *Notes.*)

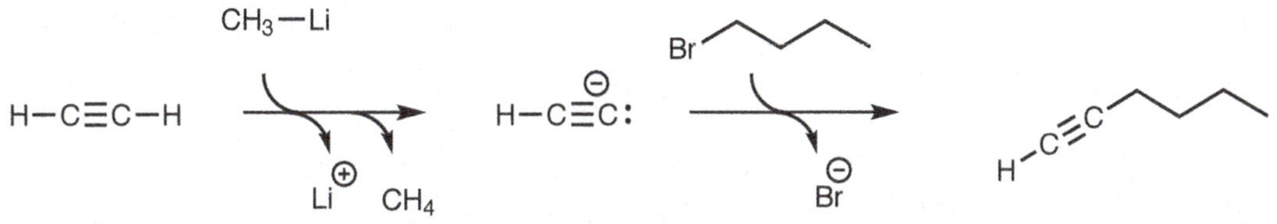

11. An S$_N$2 reaction of methyl phenylacetate enolate with iodopropane to give methyl 2-phenylpentanoate. (See *Notes.*)

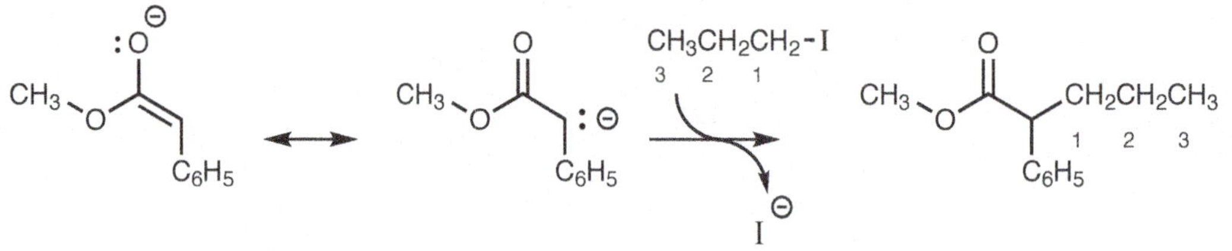

12. A Gabriel amine synthesis with formation of phthalimide anion (potassium carbonate) and alkylation with 1-bromo-2-butene. (See *Notes.*)

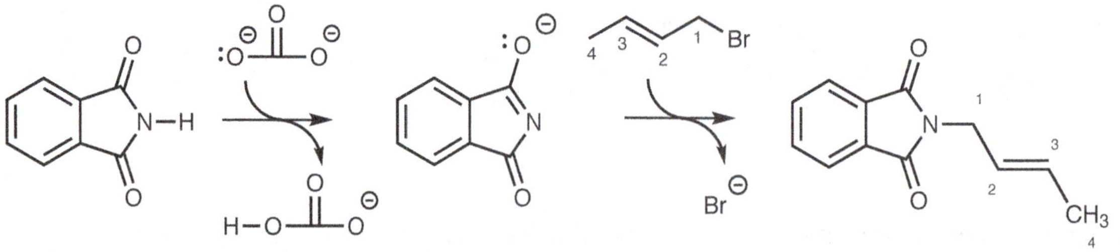

13. An S$_N$2 reaction of bromomethane with phenoxide to give methoxybenzene (anisole). (See *Notes.*)

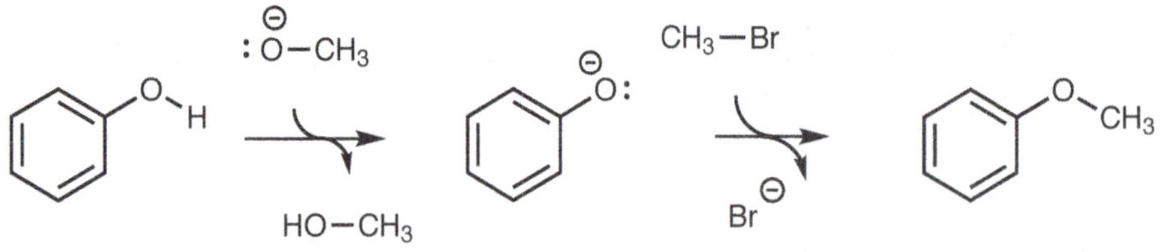

14. An S$_N$2 reaction of benzyl bromide with sodium ethoxide to give ethyl benzyl ether.

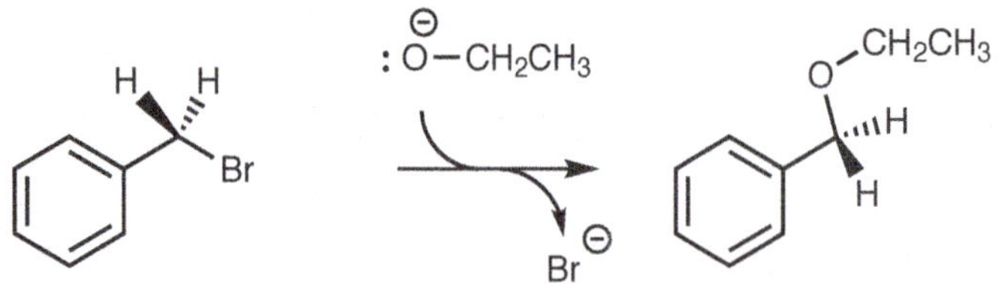

15. An S$_N$2 cleavage reaction of *p*-nitroanisole to give iodomethane and *p*-nitrophenol. (See *Notes.*)

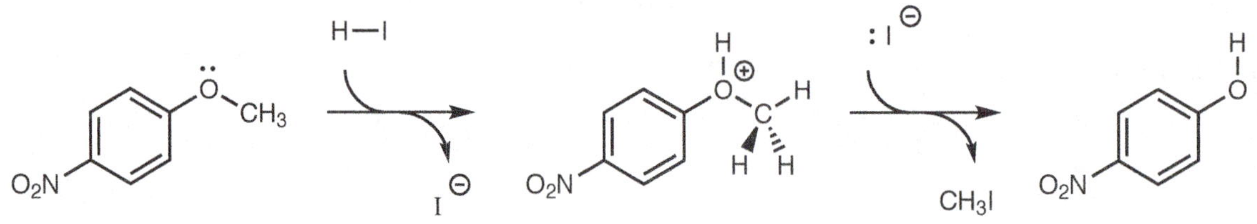

16. An S$_N$2 reaction of 1-bromo-2-butanol with sodium hydroxide to give 2-ethyloxirane (butylene oxide).

17. An S$_N$2 reaction of the anion of 1-butyne with *(R)*-2-ethyloxirane to give *(R)*-hept-5-yn-3-ol. (See *Notes*.)

Work up

18. An acid catalyzed opening of 2-ethyloxirane with methanol to give 2-methoxybutan-1-ol. (See Notes.)

cont'd

19. An S$_N$2 reaction of the enolate of acetophenone with *(R)*-2-methyloxirane (propylene oxide) to give *(R)*-4-hydroxy-1-phenylpentan-1-one. (See *Notes*.)

Step one   *Step two*

(R)-2-methyloxirane   H$_2$O   (R)-4-hydroxy-1-phenylpentan-1-one

S$_N$1 Substitution Reactions

20. An S$_N$1 solvolysis reaction of *t*-butyl iodide to give *t*-butyl alcohol. (See *Notes*.)

21. An S$_N$1 solvolysis reaction of *(R)*-(1-chloroethyl)benzene to give *rac*-1-phenylethanol.

22. An S$_N$1 reaction of 1-methylcyclohexanol with hydrogen chloride to give 1-chloro-1-methylcyclohexane. (see *Notes*)

23. An S$_N$1 solvolysis reaction of *(1S,3R)*-1-bromo-1,3-dimethylcyclohexane to give *(1S,3R)*- and *(1R,3R)*-1,3-dimethylcyclohexanol. (See *Notes*.)

Top-face attack. -

Bottom-face attack. - - - - - - - - - - - -

24. An S$_N$1 solvolysis reaction of 2-bromo-3-methylbutane to give 2-methyl-2-butanol. .(See *Notes*.)

II° carbocation   III° carbocation

25. An S_N1 reaction of 2-methylbut-3-en-2-ol with hydrogen bromide to give 1-bromo-3-methyl-2-butene. (see *Notes*)

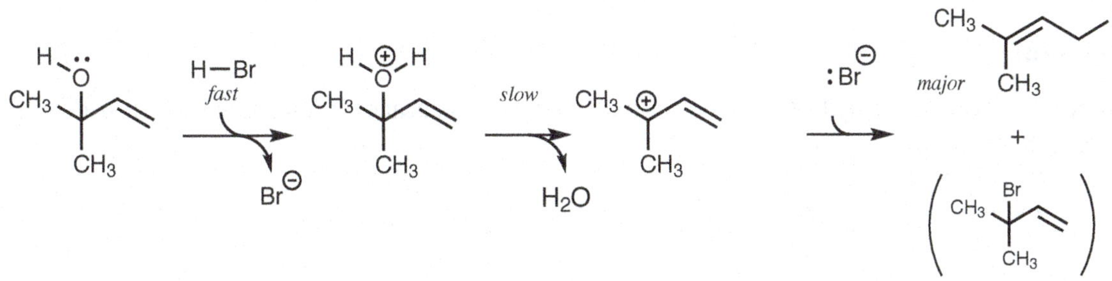

26. An S_N1 solvolysis reaction of 1-bromo-3-methyl-2-butene to give 2-methyl-3-buten-2-ol. (See *Notes*.)

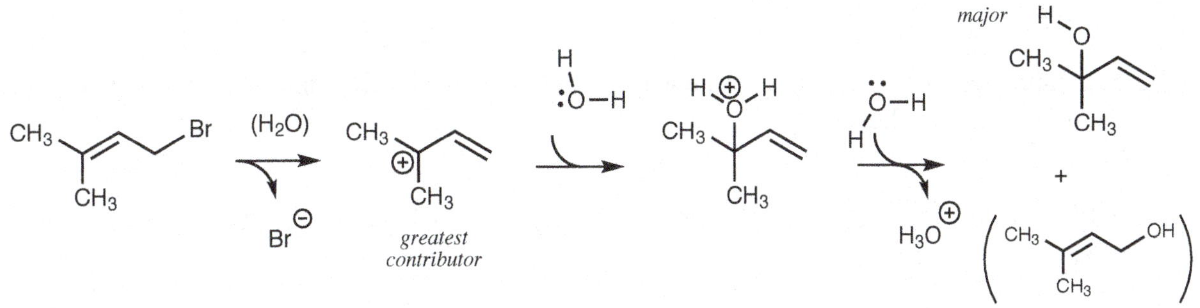

greatest
contributor

27. An S_N1 reaction of 2-methylcyclopentanol with HBr to give 1-bromo-1-methylcyclopentane.

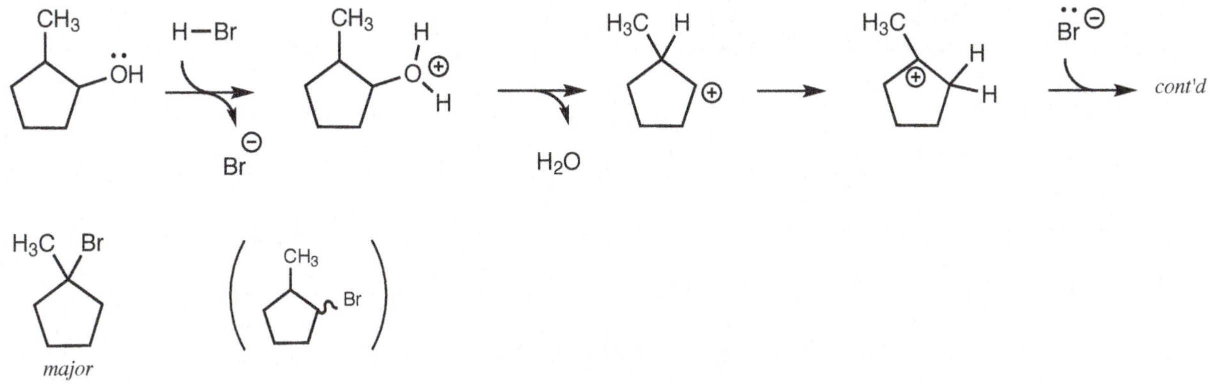

cont'd

major

28. An S_N1 solvolysis reaction of 3-chlorocyclopentene to give 3-methoxycyclopentene. (See *Notes*.)

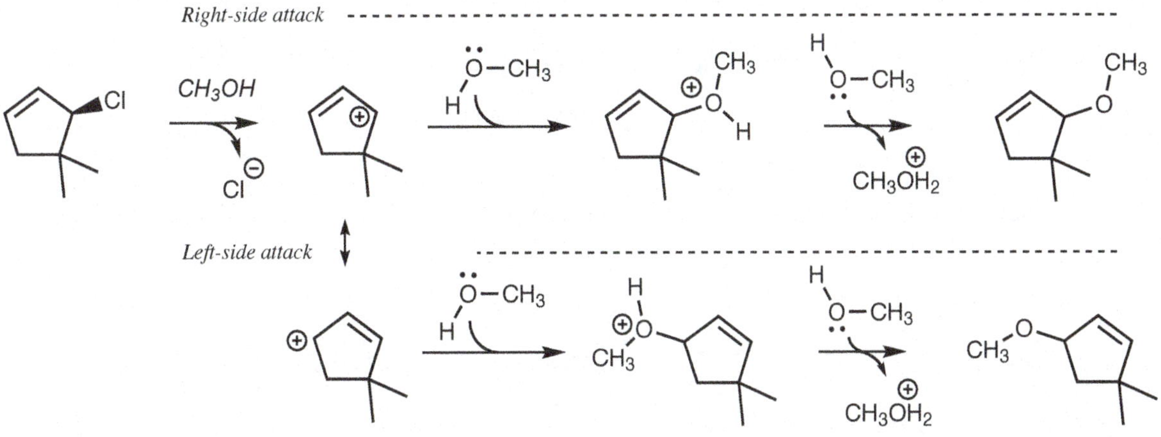

Right-side attack

Left-side attack

4 — Elimination Reactions

Alkene Formation

1. An E2 elimination reaction of hydrogen chloride from 1-chlorooctadecane with potassium *t*-butoxide to give 1-octadecene. (See *Notes*.)

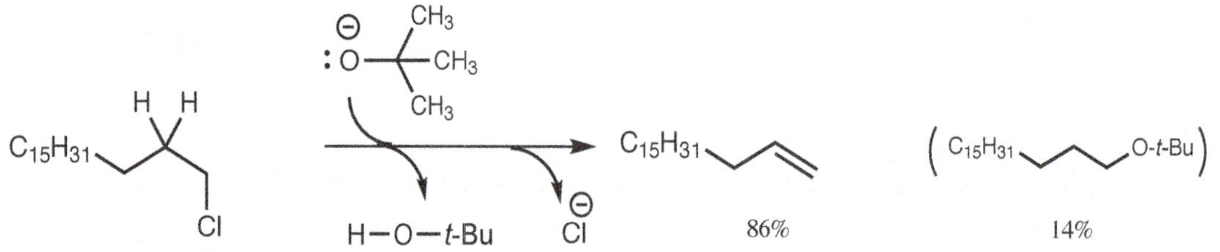

2. An E2 elimination reaction of hydrogen bromide from 2-bromobutane with sodium ethoxide to give *trans*-2-butene plus other butenes. (See *Notes*.)

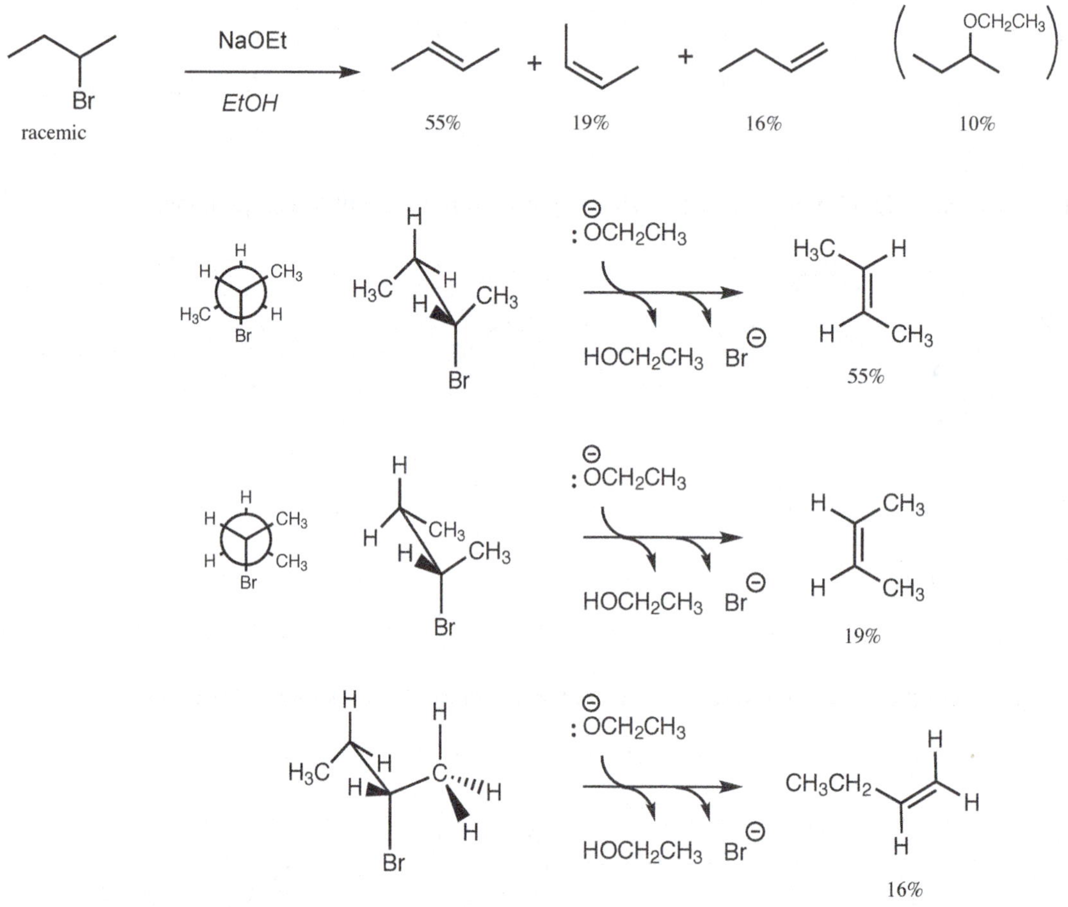

3. An E2 elimination reaction of 2-bromo-2-methylbutane to give 2-methyl-2-butene, a Zaitsev product.

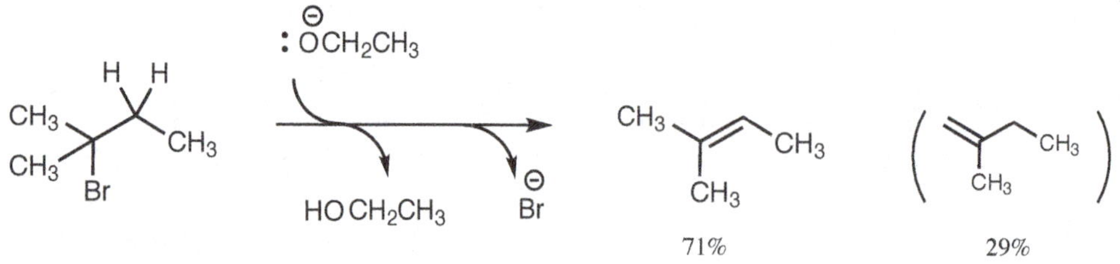

4. An E2 elimination reaction of (*1R,2R*)- or (*1S,2S*)-1-bromo-1,2-diphenylpropane to give (*Z*)-1,2-diphenylpropene. (See *Notes*.)

5. An E2 elimination reaction of *trans*-1-chloro-2-isopropylcyclohexane to give 3-isopropylcyclohexene. (see *Notes*)

stereoview

6. An E2 elimination reaction of *cis*-1-chloro-2-isopropylcyclohexane to give 3-isopropylcyclohexene. (See *Notes*.)

stereoview   *major*

7. A competitive E2 elimination reaction of *cis*- and *trans*-1-bromo-4-*t*-butylcyclohexane and one equivalent of *t*-butoxide to give 4-*t*-butylcyclohexene and unreacted bromocyclohexane. (See *Notes*.)

1 mole   *1 mole*   *1 mole*   *1 mole*

8. An E2 elimination reaction of 1-chloro-1-methylcyclohexane with ethoxide to give cyclohexene. (See *Notes*.)

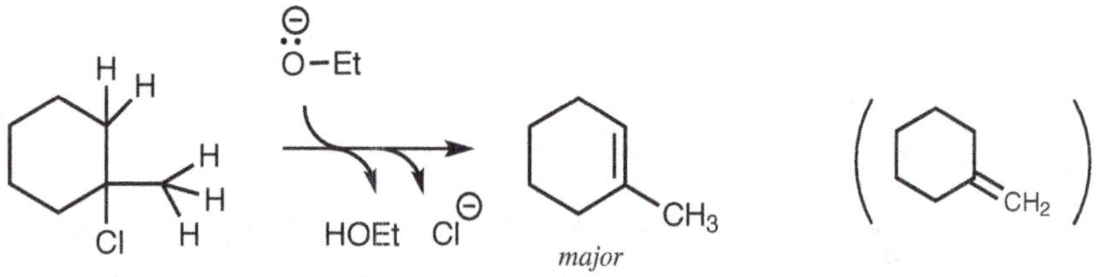

9. An E2 elimination reaction of 1-chloro-1-methylcyclohexane with *t*-butoxide to give methylenecyclohexane.

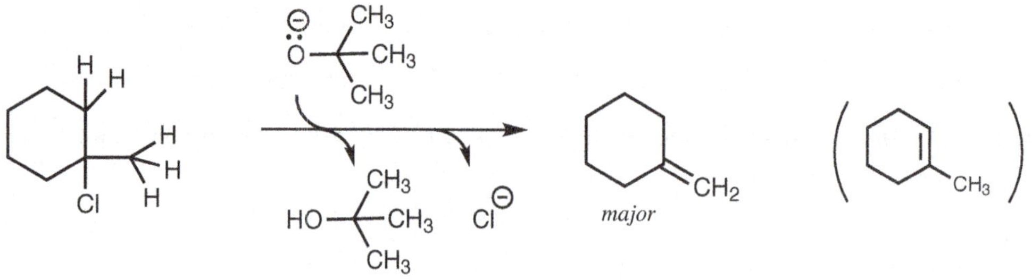

10. An E1cb elimination reaction of 3-chloro-3-methylcyclohexanone with *t*-butoxide to give 3-methyl-2-cyclohexenone. (See *Notes*.)

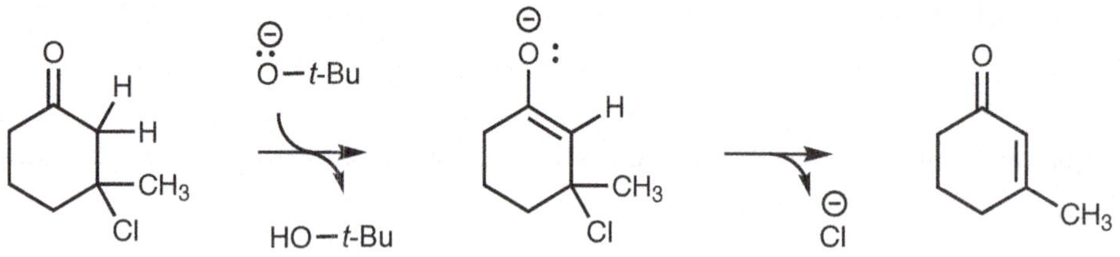

11. An E2 elimination reaction of 4-bromo-cyclohexene with *t*-butoxide to give 1,3-cyclohexadiene. (See *Notes*.)

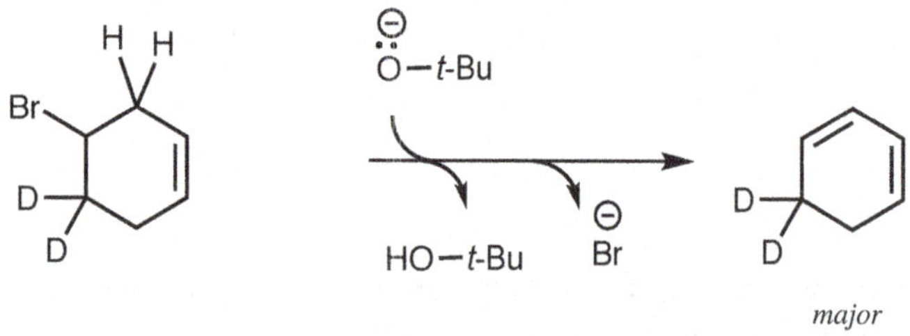

12. An E2 elimination reaction of 3-bromo-1-cyclohexene with *t*-butoxide to give 1,3-cyclohexadiene. (See *Notes*.)

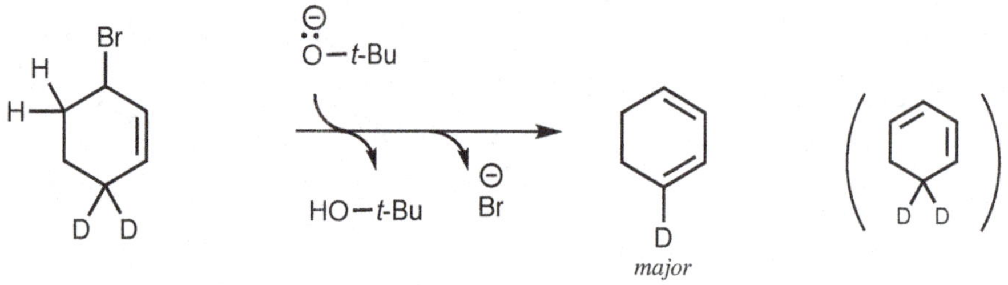

13. An E1 elimination reaction of 2-methylcyclopentanol by treatment with sulfuric acid to give methylcyclopentene. (See *Notes*.)

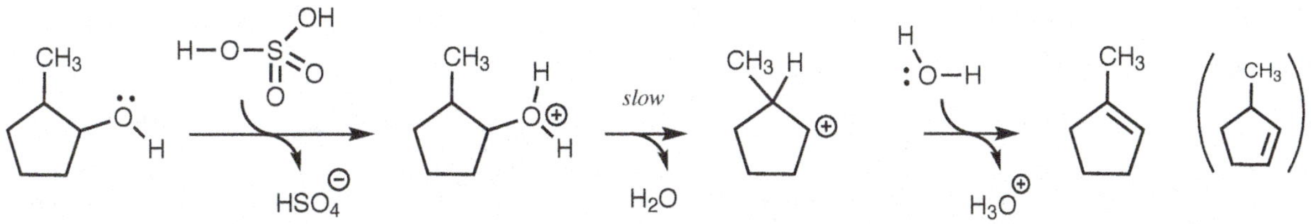

14. An E1 elimination reaction of 3-methyl-3-pentanol by treatment with acid to give pentenes. (See *Notes*.)

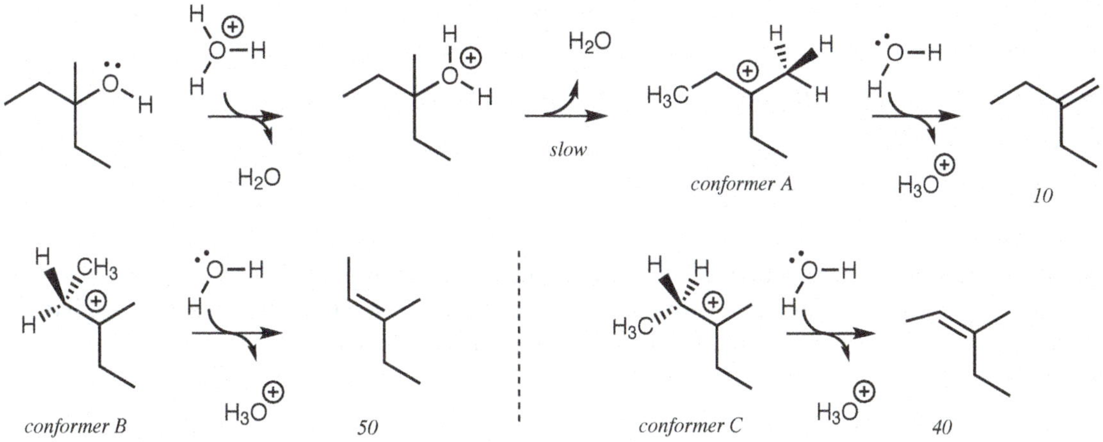

15. An E1 solvolysis-elimination reaction of a benzyl bromide in aqueous base to give methyl stilbenes. (See *Notes*.)

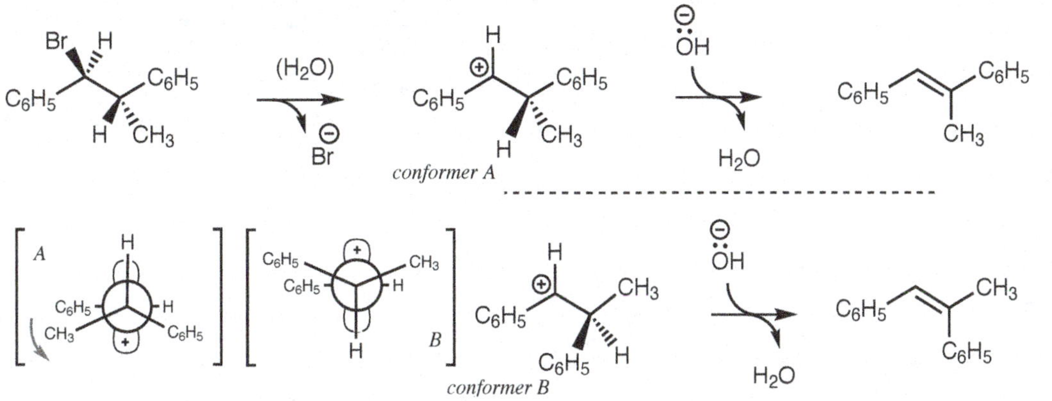

16. An E1 elimination reaction of 2-cyclobutyl-2-propanol and sulfuric acid to give 1,2-dimethylcyclopentene. (See *Notes*.)

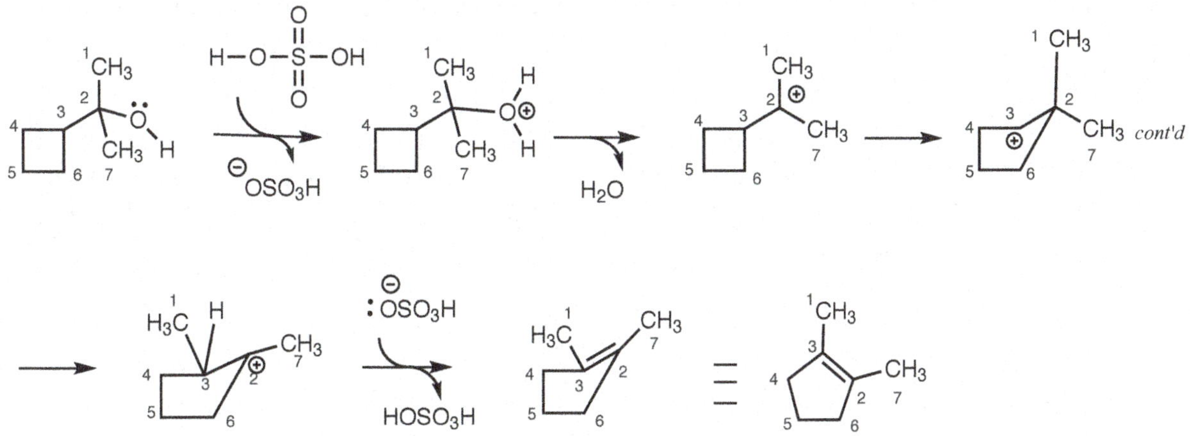

17. A Hofmann elimination reaction of a trimethylamine to give a 1-alkene. Reaction of *N,N*-dimethyl-2-pentanamine with iodomethane, silver oxide and elimination to give 1-pentene, the Hofmann elimination product. (See *Notes.*)

major

18. A Cope elimination reaction of a dimethylamine-*N*-oxide. Step 1, reaction of *N,N*-dimethyl-2,3-diphenylbutan-2-amine with hydrogen peroxide. Step 2, heating the *N*-oxide results in an elimination reaction to give *cis*-α,β-dimethylstilbene, the Zaitsev (syn) elimination product. (See *Notes.*)

Isolate and dry this product

major

19. A selenoxide elimination reaction of a cyclohexanone to a cyclohexenone, a *syn*-elimination. (See *Notes.*)

Acetylene Formation

20. A synthesis of 3-hexyne from *trans*-3-hexene by bromination and two elimination reactions. (See *Notes*.)

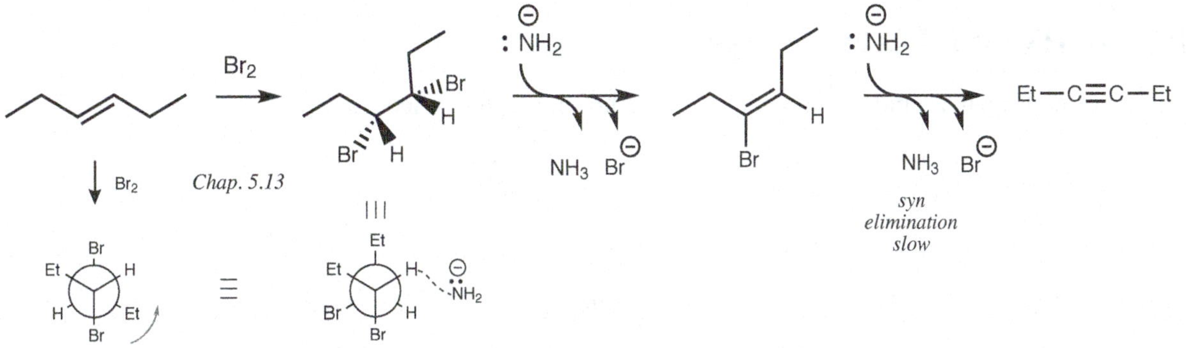

21. A synthesis of 3-hexyne from *cis*-3-hexene by bromination and two elimination reactions. (See *Notes*.)

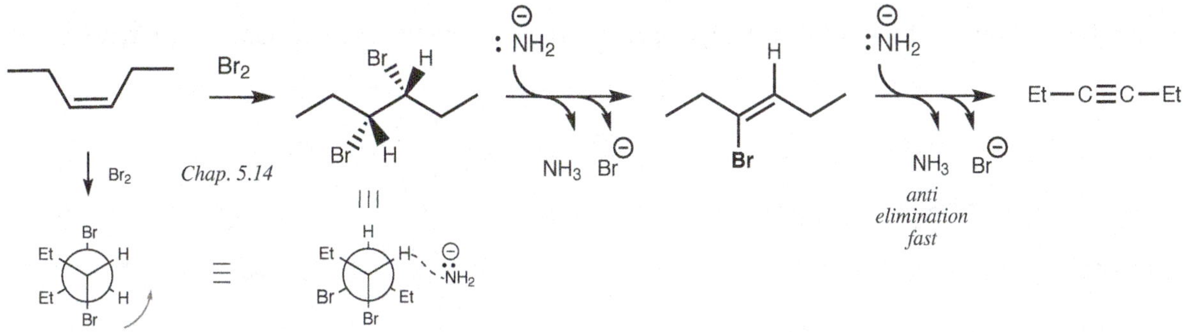

22. An E2 elimination reaction of 1,1-dibromopentane with *t*-butoxide to give 1-pentyne. (See *Notes*.)

Most stable conformer

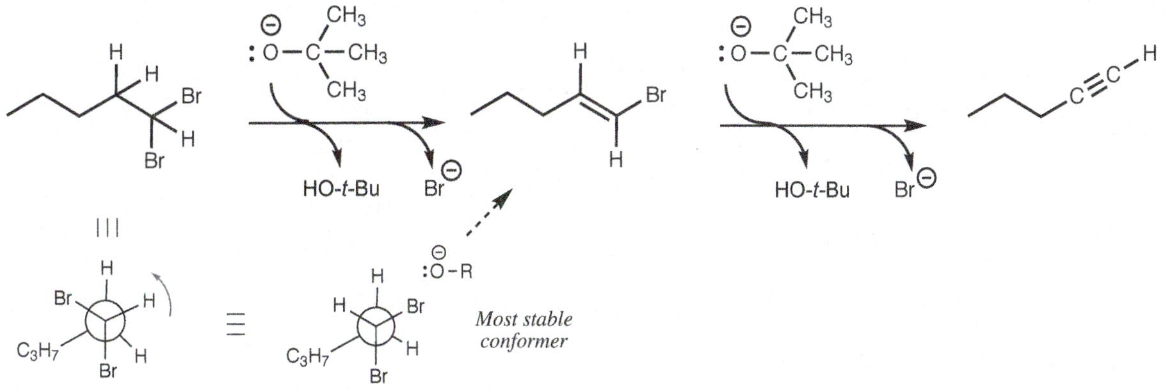

23. An E2 elimination reaction of 2,2-dibromopentane with LDA to give 1-pentyne. (See *Notes*.)

Step 1, elimination

cont'd

Step 2, work up

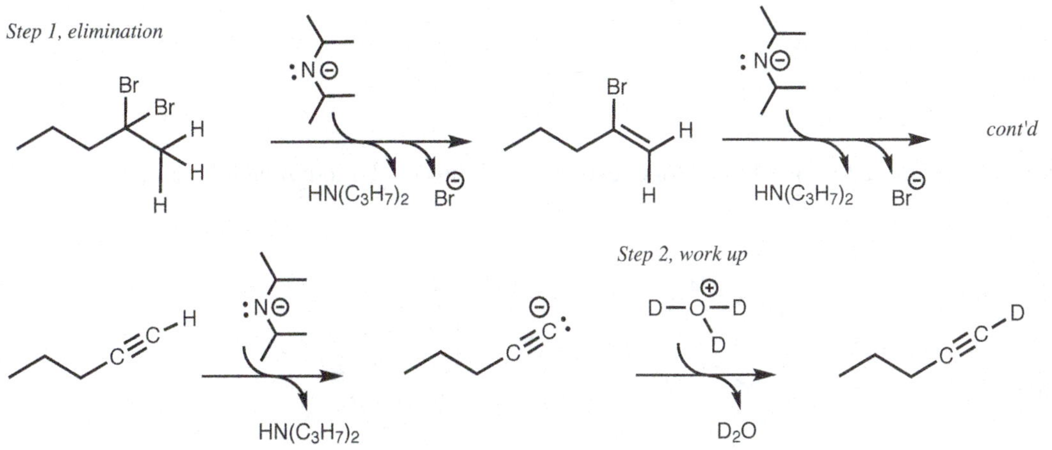

5 — Electrophilic Addition to Alkenes and Alkynes

Addition of HX and H₂O to Alkenes

1. Addition of hydrogen bromide to propene to give 2-bromopropane. (See *Notes*.)

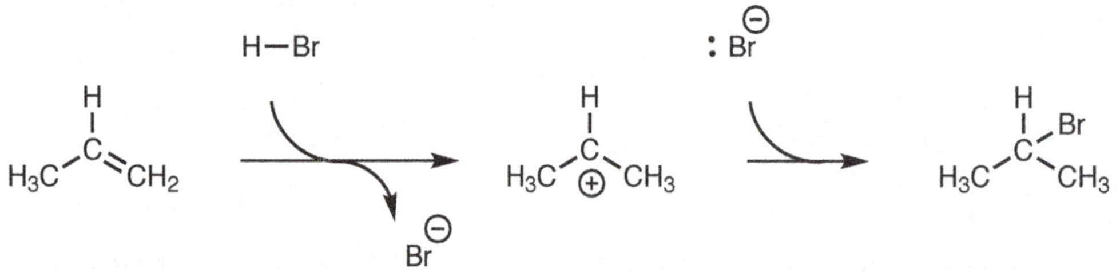

2. Acid catalyzed addition of water to methylcyclopentene to give 1-methylcyclopentanol. (See *Notes*.)

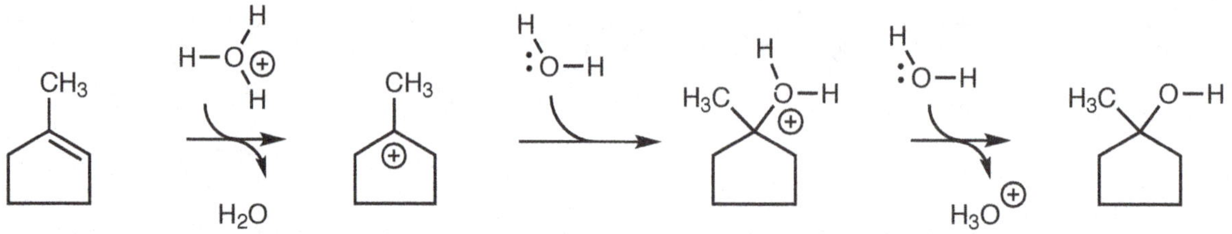

3. Addition of acetic acid to propene catalyzed by sulfuric acid to give 2-propyl ethanoate (isopropyl acetate). (See *Notes*.)

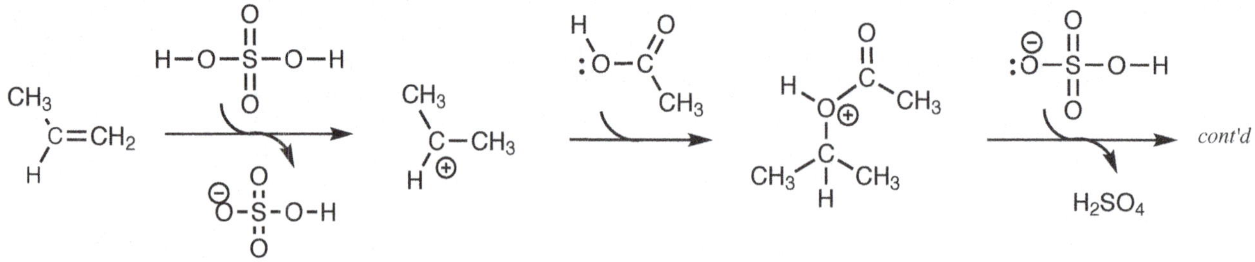

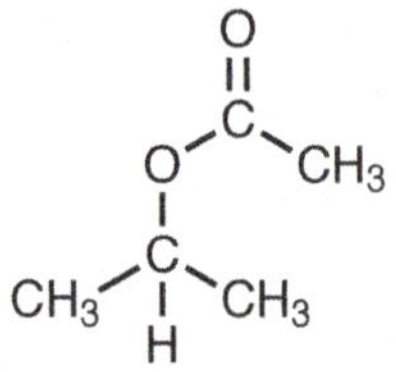

cont'd

4. Addition of hydrogen bromide to 1-methylcyclohexene to give 1-bromo-1-methylcyclohexane.

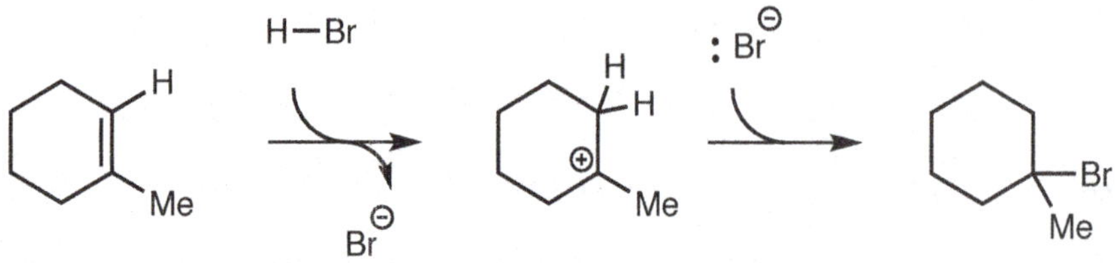

5. Addition of hydrogen chloride to *(E)*-3-hexene to give *(R)*- and *(S)*-3-chlorohexane.

(*R*)-3-chlorohexane   (*S*)-3-chlorohexane

6. Addition of hydrogen chloride to *(Z)*-3-hexene to give *(R)*- and *(S)*-3-chlorohexane.

(*S*)-3-chloro-hexane

(*R*)-3-chloro-hexane

7. Addition of hydrogen bromide to 3-methyl-1-butene to give after rearrangement, 2-bromo-2-methylbutane. (See *Notes.*)

II° carbocation   *III° carbocation*

8. Addition of HBr to 2-cyclobutylpropene to give, after rearrangement, 1-bromo-1,2-dimethylcyclopentane. (See *Notes.*)

Number the atoms to avoid confusion.

cont'd

mixture of diastereomers

9. Addition of hydrogen chloride to allylbenzene to give, after rearrangement, (1-chloropropyl)benzene. (see *Notes*)

cont'd

10. Addition of hydrogen chloride to 2-methyl-1,3-butadiene (isoprene) to give, 3-chloro-3-methyl-1-butene, the kinetic product, or 1-chloro-3-methyl-2-butene, the thermodynamic product. (See *Notes*.)

III° greater contributor

Kinetic product

Thermodynamic product

11. Addition of bromine to 2-methyl-1,3-butadiene to give 3,4-dibromo-3-methylbut-1-ene, the kinetic product and 1,4-dibromo-2-methylbut-2-ene, the thermodynamic product. (See *Notes*.)

III° greater contributor

Kinetic product

Thermodynamic product

Bromination

12. Bromination of cyclohexene to give *trans*-1,2-dibromocyclohexane. (See *Notes*.)

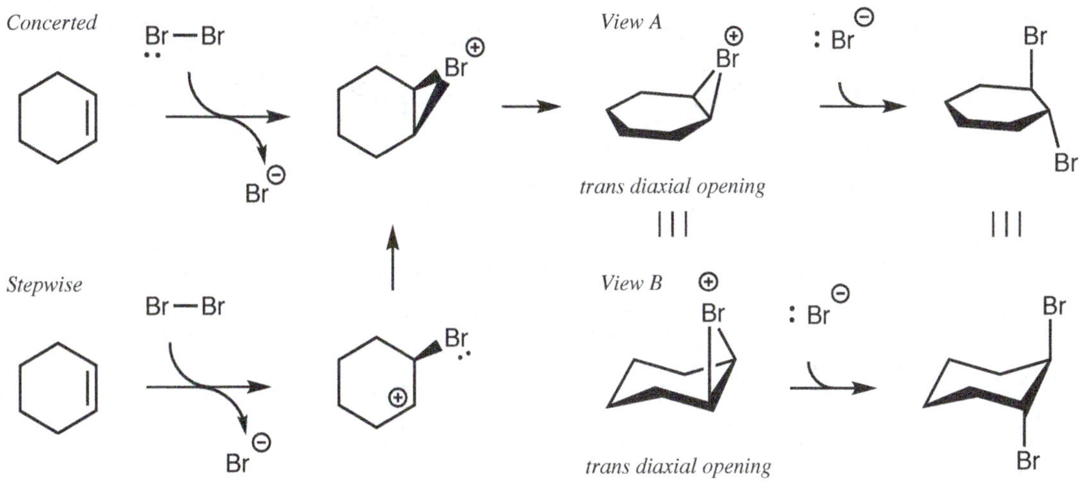

Concerted

Stepwise

View A

trans diaxial opening

View B

trans diaxial opening

13. Bromination of *trans*-2-butene to give (erythro) *(2R,3S)*- and *(2S,3R)*-2,3-dibromobutane.

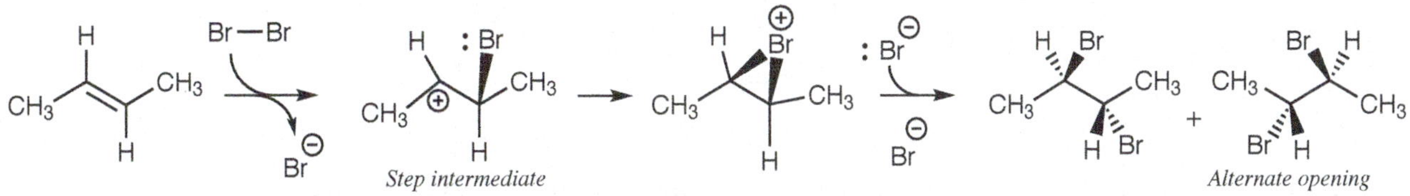

Step intermediate

Alternate opening

14. Bromination of *cis*-2-butene to give (threo) *(2R,3R)*- and *(2S,3S)*-2,3-dibromobutane.

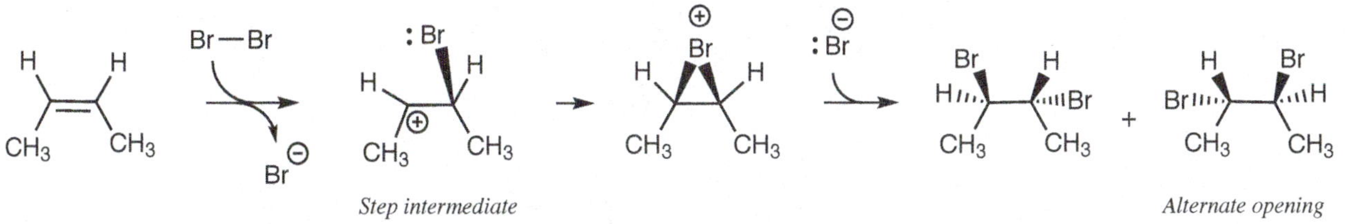

Step intermediate

Alternate opening

15. Bromination of methylcyclohexene to give *(1R,2R)*- and *(1S,2S)*-2-bromo-1-methylcyclohexanol. (See *Notes*.)

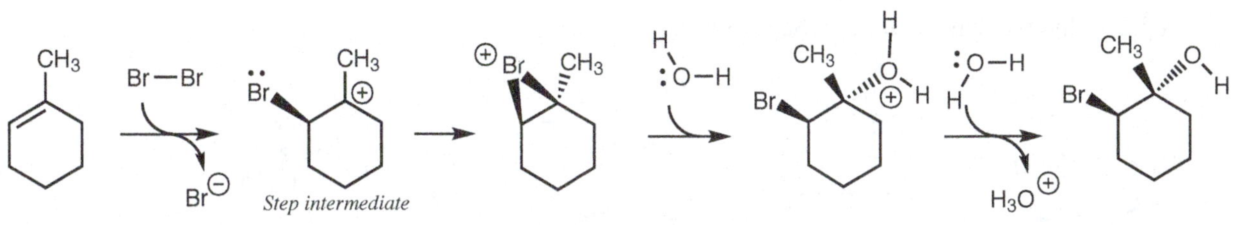

Step intermediate

Oxymercuration

16. Step 1, oxymercuration of 3-methyl-1-butene to give 3-methyl-2-butanol. (See *Notes.*)

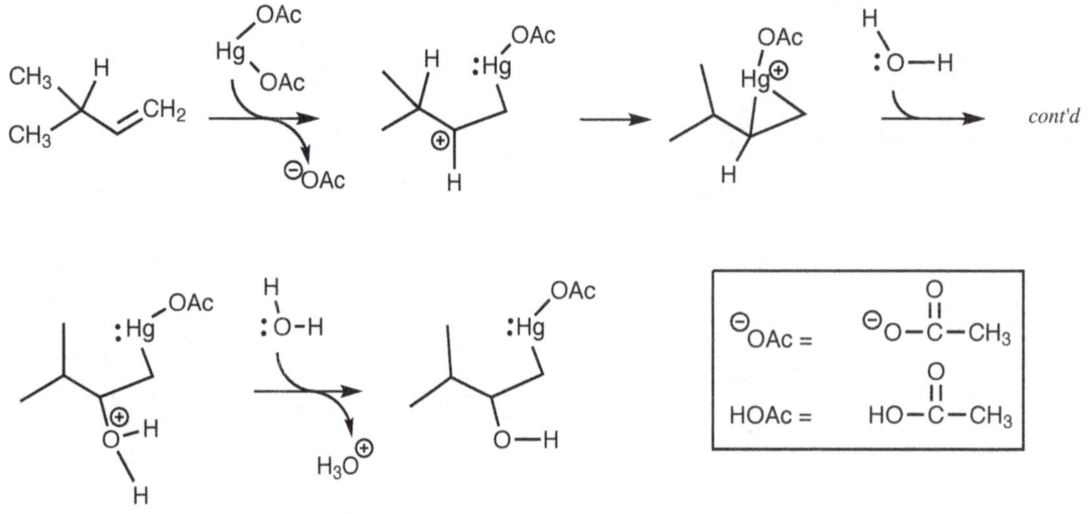

Step 2, reductive demercuration. (See *Notes.*)

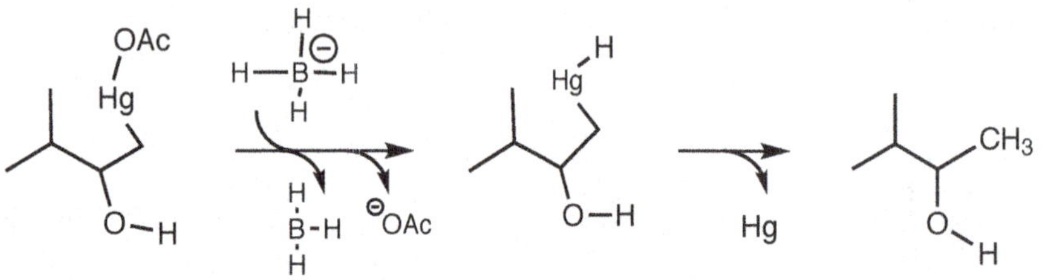

17. Step 1, oxymercuration of 1-methylcyclohexene to give 1-methylcyclohexanol. (See *Notes.*)

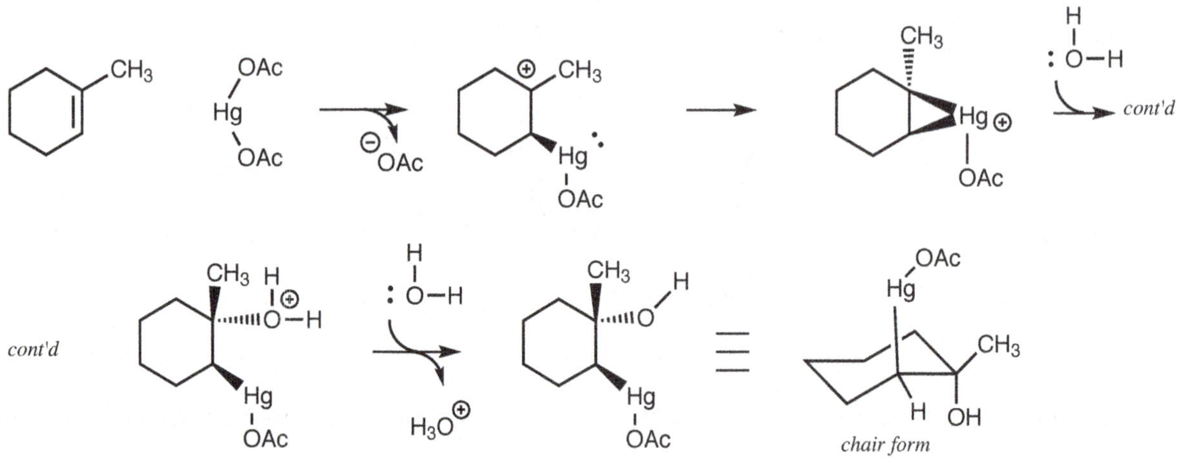

Step 2, reductive demercuration. (See *Notes.*)

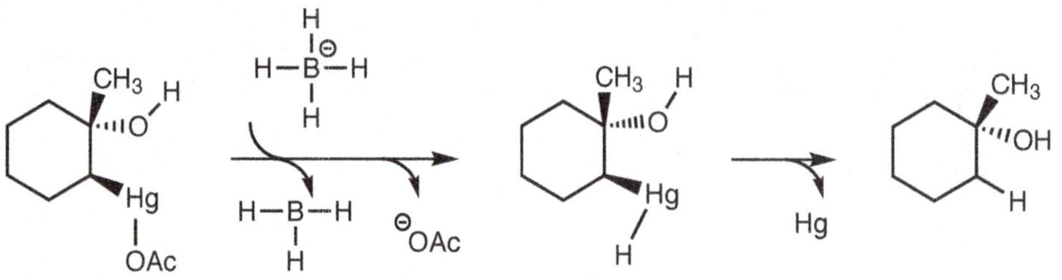

Hydroboration-Oxidation of Alkene

18. Hydroboration-oxidation of propene to give 1-propanol.

Step 1, hydroboration. Each bracket represents one of three hydroboration steps. (See *Notes*.)

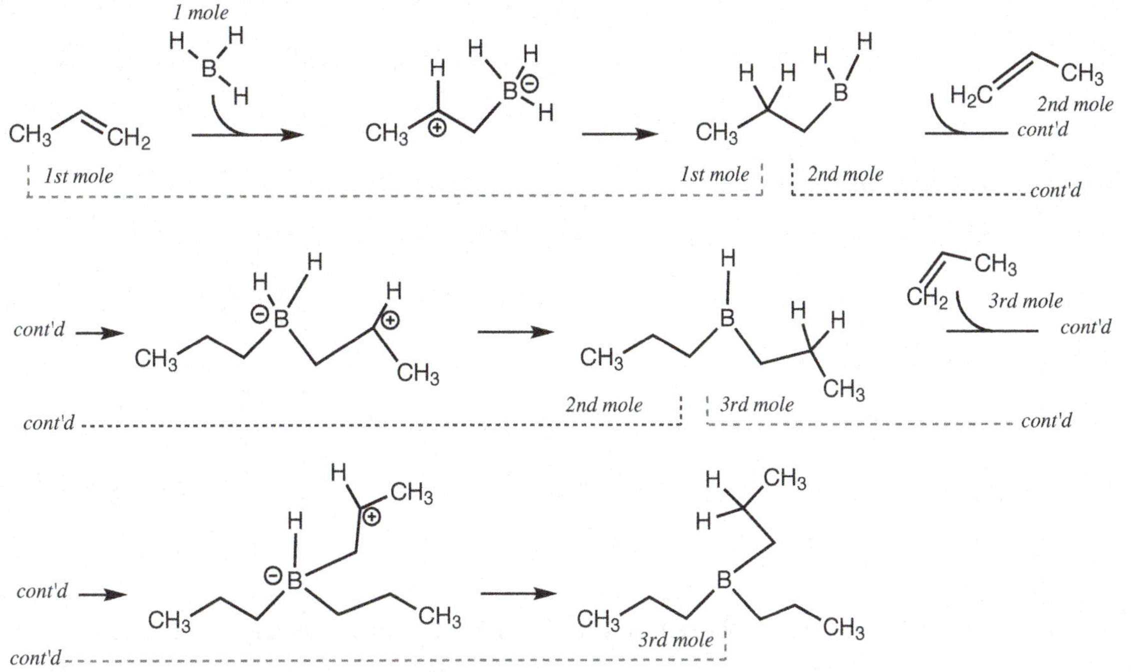

Step 2, oxidation. Each bracket represents one of three oxidation steps. (See *Notes*.)

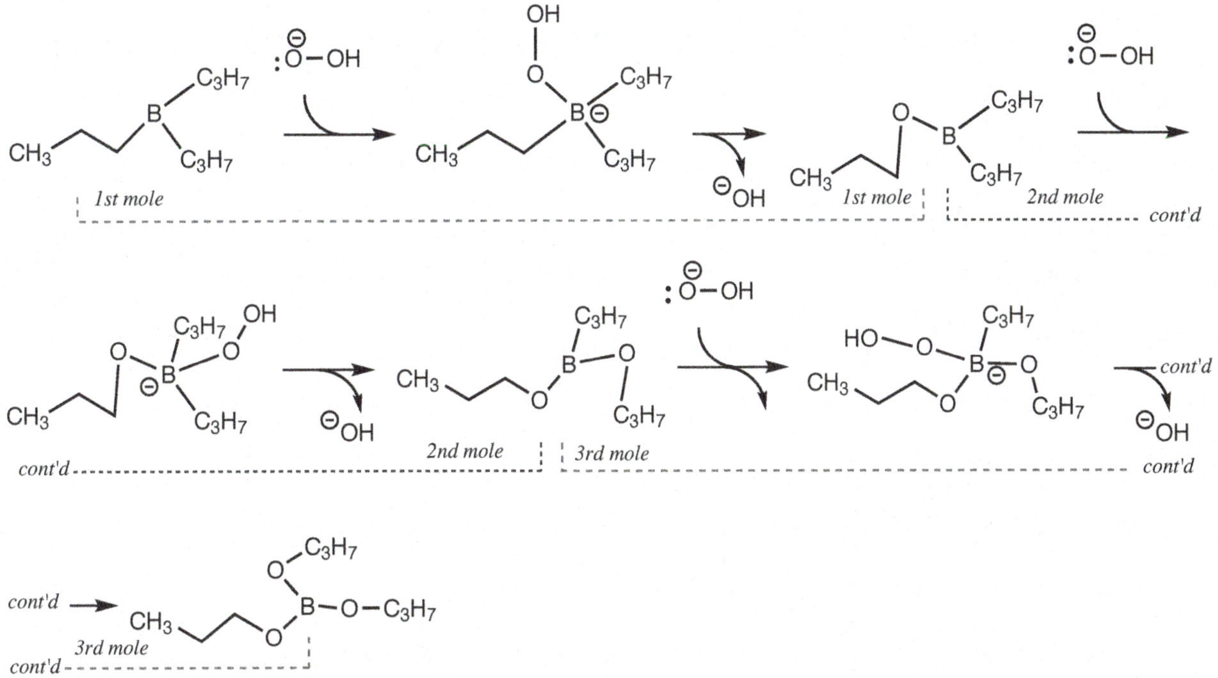

18. Step 3, borate ester hydrolysis to 1-propanol. Each bracket represents one of three hydrolysis steps. (See Notes.)

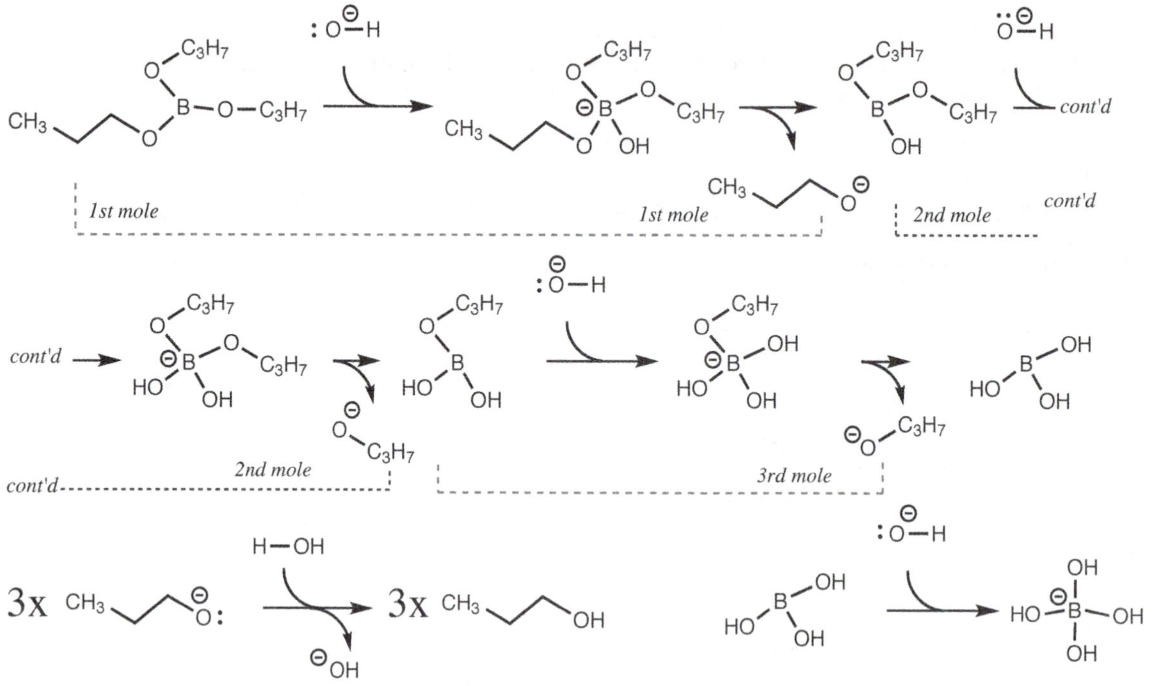

19. Hydroboration/oxidation of 1-methylcyclohexene to give *trans*-2-methylcyclohexanol. (See *Notes*.)

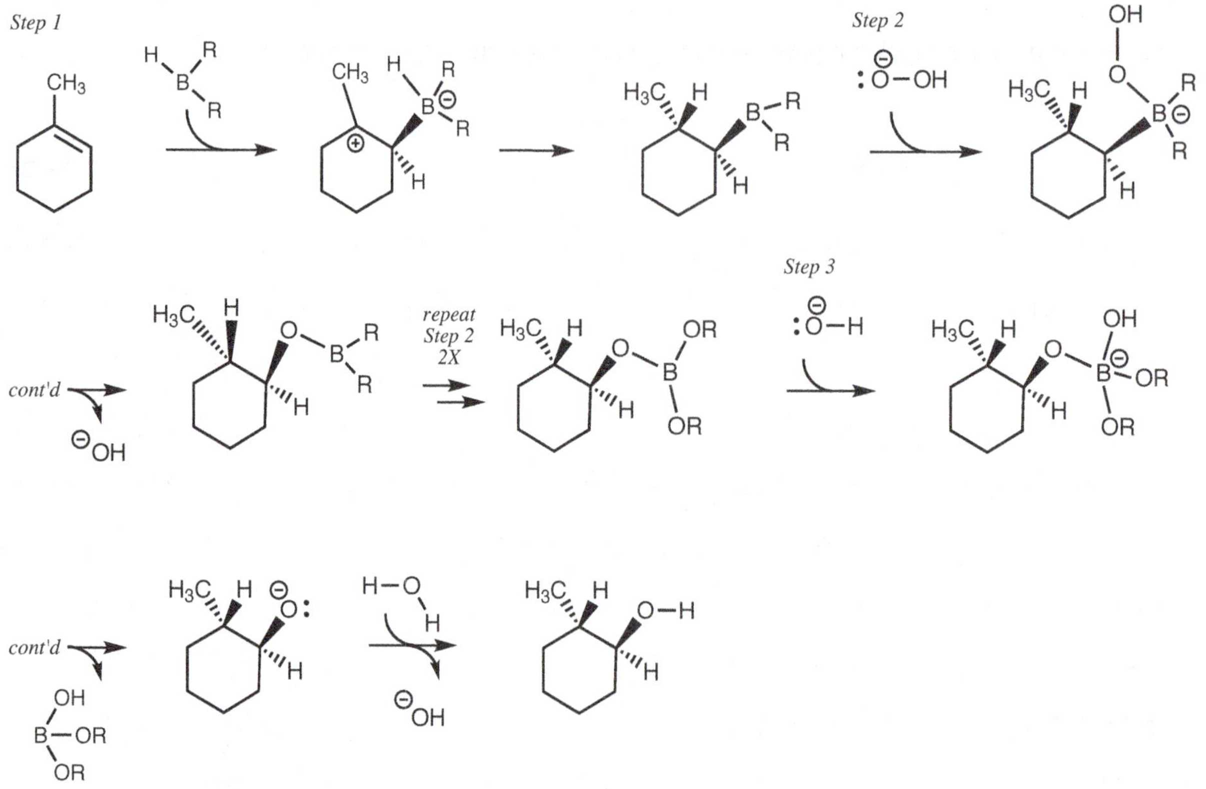

Carbon-Carbon Triple Bond Electrophilic Reactions

Addition to an Internal Acetylene

20. Addition of HCl to 2-butyne (dimethyl acetylene) to give (*E*)-2-chloro-2-butene and 2,2-dichlorobutane. (See *Notes*.)

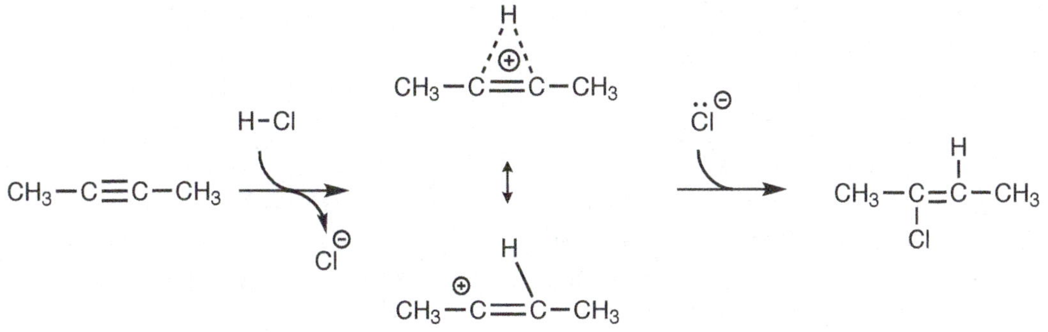

2nd Equivalent of HCl

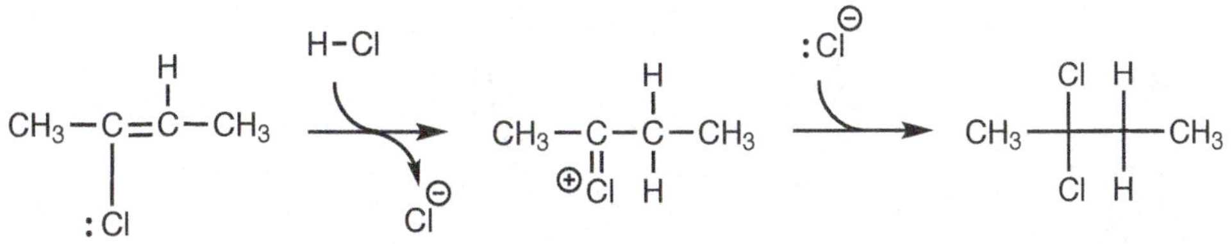

21. Addition of bromine to ethynylcyclopentane to give *(E)*-(1,2-dibromovinyl)cyclopentane and (1,1,2,2-tetrabromoethyl)cyclopentane. (See *Notes*.)

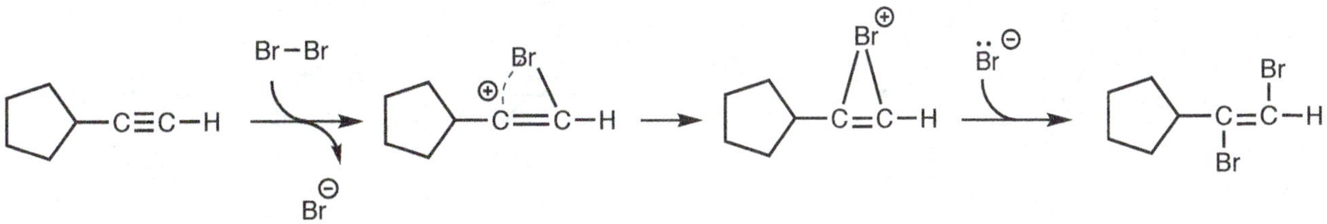

2nd Equivalent of bromine.

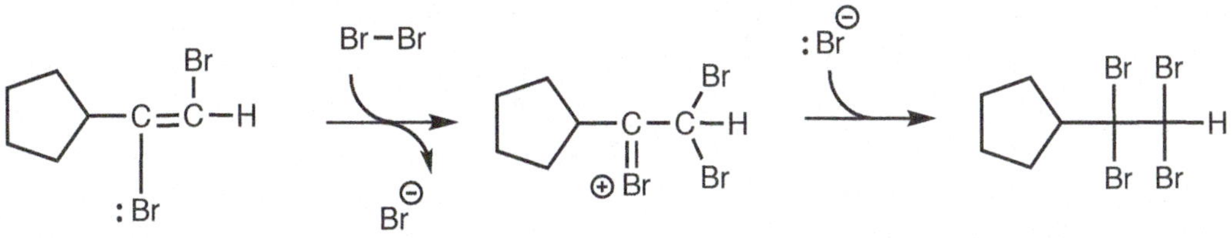

22. Sulfuric acid catalyzed hydration of 1-propynylbenzene to give 1-phenyl-1-propanone. (See *Notes*.)

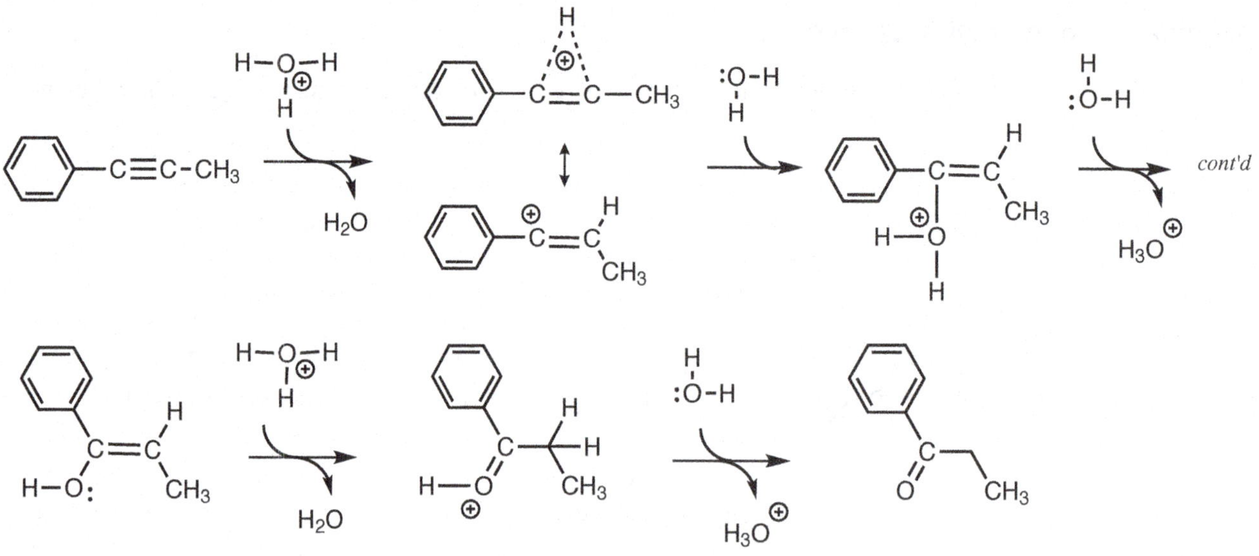

Addition to a Terminal Acetylene

23. Mercury catalyzed hydration of propyne (methyl acetylene) to give 2-propanone (acetone). (See *Notes*.)

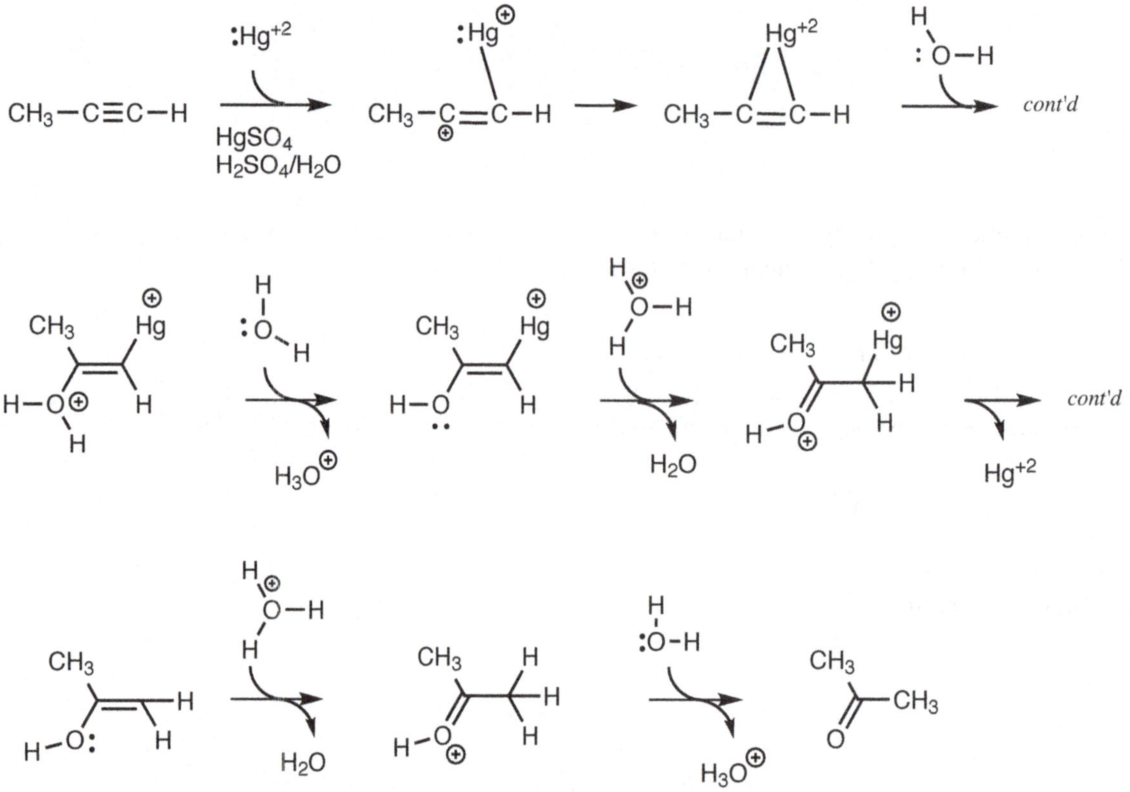

Disiamylborane Hydroboration–Oxidation of an Acetylene

24. Hydroboration-oxidation of phenylacetylene with disiamylborane to give phenylacetaldehyde. (See *Notes*.)

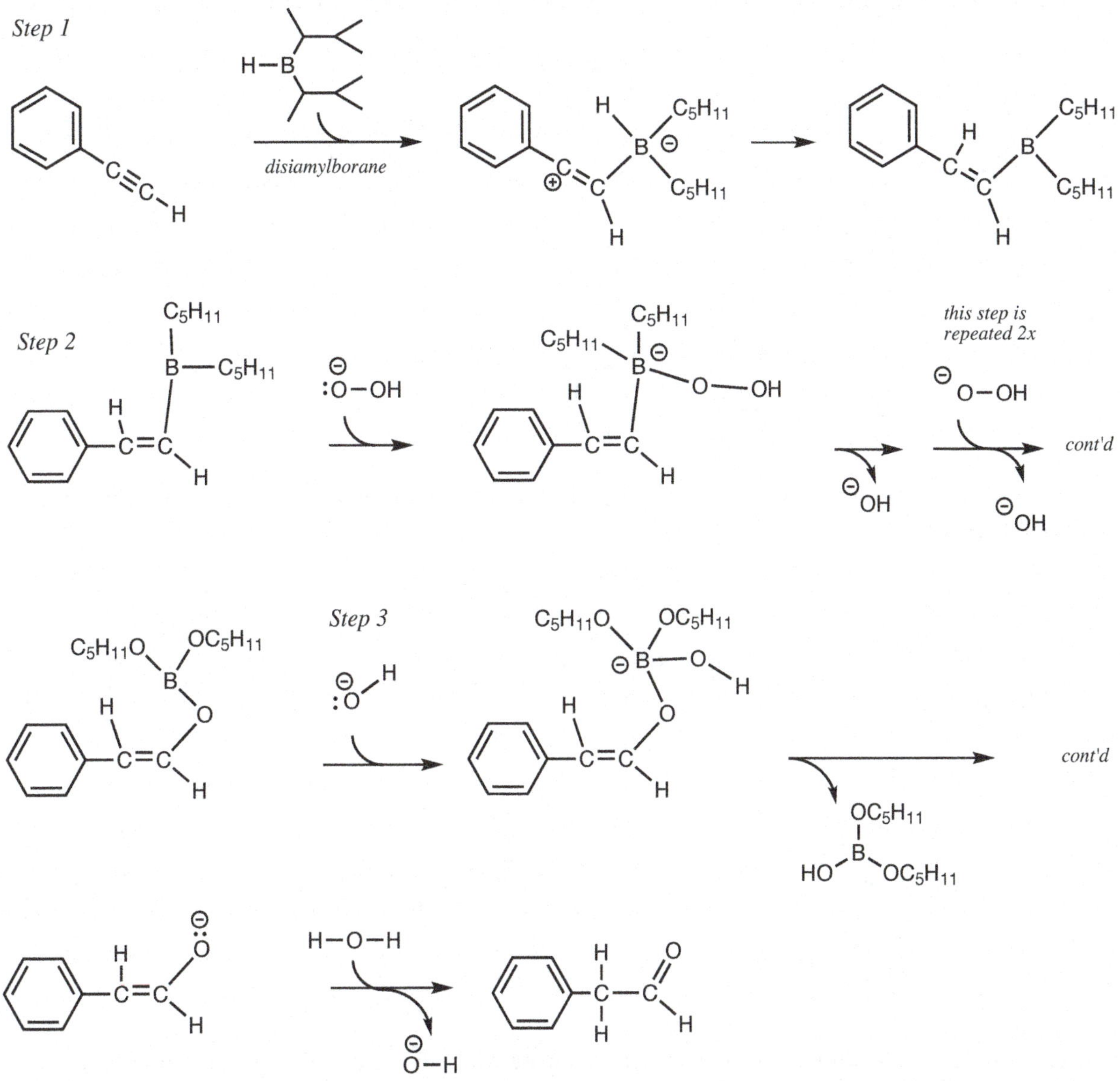

Step 1

disiamylborane

Step 2

this step is repeated 2x

cont'd

Step 3

cont'd

6 — Rearrangement Reactions

Baeyer-Villiger Oxidation

1. Acid catalyzed Baeyer-Villiger oxidation of 2,2-dimethylcyclopentanone with peracetic acid. (See *Notes.*)

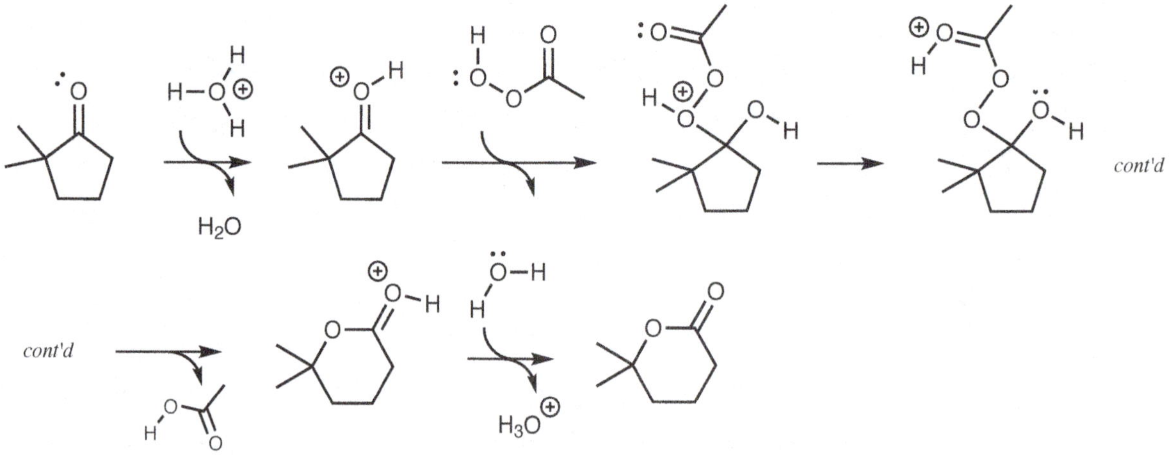

2. Baeyer-Villiger oxidation of *o*-methoxyacetophenone to *o*-methoxyphenyl acetate with peracetic acid, acid catalyzed. (See *Notes.*)

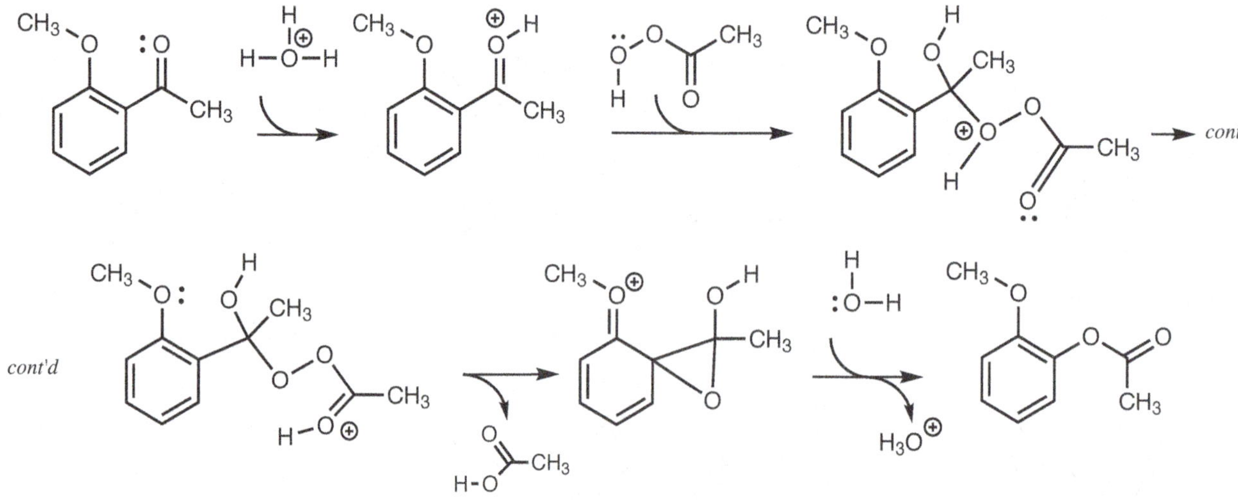

3. Acid catalyzed Baeyer-Villiger oxidation of benzaldehyde with peracetic acid to give benzoic acid. (See *Notes.*)

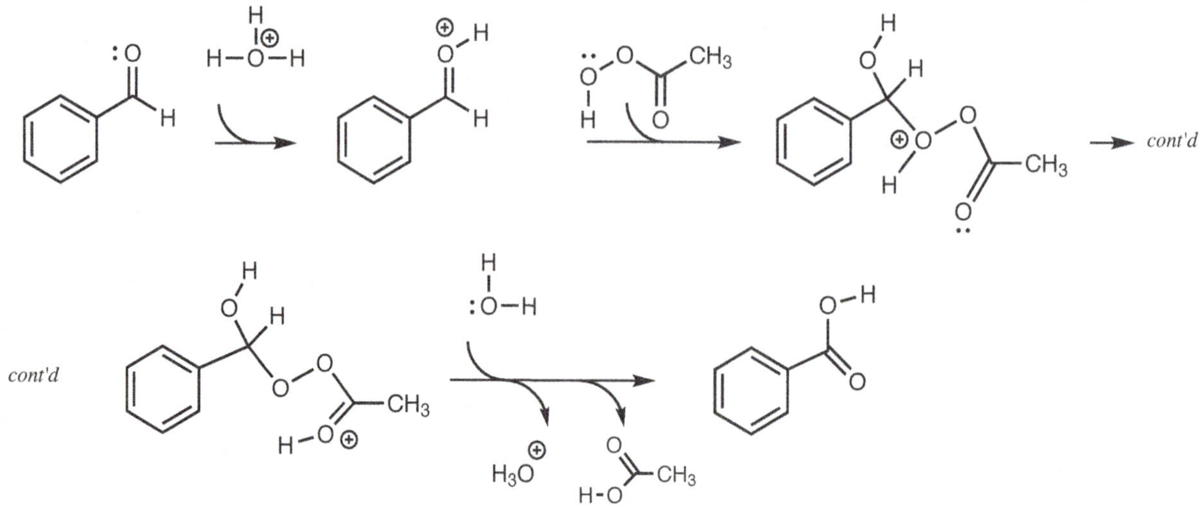

4. Baeyer-Villiger oxidation of a benzophenone with trifluoroperacetic acid. (*See Notes.*)

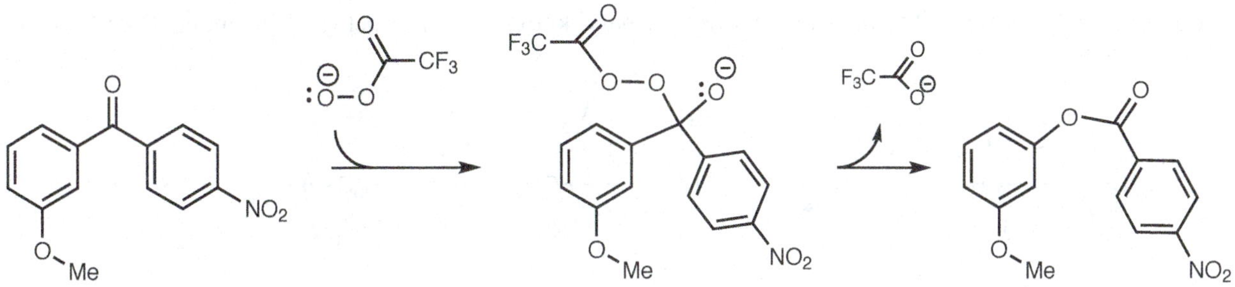

5. Baeyer-Villiger oxidation of *o*-methoxyacetophenone with *m*-chloroperoxybenzoic acid. (*See Notes.*)

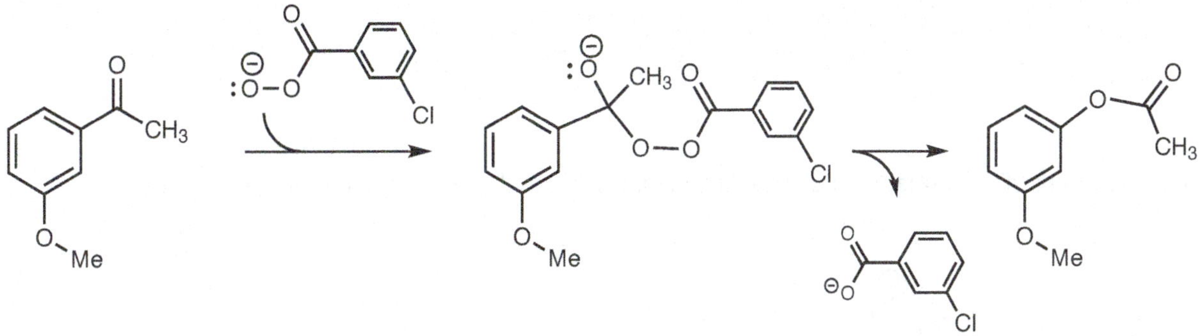

6. Baeyer-Villiger oxidation of bicyclic phenyl ketone with trifluoroperacetic acid. (*See Notes.*)

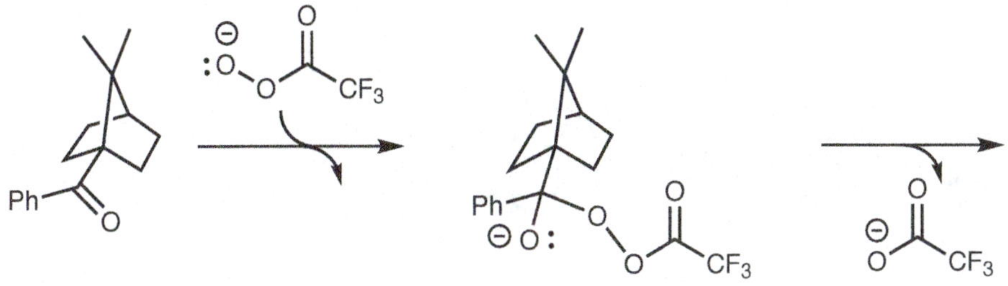

7. Baeyer-Villiger oxidation of *p*-chlorobenzaldehyde with peracetic acid to give *p*-chlorobenzoic acid. (*See Notes.*)

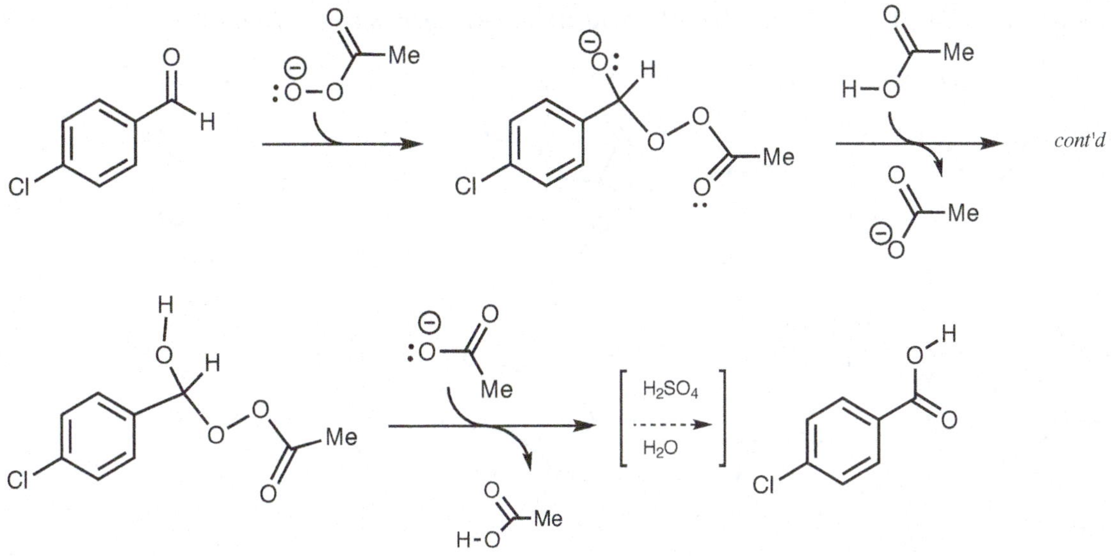

cont'd

Pinacol Rearrangement

8. Rearrangement of pinacol to pinacolone, methyl *t*-butyl ketone or 3,3-dimethyl-2-butanone. (See *Notes*.)

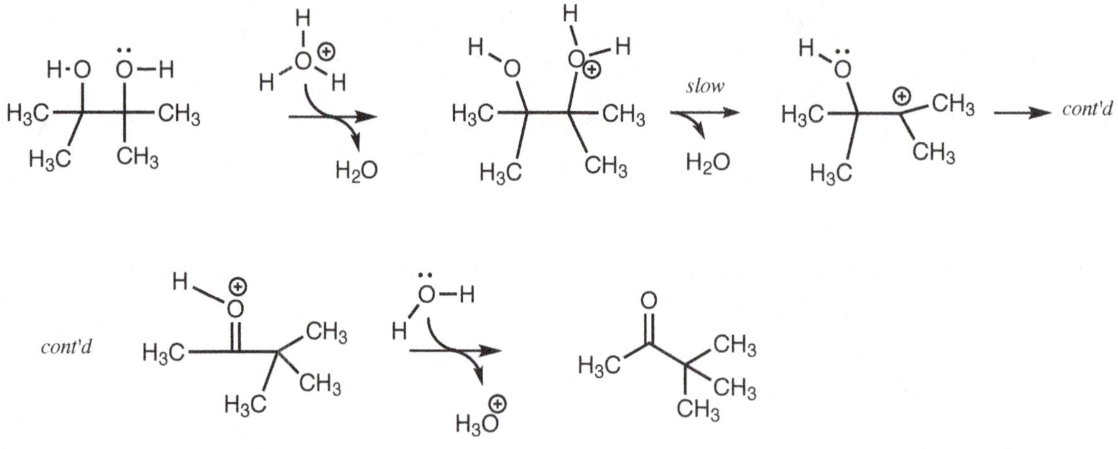

9. Rearrangement of 1,2-dimethyl-1,2-cyclohexanediol with acid to a methyl ketone. (See *Notes*.)

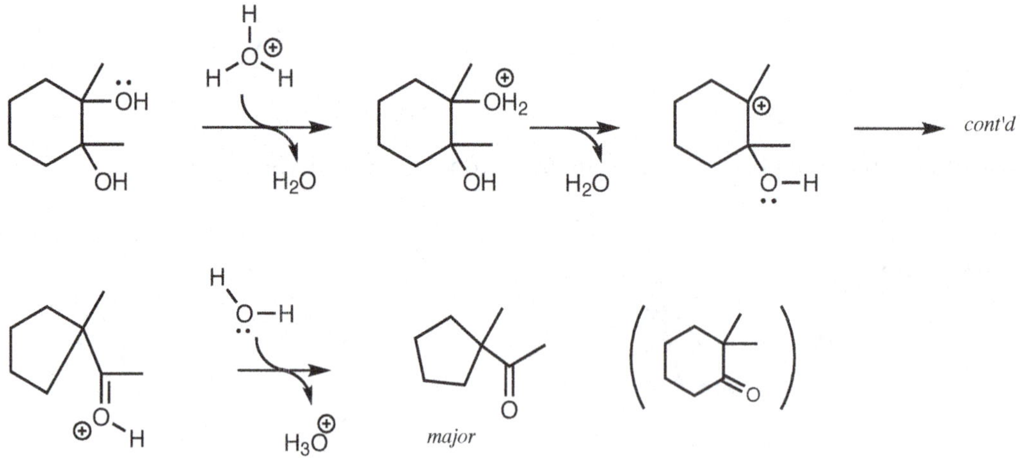

Benzilic Acid Rearrangement

10. Reaction of benzil with hydroxide to give benzilic acid after rearrangement. (See *Notes*.)

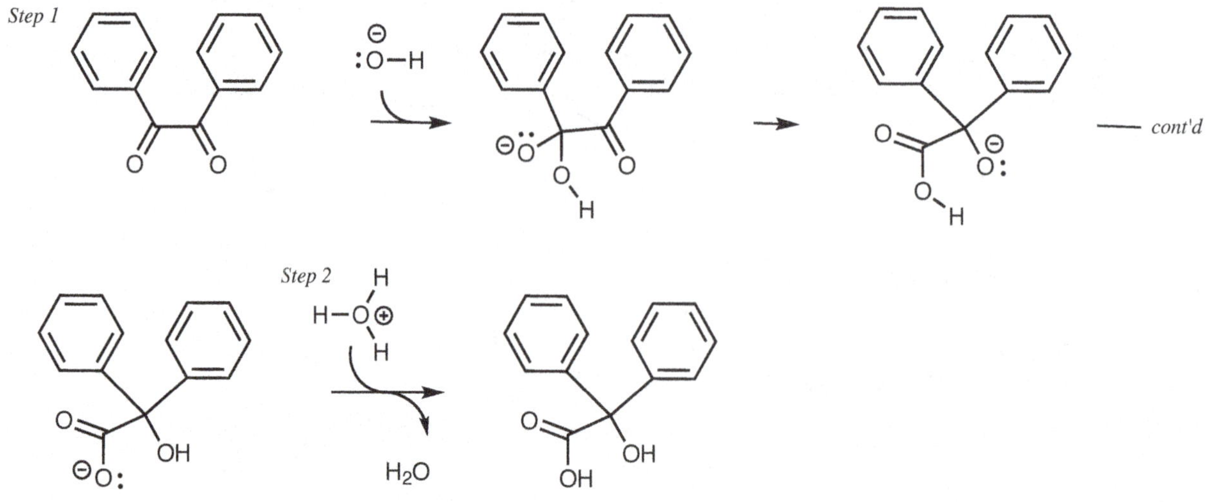

Dakin Reaction

11. Reaction of an *o*- or *p*-hydroxybenzaldehyde with basic hydrogen peroxide to give a phenol. (See *Notes.*)

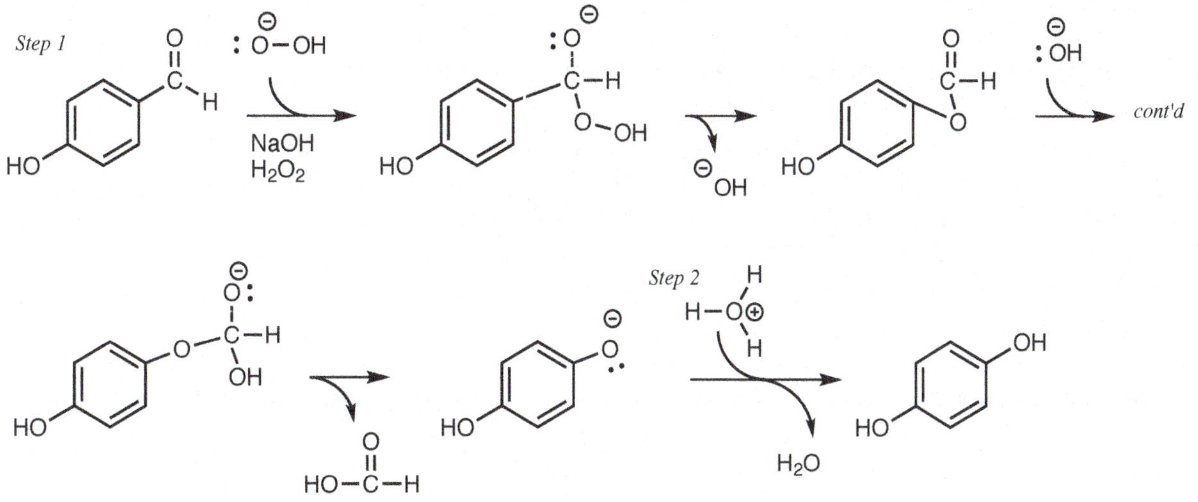

Step 1

Acetone from Cumene

12. Conversion of isopropylbenzene (cumene) to acetone and phenol.

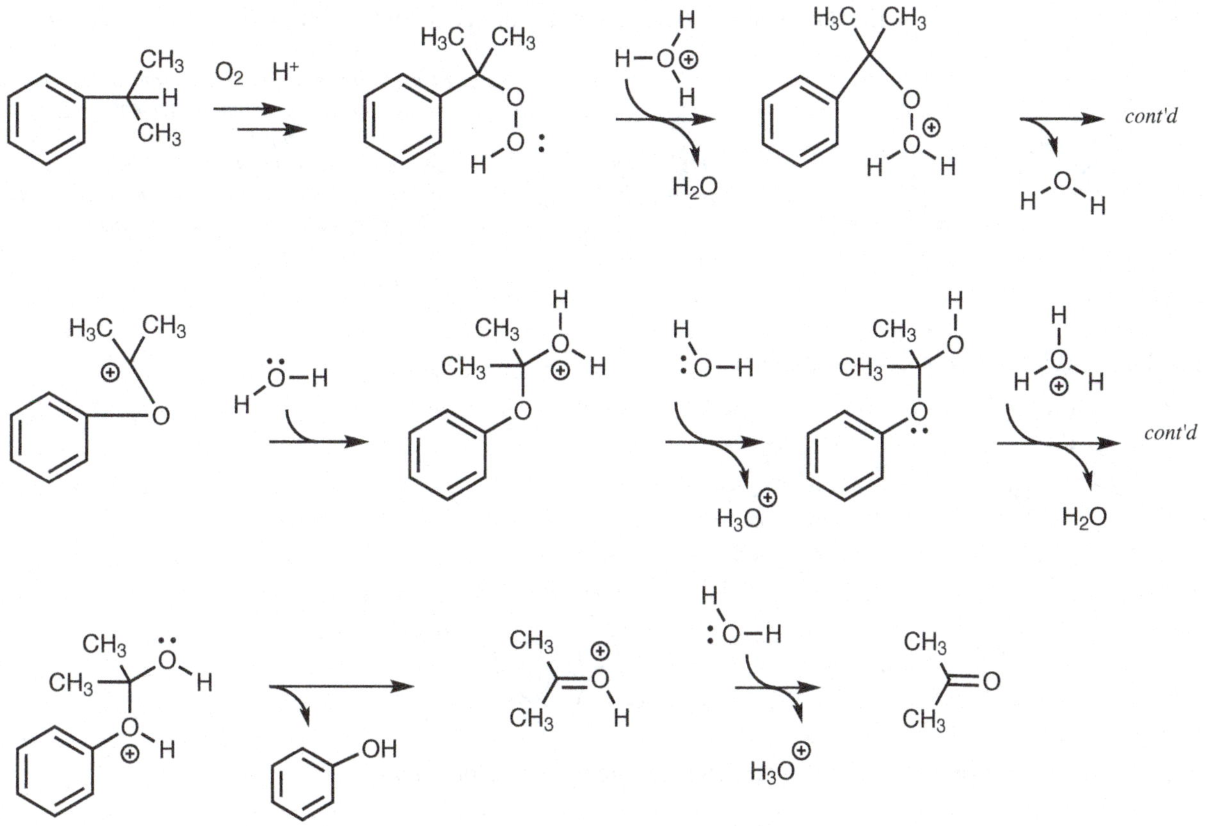

7 — Electrocyclic Reactions

Diels Alder Reactions

1. A Diels-Alder reaction between 1,3-butadiene and 2-propenal (acrolein).

2. A Diels-Alder reaction between 1,3-cyclopentadiene and (E)-2-butenal (*trans*-crotonaldehyde).

redraw

The preferred orientation of the dienophile is below. You should check with your text for further details.

This is the same example as above. It is more difficult to negotiate the atom movements, congestion, and drawing the final product all at once. I wrote the upper example in two steps as it is easier to visualize and convert it to the bicyclic product.

The preferred orientation has the electron-withdrawing group of the dienophile overlapping with the diene, if possible.

3. The Diels-Alder dimer of cyclopentadiene.

major

minor

4. A Diels-Alder reaction between 1,3-butadiene and methyl (Z)-2-butenoate.

5. A Diels-Alder reaction between 2-methyl-1,3-butadiene and *N*-methylmaleimide.

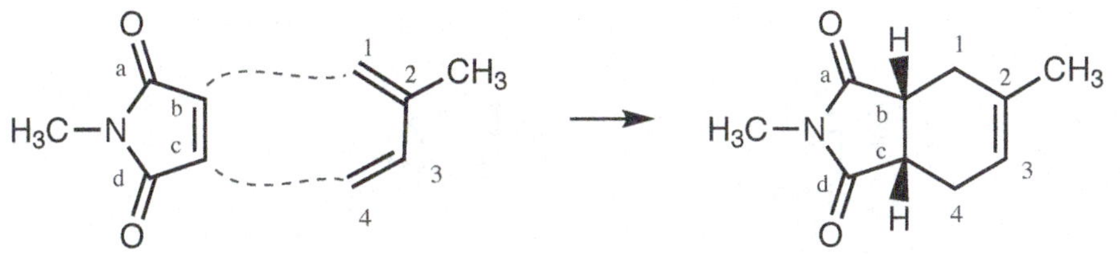

6. A Diels-Alder reaction between furan and but-3-en-2-one (methyl vinyl ketone, MVK).

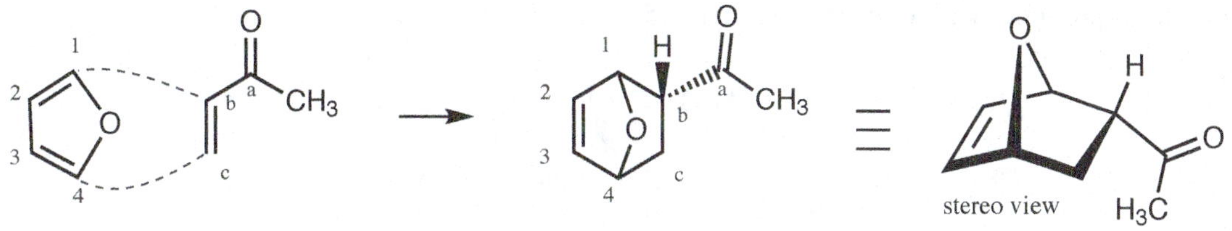

stereo view

7. A Diels-Alder reaction between cyclopentadiene and dimethyl acetylenedicarboxylate.

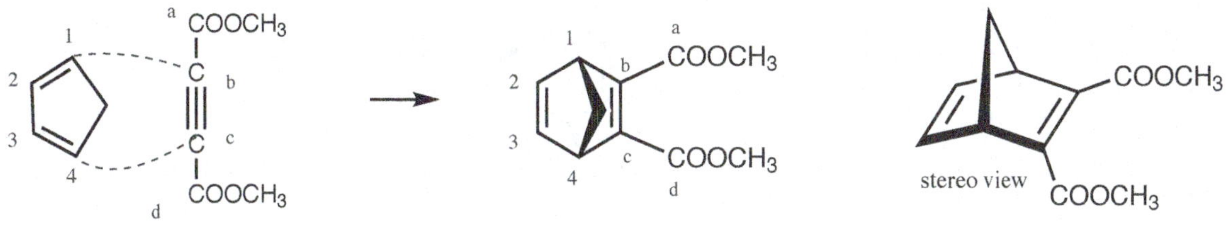

stereo view

8. A reverse-forward Diels-Alder reaction between cyclopentadiene and maleic anhydride.

heat

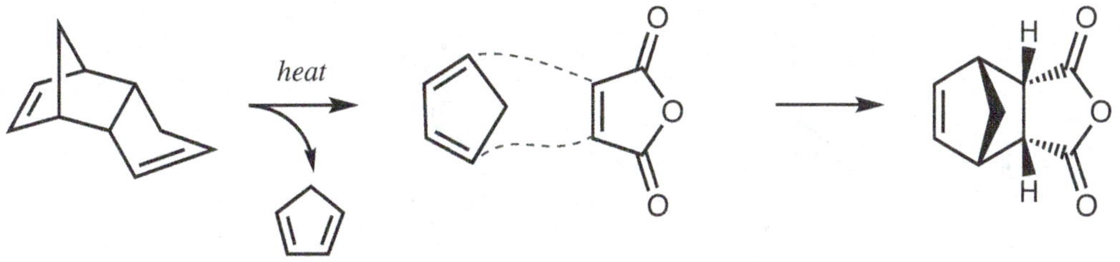

9. A reverse-forward Diels-Alder between butadiene (sulfone) and maleic anhydride.

heat

SO_2

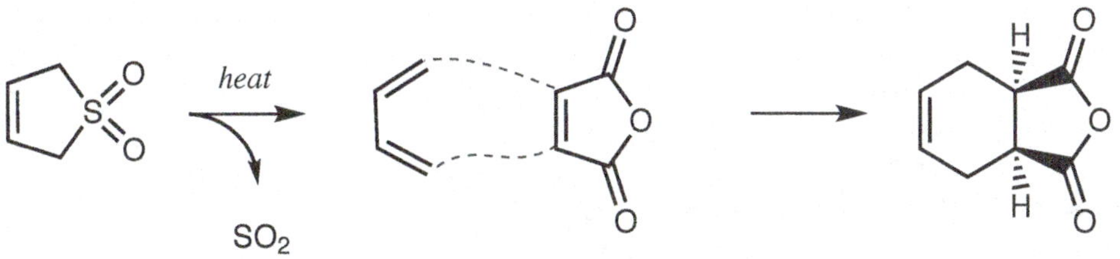

10. A Diels-Alder reaction between (*2E,4E*)-hexa-2,4-diene and maleic anhydride.

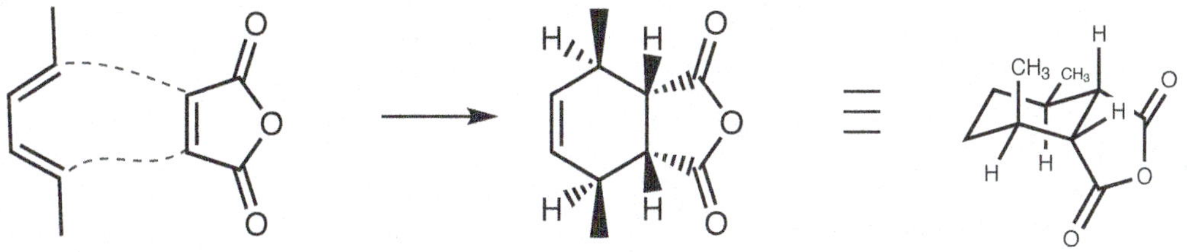

11. Draw the Diels-Alder product for the reaction of 1,3-cyclohexadiene and 2-butenal (crotonaldehyde).

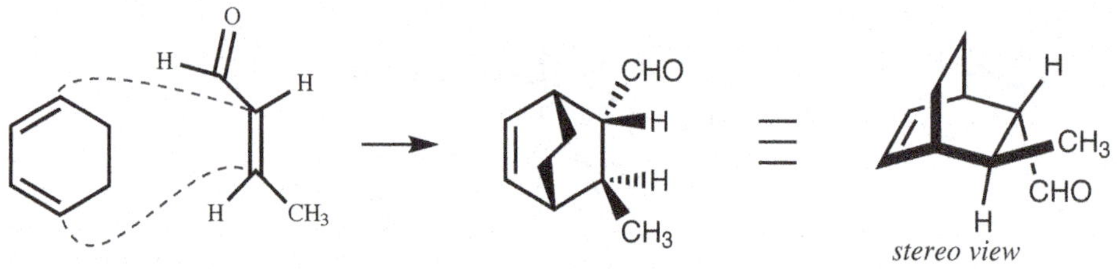

stereo view

12. A Diels-Alder reaction between 1-methoxy-1,3-butadiene and MVK gives a major product. To determine the structure of the major product, work down to Example 15.

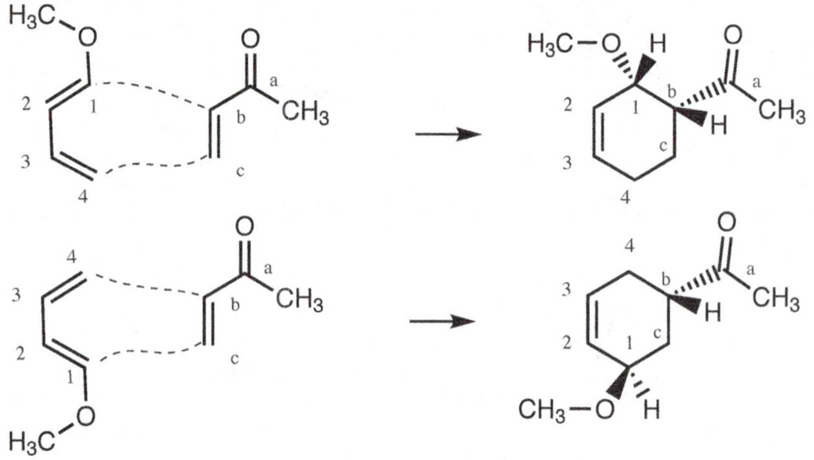

13. Examine the electrostatic charges of the starting materials by drawing the resonance structures.

Add curved arrows

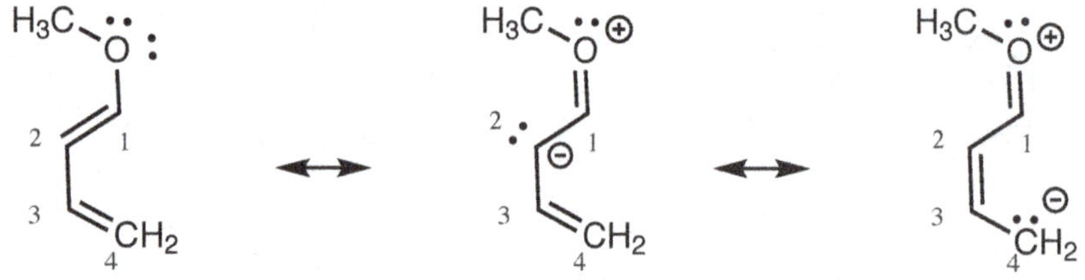

14. Examine the electrostatic charges of the starting materials by drawing the resonance structures.

Add curved arrows

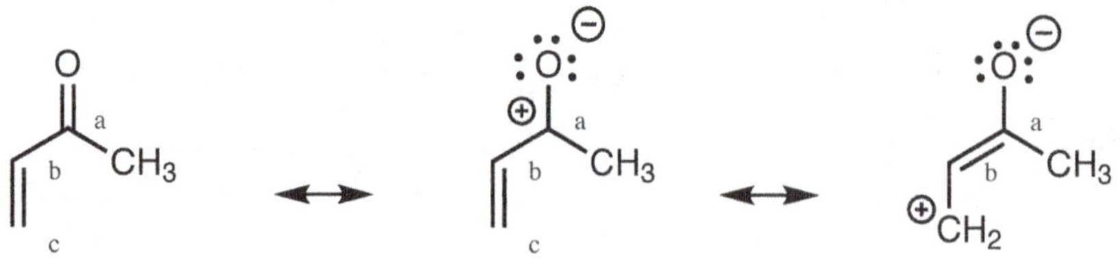

15. How do resonance structures shown in 13 and 14 predict the product for 12? Which structure will be the major product?

Other Electrocyclic Reactions (See *Notes.*)

16. A 3+2 cycloaddition between cyclopentene and benzonitrile oxide. How many pairs of electrons (curved arrows) move to complete the formation of product. Compare the number of electrons that move in this reaction with the Diels-Alder reaction. Is it the same?

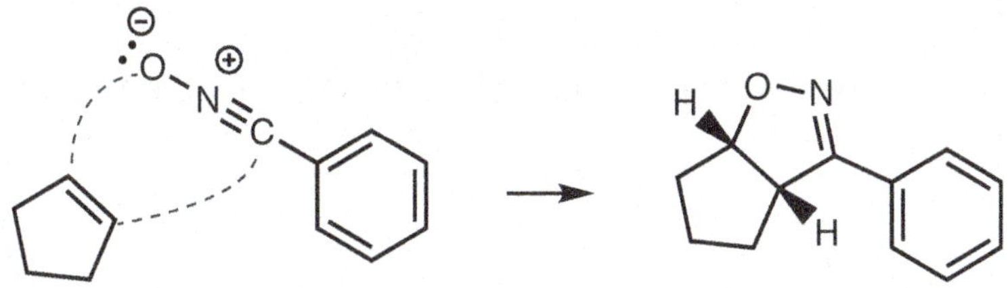

17. A Claisen rearrangement (electrocyclic) reaction to transfer a group from oxygen to carbon.

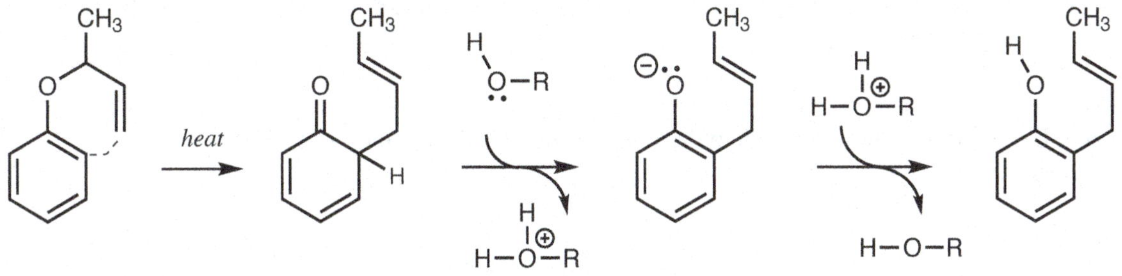

18. A double Claisen rearrangement (electrocyclic) reaction to transfer a group from oxygen to carbon to another carbon. Count the number of electrons that move in each step.

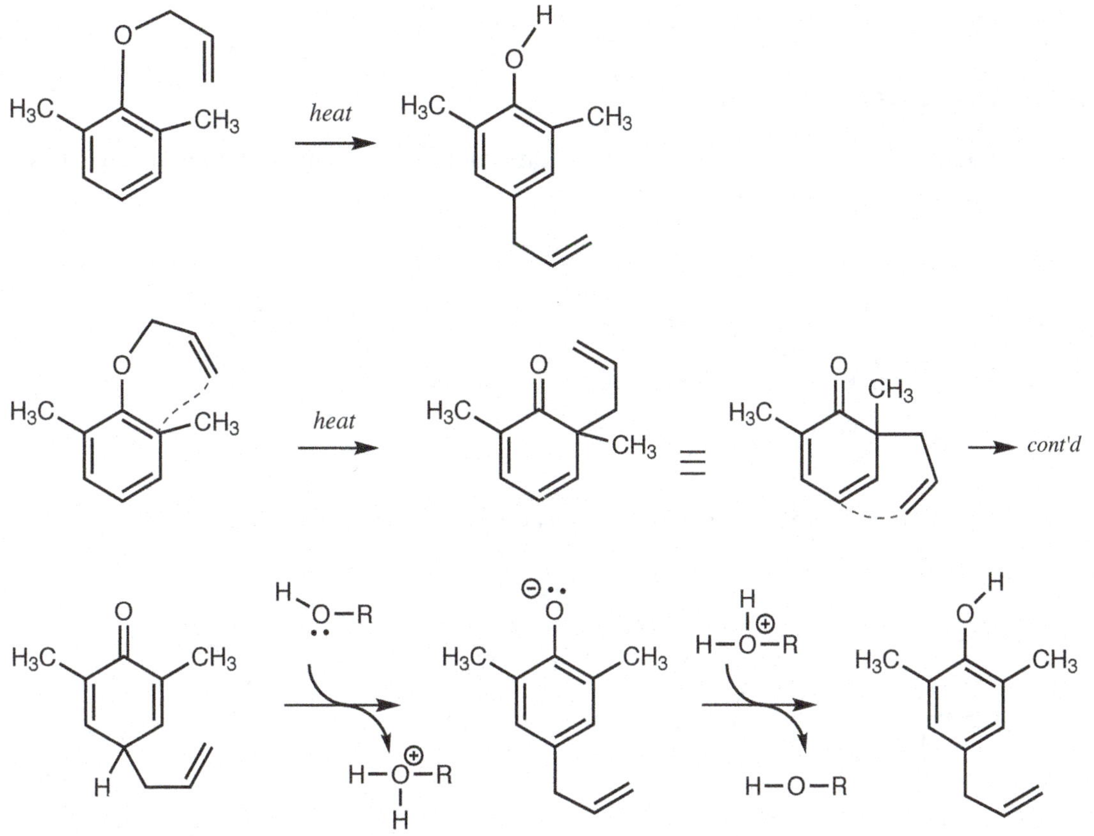

8 — Carbonyl Addition and Addition-Elimination Reactions

Additions by C, N, and O. Addition by hydrogen nucleophiles are discussed in Chapter 11, Reduction.

Grignard Addition to a Carbonyl Group

1. Addition of methyl magnesium bromide to cyclohexanone to give 1-methylcyclohexanol (for formation of Grignard reagents, see *Notes*).

2. Addition of a Grignard reagent to acetaldehyde to give 6-methyl-2-heptanol. (See *Notes*.)

Alkyllithium Addition to a Carbonyl Group

3. Addition of ethyllithium to benzaldehyde to give 1-phenylpropanol. (for formation of lithium reagent, see *Notes*).

4. Addition of propynyllithium to acetone to give 2-methylpent-3-yn-2-ol. (see Chapter 2.14)

Wittig Reaction

5. Wittig reaction, Step 1, formation of Wittig reagent. (See *Notes*.)

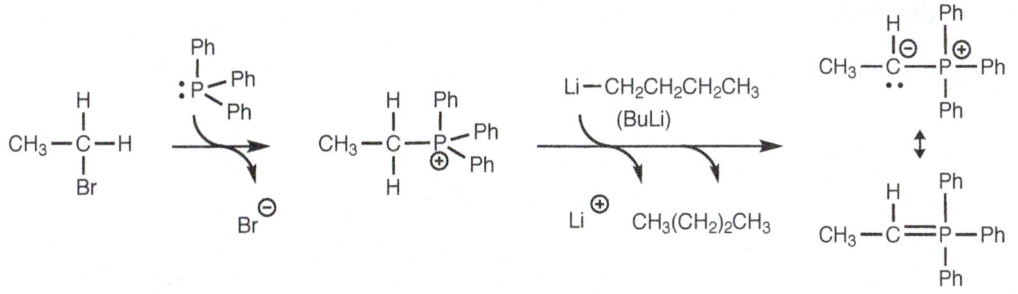

Step 2, reaction with benzaldehyde

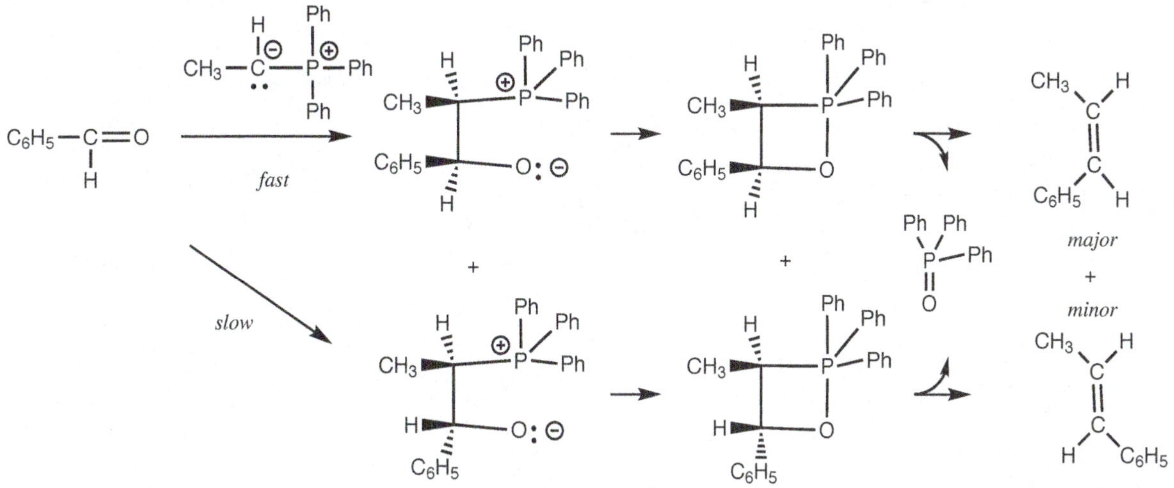

6. Wittig reaction, Horner-Wadsworth-Emmons modification, Step 1, Arbusov reaction. (See *Notes*.)

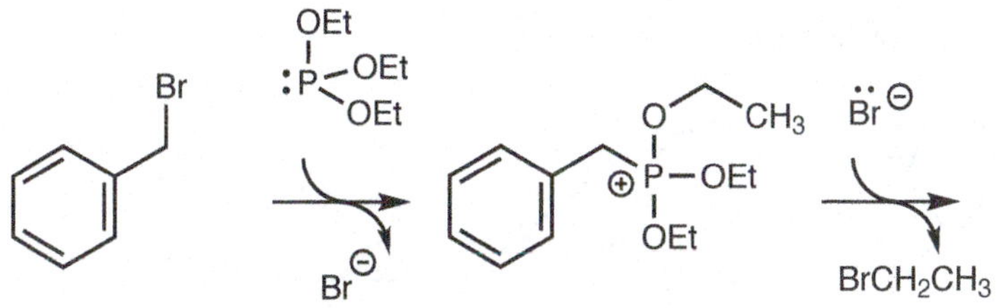

Step 2, reaction with benzaldehyde

cont'd

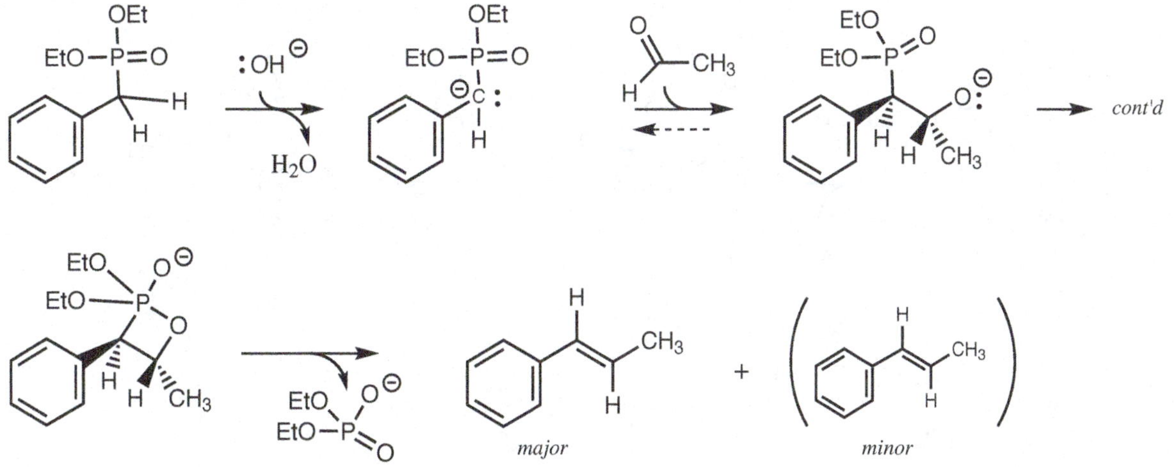

Addition-Elimination Reactions (Reversible Additions)

Ketal Formation and Hydrolysis

7. Acid catalyzed ketalization of cyclohexanone. (See *Notes*.)

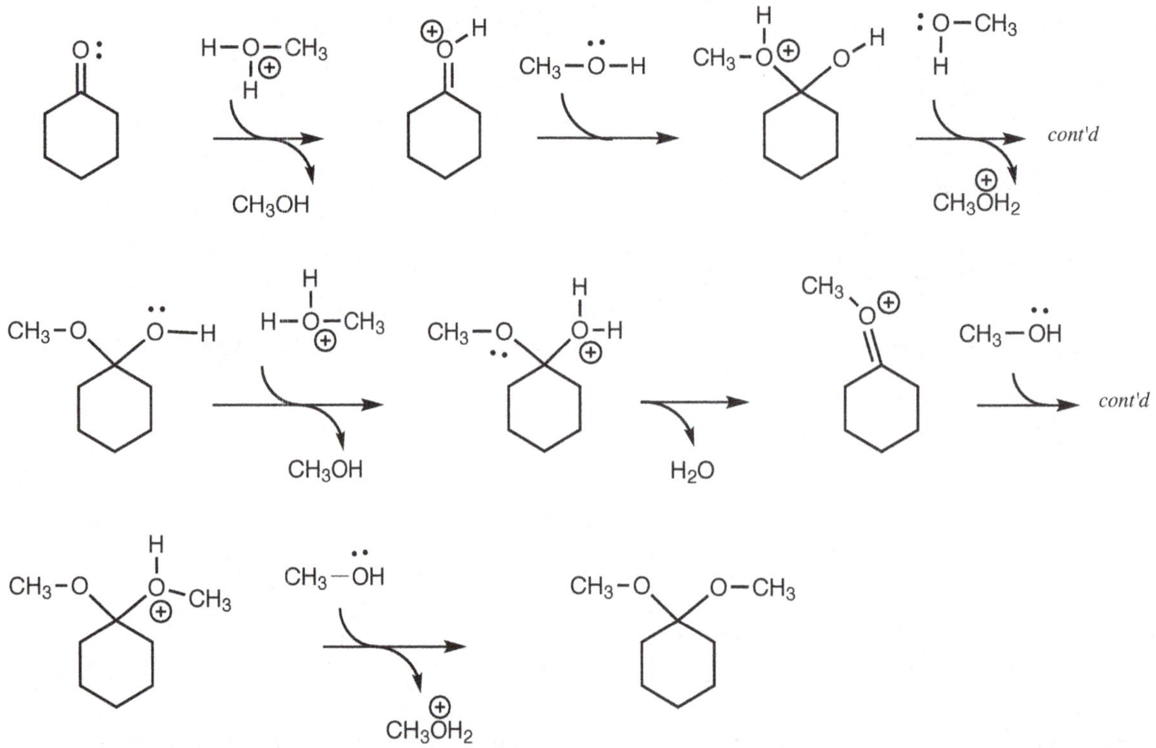

8. Acid catalyzed hydrolysis of the dioxolane acetal of benzaldehyde. (See *Notes*.)

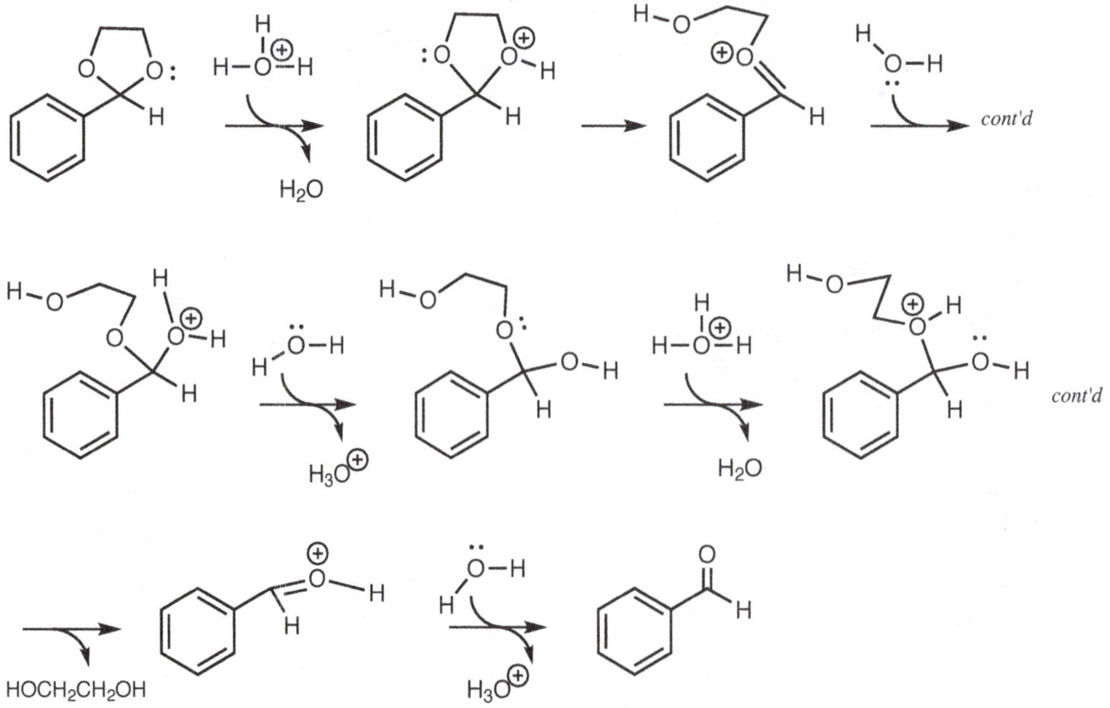

Oxime Formation

9. Formation of the oxime of cyclohexanone. (See *Notes*.)

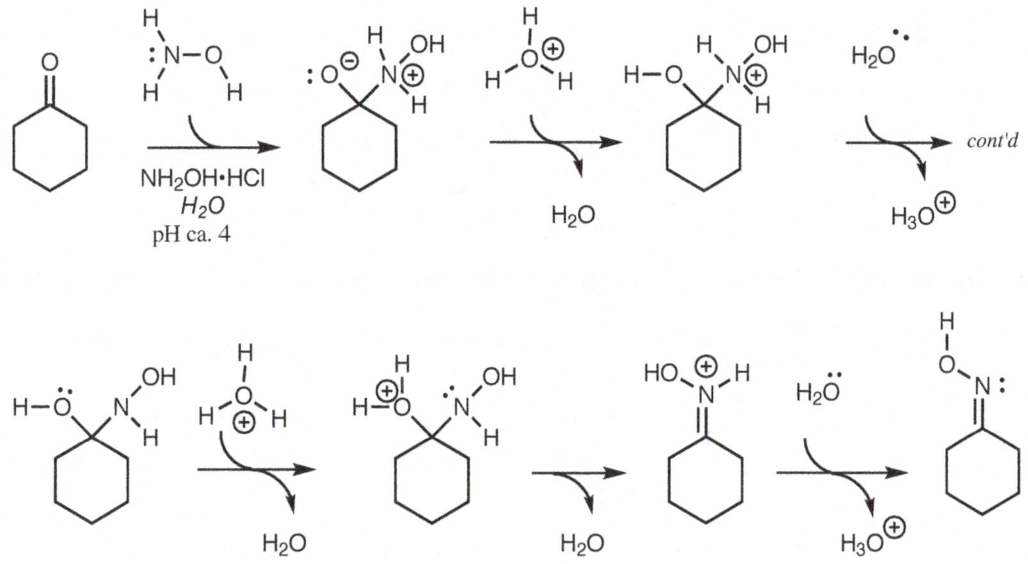

Other Additions to a Carbonyl Group

10. Formation of the cyanohydrin (2-hydroxy-2-methylpropanenitrile) from acetone. (See *Notes*.)

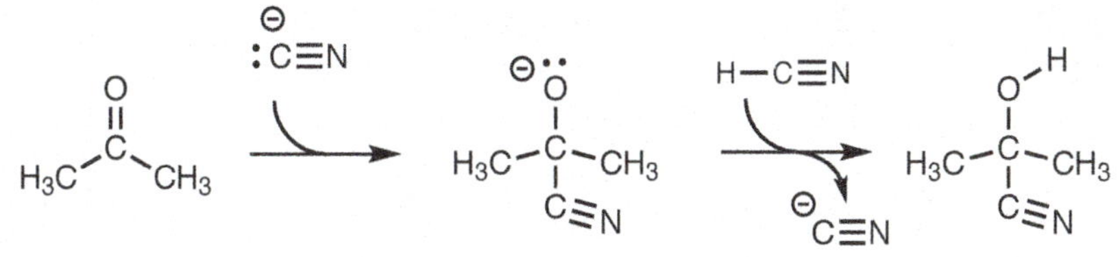

11. Reversion of the cyanohydrin (2-hydroxy-2-methylpropanenitrile) to form acetone. (See *Notes*.)

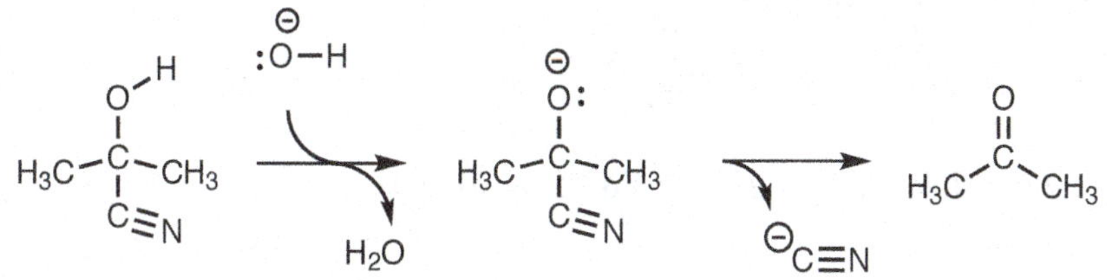

12. Addition of ethyl acetate enolate to 4-bromobenzaldehyde to give a benzyl alcohol.

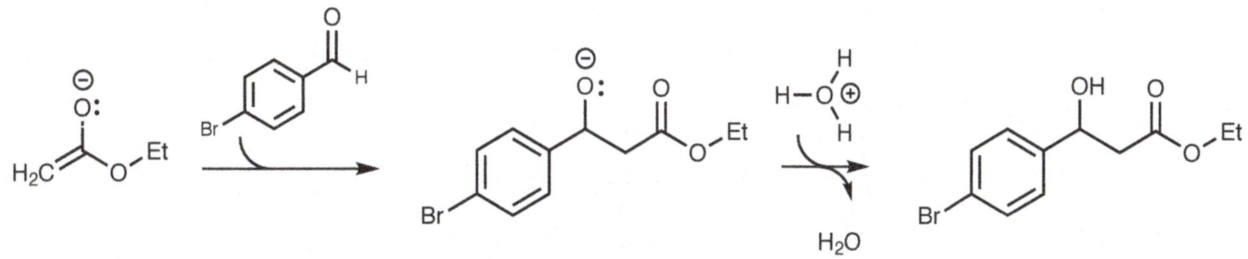

Reactions of Acyl Chlorides, Anhydrides, Esters, and Amides

Esters from Acid Chlorides or Anhydrides

13. Reaction of benzoyl chloride with ethoxide to give ethyl benzoate. (See *Notes*.)

14. Direct reaction of ethanol with acetyl chloride to give ethyl acetate. (See *Notes*.)

15. Pyridine catalyzed acylation with benzoyl chloride to give ethyl benzoate. (See *Notes*.)

16. Reaction of acetic anhydride with ethanol catalyzed by sulfuric acid. (See *Notes*.)

Amides from Acid Chlorides or Anhydrides

17. Reaction of benzoyl chloride with ethylamine to give *N*-ethylbenzamide. (See *Notes*.)

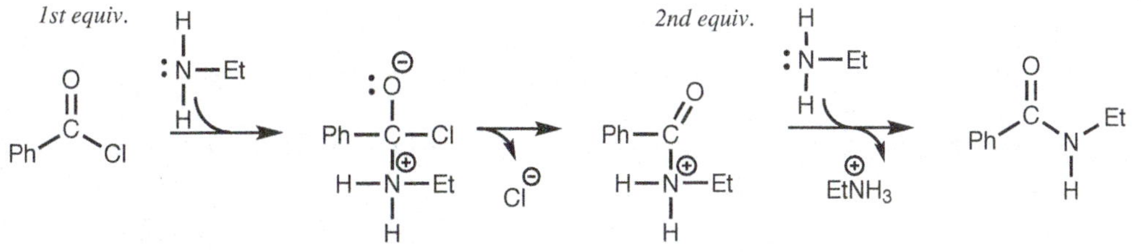

18. Reaction of acetic anhydride with aniline (and pyridine) to give acetanilide. (See *Notes*.)

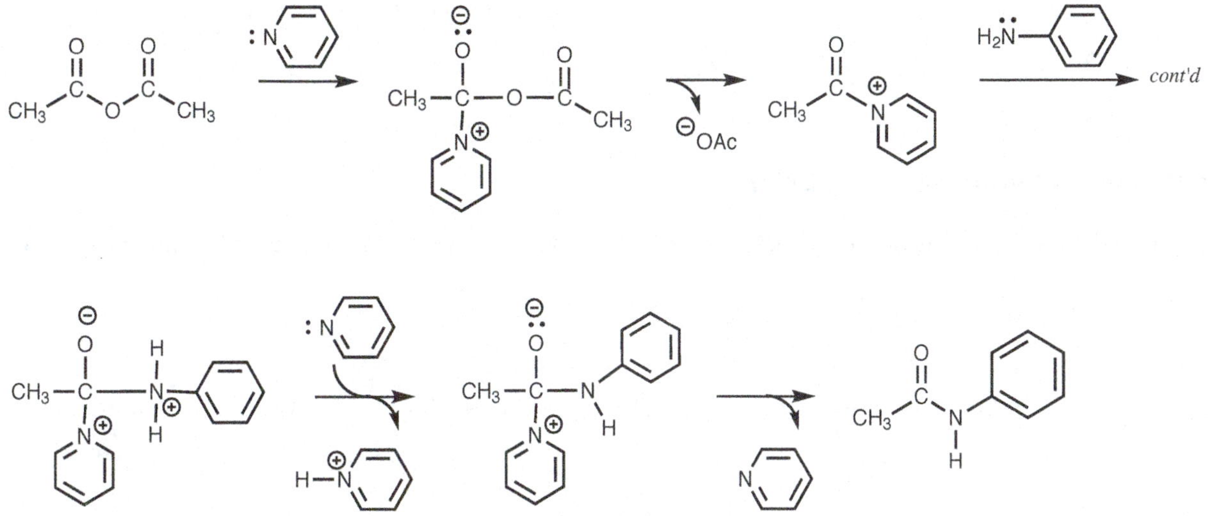

19. Reaction of acetic anhydride with aniline to give acetanilide. (See *Notes*.)

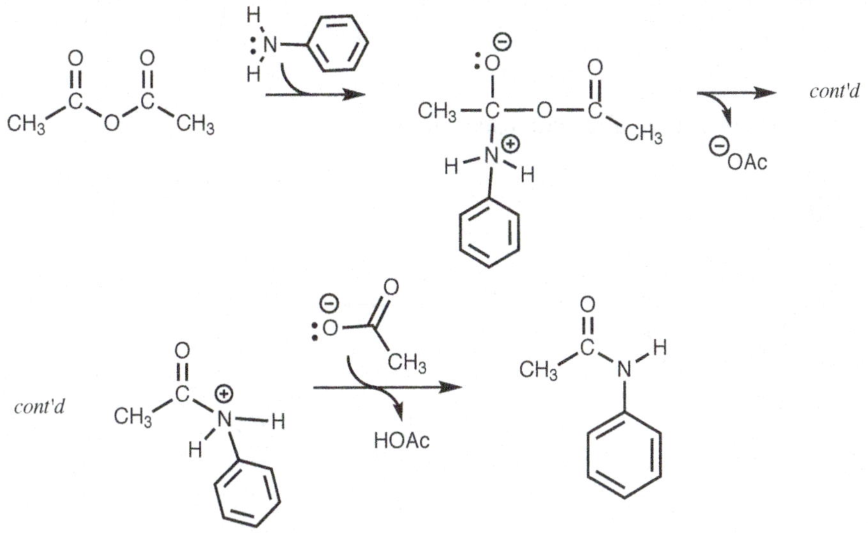

Ester from Acid with Mineral Acid Catalysis (Fischer Esterification)

20. Acid catalyzed esterfication. Formation of methyl benzoate from benzoic acid and methanol with sulfuric acid or hydrogen chloride. (See *Notes*.)

Acid Catalyzed Hydrolysis of an Ester

21. Acid catalyzed hydrolysis of methyl pentanoate (valerate) to pentanoic (valeric) acid plus methanol. (See *Notes*.)

22. Acid catalyzed conversion of *t*-butyl pivalate to pivalic acid. (See *Notes*.)

TFA

Base Hydrolysis of an Ester (Saponification)

23. Base hydrolysis of octyl isobutyrate to give octanol and isobutyric acid. Step 1, treatment with base. (See *Notes*.)

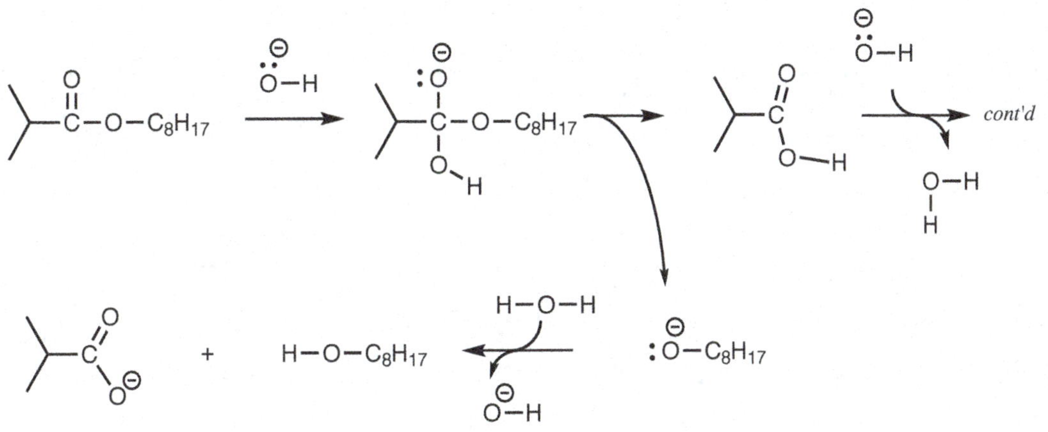

Step 2, acidification of isobutyrate and isolation of isobutyric acid.

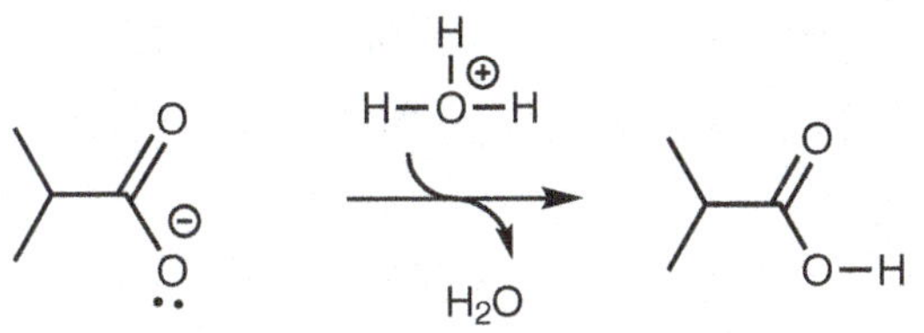

Hydrolysis of an Amide

24. Base hydrolysis of *N*-butylacetamide to give *n*-butylamine and acetic acid. (See *Notes*.)

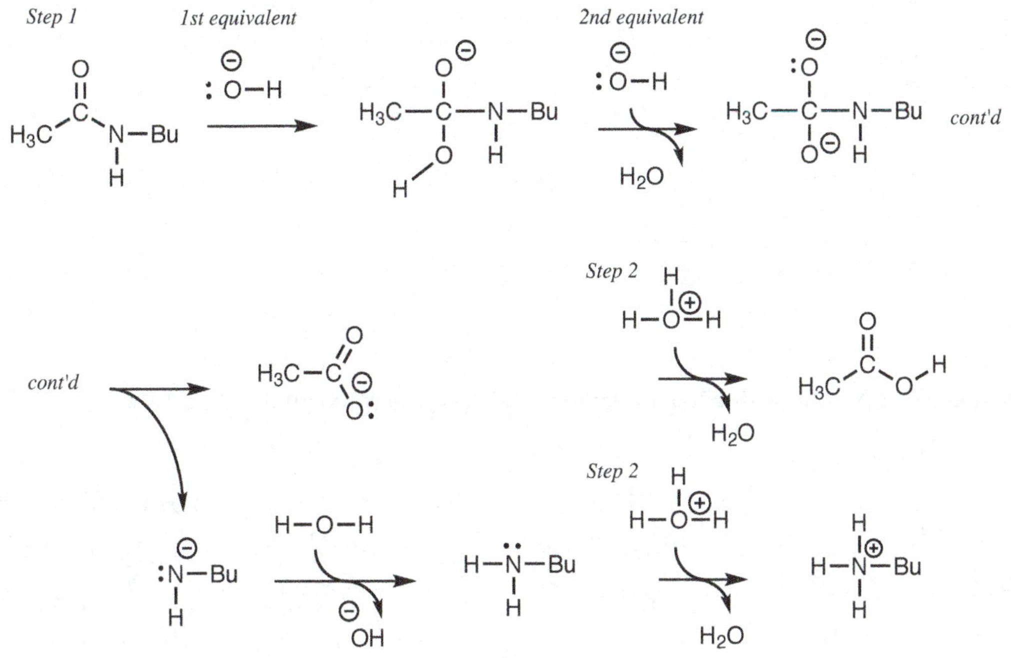

25. Acid catalyzed hydrolysis of *N,N*-dimethylacetamide to give acetic acid and dimethylammonim chloride. (See *Notes*.)

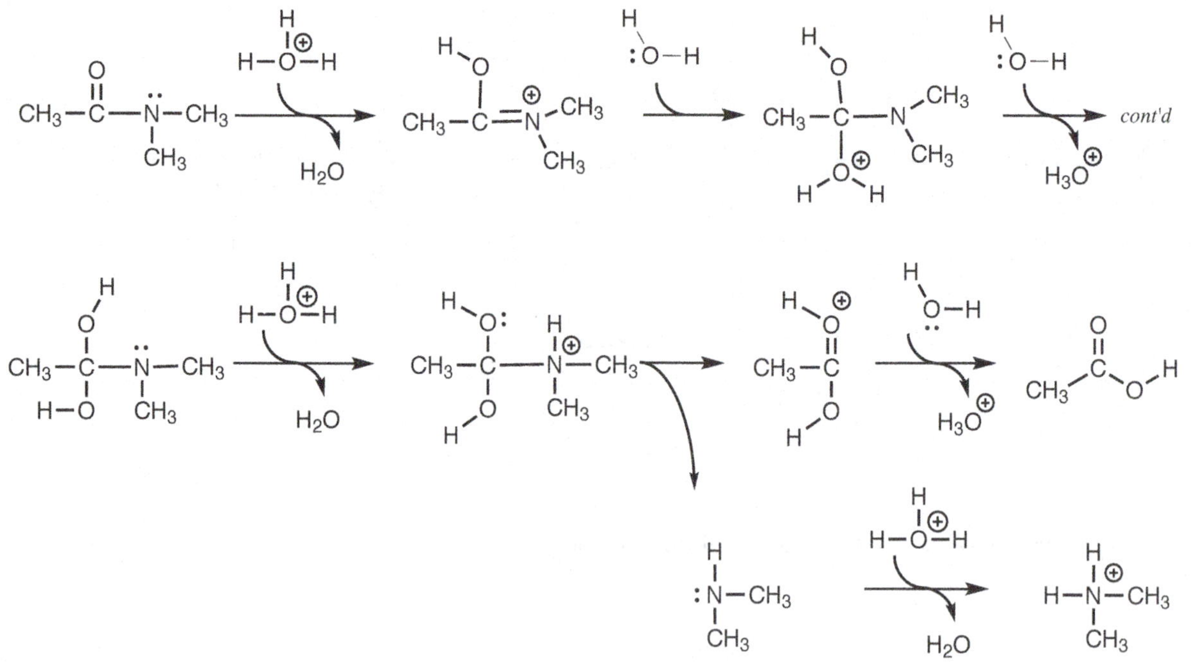

Reactions of Esters

26. Addition of methyl magnesium bromide to ethyl benzoate to give 2-phenyl-2-propanol. (See *Notes*.)

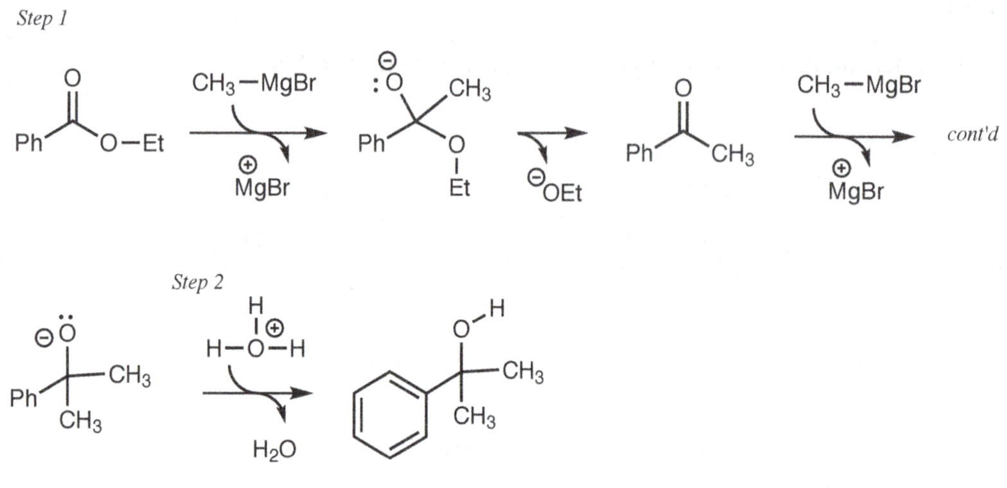

Step 1

Step 2

27. Reaction of 4-nitrophenyl propionate with ethyl amine to give *N*-ethyl propionamide. (See *Notes*.)

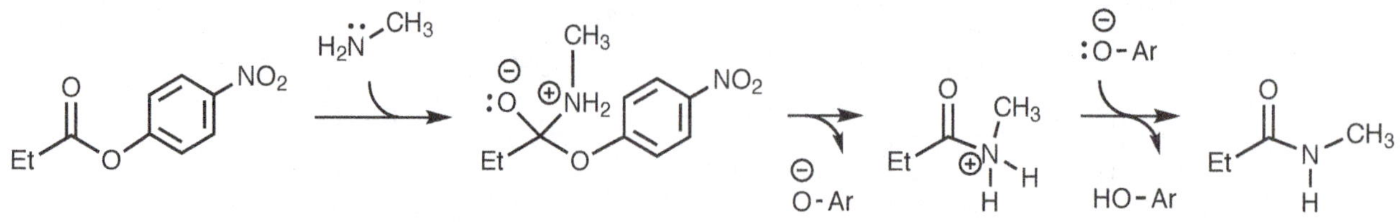

Reactions of Nitriles

28. Acid catalyzed hydrolysis of benzonitrile to give benzoic acid and ammonium chloride. (See *Notes*.)

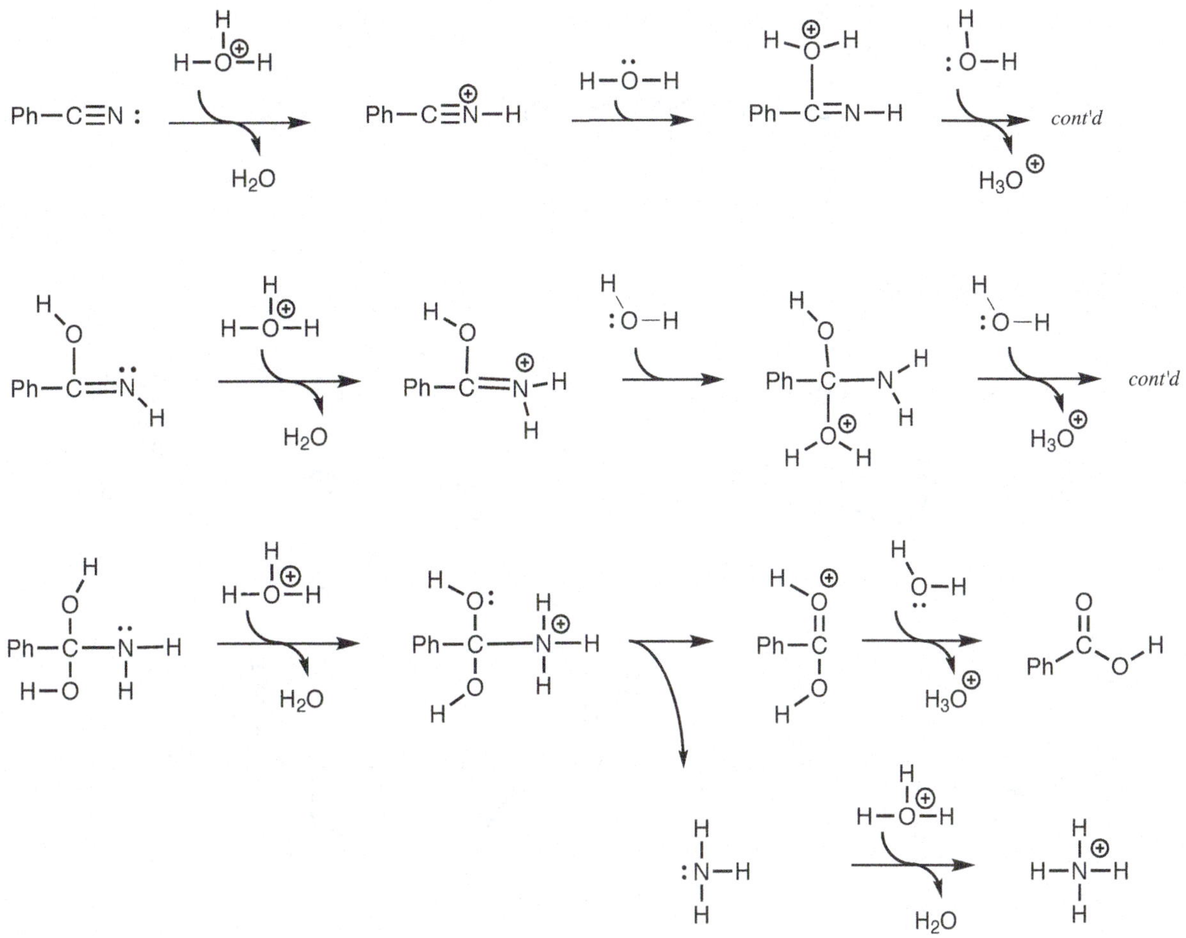

29. Base catalyzed hydrolysis of a nitrile, conversion of cyclopentanecarbonitrile to cyclopentanecarboxamide. (See *Notes*.)

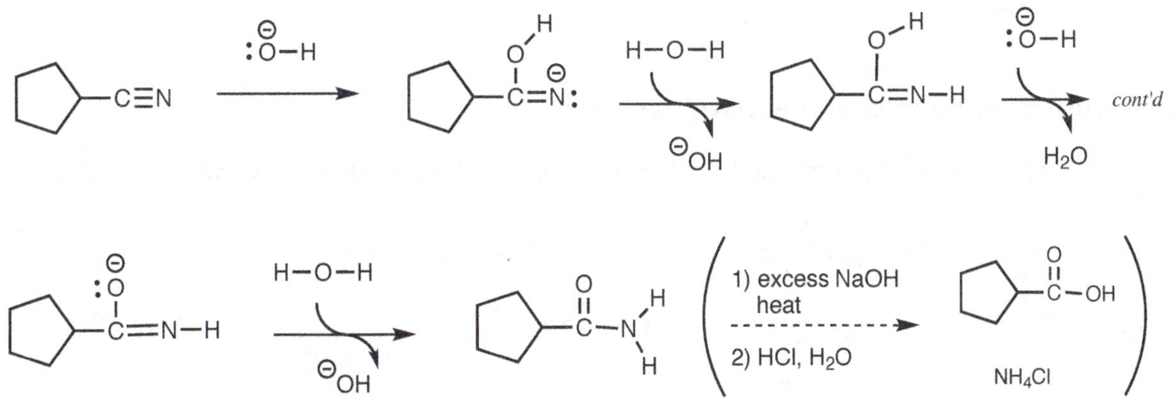

30. Addition of phenyl lithium to a nitrile to give, after hydrolysis, cyclohexyl phenyl ketone. (See *Notes*.)

Step 1, addition to the nitrile.

Step 2, hydrolysis of the imine.

Miscellaneous, Ester Formation with Diazomethane

31. Reaction of diazomethane with a carboxylic acid to form methyl 2-butenoate (methyl crotonate). (See *Notes*.)

9 — Reactions of Enols and Enolates

Aldol Reaction

1. Base catalyzed aldol condensation of butanal (butyraldehyde). (See Notes.)

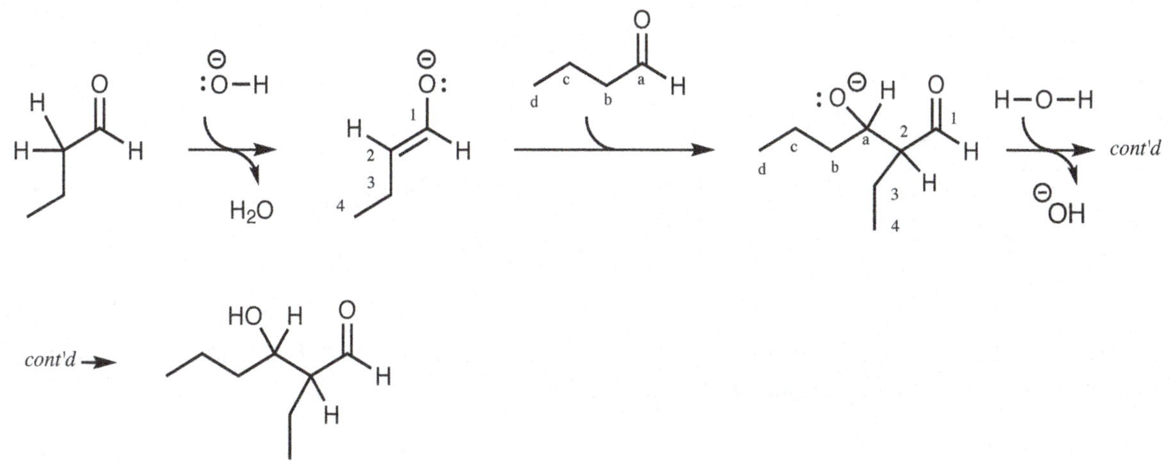

Base catalyzed dehydration step. Under concentrated base or heating, the dehydration reaction may occur spontaneously. (See Notes.)

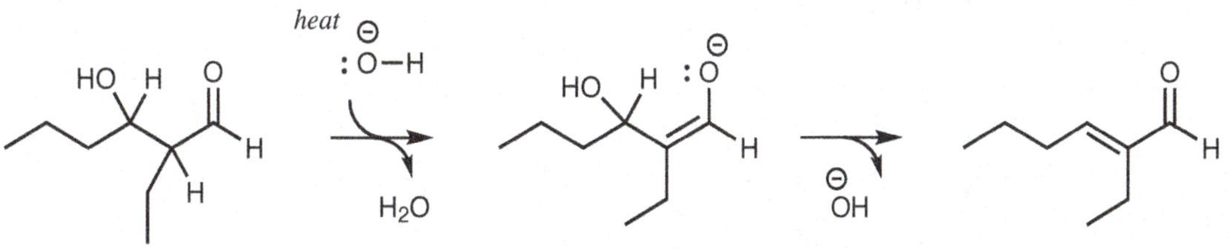

2. Directed kinetic aldol condensation of 2-methylcyclohexanone with propanal (propionaldehyde). Step 1, enolate formation; step 2, reaction with propanal; step 3, neutralization. (See Notes.)

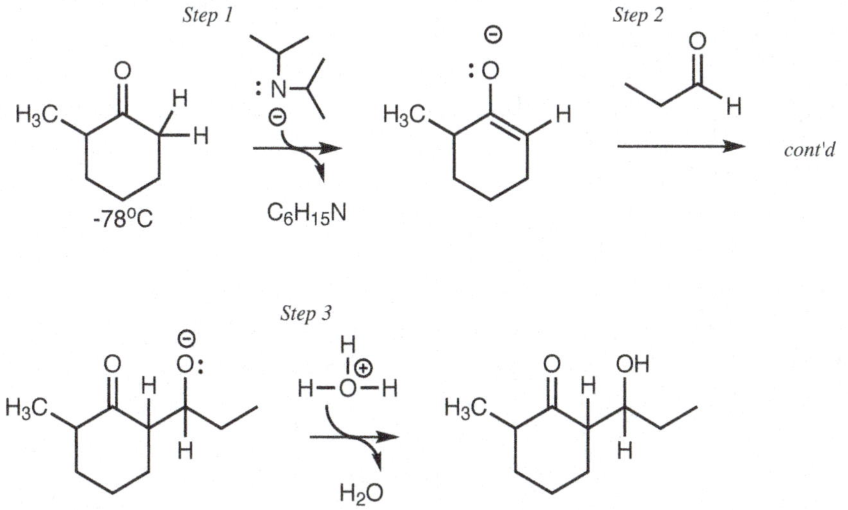

3. Base catalyzed mixed or crossed aldol condensation of acetone and benzaldehyde.

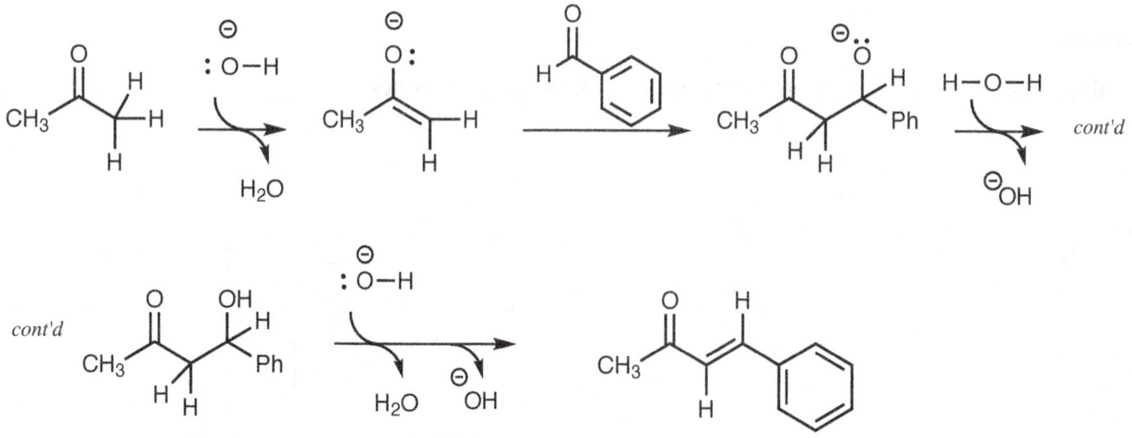

4. Mannich reaction, acid catalyzed enolization of 2-propanone in a reaction with diethylamine, formaldehyde, and 2-propanone to give 4-(diethylamino)butan-2-one. (See *Notes*.)

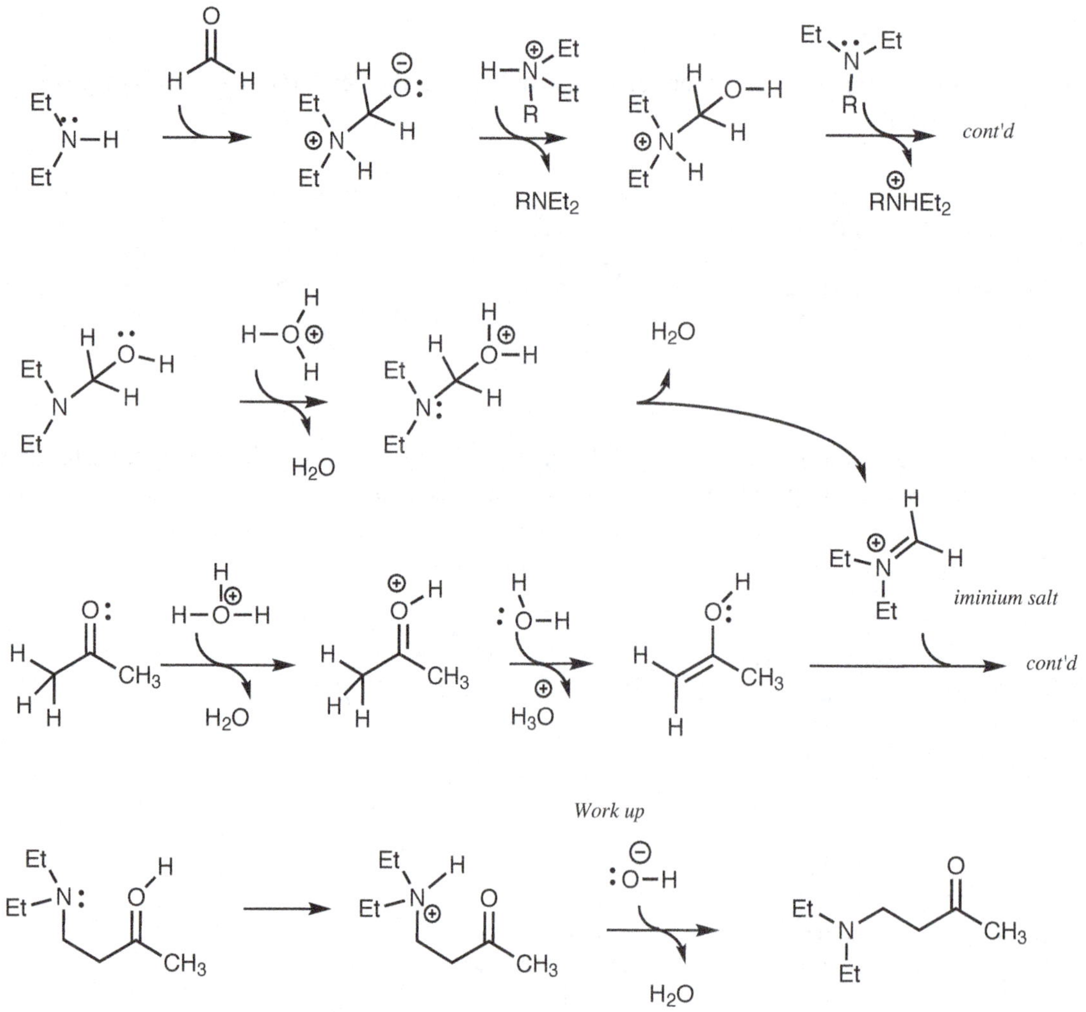

Work up

Claisen Condensation

5. Ethoxide catalyzed Claisen condensation of ethyl acetate to ethyl acetoacetate (ethyl 3-oxobutanoate). (see *Notes*)

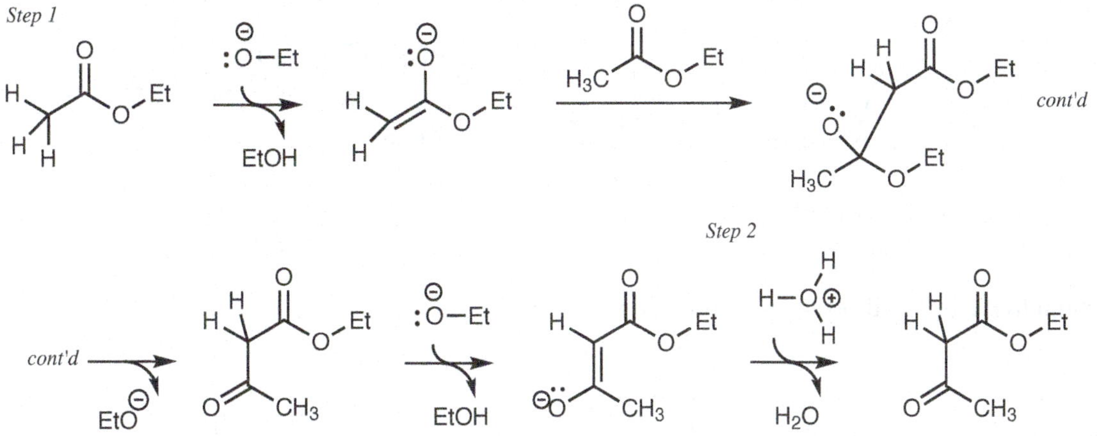

Step 1

Step 2

6. Ethoxide catalyzed mixed or crossed-Claisen condensation of cyclohexanone and ethyl formate. (See *Notes*.)

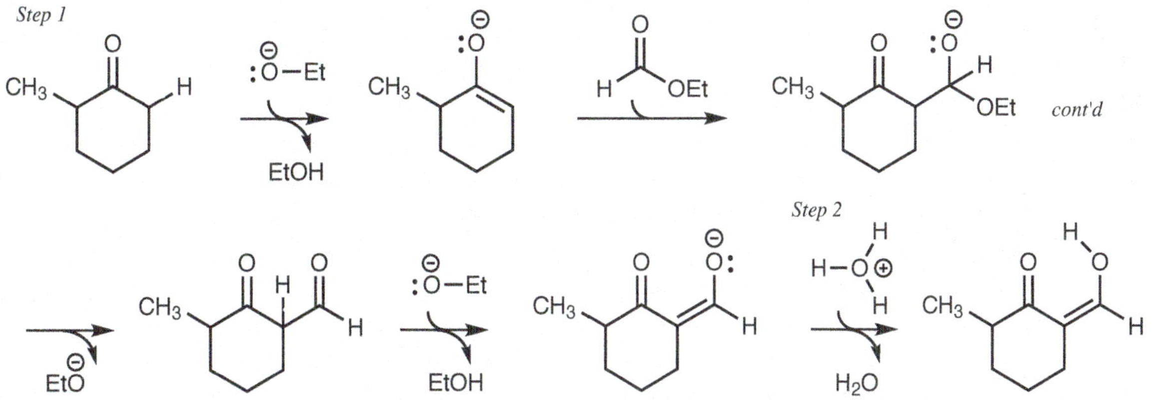

Step 1

Step 2

7. Ethoxide catalyzed mixed or crossed-Claisen condensation of ethyl acetate and ethyl benzoate. (See *Notes*.)

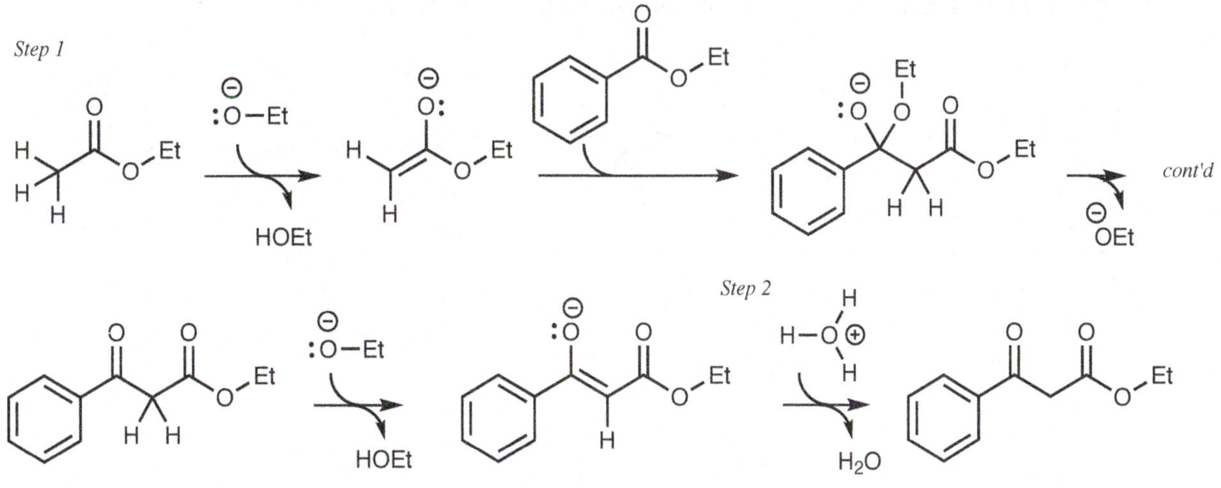

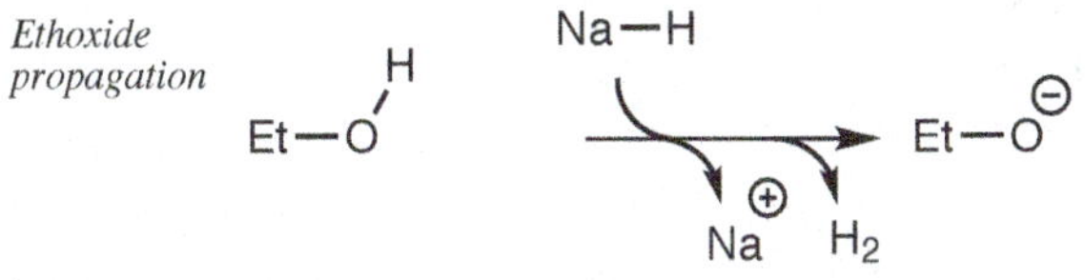

Step 1

Step 2

Ethoxide propagation

Acetoacetate Synthesis

8.   Step 1, S$_N$2 alkylation of acetoacetate.

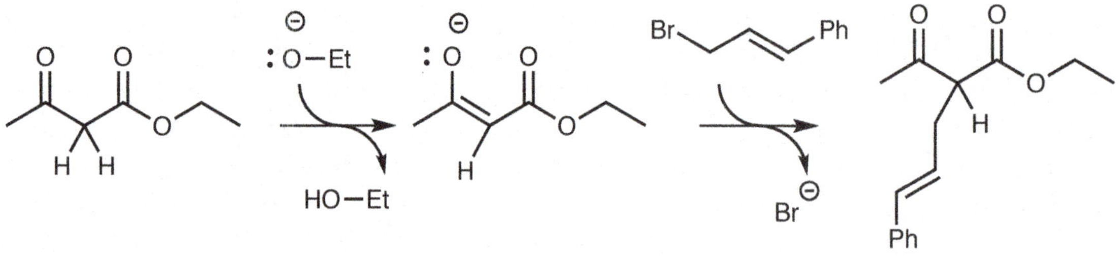

Step 2, sodium hydroxide hydrolysis of ester (saponification).

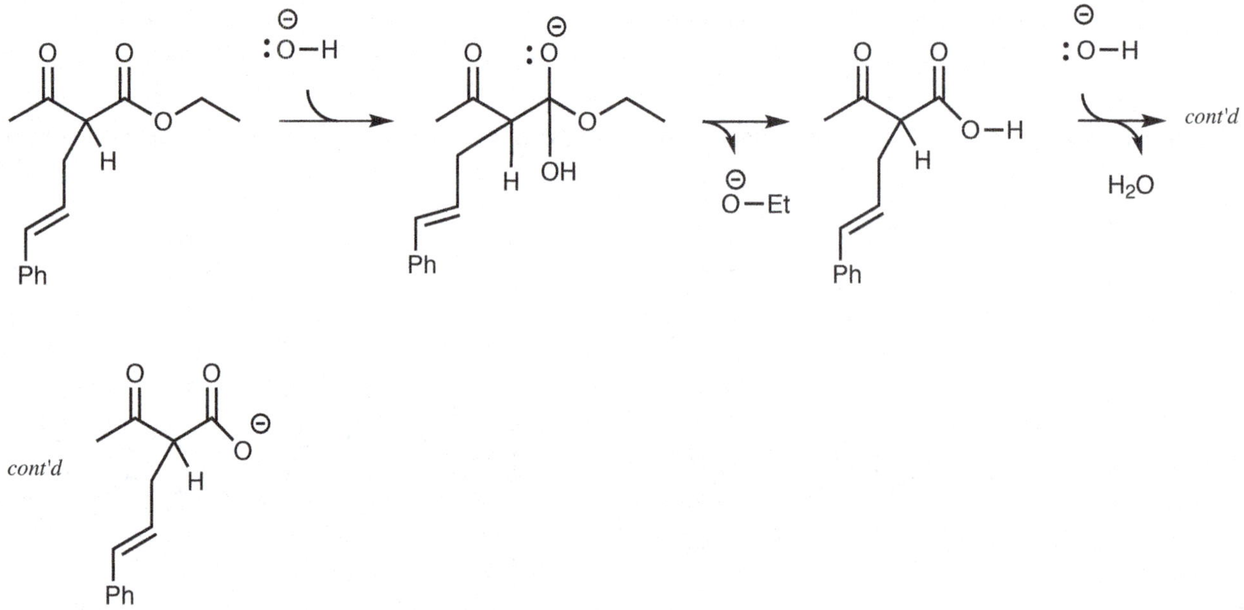

cont'd

Step 3, acidification of carboxylate, decarboxylation, and tautomerization.

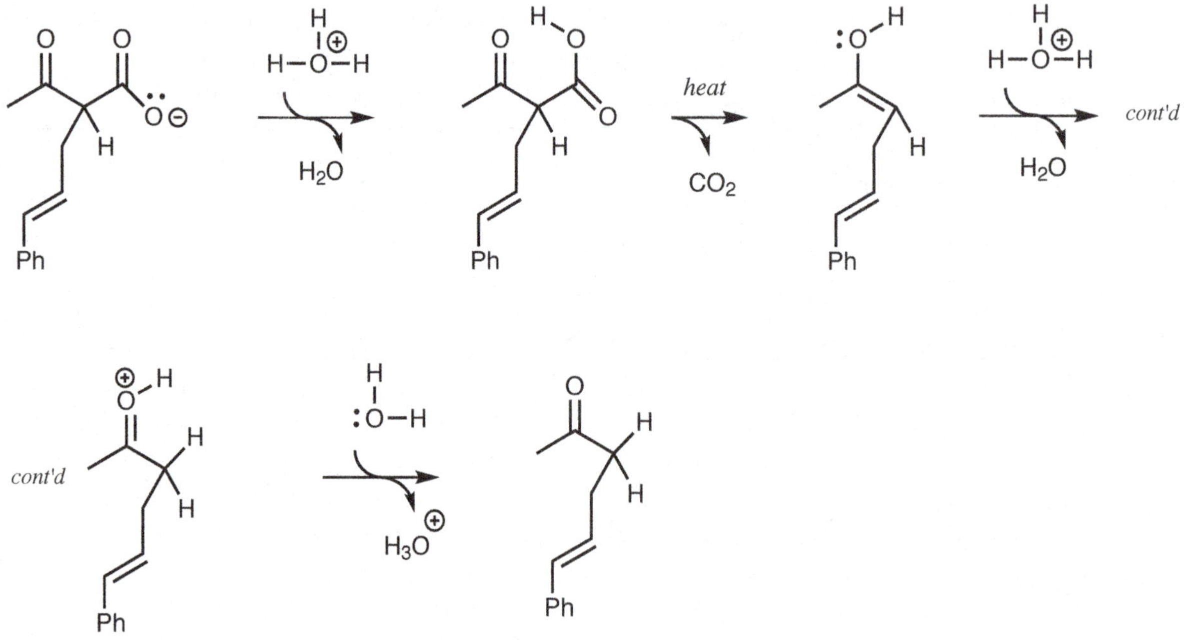

cont'd

Enolate Alkylation Reactions

9. Enolization and alkylation of ethyl propionate with benzyl bromide. (See *Notes*.)

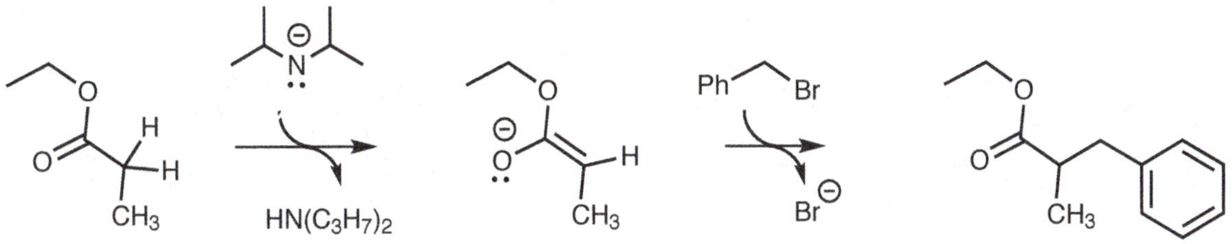

10. Sequential alkylation of a dianion of a *beta*-ketoester. (See *Notes*.)

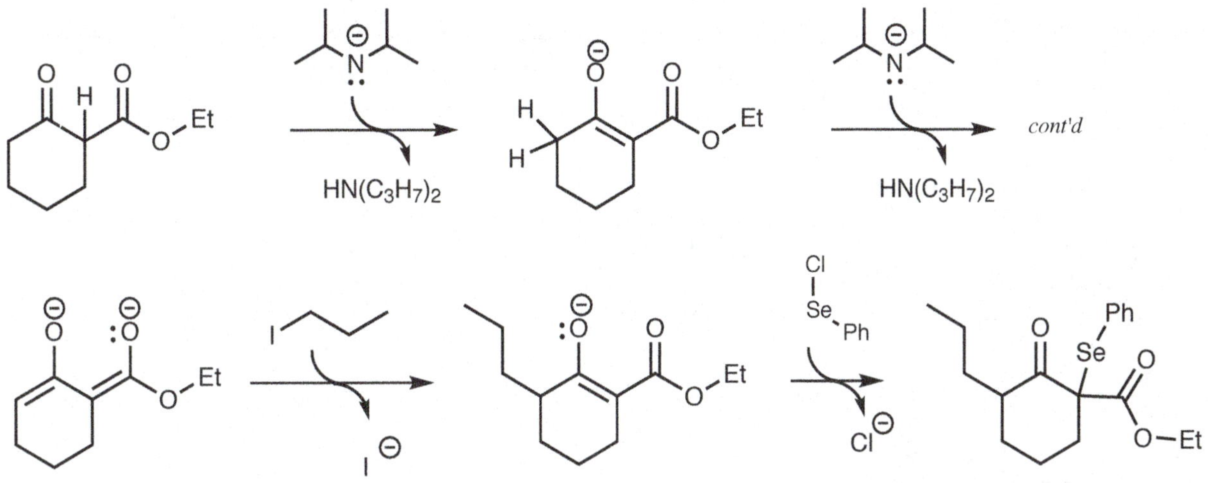

11. A retro-Claisen reaction from a *beta*-ketoester. (See *Notes*.)

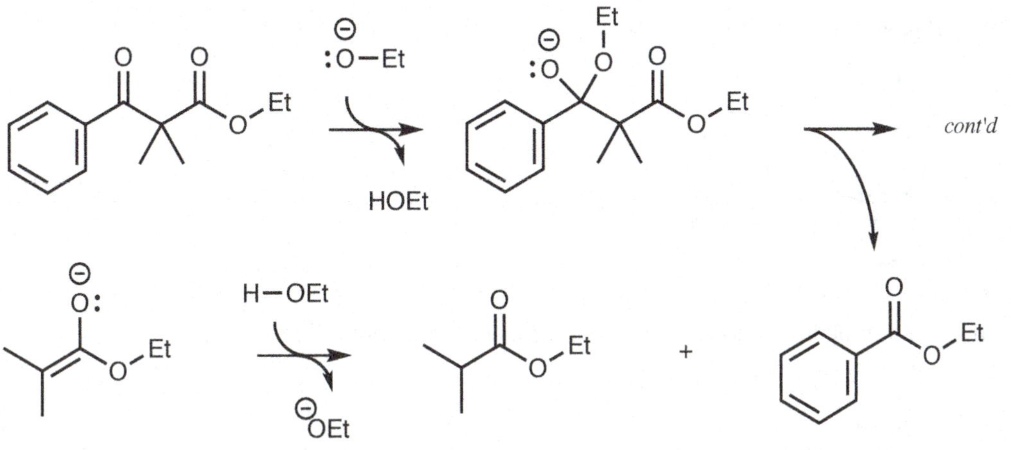

12. Enolization and alkylation of phenylacetonitrile with methyl iodide. (See *Notes*.)

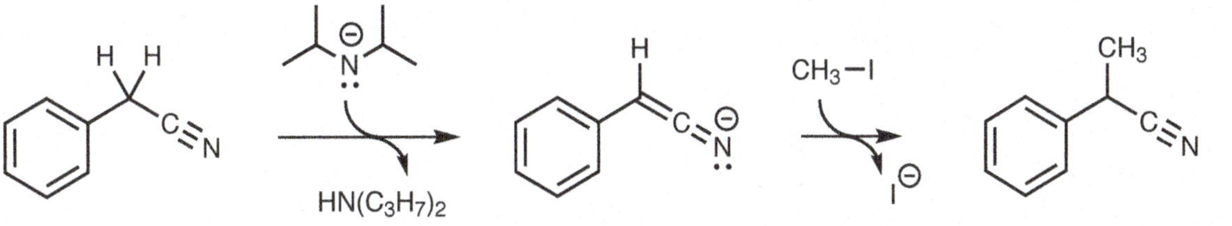

Halogenation of Carbonyl Compounds

13. Basic bromination of 3-methyl-2-butanone with sodium hydroxide and bromine, bromoform reaction. (See *Notes*.)

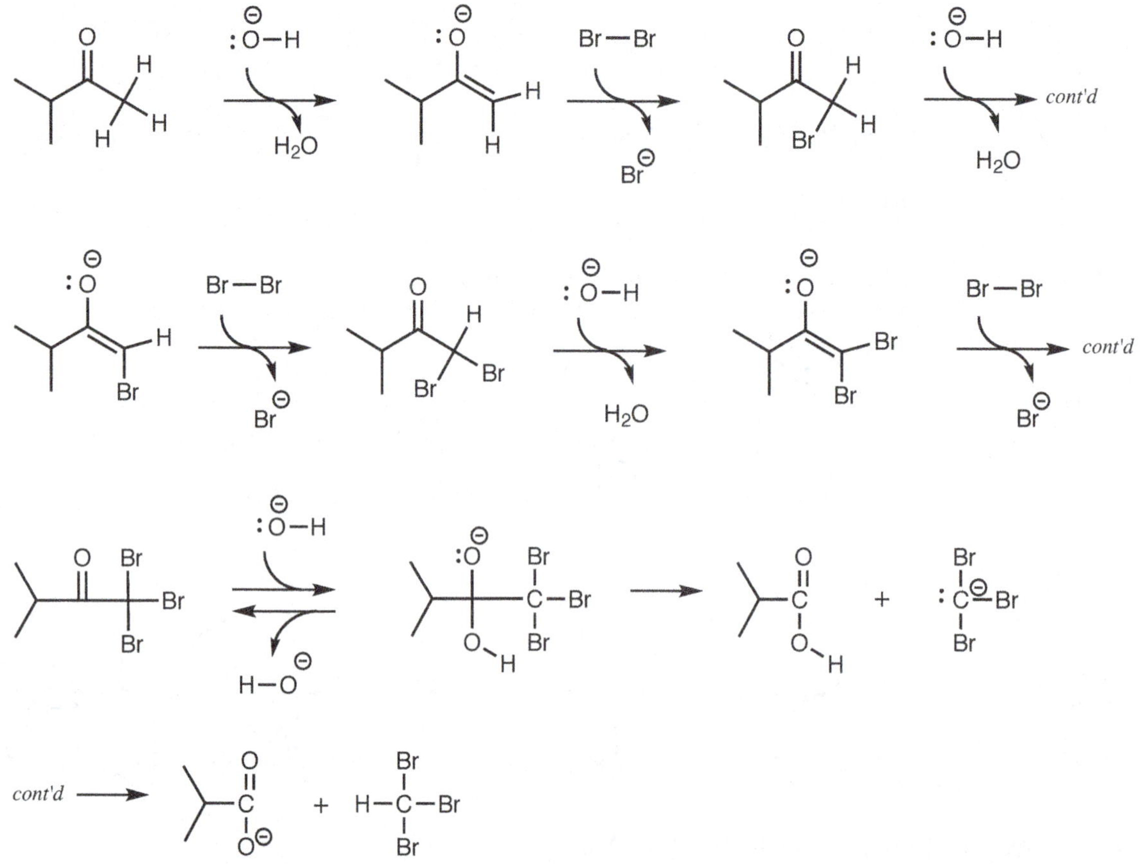

Step 2, acidification and isolation of isobutyric acid.

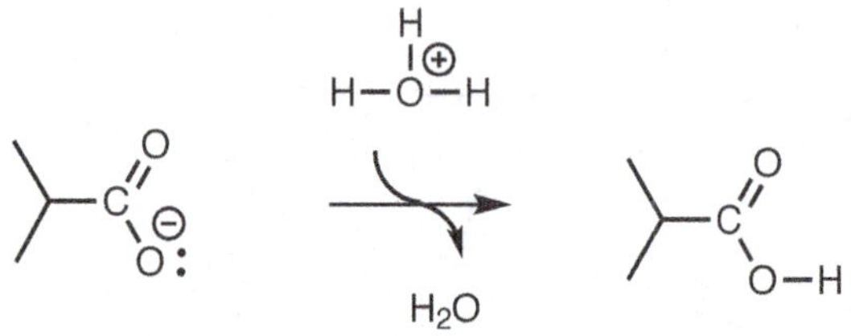

14. Acid catalyzed bromination of acetophenone to give α-bromoacetophenone. (See *Notes*.)

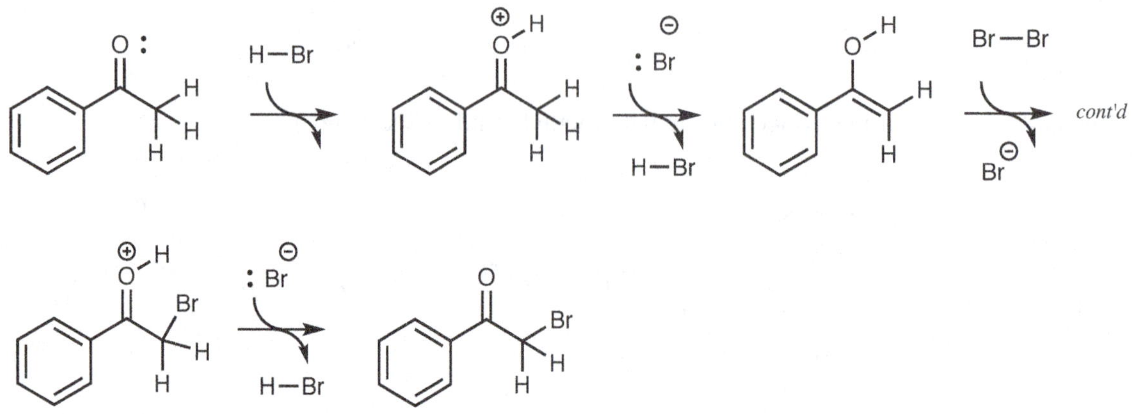

Michael Addition or 1,4-Conjugate Addition Reaction

15. Michael addition reaction of dimethylamine to methyl vinyl ketone. (See *Notes*.)

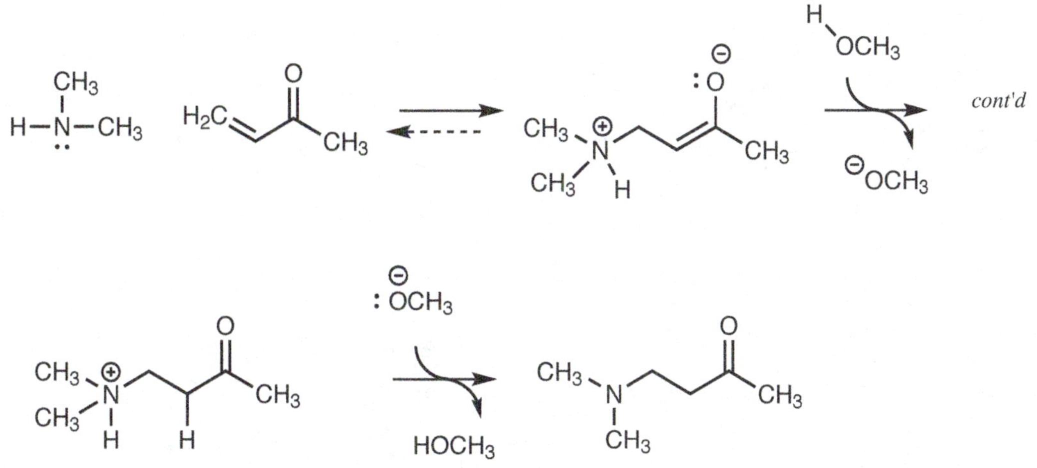

16. Michael addition reaction of ethyl acetoacetate to cyclohexenone. (See *Notes*.)

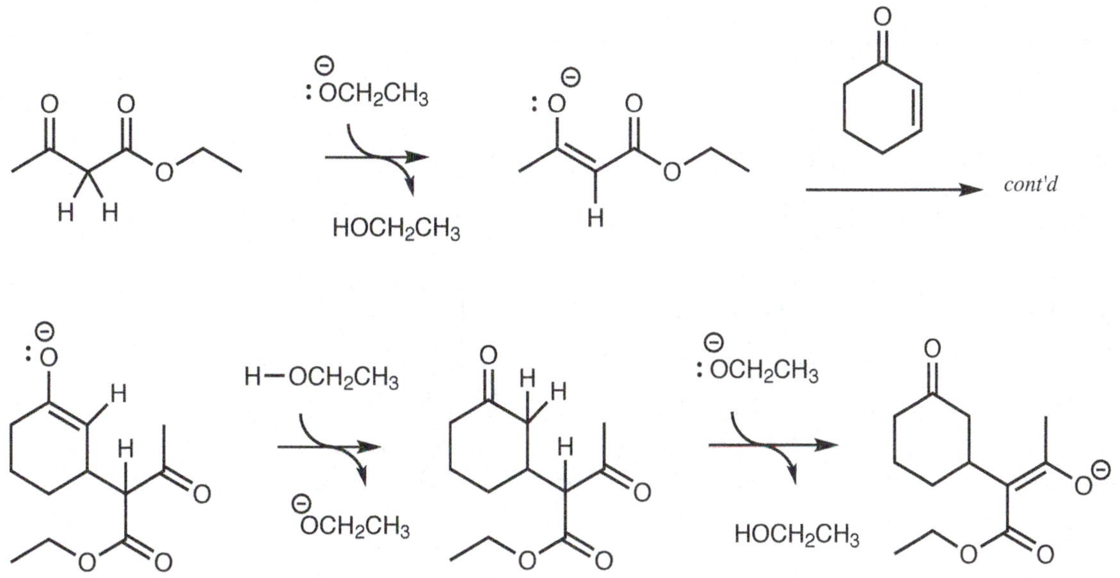

Step 2, neutralization of reaction mixture.

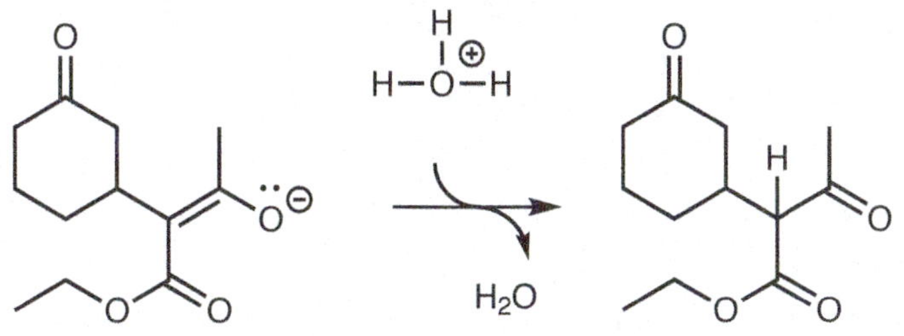

Enamine Alkylation

17. Step 1, formation of the morpholine enamine of cyclohexanone. (See *Notes*.)

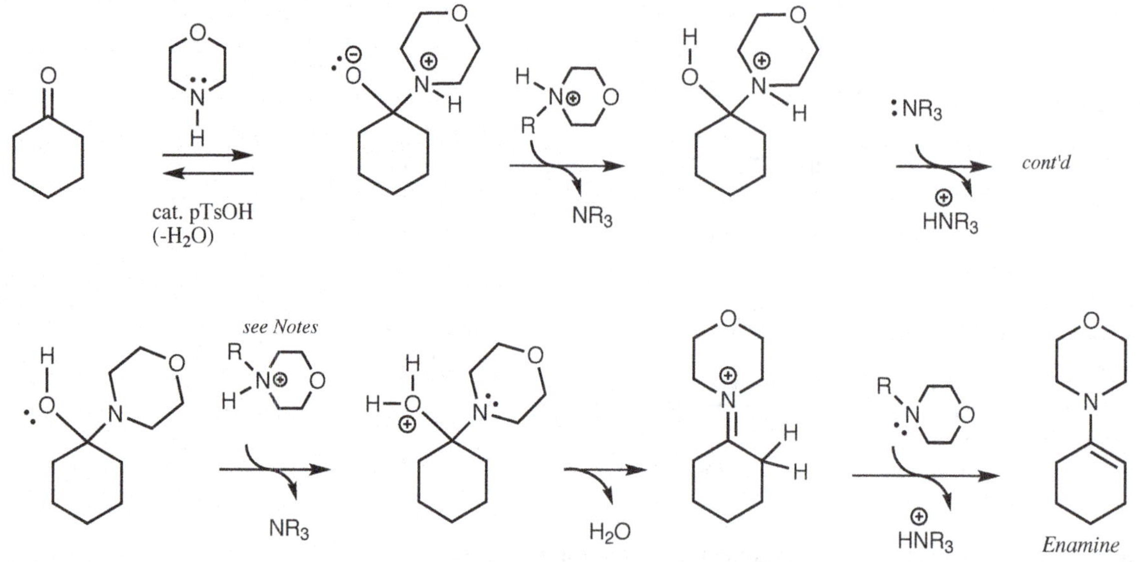

cat. pTsOH
(-H$_2$O)

:NR$_3$

cont'd

NR$_3$

HNR$_3$

see Notes

NR$_3$

H$_2$O

HNR$_3$

Enamine

Step 2, enamine alkylation with methyl iodide. (See *Notes*.)

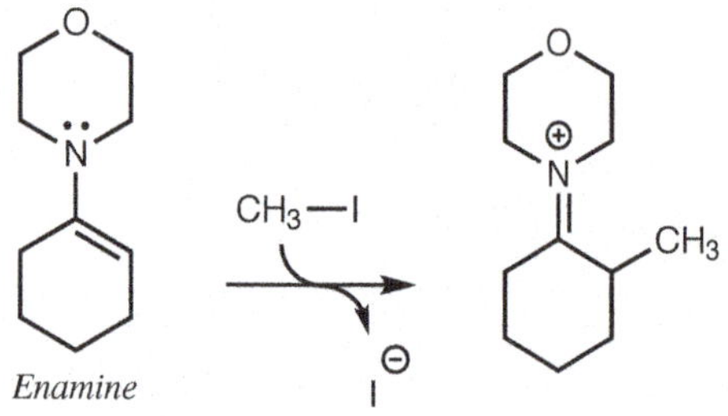

CH$_3$—I

Enamine

CH$_3$

Step 3, hydrolysis of iminium salt and isolation of substituted cyclohexanone.

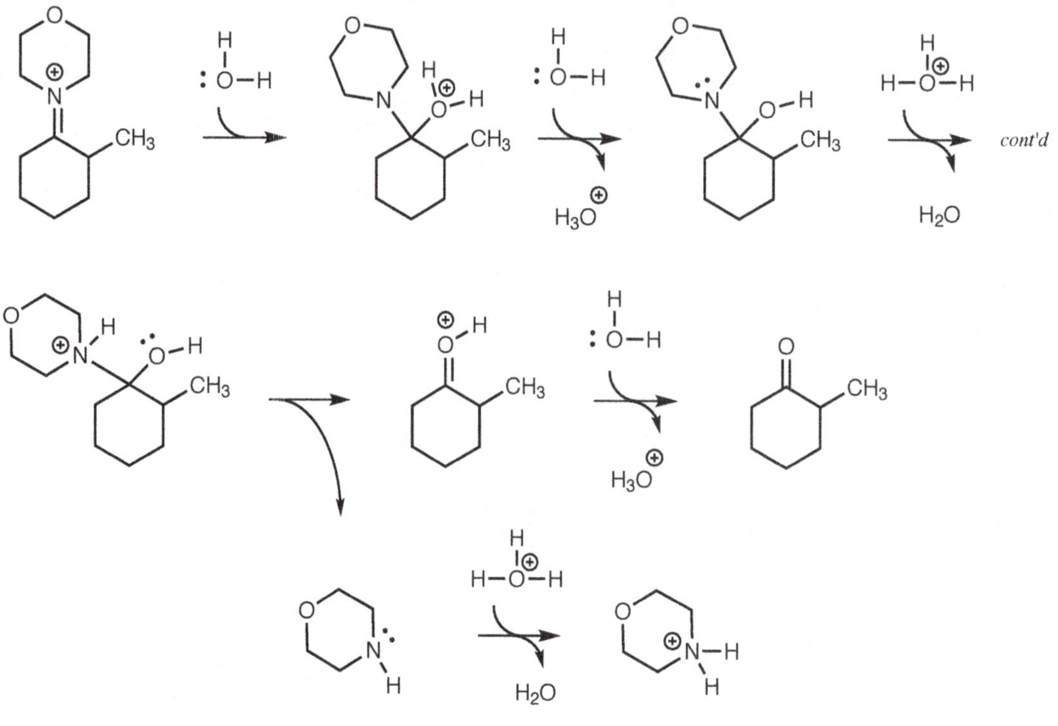

CH$_3$

CH$_3$

CH$_3$

cont'd

H$_3$O

H$_2$O

CH$_3$

CH$_3$

CH$_3$

H$_3$O

H$_2$O

10 — Dehydration/Halogenation Agents

1. Reaction of cyclohexanecarboxamide with acetic anhydride to give cyclohexanecarbonitrile. (See *Notes*.)

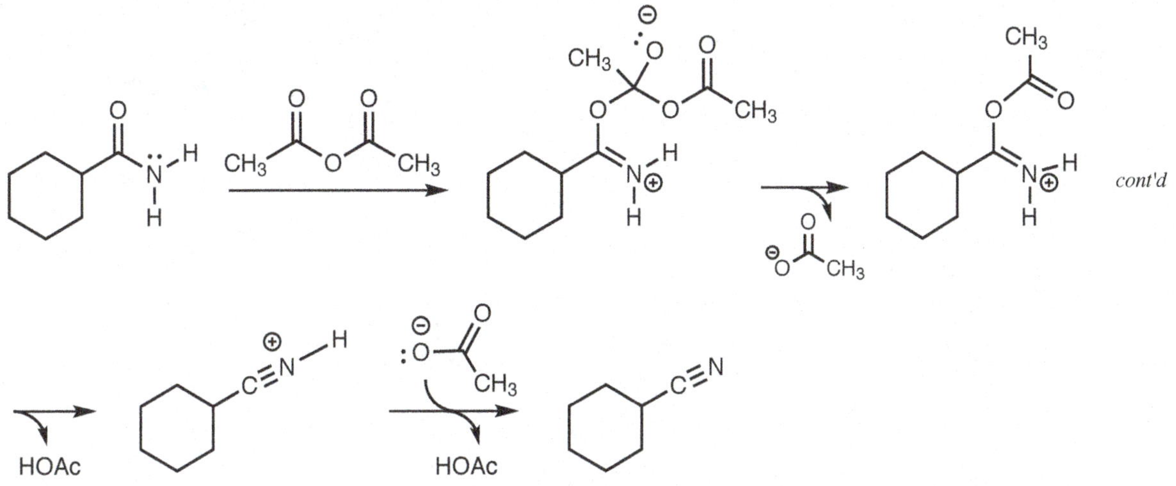
cont'd
HOAc
HOAc

2. S_N2 reaction of thionyl chloride with 1-butanol to give chlorobutane. (See *Notes*.)

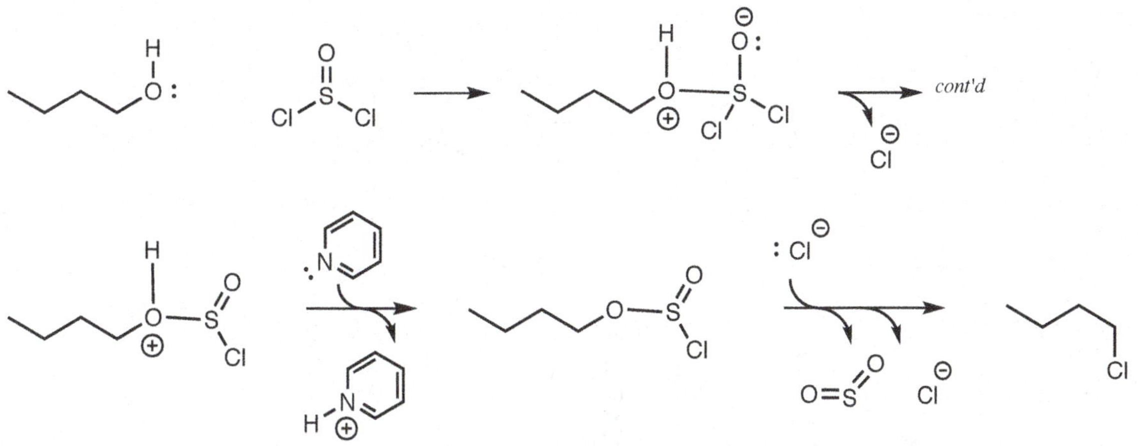
cont'd

3. Reaction of 1-butanol with tosyl chloride and pyridine to give butyl tosylate. (See *Notes*.)

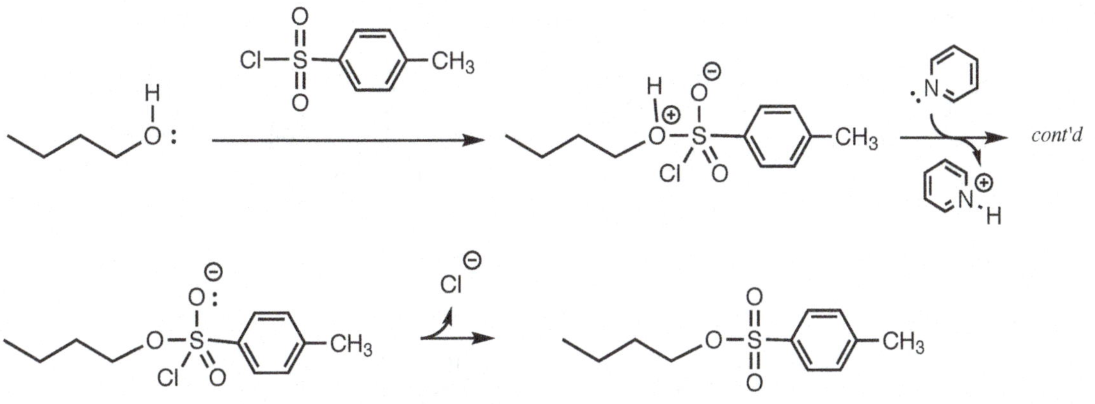
cont'd

Carboxylic Acid with Thionyl Chloride

4. Conversion of acetic acid to acetyl chloride with thionyl chloride.

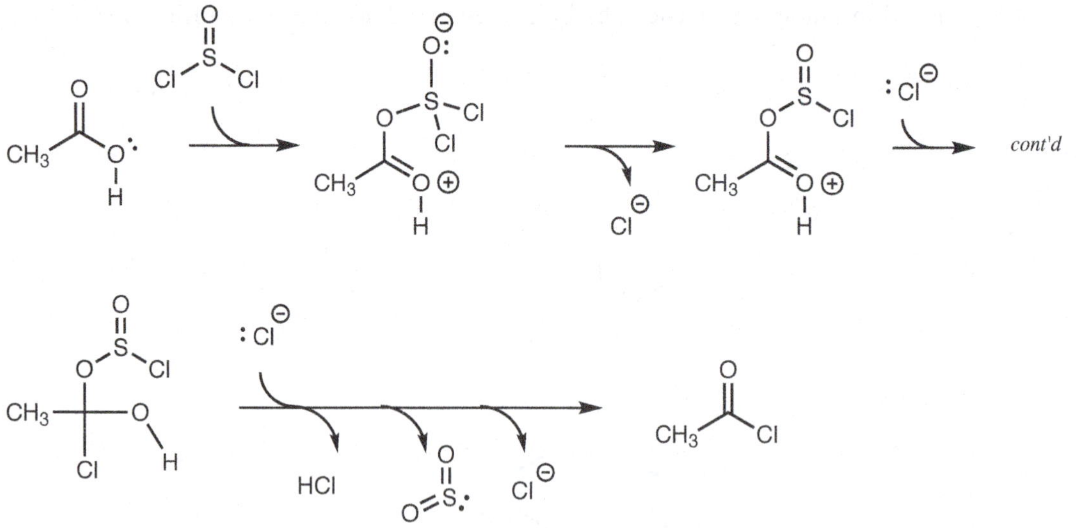

5. Conversion of acetic acid to acetyl chloride with thionyl chloride. Part 1, reaction of thionyl chloride with DMF to form iminium salt. (See Notes.)

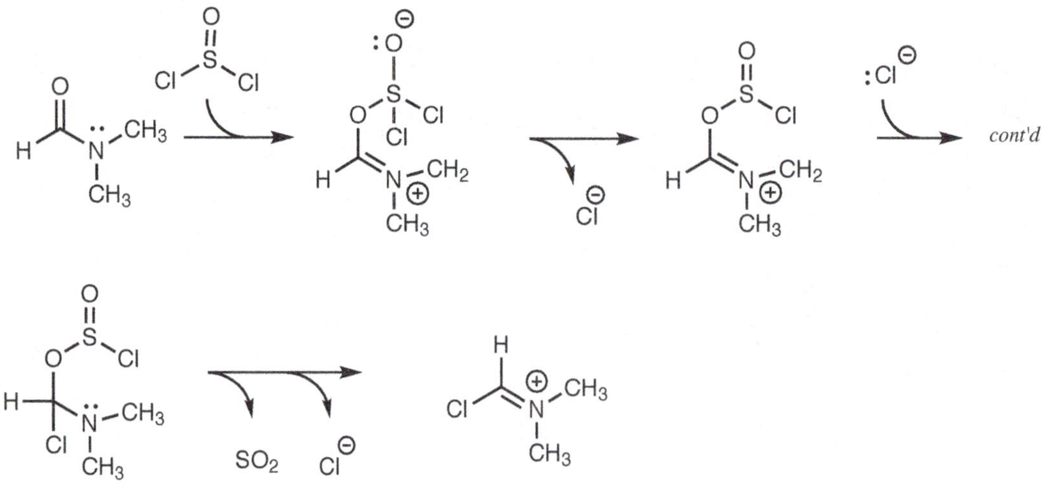

Part 2, reaction of iminium salt with carboxylic acid.

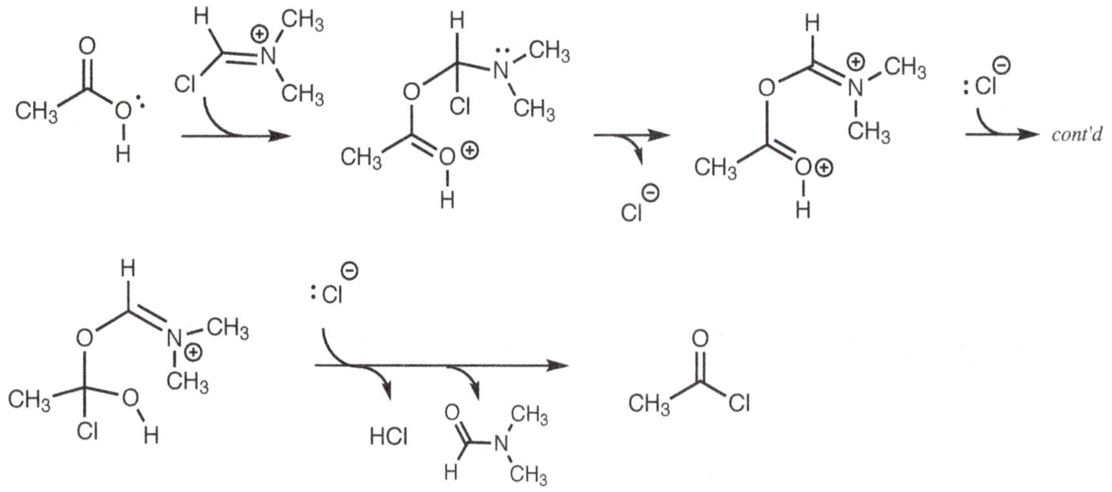

Halide from an Alcohol with a Phosphorus Reagent

6. Reaction of isobutyl alcohol with phosphorus tribromide to give isobutyl bromide. (See *Notes*.)

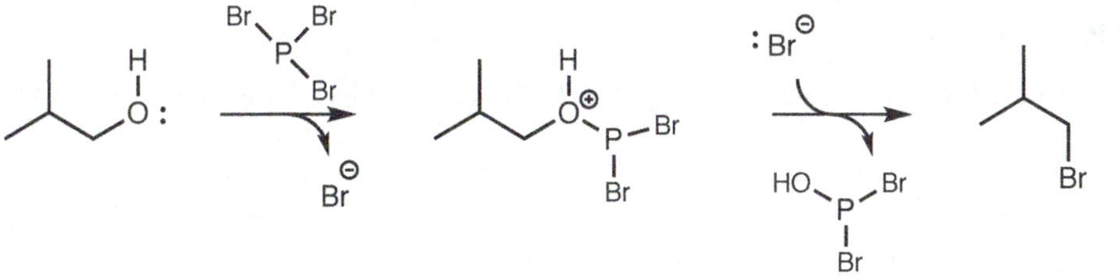

7. Reaction of triphenylphosphine, carbon tetrachloride and cyclopentanol to give chlorocyclopentane. (See *Notes*.)

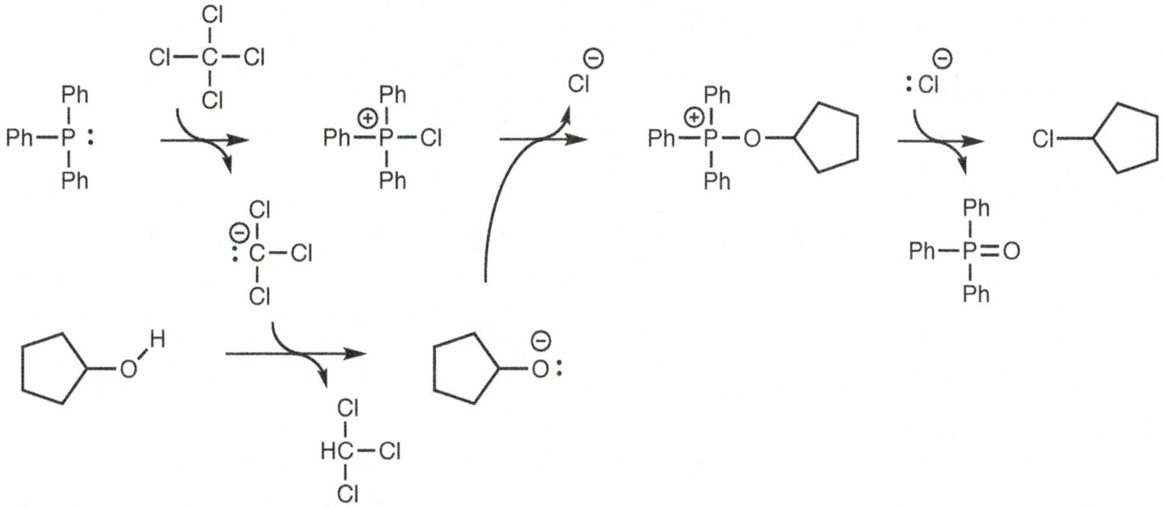

Ester from an Alcohol and Carboxylic Acid with a Phosphorus Reagent

8. Mitsunobu reaction with triphenyl phosphine and diethyl azodicarboxylate to prepare an ester from an acid and an alcohol.

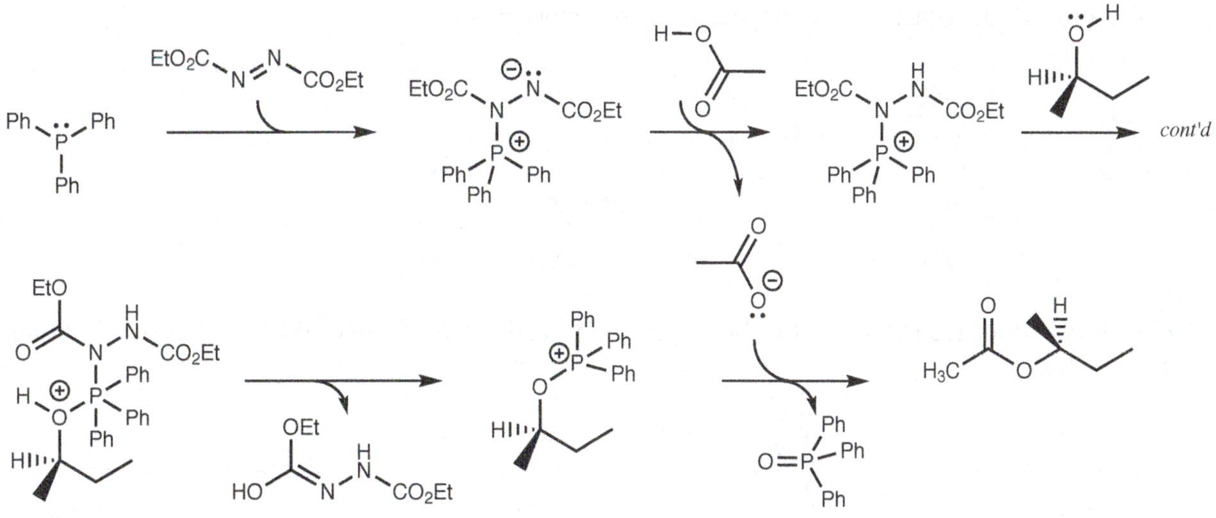

11 — Reduction Reactions

Sodium Borohydride Reductions

1. Sodium borohydride reduction of 2-methylpropanal (isobutyraldehyde) to 2-methyl-1-propanol (isobutyl alcohol).

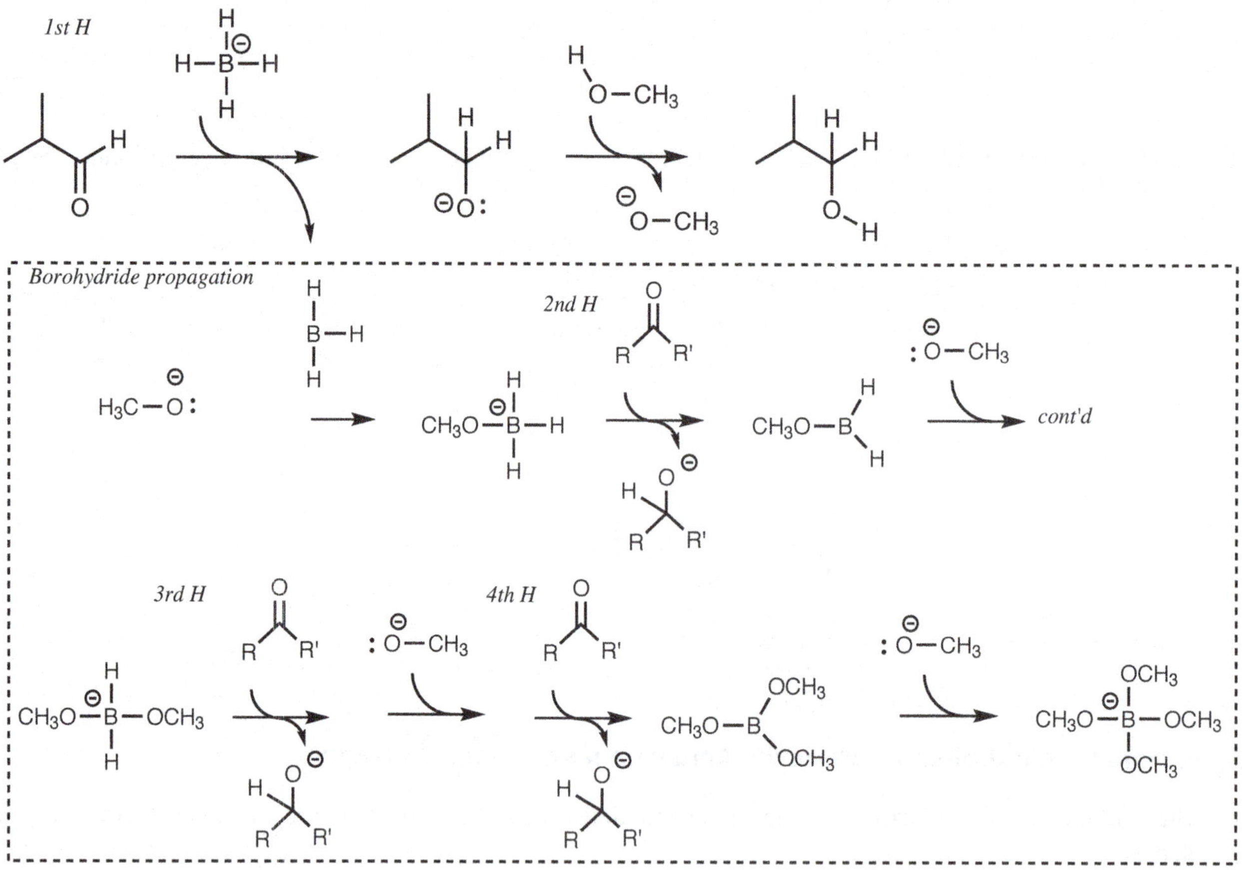

2. Sodium borohydride reduction of cyclopentanone to cyclopentanol.

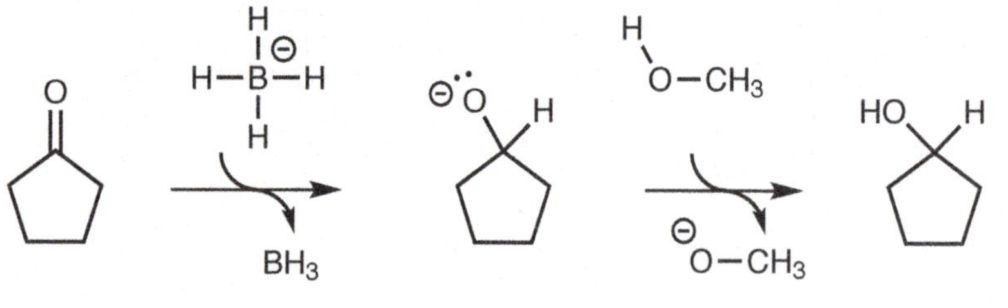

3. Sodium borohydride reduction of ethyl 4-oxocyclohexanecarboxylate to give ethyl 4-hydroxycyclohexanecarboxylate. (See *Notes*.)

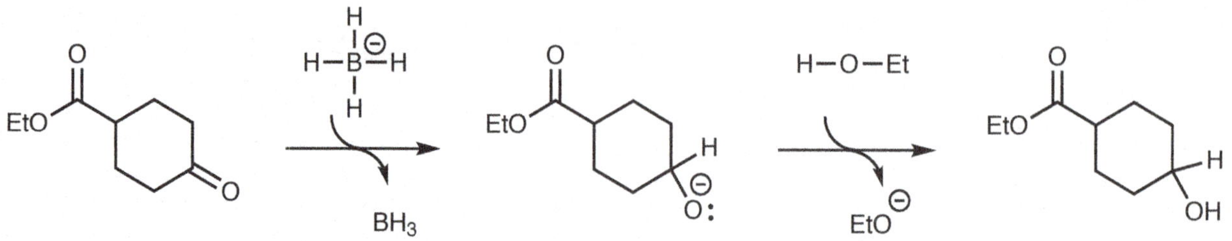

Lithium Aluminum Hydride Reductions

4. Lithium aluminum hydride reduction of acetophenone to 1-phenylethanol.

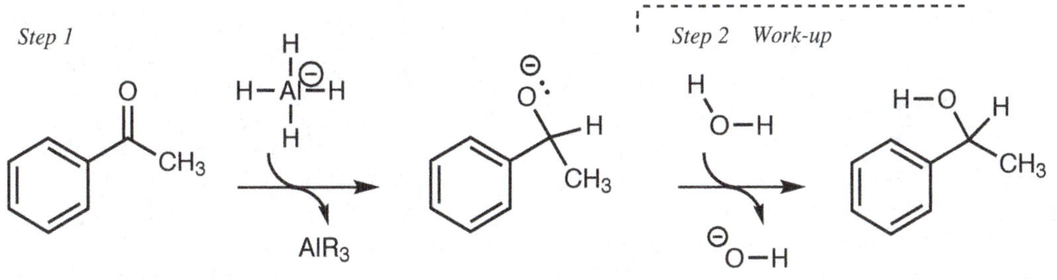

5. Lithium aluminum hydride reduction of ethyl butanoate (butyrate) to give ethanol and butanol. (See *Notes*.)

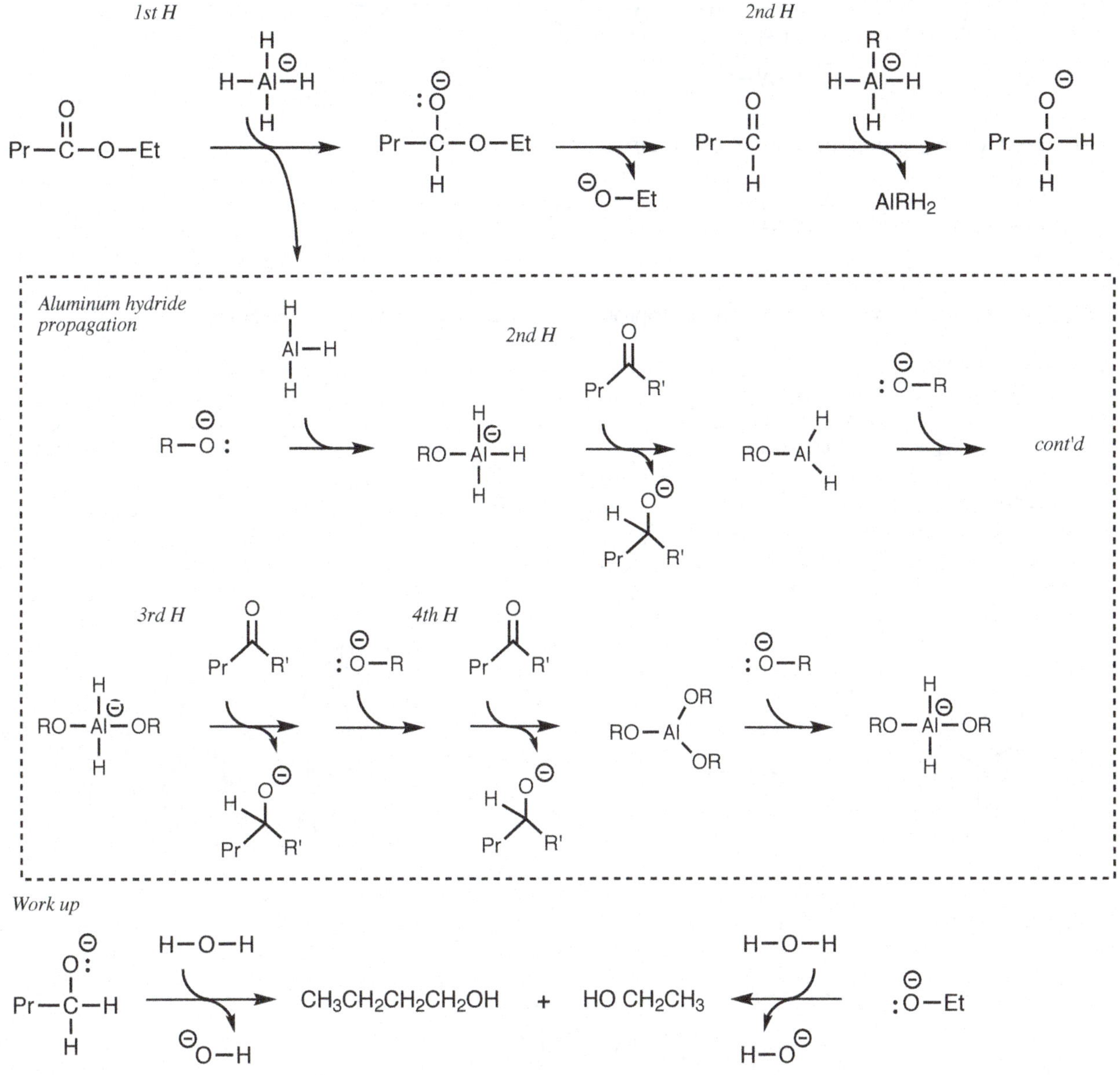

6. Lithium aluminum hydride reduction of a cyclic amide(1-methylpyrrolidin-2-one) to give a cyclic amine (1-methylpyrrolidine). (See *Notes*.) See Example 5 for regeneration of aluminum (IV) reductant.

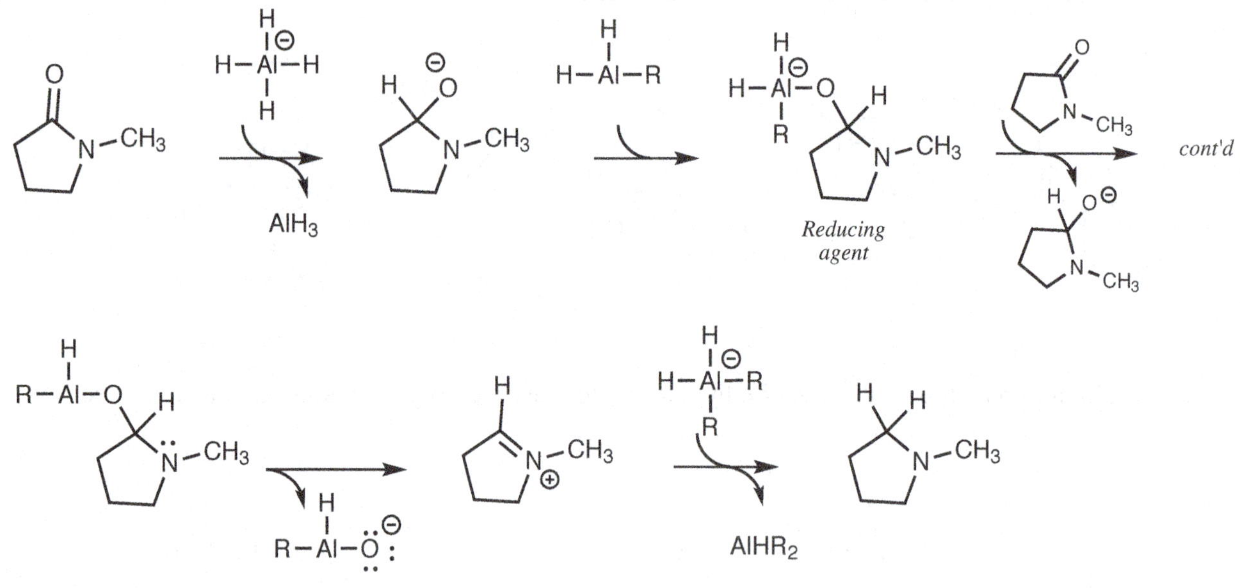

cont'd

AlH₃   Reducing
                                                                   agent

Preferred leaving group for lithium aluminum hydride reductions.

7. Lithium aluminum hydride reduction of propionyl anilide to *N*-propylaniline. (See *Notes*.)

Step 1, reduction.

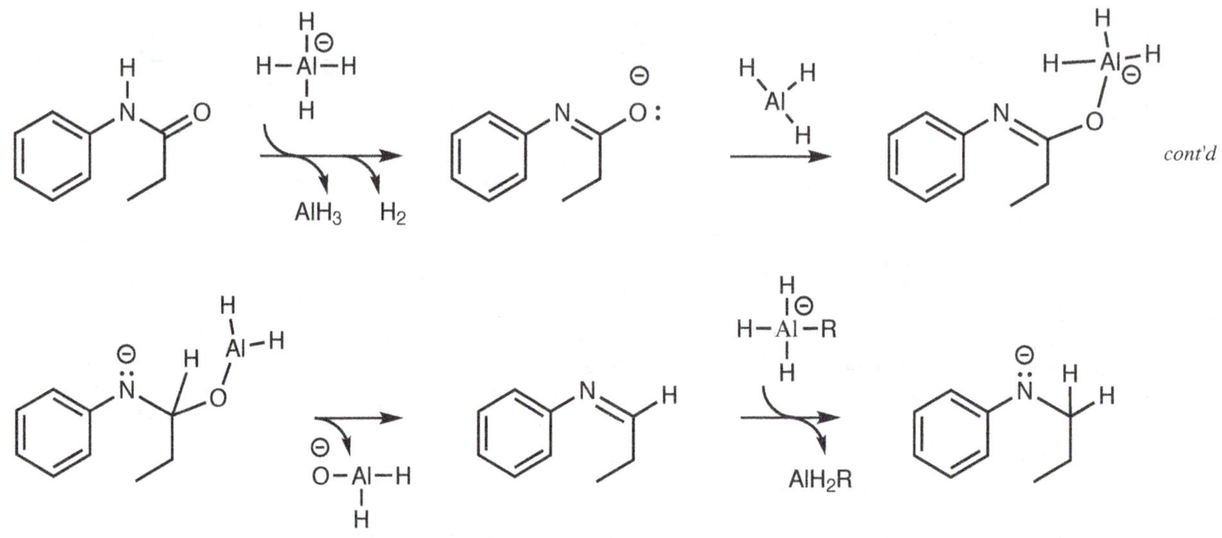

cont'd

AlH₃   H₂

Step 2, work up.

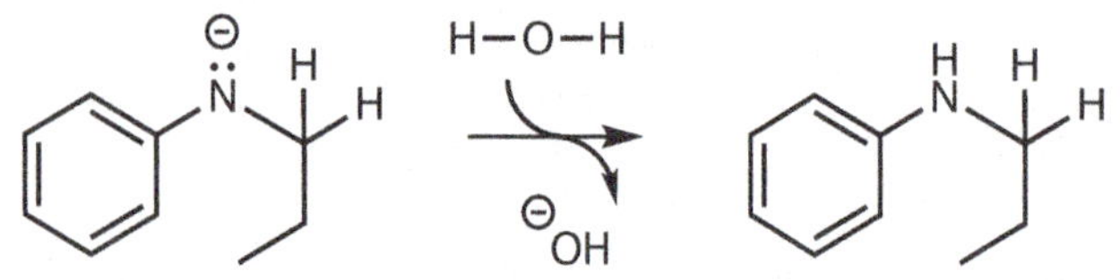

8. Lithium aluminum hydride reduction of phenylacetic acid to 2-phenylethanol. Step 1, reduction. (See *Notes*.)

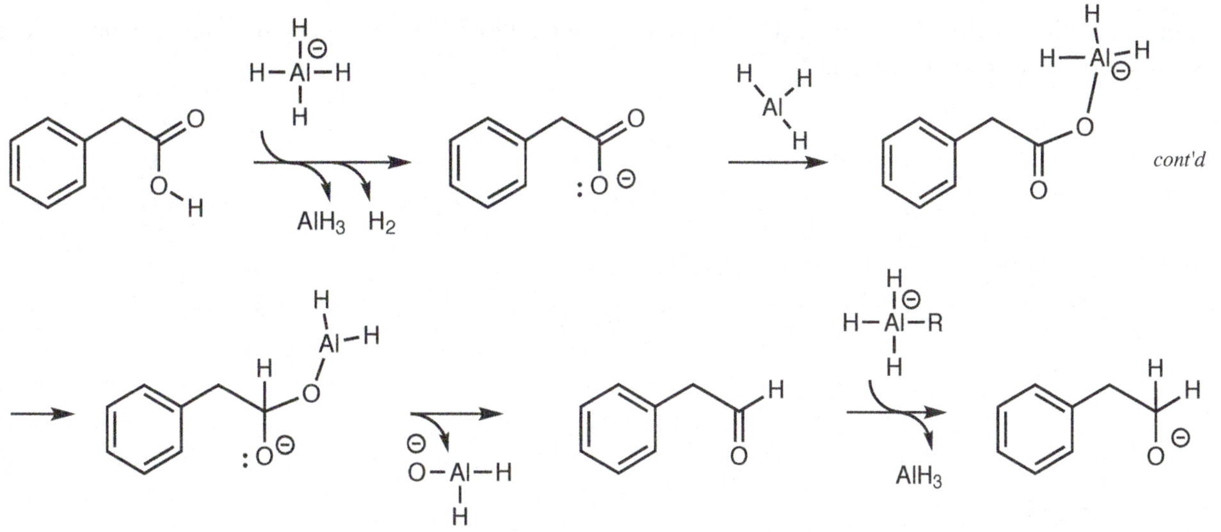

Step 2, work up.

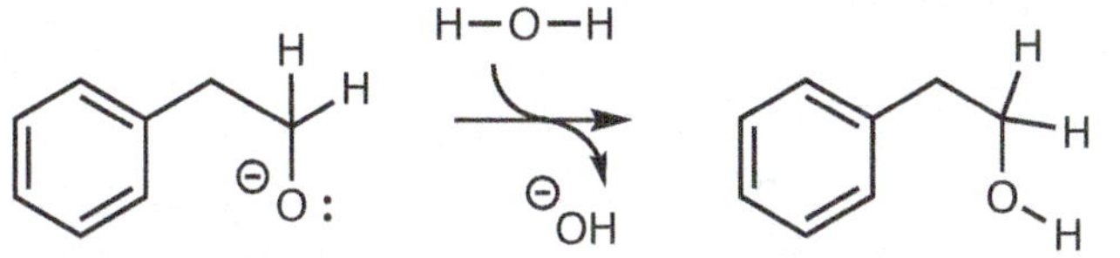

Reductive Amination

9. Step 1, formation of imine from benzaldehyde and ethylamine. (See *Notes*.)

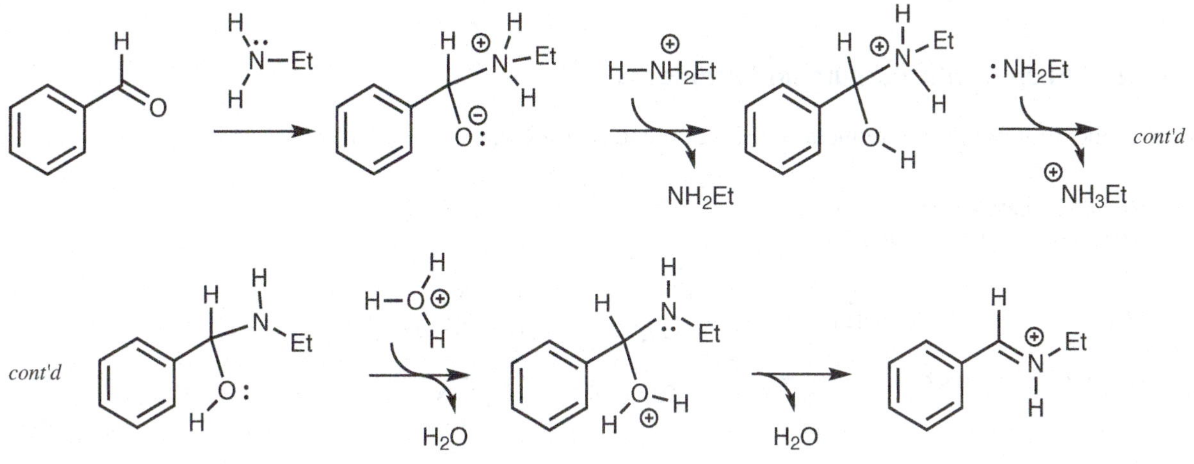

Step 2, reduction

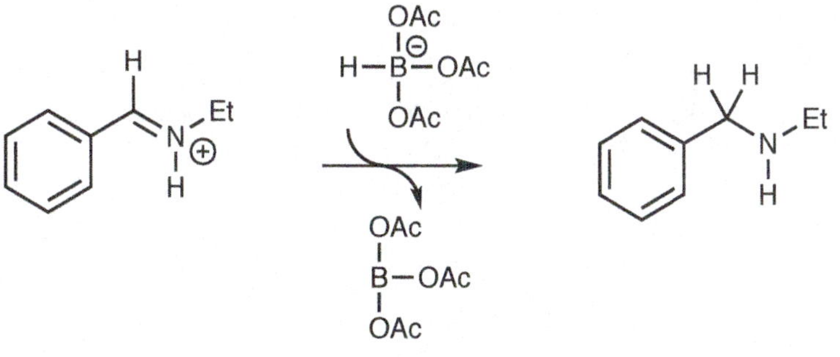

Even though this is written as a two-step process, which it is, both steps can be carried out at the same time. The imine, as it forms, can be reduced.

Triacetoxyborohydride is one of several reagents that one can use. Others are sodium cyano-borohydride (NaBH₃CN) or hydrogen with a palladium or nickel catalyst.

Diisobutylaluminum Hydride Reduction of an Ester

10. Step 1, addition of DIBAH or DIBAL(H) (diisobutylaluminum hydride) to ethyl isobutyrate. Reduction to give isobutyraldehyde (2-methylpropanal).

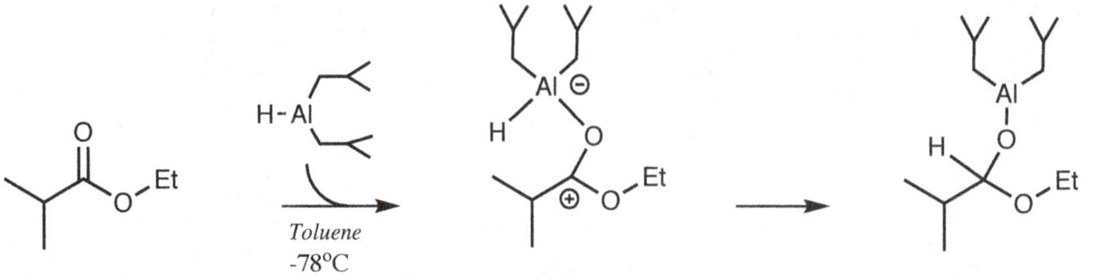

Toluene
-78°C

Step 2, work-up

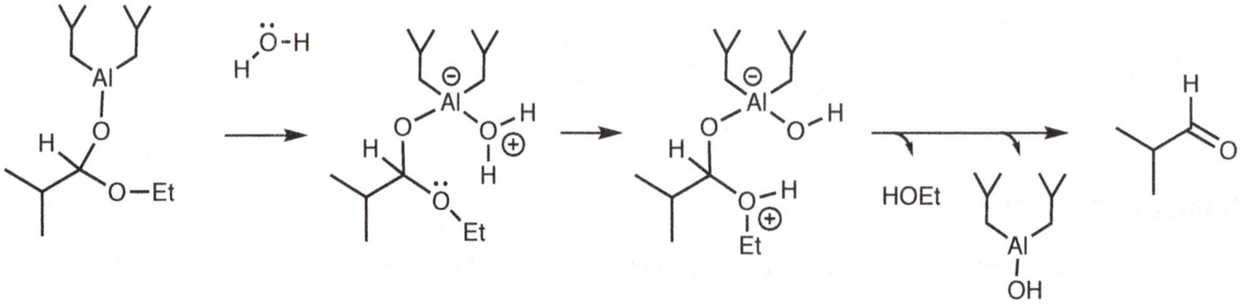

Reduction of Alkyne with Sodium and Ammonia

11. Sodium and ammonia *trans*-reduction of 2-butyne to *trans*-2-butene. (See *Notes*.)

Note, use a single barbed arrow for one-electron transfers reaction.

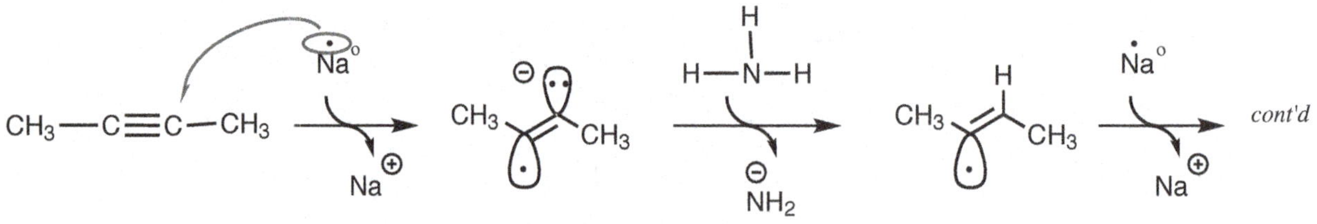

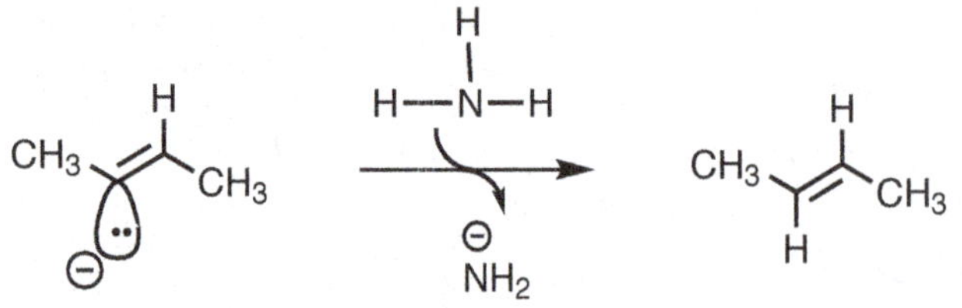

The pKa of CH₂=CH₂ is 44 and NH₃ is 36. Which conjugate base is stronger?

Wolff Kischner Reduction

12. Reaction of the ketone with hydrazine under basic conditions to form the hydrazide.

Reaction of the *in situ* formed hydrazide with KOH to form the methylene.

Catalytic Reduction of Nitrobenzene

Catalytic reduction of nitrobenzene to aniline. This isn't a mechanism. It shows that hydrogen can add across N-O single and double bonds. It is a very facile reduction. One can see that the oxygen atoms form water and the reduction takes three moles of hydrogen.

13. Catalytic reduction of nitrobenzene to aniline.

12 — Oxidation Reactions

General Form For Oxidations

See *Notes* for further discussion of oxidation reactions.

Chromic Acid Oxidation

1. Chromic acid oxidation (Jones oxidation) of 3-methyl-2-butanol to 3-methyl-2-butanone. (See *Notes*.)

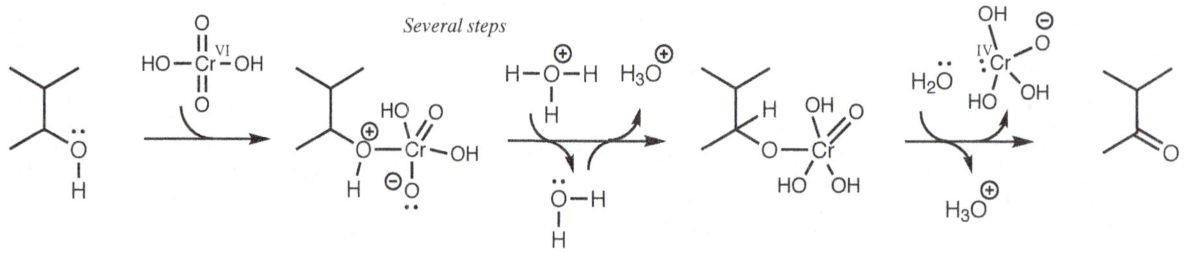

2. Chromic acid oxidation of isobutyl alcohol (2-methyl-1-propanol) to isobutyric acid (2-methylpropanoic acid). (See *Notes*.)

   Step 1, chromate ester formation and oxidation to aldehyde.

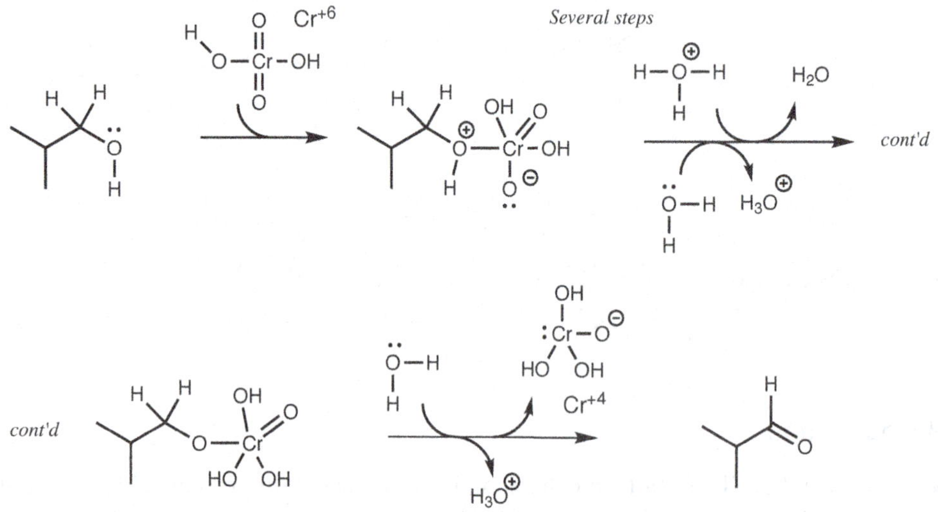

   Step 2, hydration of aldehyde and oxidation of hydrate to isobutyric acid (2-methylpropanoic acid).

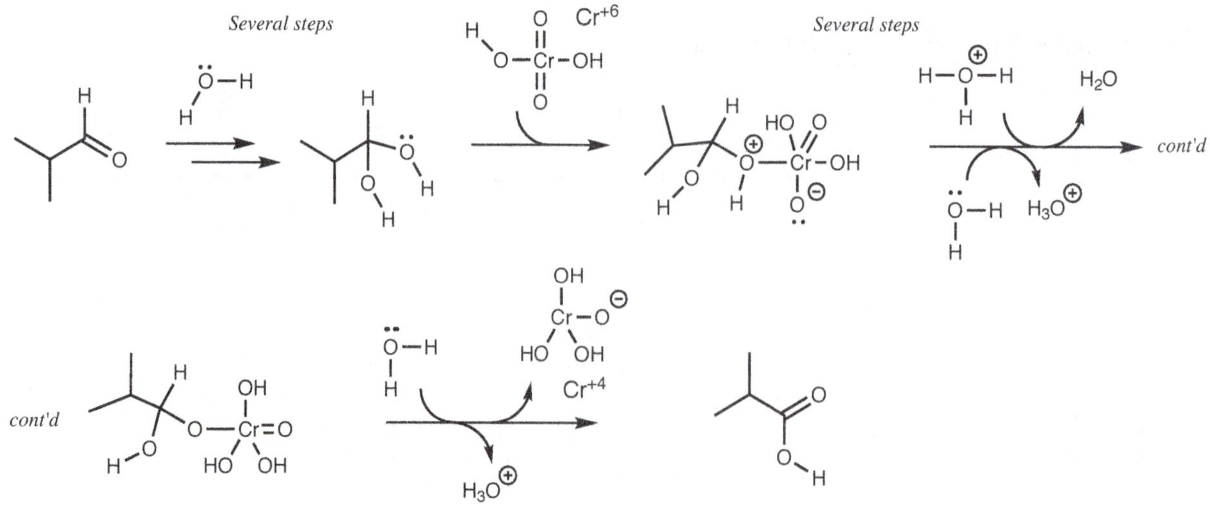

PCC, Tollens, Hypochlorite, *m*CPBA, and Sulfonium Based Oxidations

3. Pyridinium chlorochromate (PCC) oxidation of benzyl alcohol to benzaldehyde.

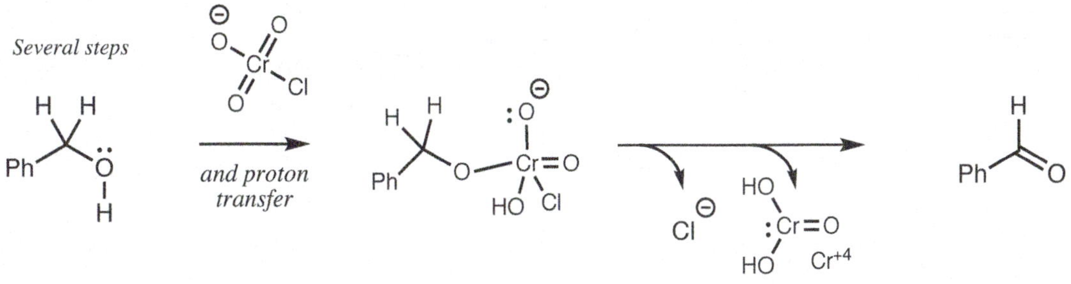

4. Tollens oxidation of 2-methylpropanal (isobutyraldehyde) with silver oxide (or hydroxide).

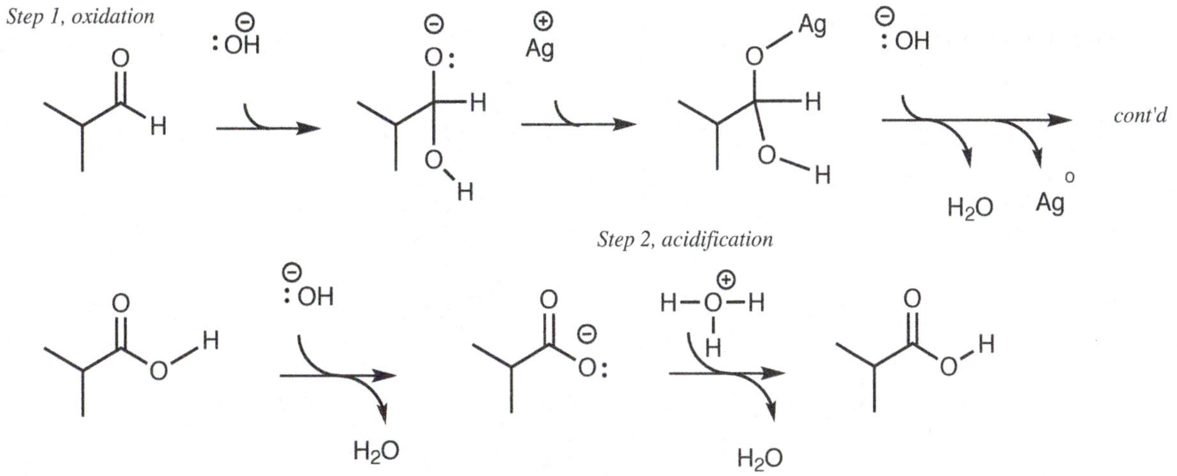

5. Oxidation of cyclohexanol to cyclohexanone with sodium hypochlorite (NaOCl, bleach).

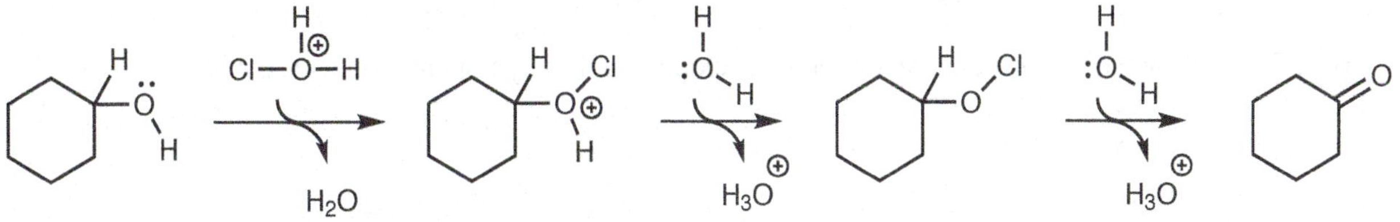

6. Peracid epoxidation of *trans*-2-butene with *m*-chloroperoxybenzoic acid (MCPBA) to give an epoxide, (*2R,3R*)-2,3-dimethyloxirane. (See *Notes*.)

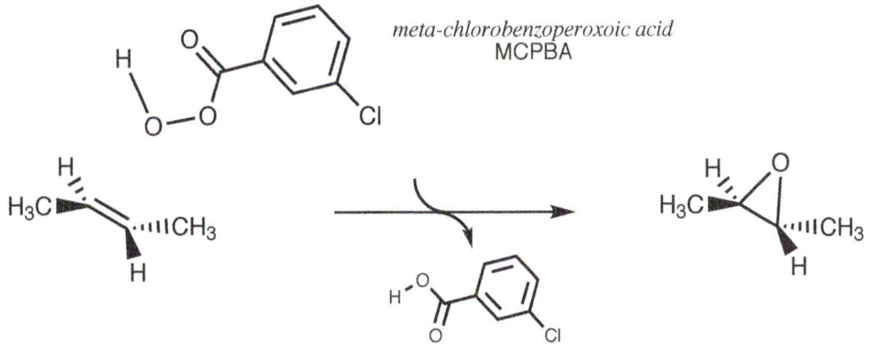

7. Step 1, Swern oxidation, preparation of chlorosulfonium salt.

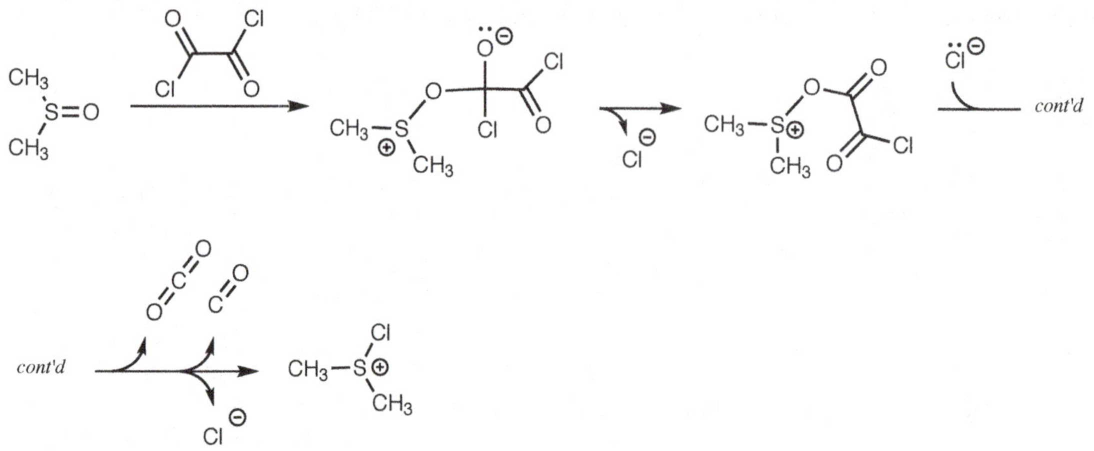

Step 2, Swern oxidation, oxidation step.

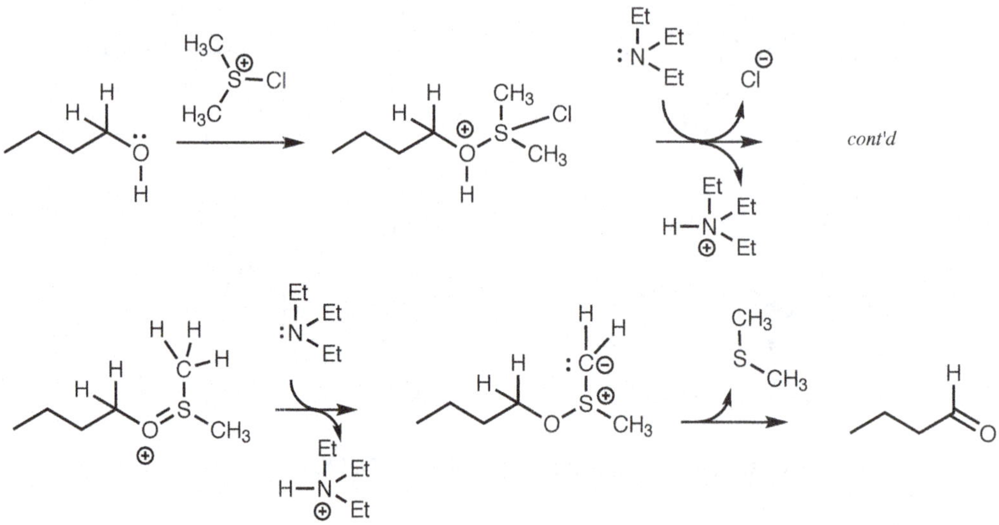

8. Corey-Kim (chlorosulfonium salt) oxidation of cyclopentanol to cyclopentanone. (See *Notes.*)

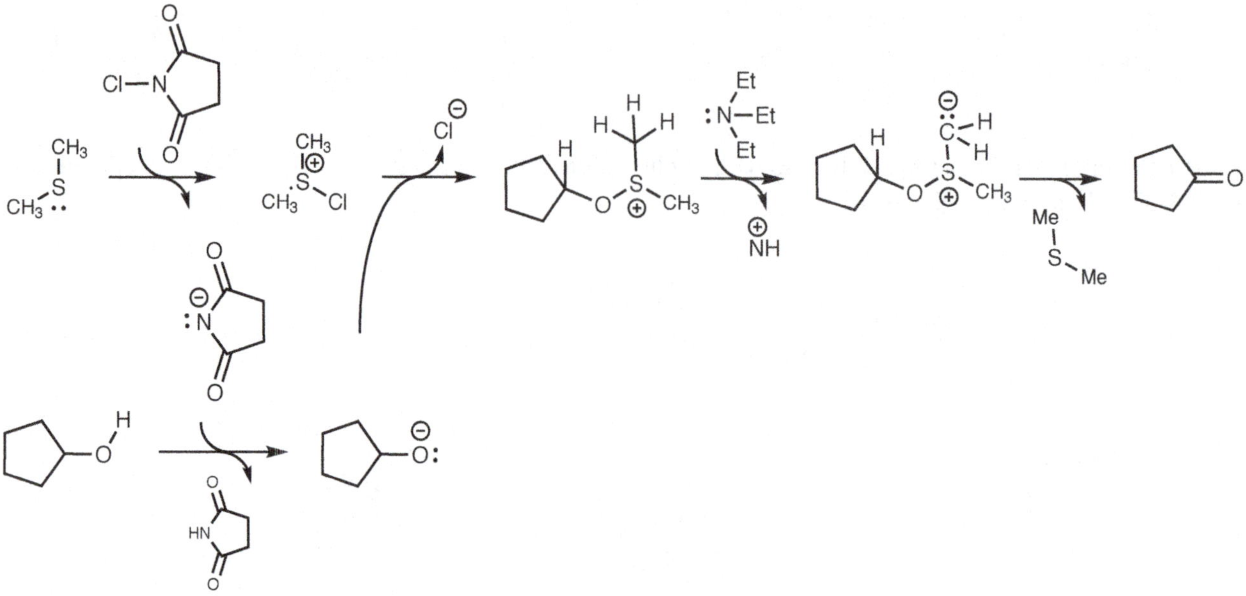

Ozone Oxidation

9. Step 1, ozone reaction with an alkene to give an ozonide.

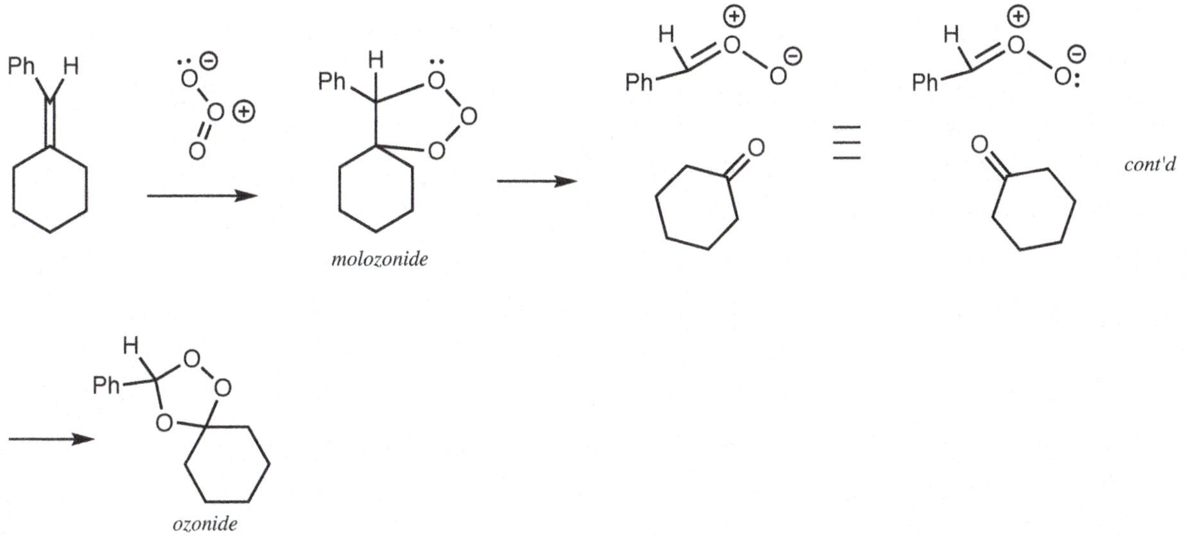

molozonide

cont'd

ozonide

Step 2, reduction of ozonide to two carbonyl compounds. (See *Notes*.)

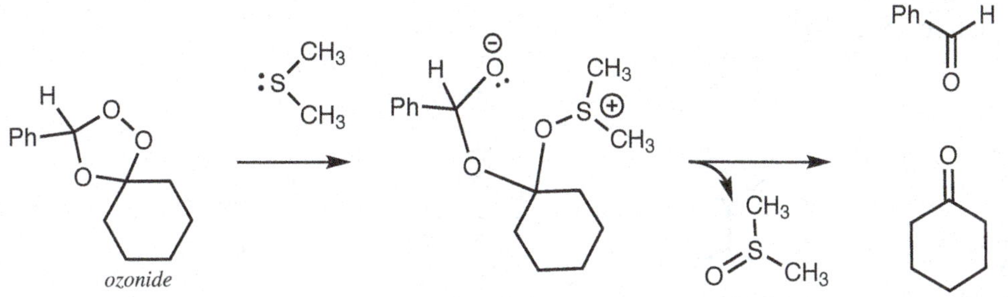

ozonide

Alternate step 2, oxidative destruction of ozonide. (See *Notes*.)

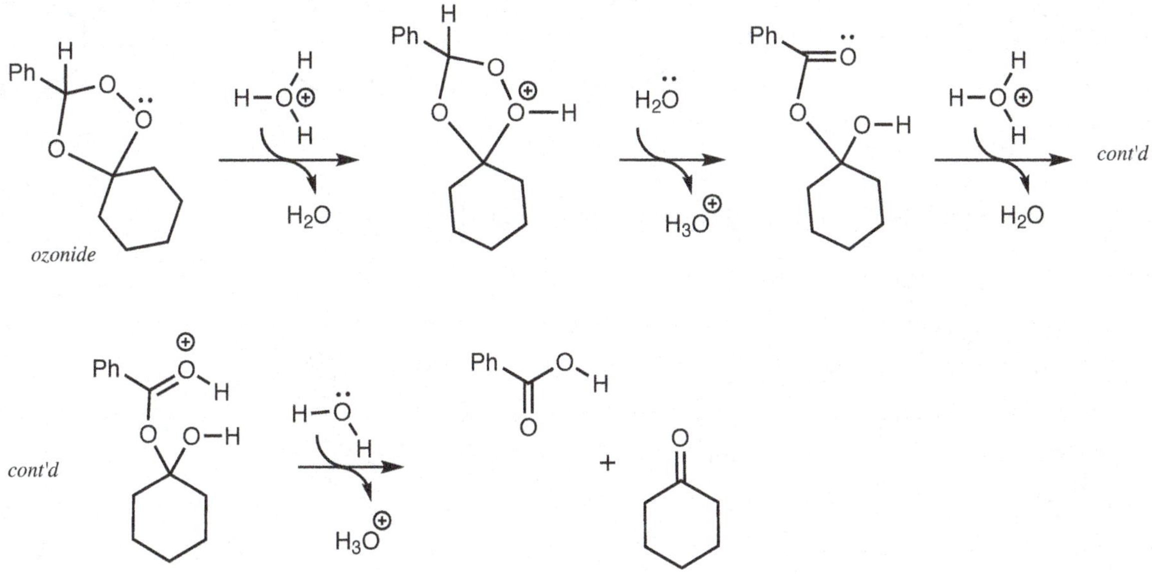

ozonide

cont'd

cont'd

Osmium Tetroxide, Potassium Permanganate, and Periodate Oxidations

10. Osmium tetroxide oxidation of cyclohexene to give *cis*-1,2-cyclohexanediol. (See *Notes*.)

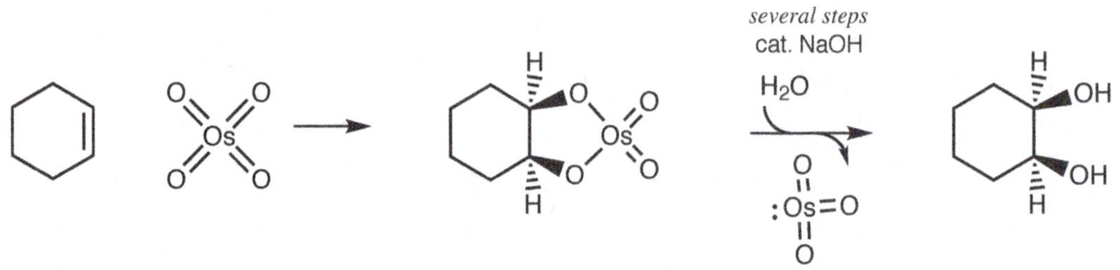

Regeneration of oxidant with *tert*-butylperoxide.

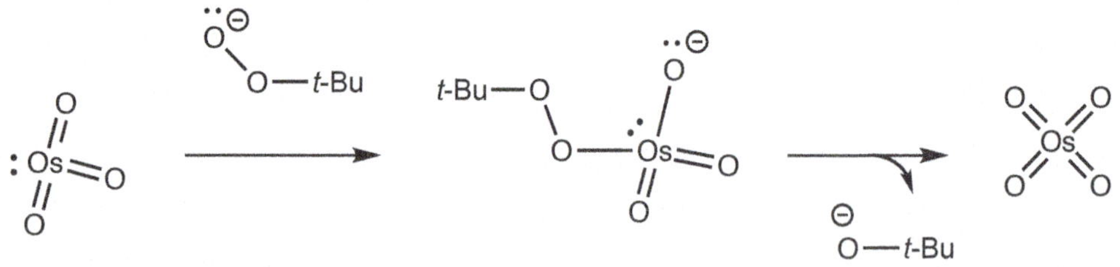

11. Potassium permanganate oxidation of methylcyclohexene to 1-methylcyclohexane-1,2-diol at low temperature or cleavage at high temperature. (See *Notes*.)

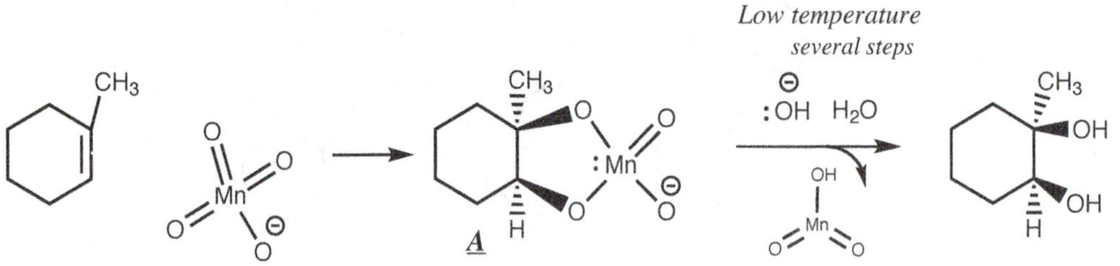

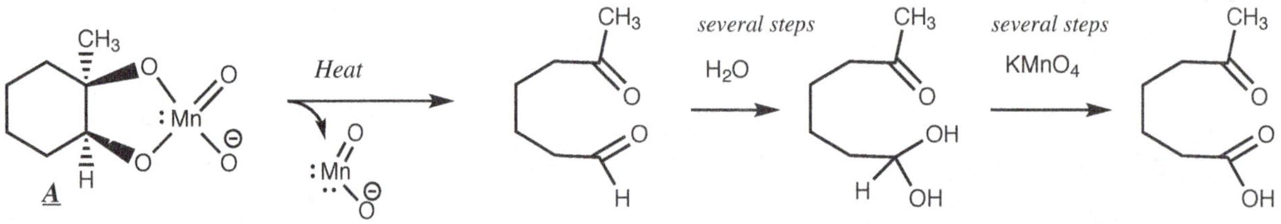

12. Periodate cleavage of a 1,2-diol to give a dicarbonyl compound. (See *Notes*.)

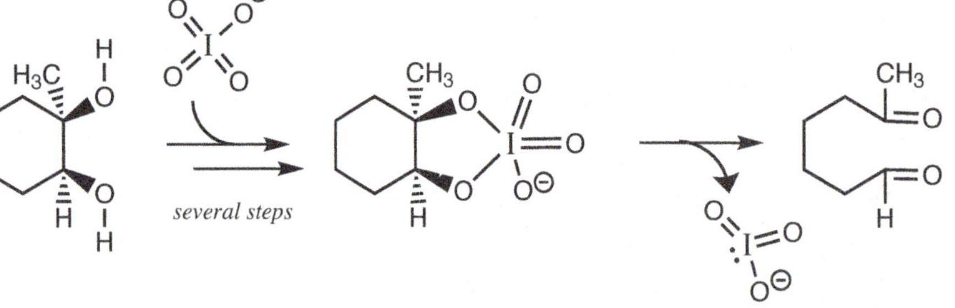

13 — Organometallic Reactions

The (transition) organometallic reactions in this chapter are of increasing importance. The reactions may recount the steps more than explain a reaction. Significantly absent is the concept of 18 valence electrons about the catalytic palladium atom.

Acyclic Heck Reaction

1. Step 1, reduction of palladium (II) to zero valent palladium with propene. (See *Notes.*)

Step 2, the catalytic cycle (oxidative-addition, *syn*-addition, *syn*-elimination, and reductive-elimination) with 2-bromopropene and propene.

The alkene pi-electrons probably complex with palladium and are donated to the palladium before the syn addition takes place.

Cyclic Heck Reaction

2. Step 1, reduction of palladium (II) to zero valent palladium with cyclopentene.

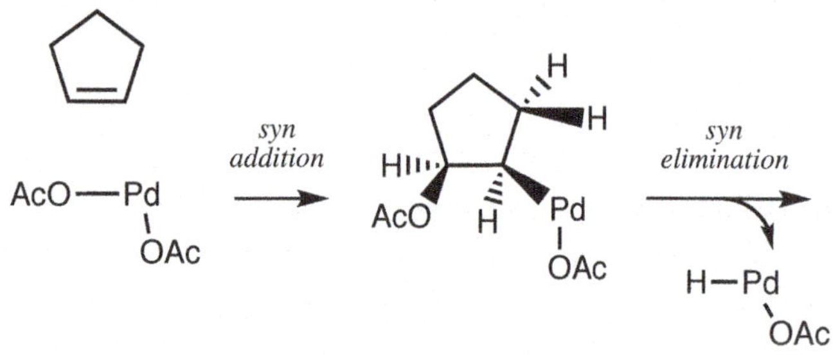

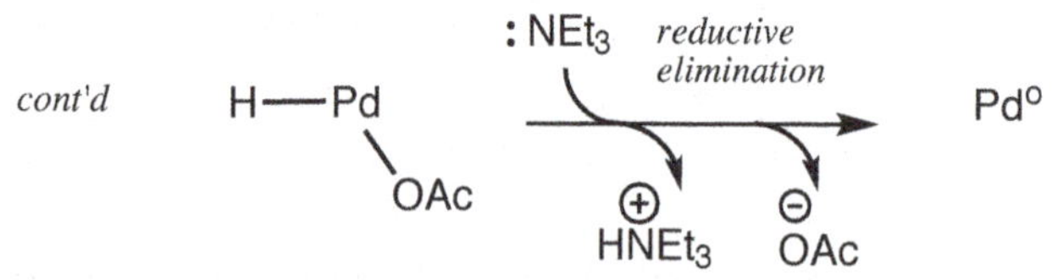

Step 2, the catalytic cycle (oxidative-addition, *syn*-addition, *syn*-elimination, and reductive-elimination) with iodobenzene and cyclopentene.

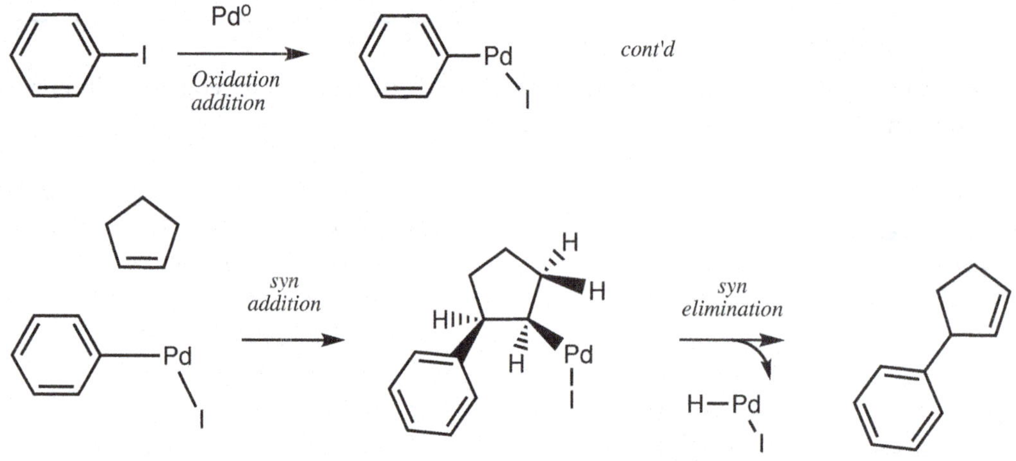

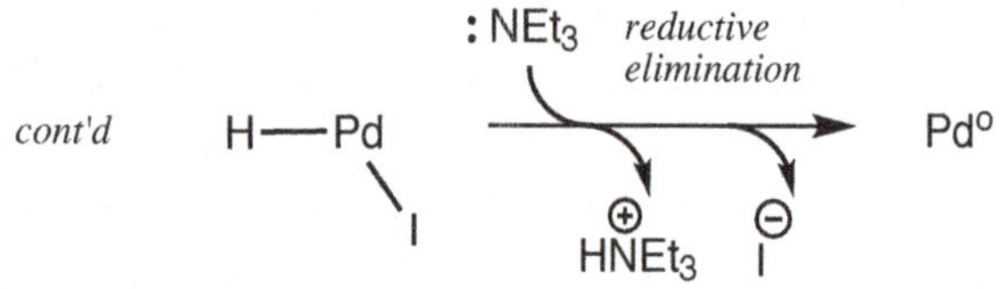

Catalytic Reduction of an Alkene (See *Notes*.)

3. Catalytic hydrogenation of *cis*-3-hexene to hexane.

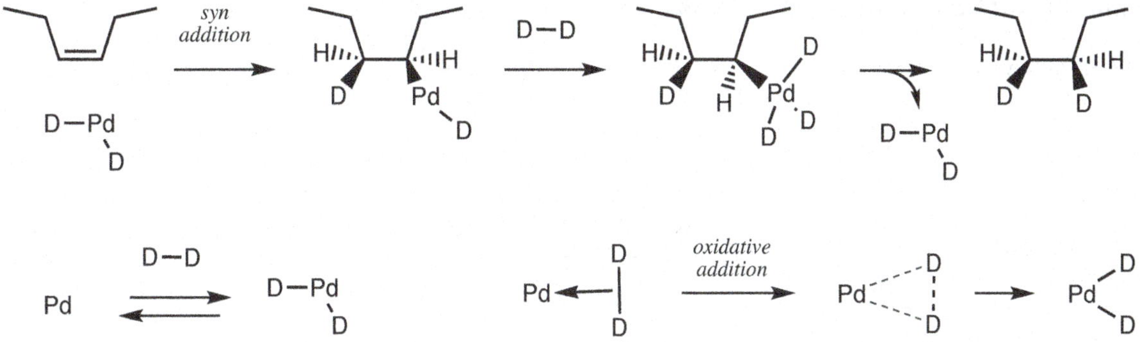

Gilman Reagent (See *Notes*.)

4. Formation of Gilman reagent, lithium dimethylcuprate.

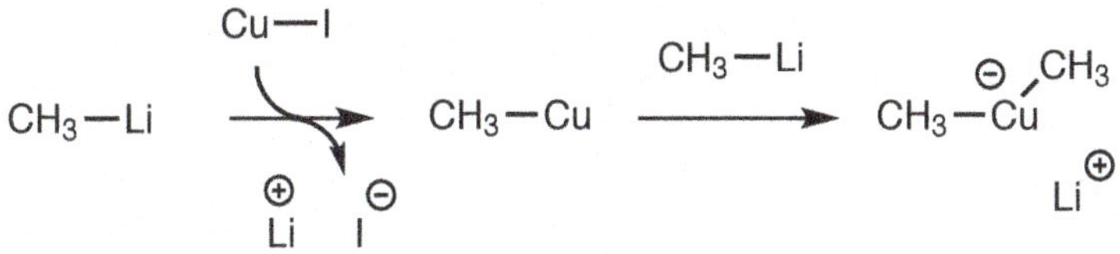

Coupling of Gilman reagent, lithium dimethylcuprate with (Z)-1-bromopent-1-ene to give *cis*-2-hexene.

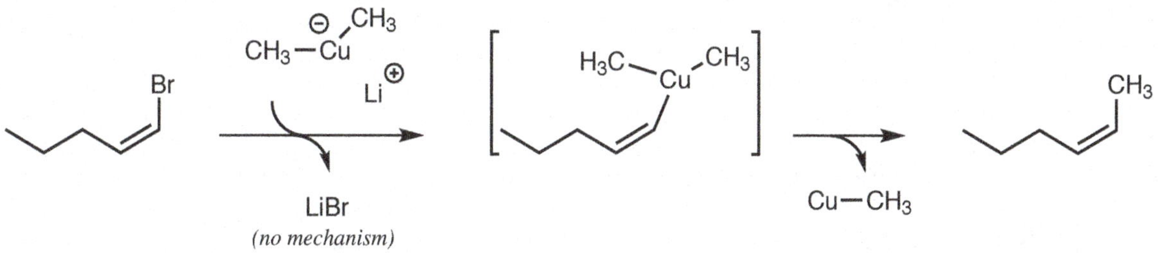

(no mechanism)

5. A 1,4-conjugate addition of lithium dimethylcuprate to 2-cyclohexenone to give 3-methylcyclohexanone.

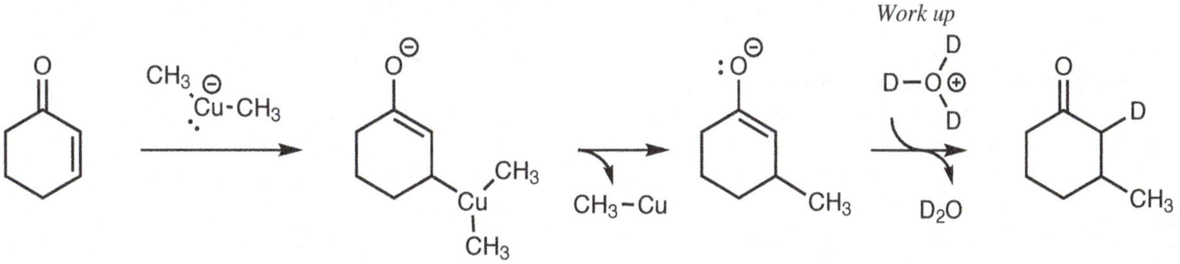

6. Reaction of lithium dimethylcuprate to benzoyl chloride to give acetophenone.

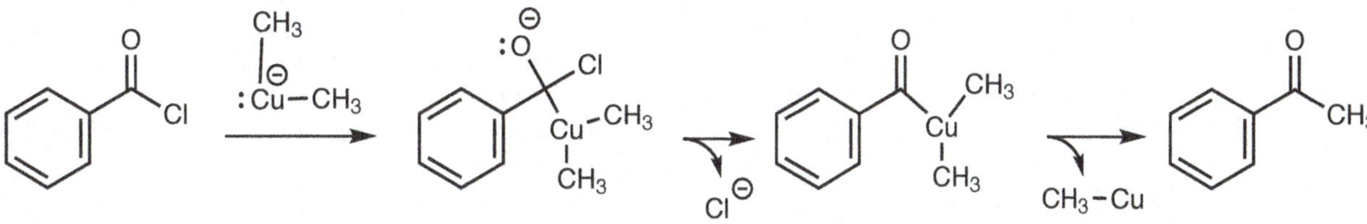

14 — Aromatic Substitution Reactions

Electrophilic Aromatic Substitution of Benzene

1. Friedel Crafts acylation of benzene.

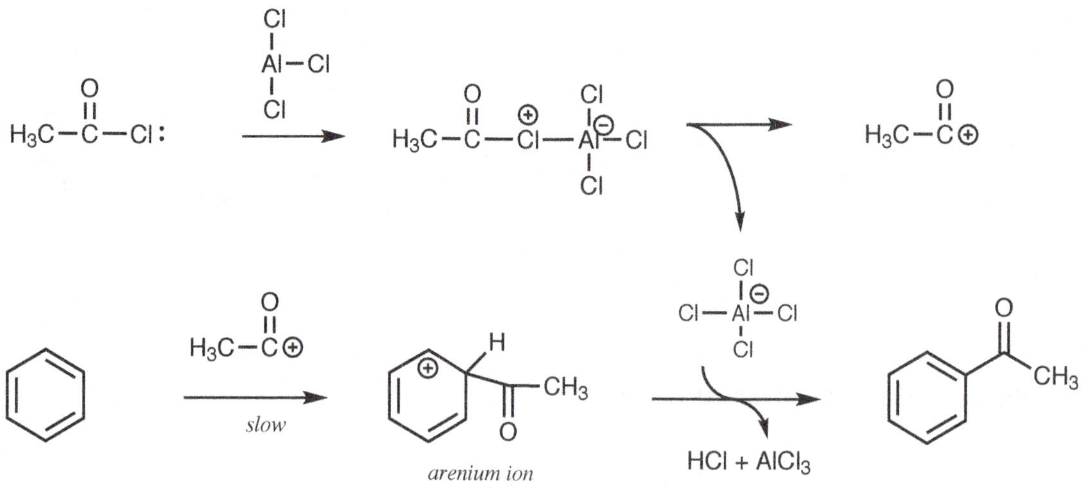

arenium ion

2. Friedel Crafts alkylation of benzene. (See *Notes*.)

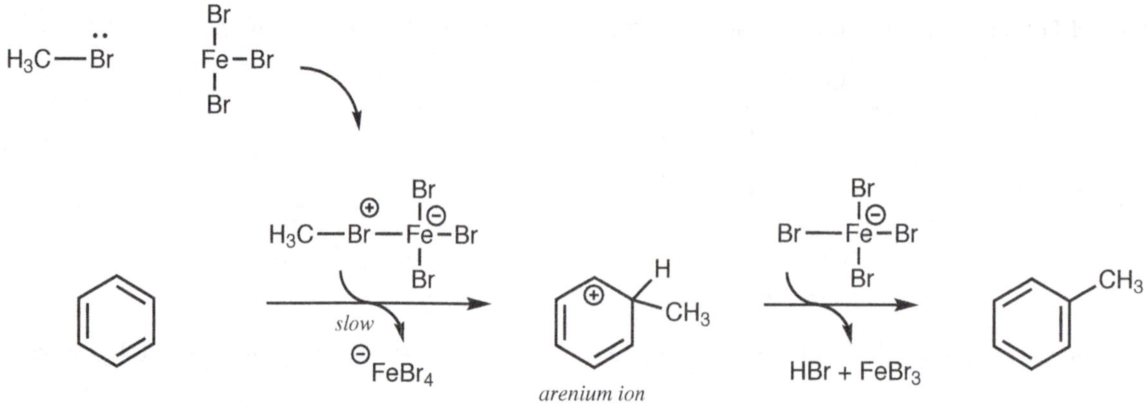

arenium ion

3. Ferric bromide bromination of benzene.

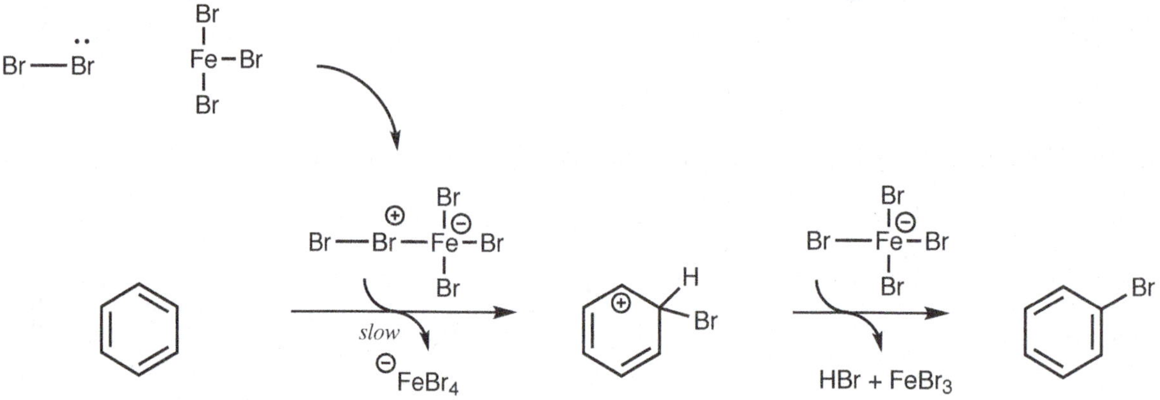

4.   Nitration of benzene with nitric and sulfuric acids.

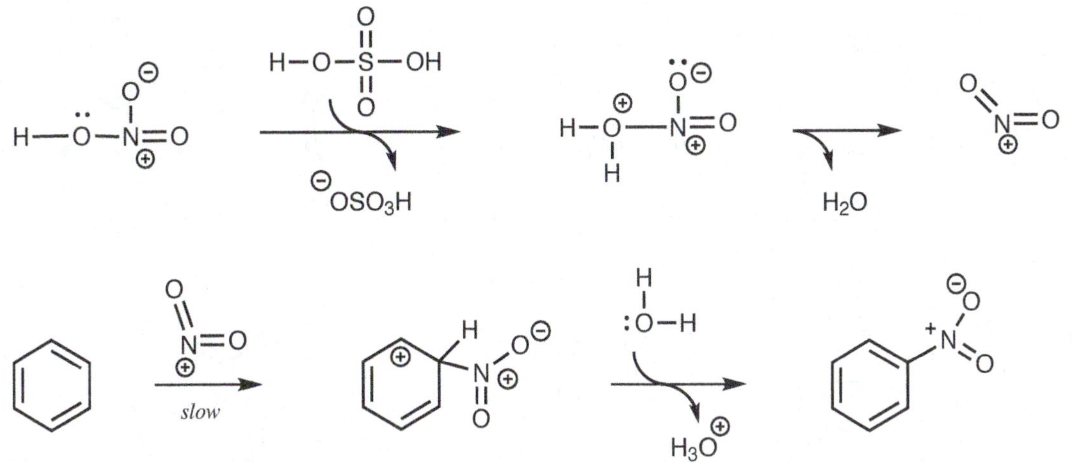

Electrophilic Substitution of Substituted Aromatic Compounds (See *Notes*.)

5.   Bromination of acetophenone to give *m*-bromoacetophenone.

   (For 5, 6, and 7, answer the question, "Will it react faster than benzene?")

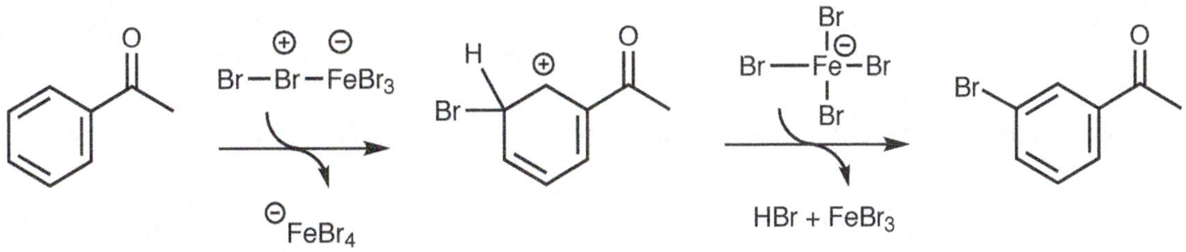

6.   Nitration of methyl *p*-chlorobenzoate to give methyl 4-chloro-3-nitrobenzoate.

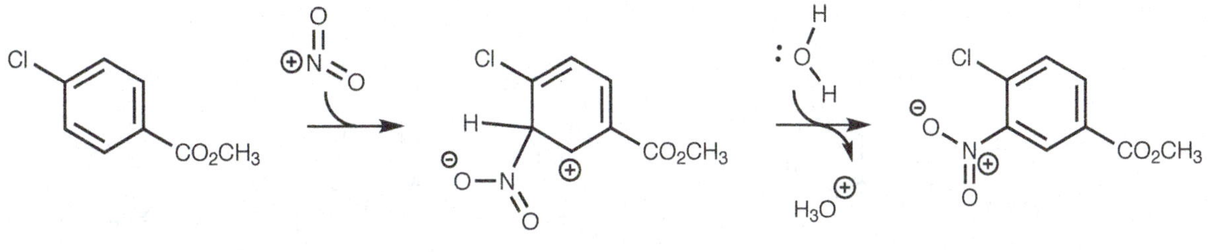

7.   Friedel-Crafts acylation of phenyl acetate with acetyl chloride and aluminum chloride. (See #1)

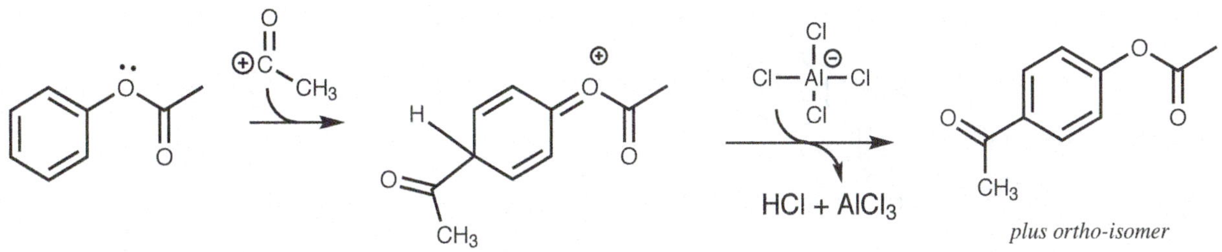

plus ortho-isomer

8. Triflic acid catalyzed acetylation of toluene to give *o*- and *p*-methylacetophenone.

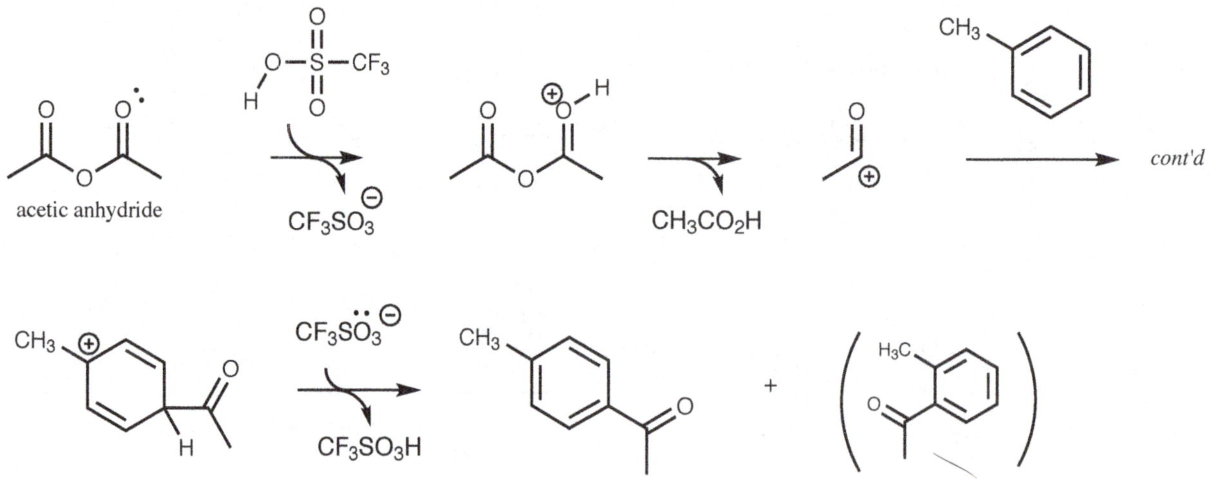

9. Aluminum chloride catalyzed chlorination of methyl 3-methoxybenzoate (methyl *m*-anisate) to give methyl 2-chloro-5-methoxybenzoate. (See *Notes*.)

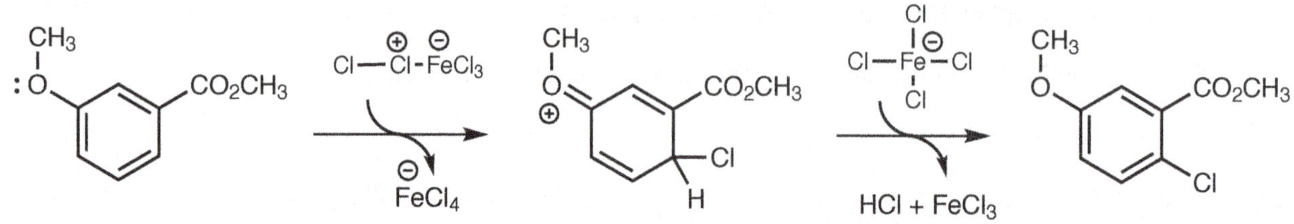

10. Friedel-Crafts alkylation of 4-nitro-*N*-*p*-tolylbenzamide with two equivalents of chloromethane. (See *Notes*.)

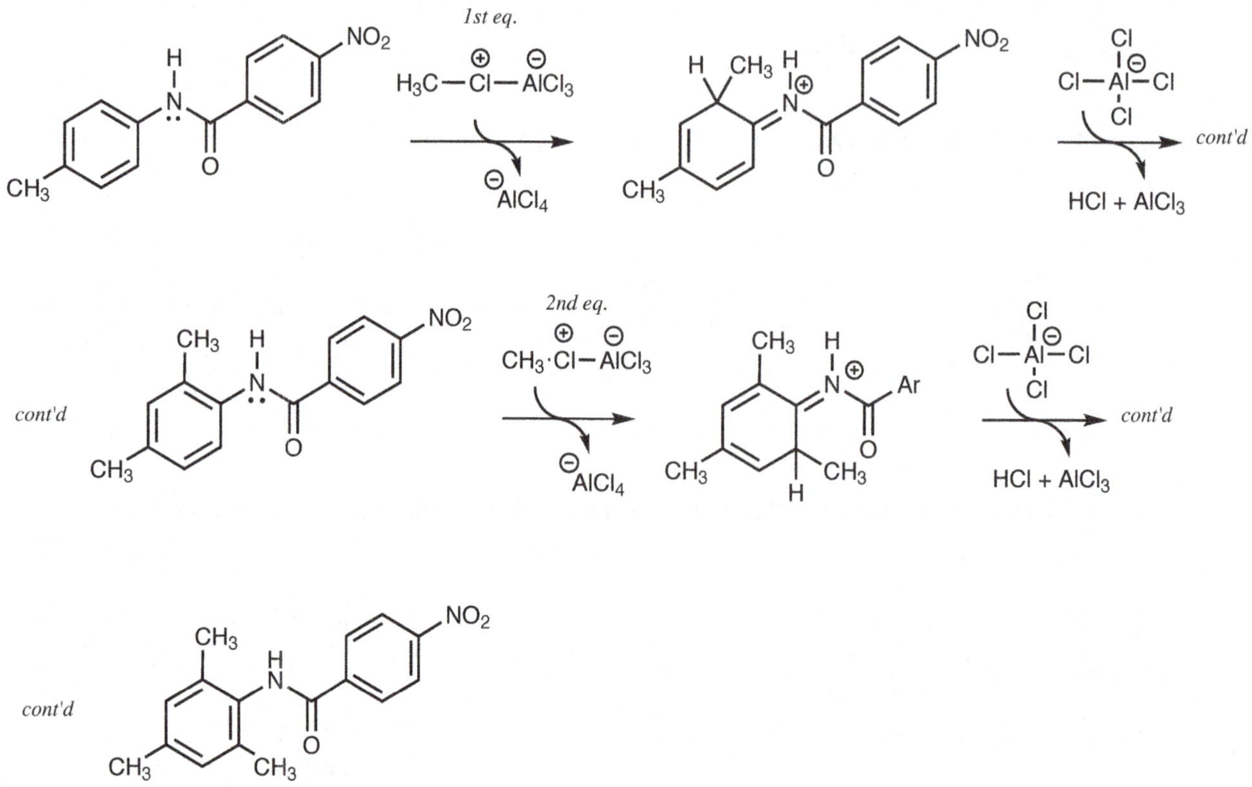

Nucleophilic Aromatic Substitution

11. Nucleophilic aromatic substitution of 1-fluoro-4-nitrobenzene with ammonia to give 4-nitroaniline. (See *Notes*.)

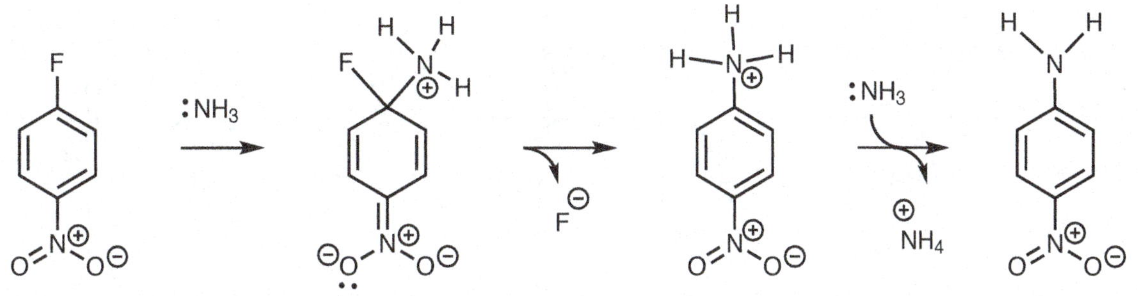

12. Nucleophilic aromatic substitution of 1,2-difluoro-4-nitrobenzene with sodium methoxide to give 2-fluoro-1-methoxy-4-nitrobenzene.

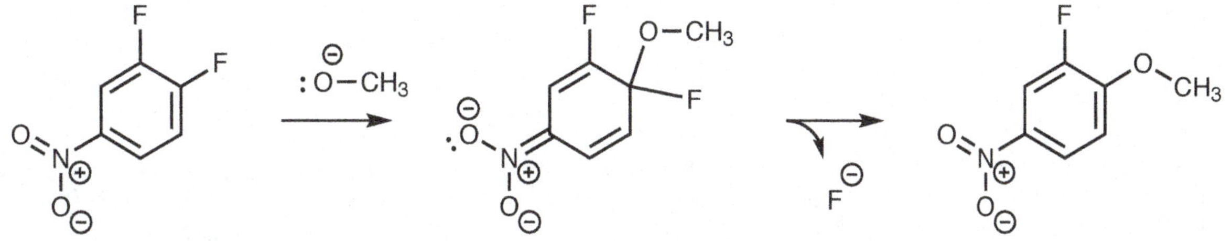

Benzyne Reaction

13. Reaction of 4-bromoanisole with sodium amide (sodamide, $NaNH_2$) to give 3- and 4-methoxyaniline via a benzyne intermediate. (See *Notes*.)

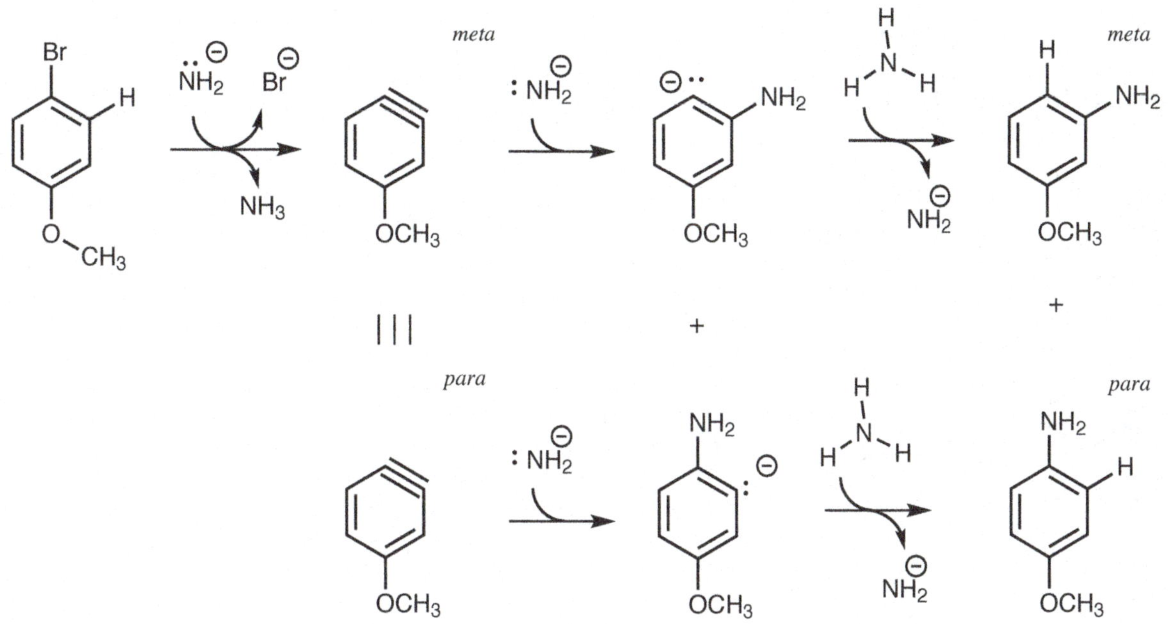

Diazonium Chemistry

14. Formation of a diazonium salt from aniline. (See *Notes*.)

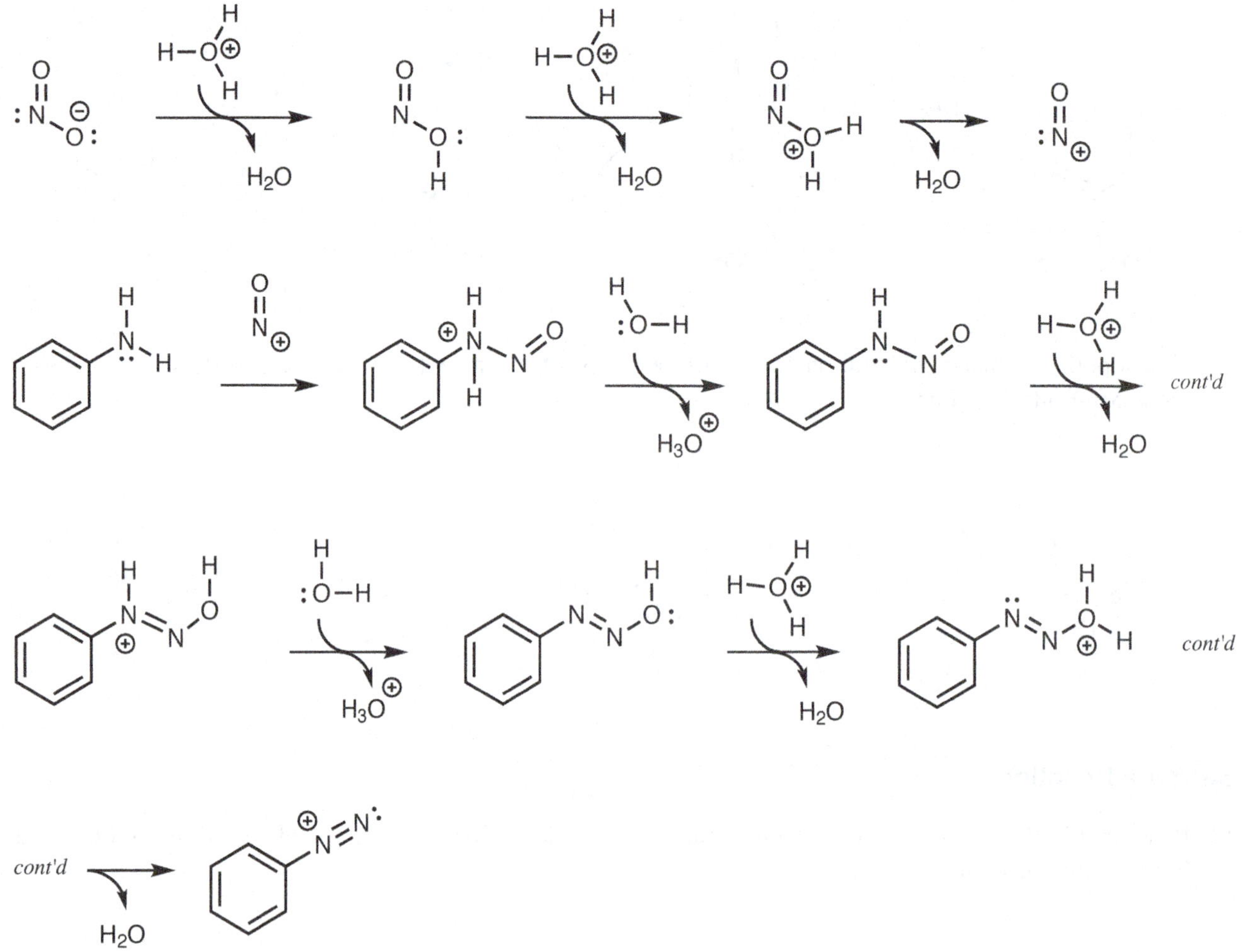

Reaction of Diazonium Salts

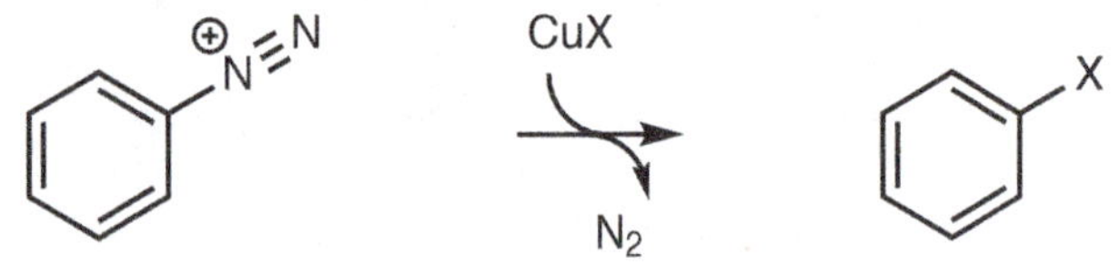

X = Cl, Br, CN, OH

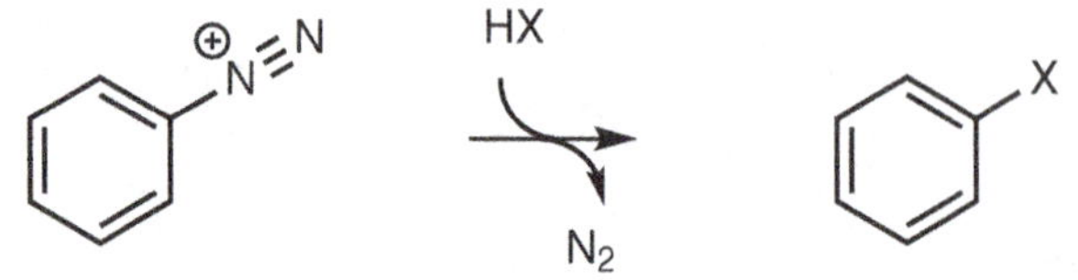

X = I, F or OH

15 — Carbene and Nitrene Reactions

Carbene Reactions

1. Simmons-Smith carbene addition to cyclohexene to give a bicyclo[4.1.0]heptane. (See *Notes*.)

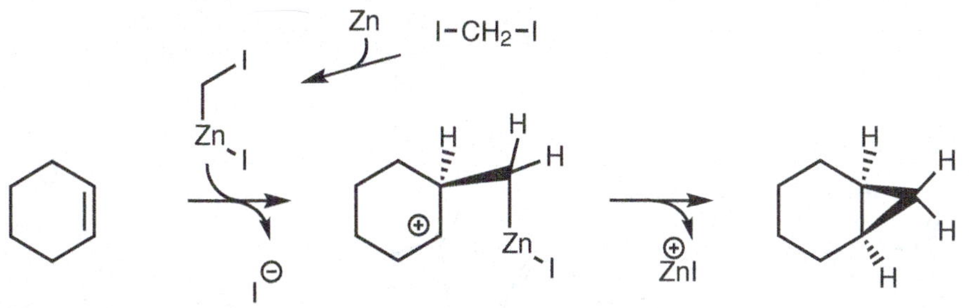

2. Dihalocarbene addition to (*E*)-1-phenylpropene to give a dibromocyclopropane.

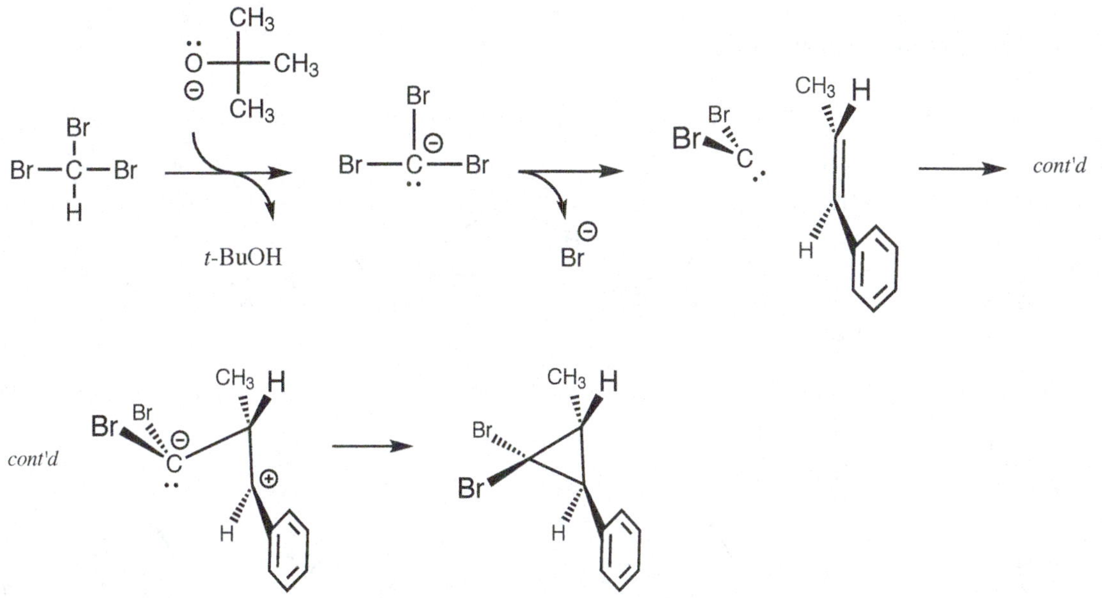

3. Dihalocarbene addition to cyclohexene to give 7,7-dichlorobicyclo[4.1.0]heptane.

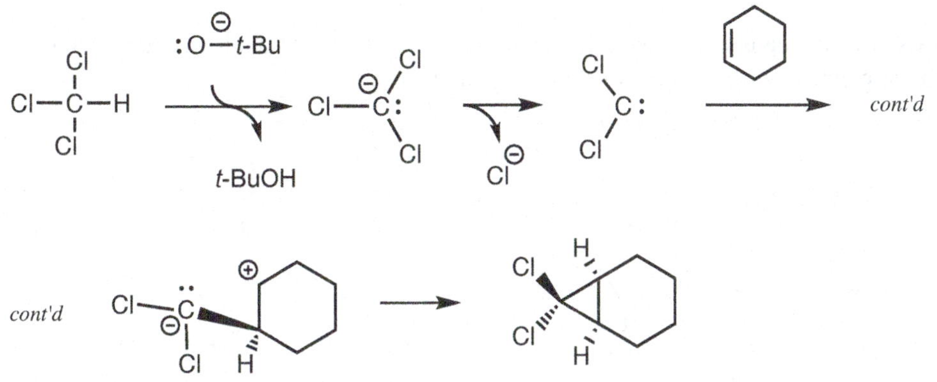

Curtius Rearrangement

4. Step 1, Curtius rearrangement, reaction of acid chloride with azide and rearrangement via a nitrene intermediate.

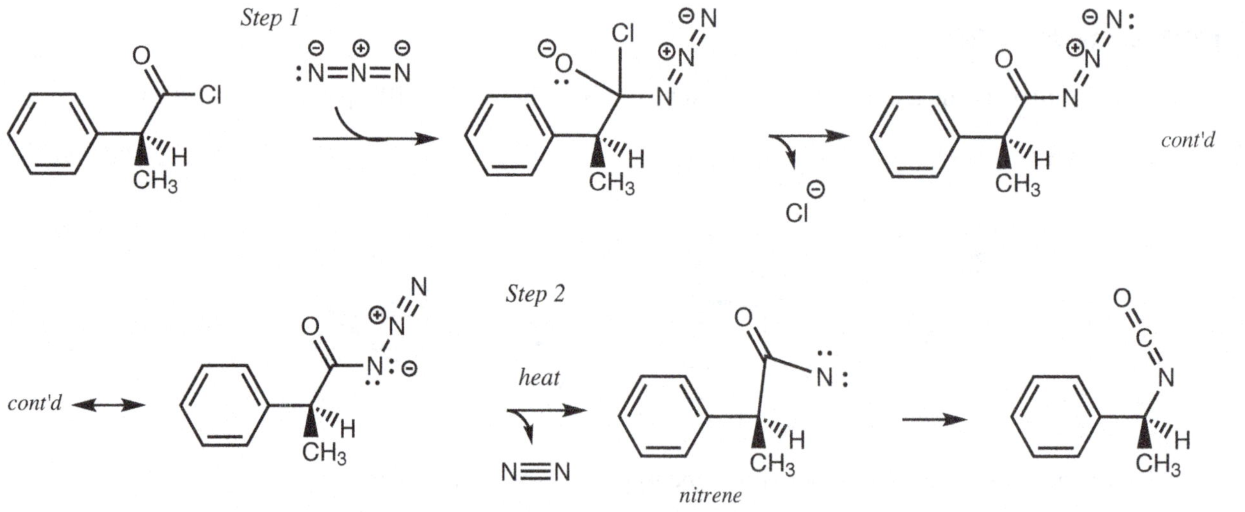

Step 1

cont'd

cont'd

Step 2

heat

nitrene

Step 2, hydrolysis of isocyanate to an amine.

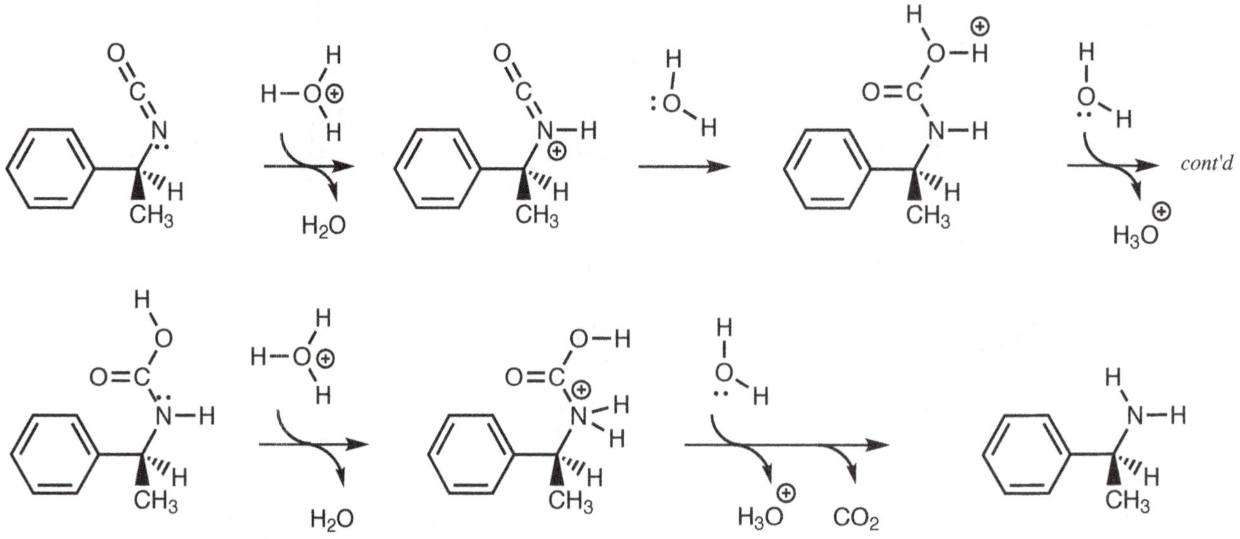

cont'd

must be solated from its ammonium salt

Alternate Step 2, hydrolysis of isocyanate to a carbamate. A carbamate is a nitrogen analog of a carbonate ester. The steps are similar to the above reaction with an alcohol replacing the water.

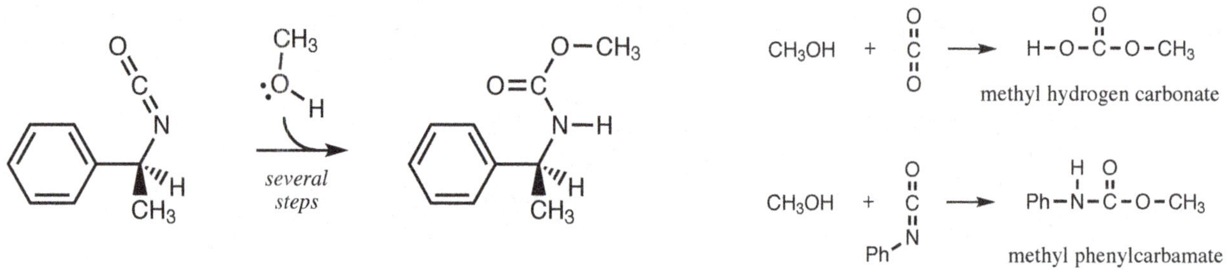

several steps

CH_3OH +   ⟶ $H-O-C-O-CH_3$

methyl hydrogen carbonate

CH_3OH +   ⟶ $Ph-N-C-O-CH_3$

methyl phenylcarbamate

Hoffmann Rearrangement

5. Step 1, Hoffmann rearrangement of a primary amide to give the corresponding amine. The reaction of the amide with sodium hydroxide and bromine.

This intermediate may be skipped

Step 2, Acidification and decarboxylation to give the amine.

carbamic acid (nitrogen analog of carbonic acid)

16 — Radical Reactions

For discussion of radical reactions, see *Notes*.

Bromination (or Chlorination) Reaction

1. Free radical bromination of cyclohexane to give bromocyclohexane.

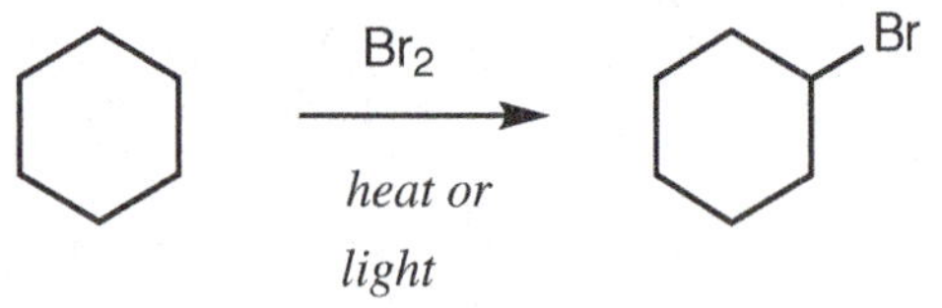

Initiation

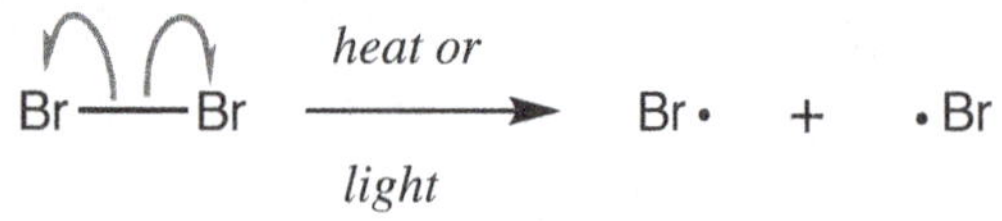

Propagation

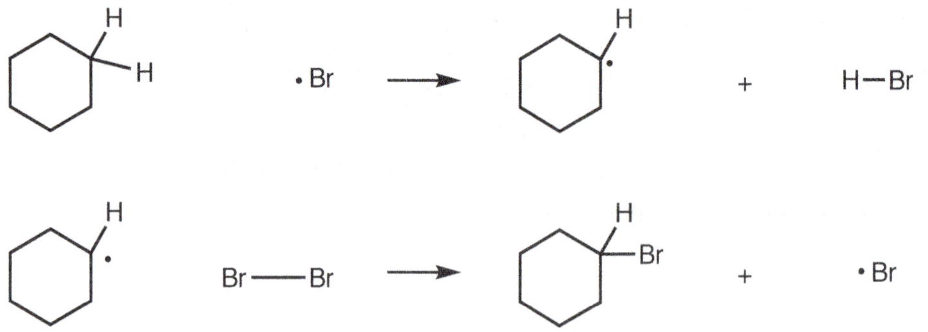

Termination

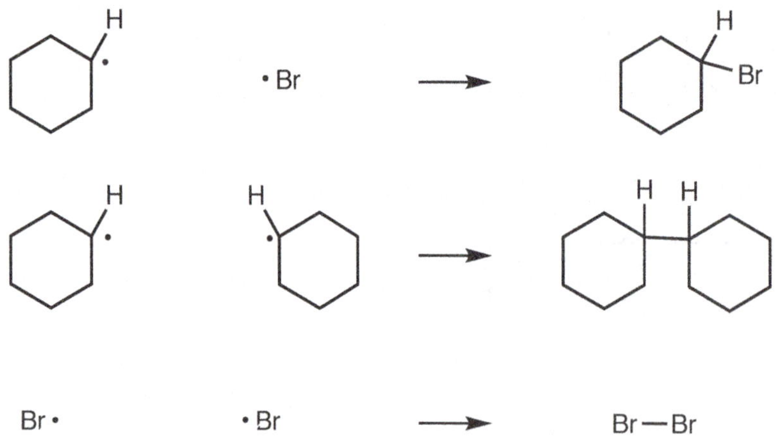

Allylic Bromination with NBS

2. Free radical bromination of cyclohexene with *N*-bromosuccinimide, an allylic bromination.

Overall reaction

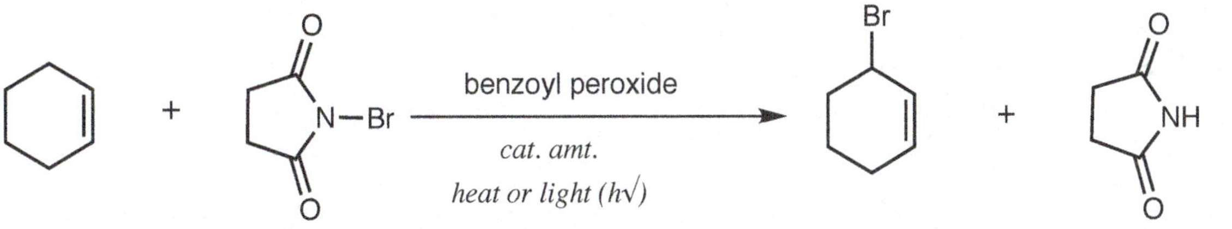

NBS

Initiation (See *Notes*.)

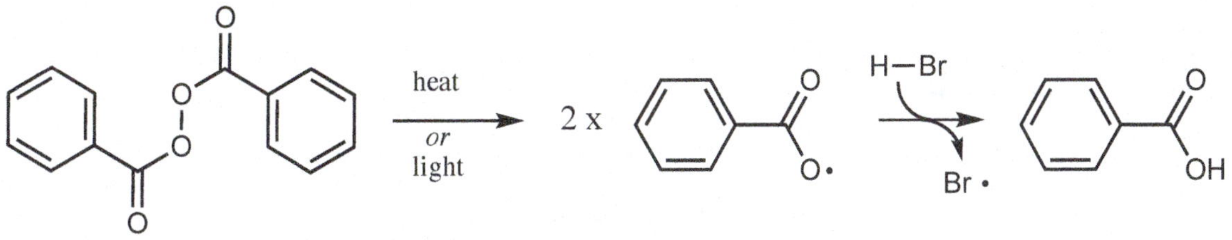

Propagation

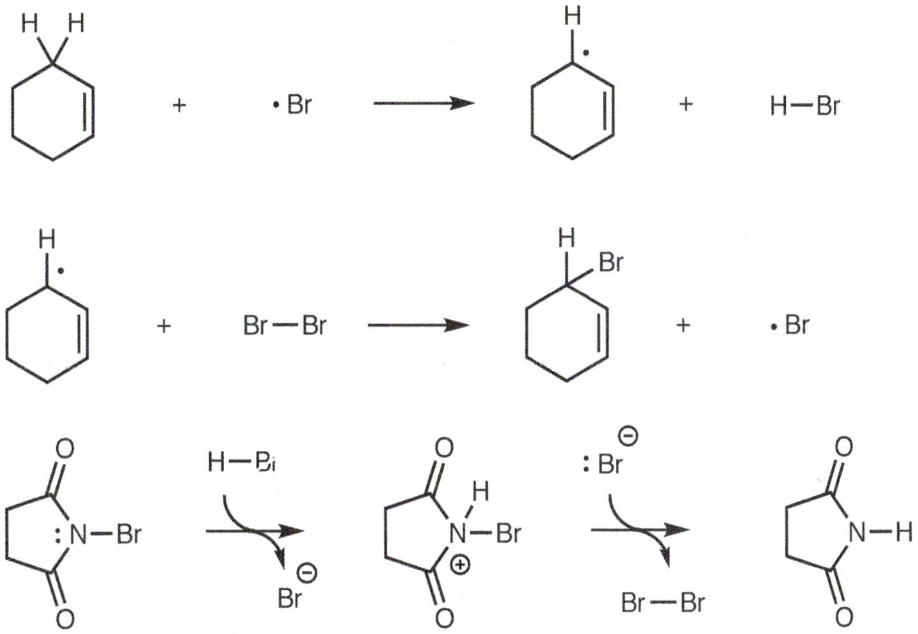

Termination

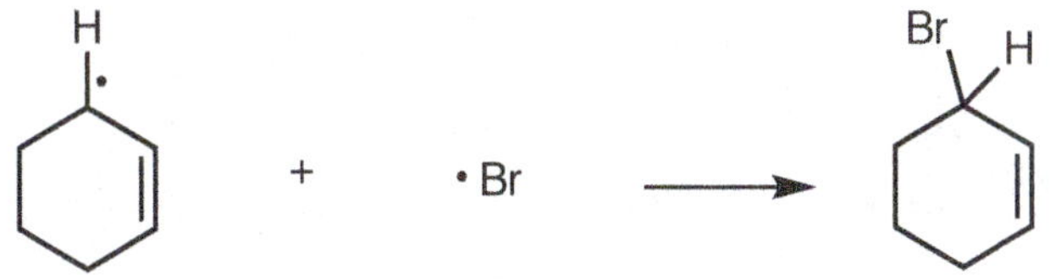

+ *others*

Radical Addition of Hydrogen Bromide

3. Free Radical Addition of HBr/H_2O_2 to Alkene (See *Notes.*)

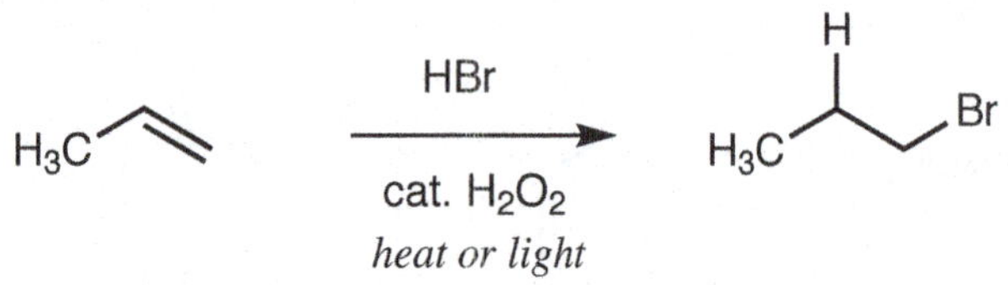

Initiation

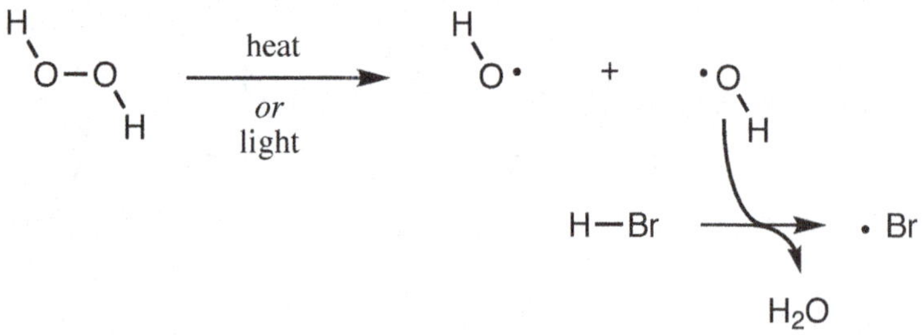

Propagation

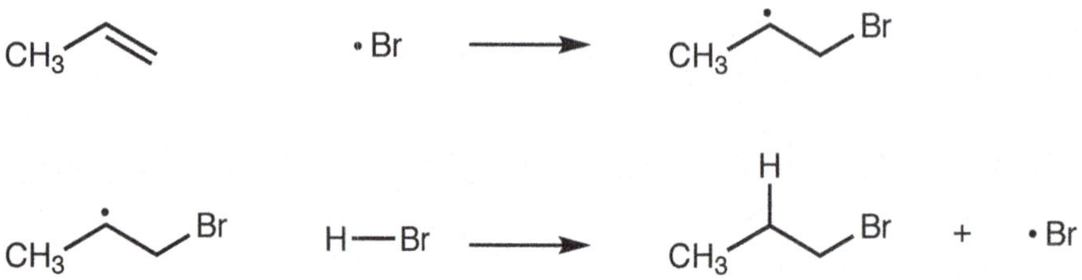

Termination

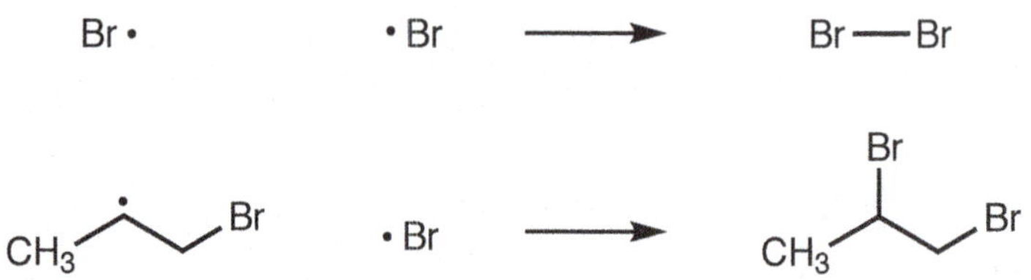

others

Benzylic Bromination with NBS

4. Benzylic bromination of ethylbenzene with NBS to give 1-bromo-1-phenylethane.

Overall reaction

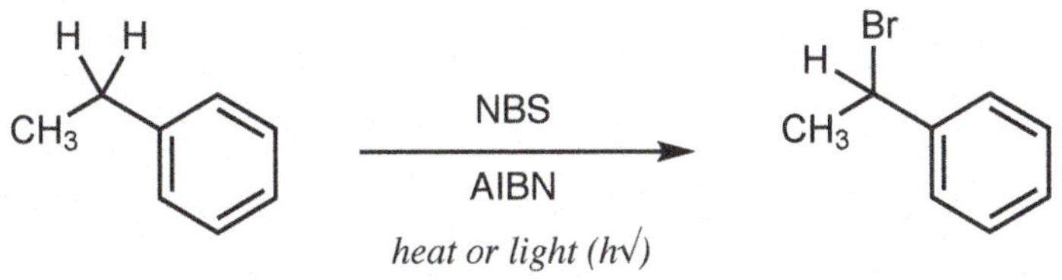

Initiation with azobisisobutyronitrile (AIBN)

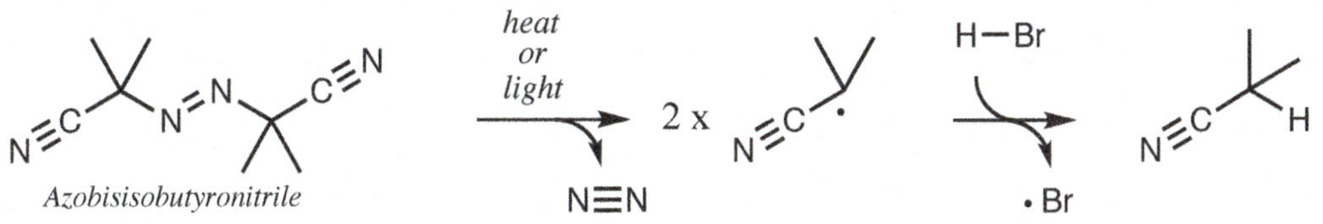

Propagation

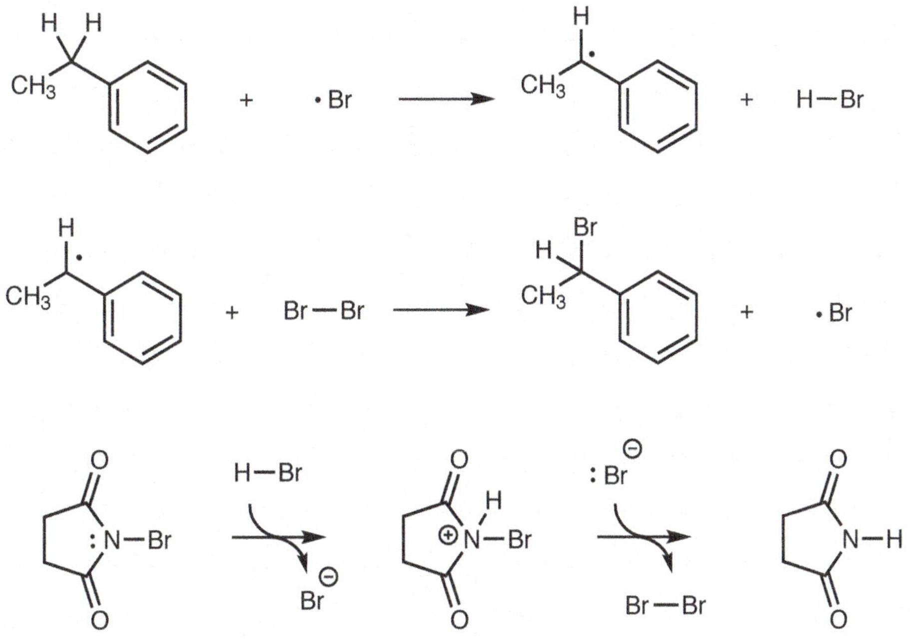

Termination

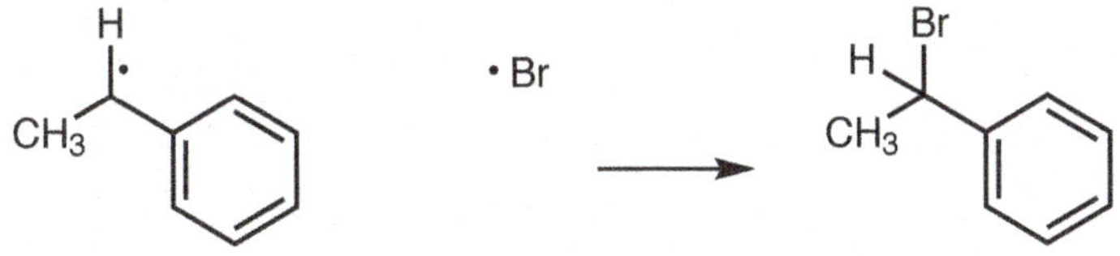

+ *others*

1 — Getting Ready for Reactions

About the Atom

In the following table, the relative stabilities of some carbanions and carbocations are listed. In this instance, we are looking at how the substituents affect electron deficient or electron rich carbon atoms.

Carbanion Structure	Relative stability	Carbocation Structure	Relative stability
$H-\overset{\ominus}{\underset{H}{C}}-H$	1 (most)	$H-\overset{\oplus}{\underset{H}{C}}-H$	4
$H_3C-\overset{\ominus}{\underset{H}{C}}-H$	2	$H_3C-\overset{\oplus}{\underset{H}{C}}-H$	3
$H_3C-\overset{\ominus}{\underset{CH_3}{C}}-H$	3	$H_3C-\overset{\oplus}{\underset{CH_3}{C}}-H$	2
$H_3C-\overset{\ominus}{\underset{CH_3}{C}}-CH_3$	4	$H_3C-\overset{\oplus}{\underset{CH_3}{C}}-CH_3$	1 (most)

1. If you compare the carbanion stabilities, what effect does replacing a hydrogen atom with a methyl group have on the stability of the carbanion?

2. Comparing the carbocation stabilities, what effect does replacing a hydrogen atom with a methyl group have on the stability of the carbocation?

3. Many textbooks state that carbon is a better electron donor than hydrogen. Are the carbocation and carbanion stabilities consistent with carbon being a better electron donor?

4. The pK_a of trifluoromethane (CF_3H) is approximately 25 while that of methane (CH_4) is 50. Which hydrogen-electron pair distance will be greater? [methane or trifluoromethane] How might the fluorine atoms have affected that distance?

5. From your predictions above, which electrons would be held closer to the carbon nucleus, the methyl or the trifluoromethyl carbanion? Methyl versus tertiary butyl carbanion?

6. In the following problem, if an oxygen atom is attached to another (electronegative) atom; predict which compound would be more acidic, HOH (water), HOOH (hydrogen peroxide), or HOCl (hypochlorous acid)? What effect will the atom attached to the oxygen (H, OH, and Cl) have on the proton-oxygen electron pair?

7. An analogy that I suggest for predicting chemical reactivity is a boxer. The boxer with a longer reach would have an easier time to hit the nucleus of a neighboring atom. Which do you think would have a longer reach, CH_3^- or HO^-?

Guide to Drawing Resonance Structures

Resonance Structures

If you are unfamiliar with the use of the curved arrow, refer to the discussion in the *Notes* section. A good place to start pushing electrons is in drawing resonance structures. They have the elements of electron movement, but the problems will be more limited in scope.

Resonance Structures of Anions

Add the missing curved arrows for the following resonance structures. For 1-6, the first and last structures are the same. In that case, you are converting it back to the starting structure.

1.

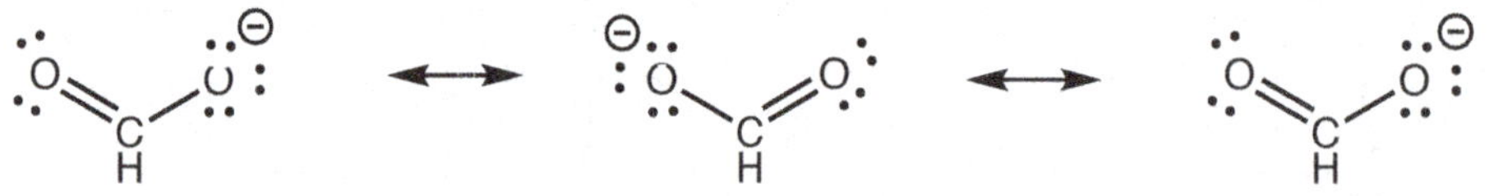

2.

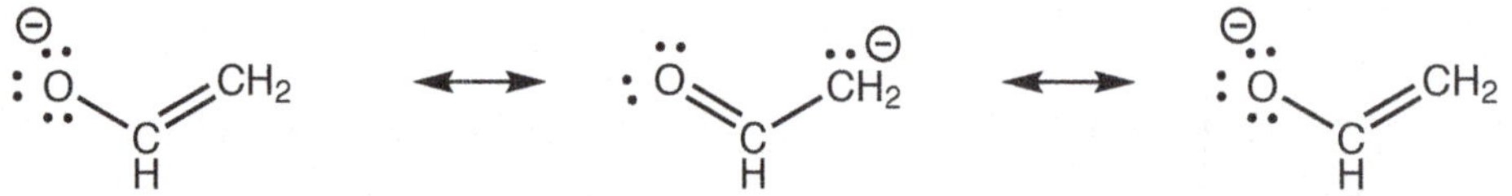

3.

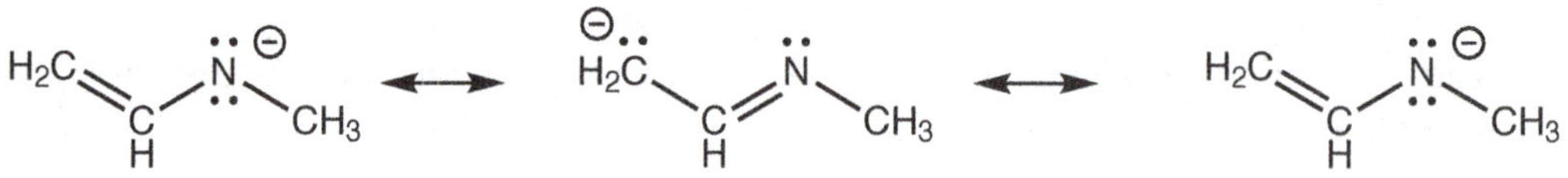

4.

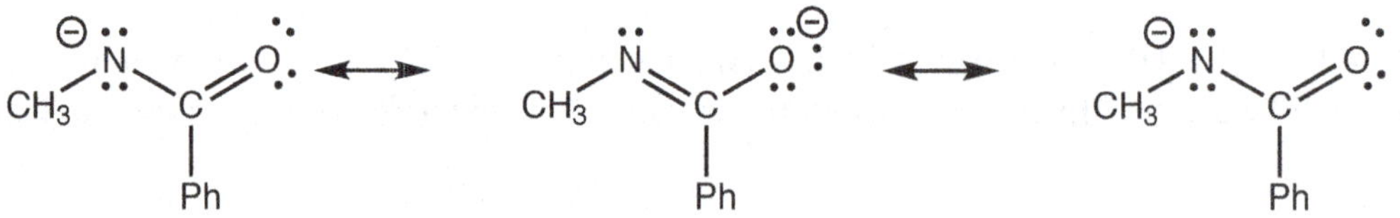

5.

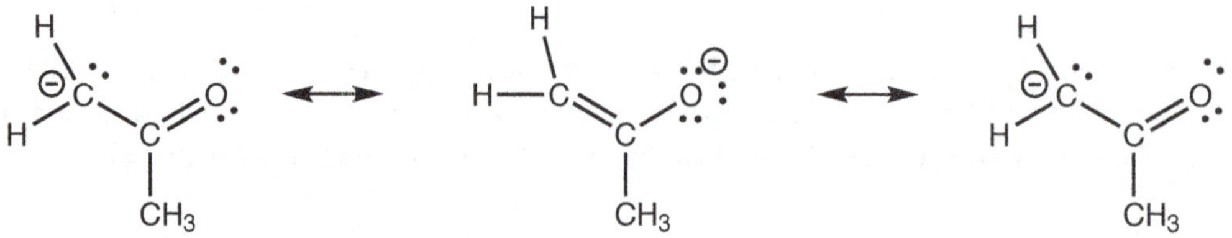

6.

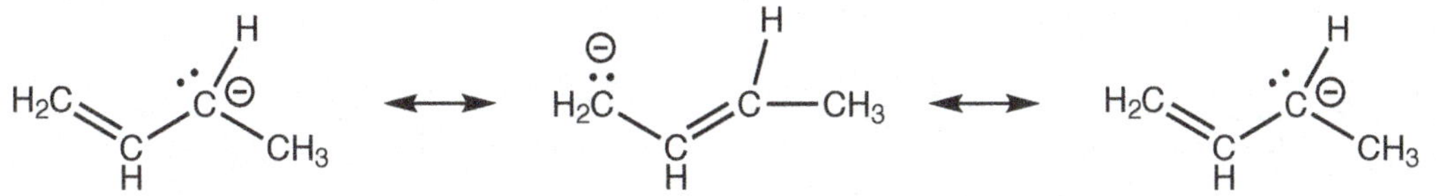

7.

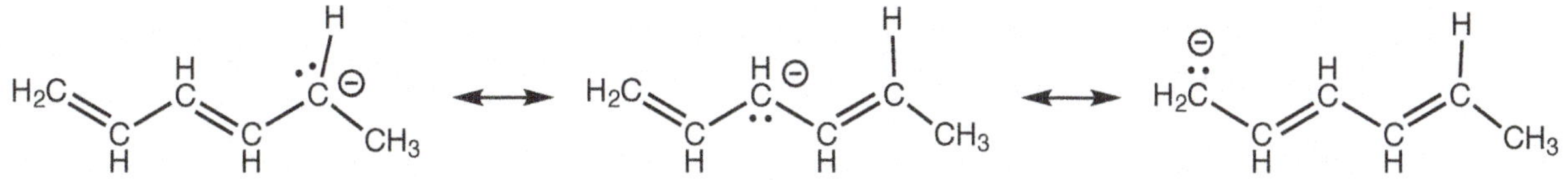

8.

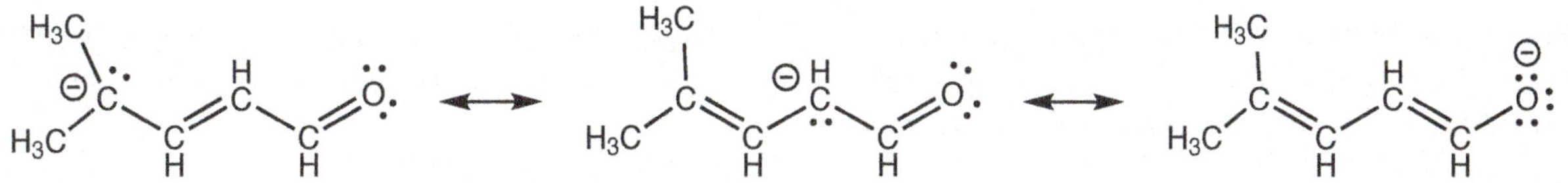

These examples show how the electrons can move, where the resulting charge will form, and how the charge can be distributed. It does not tell you on which atoms the greater charge density might exist nor upon which atom a reaction might next occur. You will note that since the original structure had a negative charge, the only charge that exists on any of the resulting resonance structures is a negative charge.

Resonance Structures of Cations

Conversely, if there is a positive charge, it is the positive charge that will attract (pull) electrons. Start a curved arrow from a neighboring pi or pair of non-bonded electrons and bring it toward the positive charge. You will note that only one curved arrow is necessary to create a new resonance structure. Because the original structure is a cation, completing its octet should not require further electron movement.

Add curved arrows to the following structures. For 9 and 10, the first and last structures are the same. In that case, you are converting it back to the starting structure.

9.

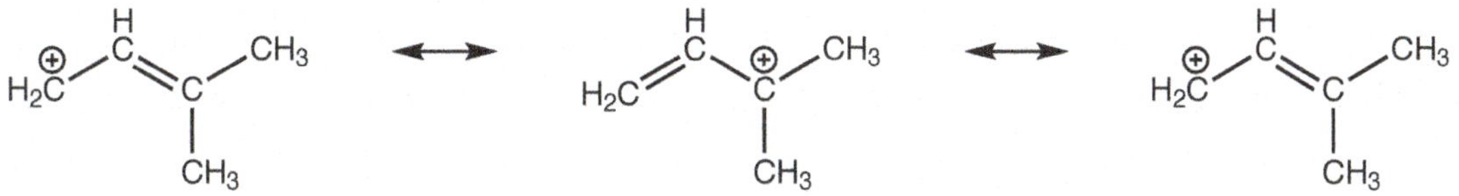

10.

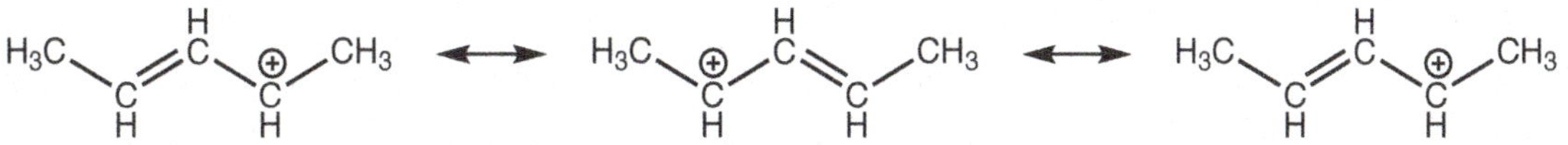

11.

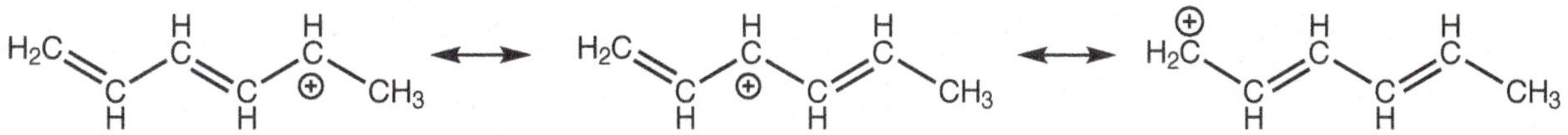

12.

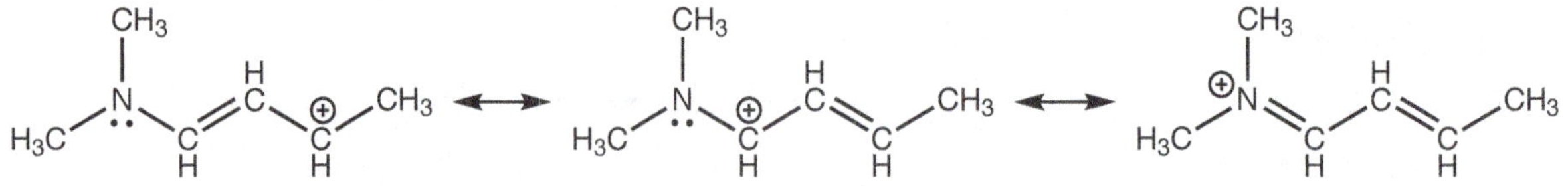

13.

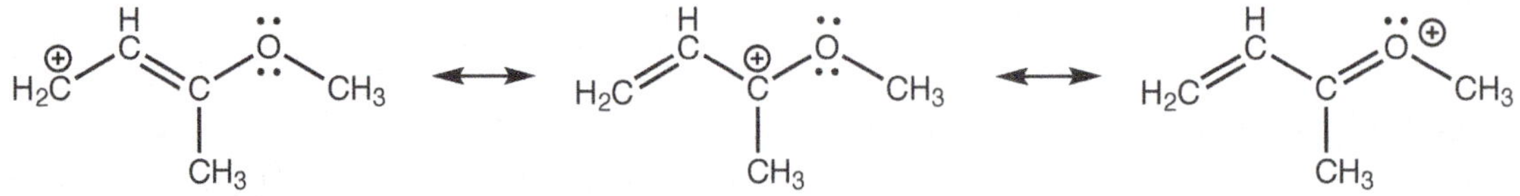

Again, these examples show how the electrons can move, where the resulting charge will form, and how the charge can be distributed. It does not tell you on which atoms the greater charge density might exist nor upon which atom a reaction might next occur. You will note that since the original structure had a positive charge, the only charges that exist on any of the resulting resonance structures are positive charges.

Resonance Structures of Neutral Compounds with Non-Bonded Electrons

If there isn't a charge and there are adjacent non-bonded electrons, then it will be the non-bonded electrons that will move (push) toward a neighboring pi bond. Start a curved arrow with the non-bonded electrons and direct them to the neighboring double bond.

Add curved arrows to the following structures. For 14-17, the first and last structures are the same. In that case, you are converting it back to the starting structure.

14.

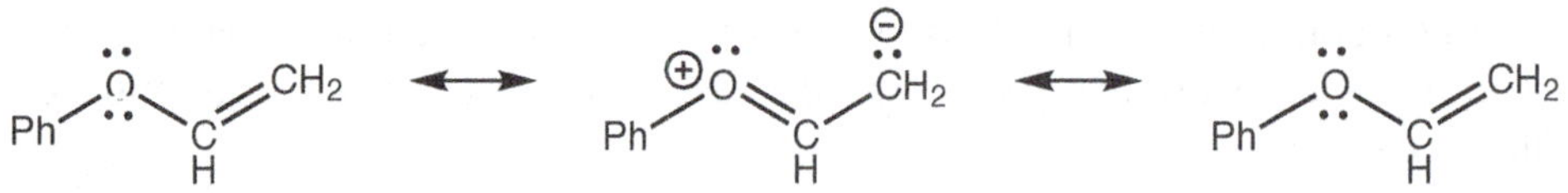

15.

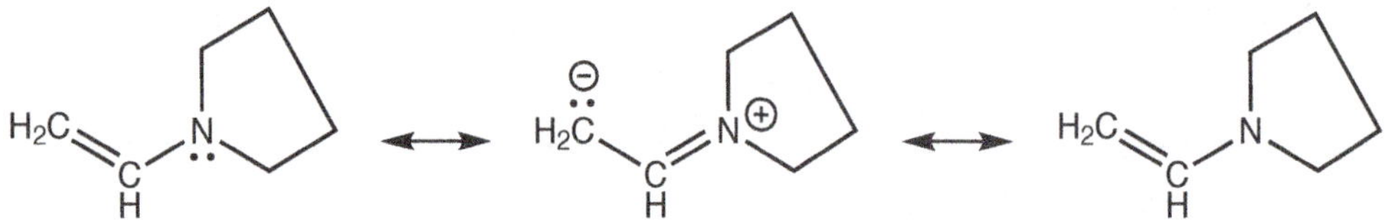

16.

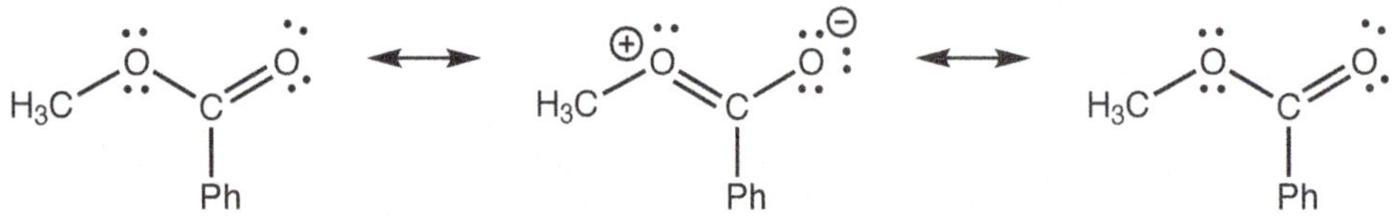

17.

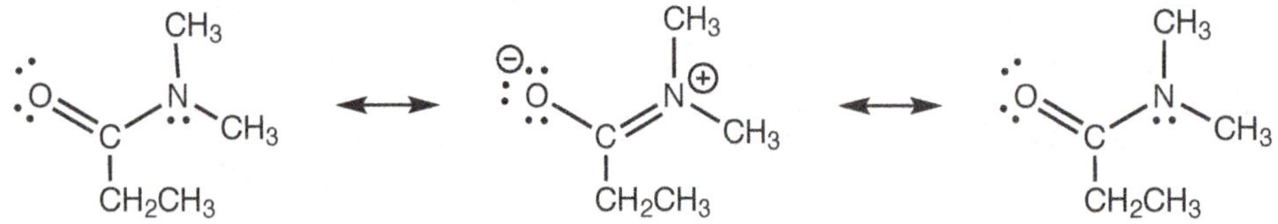

18.

19.

Again, these examples show how the electrons can move, where the resulting charges will form, and how the charges can be distributed. It does not tell you on which atoms the greater charge densities might exist nor upon which atoms a reaction might next occur. You will note that since the original structure was neutral, the net charges that exist on any of the resulting resonance structures are also neutral and only two atoms have a charge.

Resonance Structures of Neutral Compounds without Non-Bonded Electrons

What if there is no charge and there are no neighboring non-bonded electrons? Then we will push the pi electrons of a double bond toward the least substituted carbon or the most electron withdrawing atom, see Example 26. Start a curved arrow with the pi electrons of a double bond and direct them from the most to the least substituted carbon or for a C=O bond, toward the oxygen atom.

20.

21.

This is the same example as above, however the arrows are pointing in the opposite direction. If you are uncertain in which direction the electrons might move, a good strategy is to draw an arrow in the opposite direction and then to compare the results of the two possibilities. Compare the result below with the one above. Which is the more stable? If you do not recognize the lower arrangement as a lesser contributor, you may need to refer to your textbook for the rules of carbanion, carbocation, and resonance stability. Examples 23 and 24 are similar, which is more stable?

22.

23.

24. Compare Examples 23 and 24, which is preferred?

25.

26.

In these examples, we have incorporated chemical principles. The carbocations that are the most substituted are the most stable. Coinciding with this principle is that carbanions with the least substitution (or a heteroatom, last example) are the most stable.

Resonance Structures of Radicals

Radicals, compounds with unpaired electrons, are less stable than those with paired electrons. The fate of radical reactions is to form a paired-electron bond. However, sharing unpaired electrons with neighboring non-bonded electrons or pi-bonds can attain added stability. Also note the curved arrow has a single barb indicating the movement of a single electron. Two arrows are required for a pair of electrons.

27.

28.

2 — Acid-Base Chemistry

Bronsted-Lowrey Acids and Bases

Acid-base reactions are often the first intermolecular reaction you will encounter. A proton will be exchanged from the strongest acid to the strongest base.

In these examples, you must note the conjugate acid and conjugate base that result in each reaction. You should note the use and meaning of the curved arrows. If the example does not contain a curved arrow, you must supply one. The meaning of the curved arrow is important for you to understand.

- The rule for predicting the product of an acid-base reaction is simple. A reaction will generally give the product that is the weakest base (or conjugate base). The base strength of a compound is related to the acidity of the acid, the stronger the acid, the weaker the base, or the corollary, the weaker the acid, the stronger the base. In order to compare the base strengths, the acidities of the acid and conjugate acid must be determined first.

- Look at each example and write the pK_a under each acid (on the left) and conjugate acid (on the right). Be careful that you correctly identify the acid and the corresponding pK_a. Strong acids have a small pK_a and weak acids have a large pK_a. You may need to use a table to find some values.

- For each acid (or conjugate acid), write the corresponding pK_a beneath it. Write the label "B(ase)" (or "CB") under each base (or conjugate base). Look at the pK_a of the acid or conjugate acid. The strongest acid corresponds with the weakest base. Label it, "weakest base". In example 1, the pK_a of HF is 3.2. It is placed under HF. The pK_a of acetic acid is 4.75 and 4.75 is placed under it. Because HF is the stronger acid (lowest pK_a), its conjugate base will be the weakest base. The equilibrium will shift to the right.

- Label the equilibrium of each reaction, L(eft) or R(ight). Example 1, R.

1. What is the weakest base?

Base	Acid	Conjugate Acid	Conjugate Base

L/**R**   3.2   4.75

Write sentences to describe the curved arrows.

2. Label acids, bases, and conjugate acids and bases.

L/**R**   base   3.2   15.7   conj. Base

Add the curved arrows and write sentences to describe the curved arrows.

Continue by completing the equation, adding curved arrows, pK$_a$ values, and direction of equilibrium.

3. Write a sentence(s) describing any bonds being made or broken.

$$H-\overset{..}{\underset{..}{O}}-H \quad + \quad H-\overset{..}{\underset{..}{F}}: \quad \rightleftharpoons \qquad\qquad + \qquad :\overset{..}{\underset{..}{F}}:^{\ominus}$$

L/R   base   3.2   -1.7   conj. Base

4. Write a sentence(s) describing any bonds being made or broken.

$$H-\overset{..}{\underset{..}{S}}-H \quad + \quad H-\overset{..}{\underset{\underset{H}{|}}{N}}-H \quad \rightleftharpoons \qquad\qquad + \quad H-\overset{\overset{H}{|}}{\underset{\underset{H}{|}}{N}}\overset{\oplus}{-}H$$

L/**R**   7.0   conj. base   9.2

5. Write a sentence(s) describing any bonds being made or broken.

$$H-\overset{..}{\underset{..}{F}}: \quad + \quad :\overset{..}{\underset{..}{Cl}}:^{\ominus} \quad \rightleftharpoons \qquad\qquad +$$

L/R   3.2   conj. base   -8

6. Write a sentence(s) describing any bonds being made or broken.

$$\text{C}_6\text{H}_5-\overset{..}{\underset{..}{O}}-H \quad + \quad {}^{\ominus}:\overset{..}{\underset{..}{O}}-H \quad \rightleftharpoons \qquad\qquad +$$

L/R   10.0   conj. Base   15.7

7. Write a sentence(s) describing any bonds being made or broken.

$$H_3C-\overset{\overset{O}{\|}}{C}-\overset{..}{\underset{..}{O}}-H \quad + \quad H_3C-\overset{..}{\underset{\underset{H}{|}}{N}}-H \quad \rightleftharpoons \quad H_3C-\overset{\overset{O}{\|}}{C}-\overset{..}{\underset{..}{O}}:^{\ominus} \quad + \quad H_3C-\overset{\overset{H}{|}}{\underset{\underset{H}{|}}{N}}\overset{\oplus}{-}H$$

L/R   4.75   conj. Base   10.5

8. Write a sentence(s) describing any bonds being made or broken.

$$H_3C \cdot CH_2 \cdot O-H \quad + \quad H_3C-\overset{..}{\underset{\underset{H}{|}}{N}}-H \quad \rightleftharpoons \quad H_3C \cdot CH_2 \cdot \overset{..}{\underset{..}{O}}:^{\ominus} \quad + \quad H_3C-\overset{\overset{H}{|}}{\underset{\underset{H}{|}}{N}}\overset{\oplus}{-}H$$

L/R   16.0   conj. Base   10.5

Continue by completing the equation, adding curved arrows, pK$_a$ values, and direction of equilibrium.

9.

L/R   22   conj. Base   19

10.

L/R   38   50   conj. Base

11.

L/R   9.2   -1.7   conj. Base

12.

L/R   4.6   9.2   conj. Base

13.

L/R   4.75   conj. Base   15.7

14.

L/R   24   conj. Base   50

Continue by completing the equation, adding curved arrows, pK$_a$ values, bases, and label the direction of equilibrium.

15.

$$H-\underset{\underset{H}{|}}{\overset{\overset{H}{|}}{C}}-\overset{\overset{O}{||}}{C}-O-CH_2CH_3 \quad + \quad :\overset{\ominus}{\underset{..}{O}}-CH_2CH_3 \quad \rightleftharpoons \qquad\qquad +$$

L/R   24   16

16.

$$CH_3-\overset{\overset{O}{||}}{C}-\underset{\underset{H}{|}}{\overset{\overset{H}{|}}{C}}-\overset{\overset{O}{||}}{C}-O-CH_3 \quad + \quad :\overset{\ominus}{\underset{..}{O}}-CH_3 \quad \rightleftharpoons \qquad\qquad +$$

L/R   11.7   16

17. Sometimes we may be unsure how a reaction might proceed. Will H$_2$S and (CH$_3$)$_2$NH react together? What will the products be if they do? In that case, draw ALL of the possible products and analyze the results. Complete this problem as before.

From the individual equilibria, can you predict the overall result?

$$H-\overset{..}{\underset{..}{S}}-H \quad + \quad H_3C-\overset{\overset{H}{|}}{N}-CH_3 \quad \rightleftharpoons \qquad +$$

38   <-5

$$H-S-H \quad + \quad H_3C-\underset{..}{\overset{\overset{H}{|}}{N}}-CH_3 \quad \rightleftharpoons \qquad +$$

7.0   10.7

$$+ \quad H_3C-\overset{\overset{H}{|}}{\underset{\underset{H}{|}}{N}}\overset{\oplus}{-}CH_3 \quad \rightleftharpoons \quad H-\overset{..}{\underset{..}{S}}-H \quad + \quad H_3C-\underset{..}{\overset{\overset{H}{|}}{N}}-CH_3 \quad \rightleftharpoons \quad H-\overset{H}{\underset{..}{\overset{\oplus}{S}}}H \quad +$$

18. What is the equilibrium between HOCH$_3$ and CH$_3$NHCH$_3$? First, determine the individual equilibria and then predict the overall result.

$$H-\overset{..}{\underset{..}{O}}-CH_3 \quad + \quad H_3C-\overset{\overset{H}{|}}{N}-CH_3 \quad \rightleftharpoons \qquad +$$

38   -2.2

$$H-O-CH_3 \quad + \quad H_3C-\underset{..}{\overset{\overset{H}{|}}{N}}-CH_3 \quad \rightleftharpoons \qquad +$$

15.5   10.7

$$+ \qquad \rightleftharpoons \quad H-\overset{..}{\underset{..}{O}}-CH_3 \quad + \quad H_3C-\underset{..}{\overset{\overset{H}{|}}{N}}-CH_3 \quad \rightleftharpoons \qquad +$$

Lewis Acids and Bases

The prior exercise showed a hydrogen atom accepting electrons. With Lewis acids, other atoms can also accept electrons. See your text for further discussion.

Add structures, non-bonded electrons, curved arrows, and formal charges as needed to complete the following.

19.

$$H_3C \underset{H_3C}{\overset{CH_3}{\diagdown}} \ddot{N} \qquad H \underset{H}{\overset{H}{\diagdown}} B \qquad \rightleftarrows$$

20.

$$Br \underset{Br}{\overset{Br}{\diagdown}} B \qquad :\ddot{O}: \underset{CH_3}{\overset{CH_3}{\diagdown}} \qquad \rightleftarrows$$

21.

$$Cl - \ddot{C}l \quad + \quad Cl \underset{Cl}{\overset{Cl}{\diagdown}} Fe \underset{}{\overset{Cl}{\diagup}} \quad \rightleftarrows \quad Cl - \overset{\oplus}{C}l - \overset{Cl}{\underset{Cl}{\overset{|}{Fe}}} \overset{\ominus}{-} Cl \quad \rightleftarrows \quad \overset{\oplus}{C}l \quad +$$

22.

$$:\ddot{Br} - \ddot{Br}: \quad + \quad Br \underset{Br}{\overset{Br}{\diagdown}} B \qquad \rightleftarrows$$

23.

$$H_3C - \overset{CH_3}{\underset{CH_3}{\overset{|}{N}}}: \quad + \quad Cl \underset{Cl}{\overset{Cl}{\diagdown}} Al \underset{}{\overset{}{\diagup}} Cl \qquad \rightleftarrows$$

24.

$$H - \overset{H}{\underset{H}{\overset{|}{C}}} - \overset{O}{\overset{||}{C}} - \ddot{C}l \quad + \quad Cl \underset{Cl}{\overset{Cl}{\diagdown}} Al \underset{}{\overset{}{\diagup}} Cl \quad \rightleftarrows \quad H - \overset{H}{\underset{H}{\overset{|}{C}}} - \overset{O}{\overset{||}{C}} - \overset{\oplus}{C}l - \overset{Cl}{\underset{Cl}{\overset{|}{Al}}} \overset{\ominus}{=} Cl \quad \rightleftarrows \quad + \quad Cl - \overset{Cl}{\underset{Cl}{\overset{|}{Al}}} \overset{\ominus}{=} Cl$$

25.

$$H_3CH_2C \underset{H_3CH_2C}{\diagdown} :\ddot{O}: \quad + \quad F - B \underset{F}{\overset{F}{\diagdown}} \qquad \rightleftarrows$$

26.

$$H_3C - \overset{H}{\underset{H}{\overset{|}{C}}} = \overset{H}{\underset{H}{\overset{}{C}}} \qquad \overset{\oplus}{Hg} - Cl \qquad \rightleftarrows$$

3 — Substitution Reactions

S$_N$2 Substitution Reactions

Add curved arrows to the following reactions.

1. An S$_N$2 reaction of 1-chlorobutane with sodium iodide to give 1-iodobutane. (See *Notes*.)

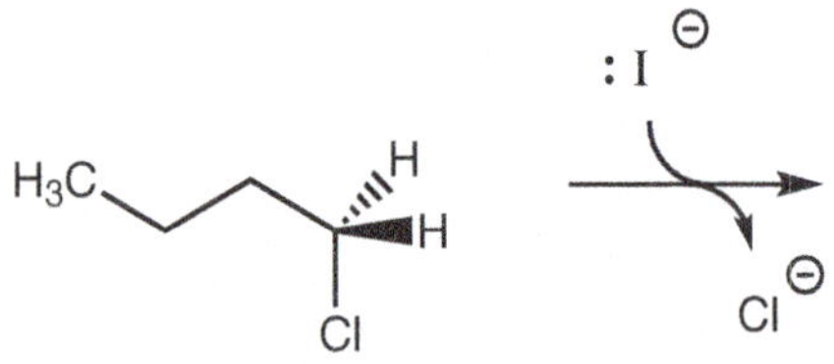

2. An S$_N$2 reaction of 1-bromobutane with ethoxide to give 1-ethoxybutane (butyl ethyl ether). (See *Notes*.)

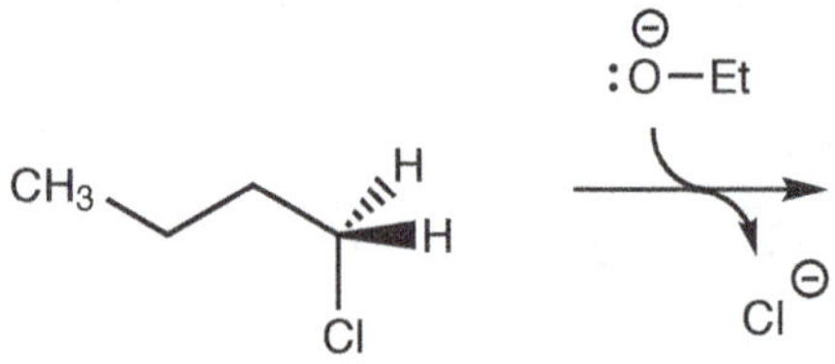

3. An S$_N$2 reaction of *(R)*-2-bromobutane with thiocyanate to give *(S)*-2-thiocyanatobutane. (See *Notes*.)

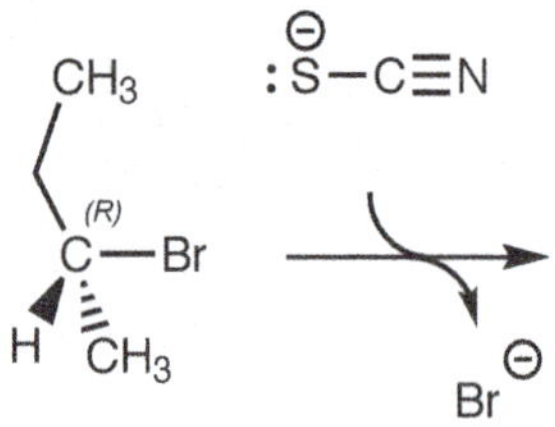

4. An S$_N$2 reaction of *(S)*-2-bromobutane with acetate to give *(R)*-*sec*-butyl acetate. (See *Notes*.)

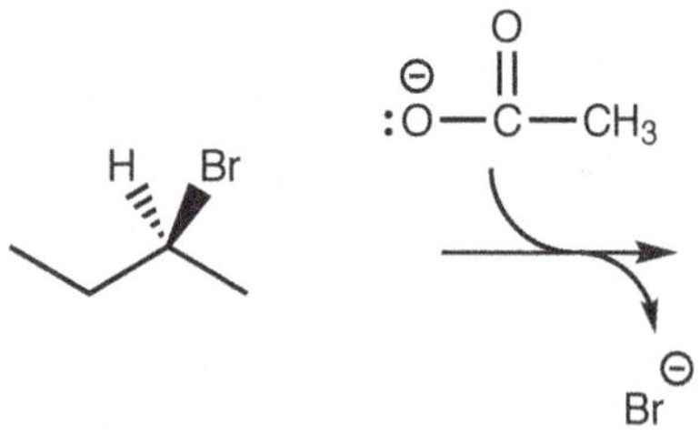

5. An S$_N$2 reaction of 1-butanol with hydrogen bromide to give 1-bromobutane. (See *Notes*.)

6. An S$_N$2 reaction of a triflate with sodium methanethiolate to give a thioether with inversion. (See *Notes.*)

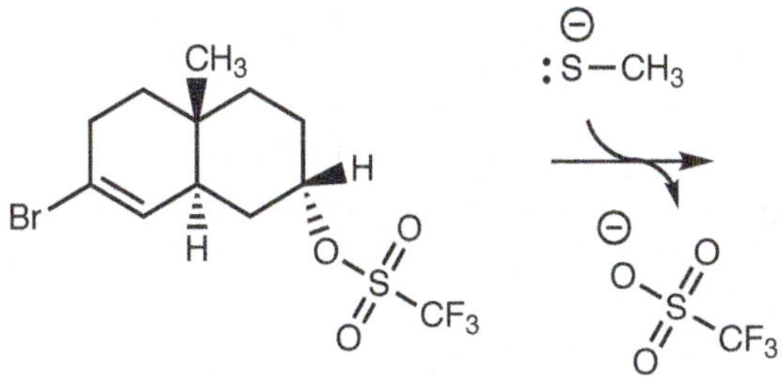

7. An S$_N$2 reaction of *cis*-1-chloro-4-methylcyclohexane with azide to give *trans*-1-azido-4-methylcyclohexane.

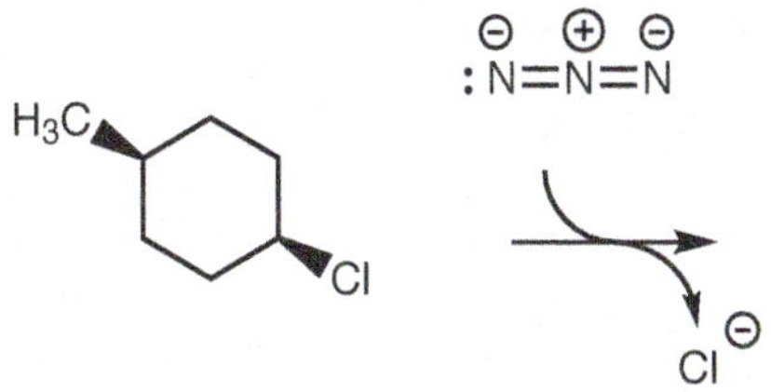

8. An S$_N$2 reaction of a ditosylate with one equivalent of cyanide to give a mono-nitrile. (See *Notes.*)

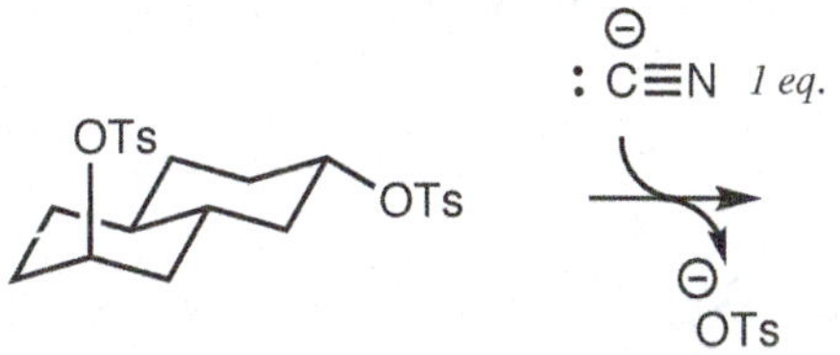

9. An S$_N$2 reaction of 1-bromo-3-chloropropane with cyanide to give 4-chlorobutanenitrile. (See *Notes.*)

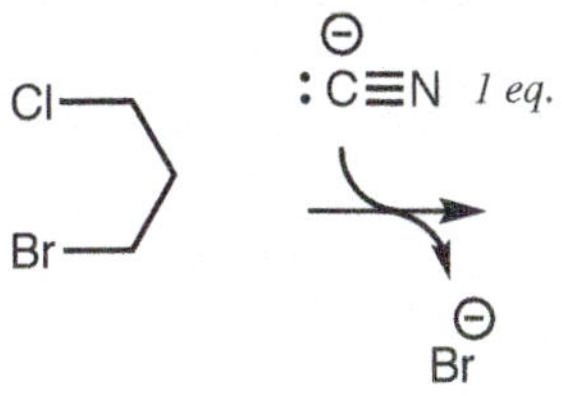

10. An S$_N$2 reaction of 1-bromobutane with acetylide anion to give 1-hexyne. (See *Notes.*)

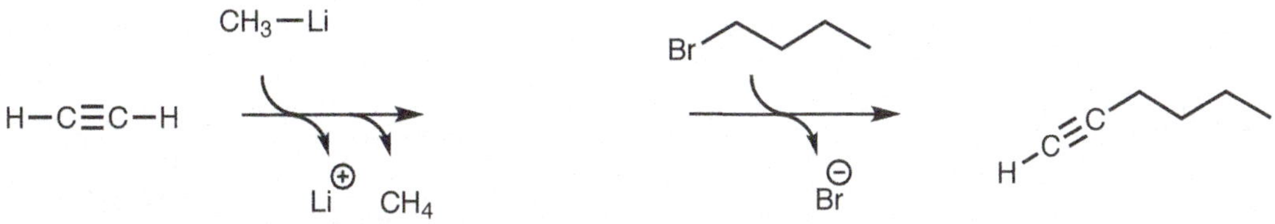

11. An S$_N$2 reaction of methyl phenylacetate enolate with iodopropane to give methyl 2-phenylpentanoate. (See *Notes*.)

12. A Gabriel amine synthesis with formation of phthalimide anion (potassium carbonate) and alkylation with 1-bromo-2-butene. (See *Notes*.)

13. An S$_N$2 reaction of bromomethane with phenoxide to give methoxybenzene (anisole). (See *Notes*.)

14. An S$_N$2 reaction of benzyl bromide with sodium ethoxide to give ethyl benzyl ether.

15. An S$_N$2 cleavage reaction of *p*-nitroanisole to give iodomethane and *p*-nitrophenol. (See *Notes*.)

16. An S$_N$2 reaction of 1-bromo-2-butanol with sodium hydroxide to give 2-ethyloxirane (butylene oxide).

17. An S$_N$2 reaction of the anion of 1-butyne with *(R)*-2-ethyloxirane to give *(R)*-hept-5-yn-3-ol. (See *Notes.*)

18. An acid catalyzed opening of 2-ethyloxirane with methanol to give 2-methoxybutan-1-ol. (See *Notes.*)

cont'd

19. An S$_N$2 reaction of the enolate of acetophenone with *(R)*-2-methyloxirane (propylene oxide) to give *(R)*-4-hydroxy-1-phenylpentan-1-one. (See *Notes.*)

(R)-4-hydroxy-1-phenylpentan-1-one

S_N1 Substitution Reactions

20. An S_N1 solvolysis reaction of *t*-butyl iodide to give *t*-butyl alcohol. (See *Notes.*)

21. An S_N1 solvolysis reaction of *(R)*-(1-chloroethyl)benzene to give rac-1-phenylethanol.

22. An S_N1 reaction of 1-methylcyclohexanol with hydrogen chloride to give 1-chloro-1-methylcyclohexane. (See *Notes.*)

23. An S_N1 solvolysis reaction of *(1S,3R)*-1-bromo-1,3-dimethylcyclohexane to give *(1S,3R)*- and *(1R,3R)*-1,3-dimethylcyclohexanol. (See *Notes.*)

24. An S_N1 solvolysis reaction of 2-bromo-3-methylbutane to give 2-methyl-2-butanol. .(See *Notes.*)

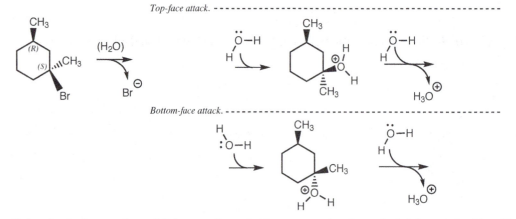

25. An S_N1 reaction of 2-methylbut-3-en-2-ol with hydrogen bromide to give 1-bromo-3-methyl-2-butene. (See *Notes.*)

26. An S_N1 solvolysis reaction of 1-bromo-3-methyl-2-butene to give 2-methyl-3-buten-2-ol. (See *Notes.*)

27. An S_N1 reaction of 2-methylcyclopentanol with HBr to give 1-bromo-1-methylcyclopentane.

28. An S_N1 solvolysis reaction of 3-chlorocyclopentene to give 3-methoxycyclopentene. (See *Notes.*)

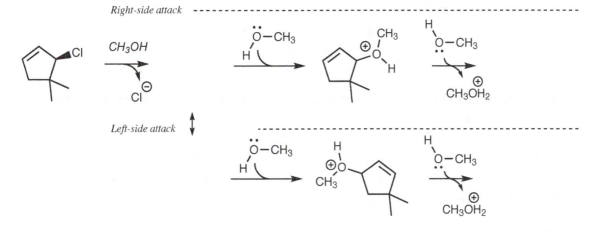

4 — Elimination Reactions

Alkene Formation

1. An E2 elimination reaction of hydrogen chloride from 1-chlorooctadecane with potassium *t*-butoxide to give 1-octadecene. (See *Notes*.)

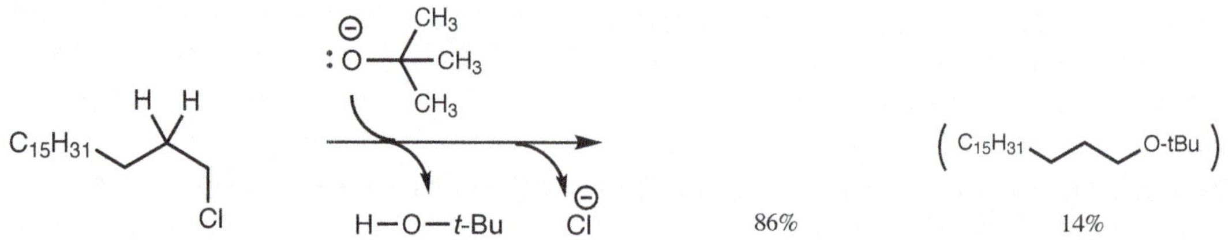

86%   14%

2. An E2 elimination reaction of hydrogen bromide from 2-bromobutane with sodium ethoxide to give *trans*-2-butene plus other butenes. (See *Notes*.)

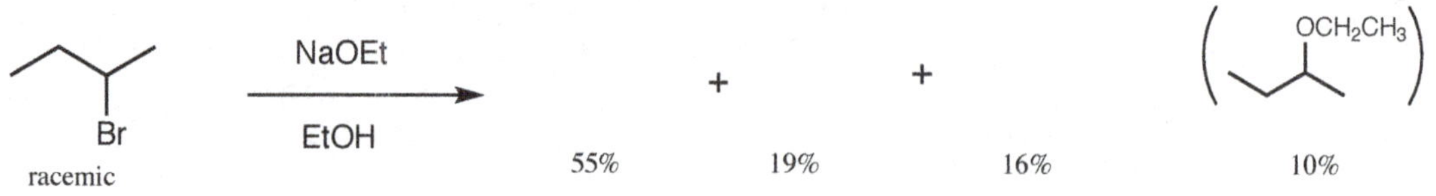

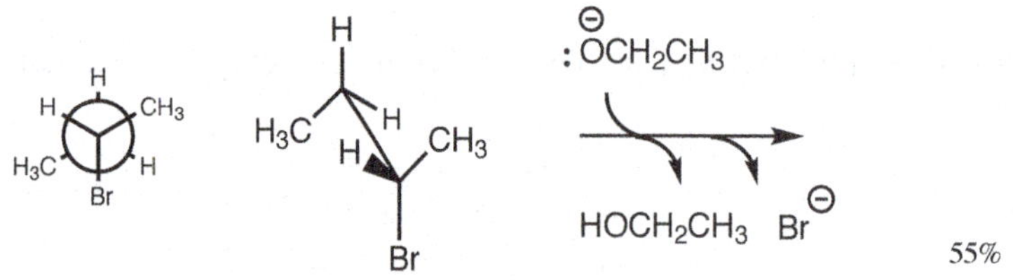

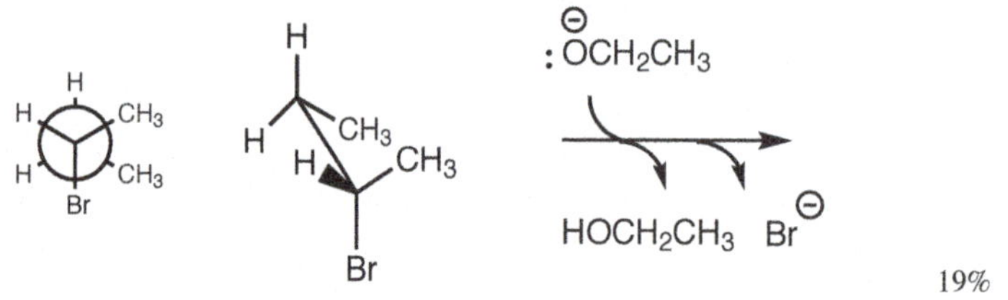

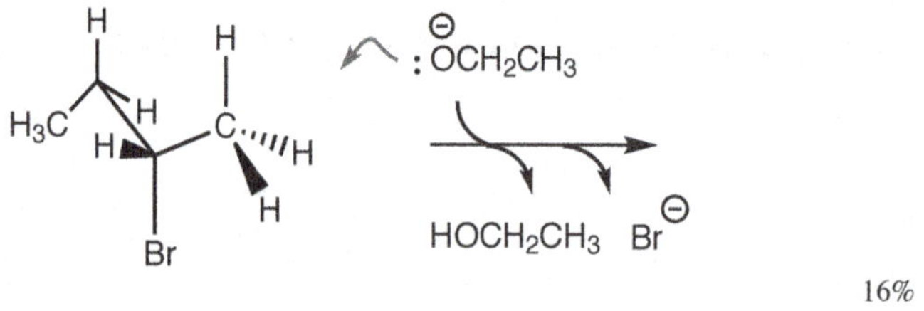

55%   19%   16%   10%

55%

19%

16%

3. An E2 elimination reaction of 2-bromo-2-methylbutane to give 2-methyl-2-butene, a Zaitsev product.

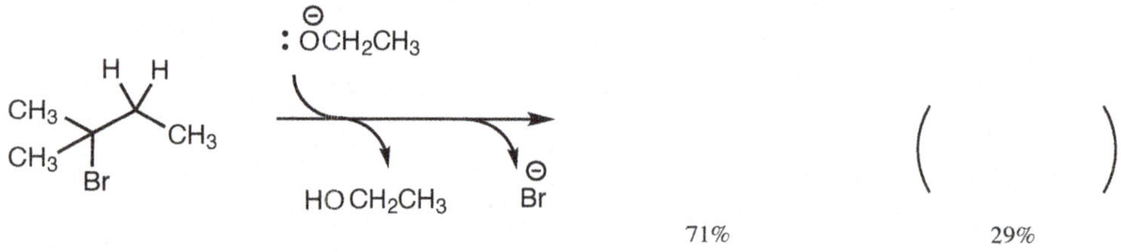

71%   29%

4. An E2 elimination reaction of (*1R,2R*)- or (*1S,2S*)-1-bromo-1,2-diphenylpropane to give (*Z*)-1,2-diphenylpropene. (See *Notes*.)

5. An E2 elimination reaction of *trans*-1-chloro-2-isopropylcyclohexane to give 3-isopropylcyclohexene. (See *Notes*.)

stereoview

6. An E2 elimination reaction of *cis*-1-chloro-2-isopropylcyclohexane to give 3-isopropylcyclohexene. (See *Notes*.)

stereoview   *major*

7. A competitive E2 elimination reaction of *cis*- and *trans*-1-bromo-4-*t*-butylcyclohexane and one equivalent of *t*-butoxide to give 4-*t*-butylcyclohexene and unreacted bromocyclohexane. (See *Notes*.)

1 mole   *1 mole*   *1 mole*   *1 mole*

8. An E2 elimination reaction of 1-chloro-1-methylcyclohexane with ethoxide to give cyclohexene. (See *Notes*.)

major

9. An E2 elimination reaction of 1-chloro-1-methylcyclohexane with *t*-butoxide to give methylenecyclohexane.

major

10. An E1cb elimination reaction of 3-chloro-3-methylcyclohexanone with *t*-butoxide to give 3-methyl-2-cyclohexenone. (See *Notes*.)

11. An E2 elimination reaction of 4-bromo-cyclohexene with *t*-butoxide to give 1,3-cyclohexadiene. (See *Notes*.)

major

12. An E2 elimination reaction of 3-bromo-1-cyclohexene with *t*-butoxide to give 1,3-cyclohexadiene. (See *Notes*.)

major

13. An E1 elimination reaction of 2-methylcyclopentanol by treatment with sulfuric acid to give methylcyclopentene. (See *Notes*.)

14. An E1 elimination reaction of 3-methyl-3-pentanol by treatment with acid to give pentenes. (See *Notes*.)

conformer A   10

conformer B   50   *conformer C*   40

15. An E1 solvolysis-elimination reaction of a benzyl bromide in aqueous base to give methyl stilbenes. (See *Notes*.)

conformer A

conformer B

16. An E1 elimination reaction of 2-cyclobutyl-2-propanol and sulfuric acid to give 1,2-dimethylcyclopentene. (See *Notes*.)

cont'd

17. A Hofmann elimination reaction of a trimethylamine to give a 1-alkene. Reaction of *N,N*-dimethyl-2-pentanamine with iodomethane, silver oxide and elimination to give 1-pentene, the Hofmann elimination product. (See *Notes*.)

step 1   *step 2*   *step 3*

major

18. A Cope elimination reaction of a dimethylamine-*N*-oxide. Step 1, reaction of *N,N*-dimethyl-2,3-diphenylbutan-2-amine with hydrogen peroxide. Step 2, heating the *N*-oxide results in an elimination reaction to give *cis*-α,β-dimethylstilbene, the Zaitsev (syn) elimination product. (See *Notes*.)

Step 1   *Step 2*

Isolate and dry this product

cont'd →

major

19. A selenoxide elimination from a cyclohexanone to a cyclohexenone, a *syn*-elimination. (See *Notes*.)

Acetylene Formation

20. A synthesis of 3-hexyne from *trans*-3-hexene by bromination and two elimination reactions. (See *Notes*.)

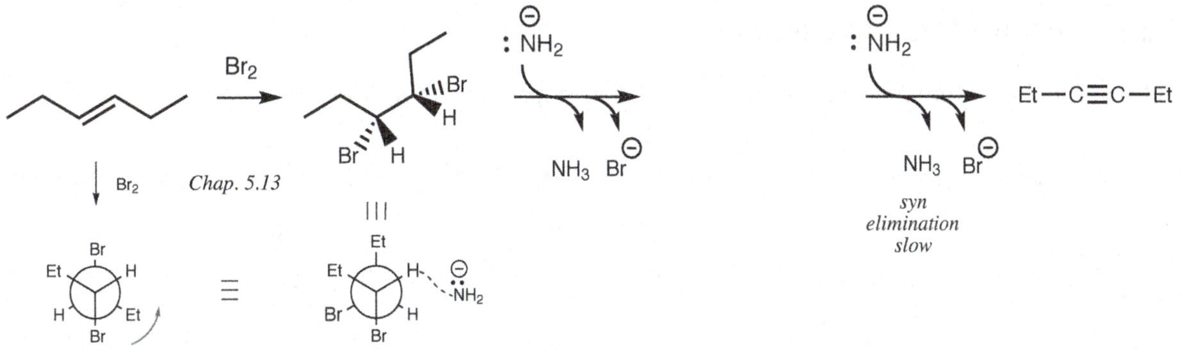

21. A synthesis of 3-hexyne from *cis*-3-hexene by bromination and two elimination reactions. (See *Notes*.)

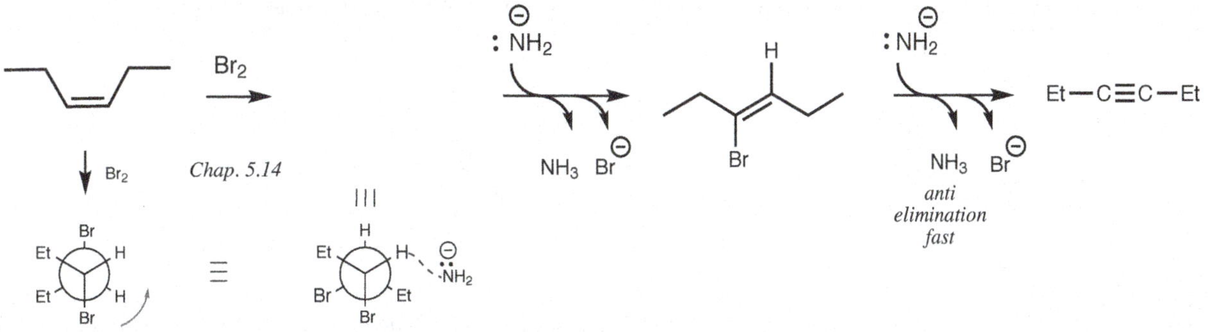

22. An E2 elimination reaction of 1,1-dibromopentane with *t*-butoxide to give 1-pentyne. (See *Notes*.)

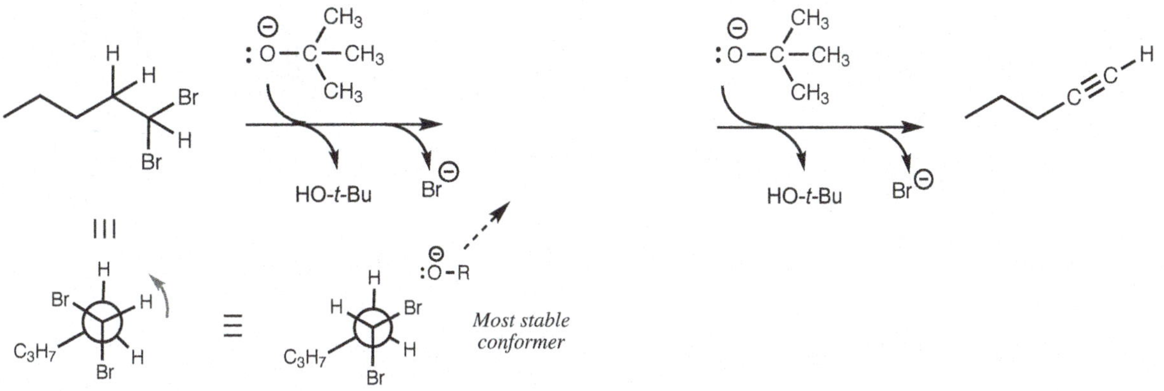

23. An E2 elimination reaction of 2,2-dibromopentane with LDA to give 1-pentyne. (See *Notes*.)

Step 1, elimination

cont'd

Step 2, work up

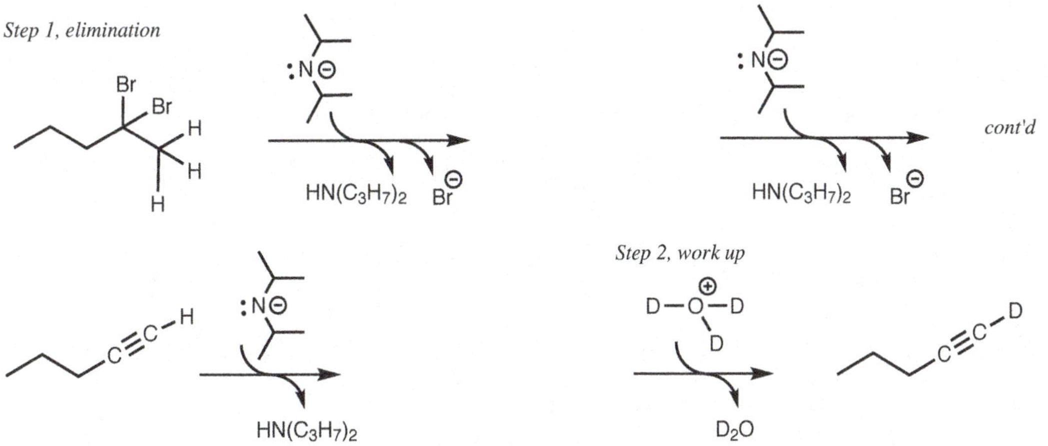

5 — Electrophilic Addition to Alkenes and Alkynes

Addition of HX and H₂O to Alkenes

1. Addition of hydrogen bromide to propene to give 2-bromopropane. (See *Notes*.)

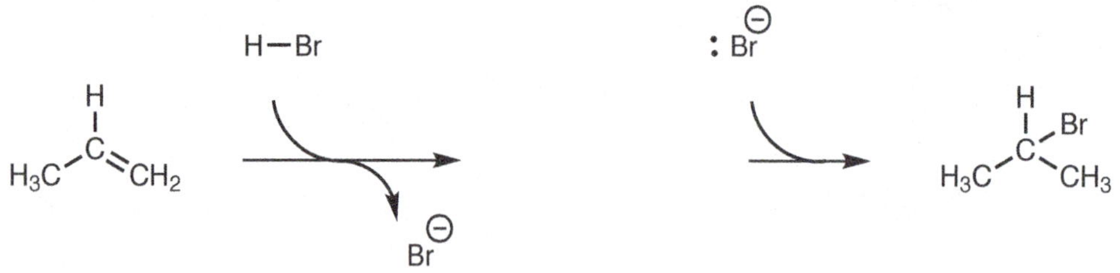

2. Acid catalyzed addition of water to methylcyclopentene to give 1-methylcyclopentanol. (See *Notes*.)

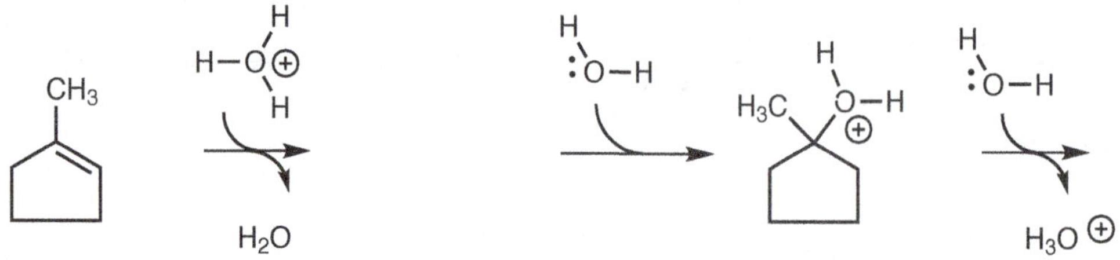

3. Addition of acetic acid to propene catalyzed by sulfuric acid to give 2-propyl ethanoate (isopropyl acetate). (See *Notes*.)

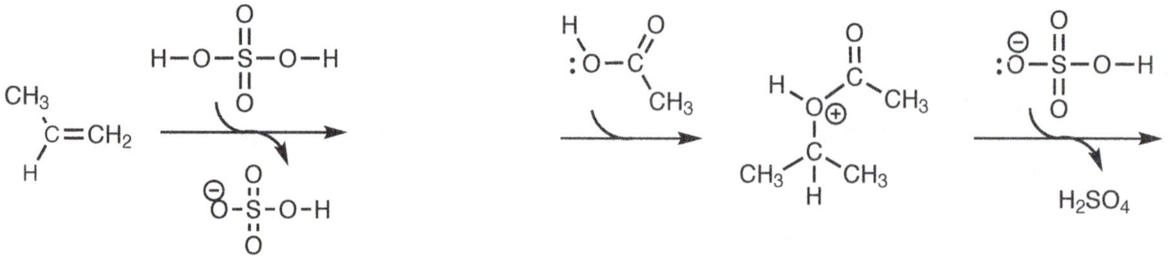

cont'd

4. Addition of hydrogen bromide to 1-methylcyclohexene to give 1-bromo-1-methylcyclohexane.

5. Addition of hydrogen chloride to *(E)-3-hexene* to give *(R)-* and *(S)-3-chlorohexane.*

(R)-3-chlorohexane   (S)-3-chlorohexane

6. Addition of hydrogen chloride to *(Z)-3-hexene* to give *(R)-* and *(S)-3-chlorohexane.*

(S)-3-chloro-hexane

(R)-3-chloro-hexane

7. Addition of hydrogen bromide to 3-methyl-1-butene to give after rearrangement, 2-bromo-2-methylbutane. (See *Notes.*)

II° carbocation

III° carbocation

55%

45%

8. Addition of HBr to 2-cyclobutylpropene to give, after rearrangement, 1-bromo-1,2-dimethylcyclopentane. (See *Notes.*)

Number the atoms to avoid confusion.

cont'd

mixture of diastereomers

9. Addition of hydrogen chloride to allylbenzene to give, after rearrangement, (1-chloropropyl)benzene. (See *Notes*.)

cont'd ⟶

10. Addition of hydrogen chloride to 2-methyl-1,3-butadiene (isoprene) to give, 3-chloro-3-methyl-1-butene, the kinetic product, or 1-chloro-3-methyl-2-butene, the thermodynamic product. (See *Notes*.)

III° greater contributor

Kinetic product

Thermodynamic product

11. Addition of bromine to 2-methyl-1,3-butadiene to give 3,4-dibromo-3-methylbut-1-ene, the kinetic product and 1,4-dibromo-2-methylbut-2-ene, the thermodynamic product. (See *Notes*.)

III° greater contributor

Kinetic product

Thermodynamic product

Bromination

12. Bromination of cyclohexene to give *trans*-1,2-dibromocyclohexane. (See *Notes*.)

Concerted

Br—Br

trans diaxial opening

View A

: Br

Stepwise

Br—Br

View B

Br   : Br

trans diaxial opening

13. Bromination of *trans*-2-butene to give (erythro) *(2R,3S)*- and *(2S,3R)*-2,3-dibromobutane.

Br—Br

Step intermediate

H   Br   : Br

CH₃   CH₃

H

+

14. Bromination of *cis*-2-butene to give (threo) *(2R,3R)*- and *(2S,3S)*-2,3-dibromobutane.

Br—Br

Step intermediate

Br   : Br

H   H

CH₃   CH₃

+

15. Bromination of methylcyclohexene to give *(1R,2R)*- and *(1S,2S)*-2-bromo-1-methylcyclohexanol. (See *Notes*.)

CH₃

Br—Br

Step intermediate

Br   CH₃

H
: O—H

: O—H
H

H₃O

CH₃   O
Br   H

Oxymercuration

16. Step 1, oxymercuration of 3-methyl-1-butene to give 3-methyl-2-butanol. (See *Notes*.)

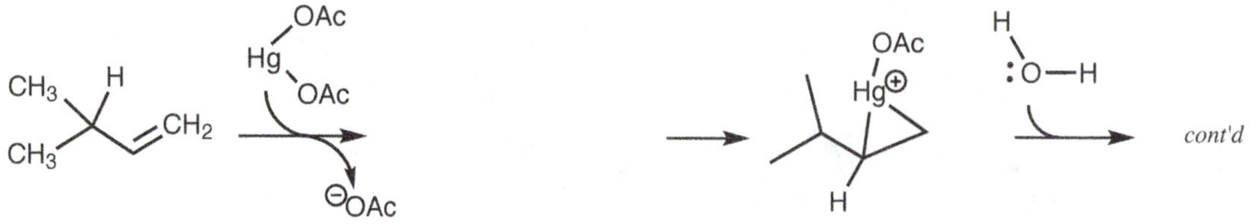

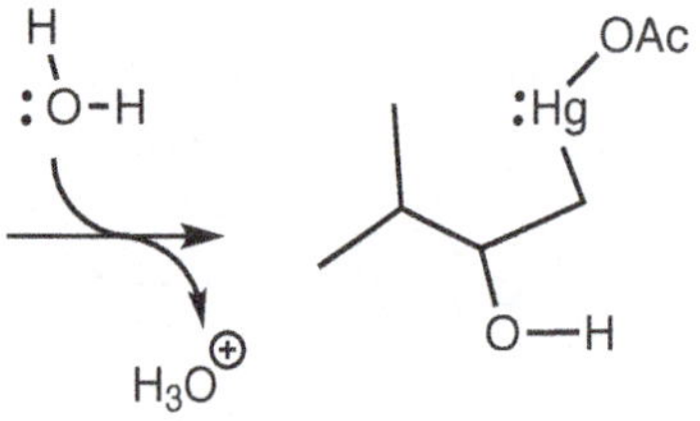

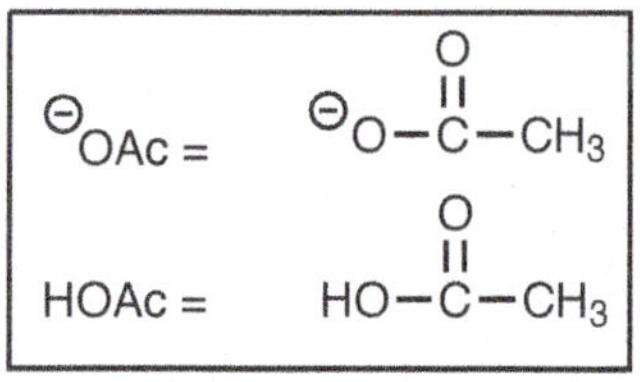

Step 2, reductive demercuration. (See *Notes*.)

17. Step 1, oxymercuration of 1-methylcyclohexene to give 1-methylcyclohexanol. (See *Notes*.)

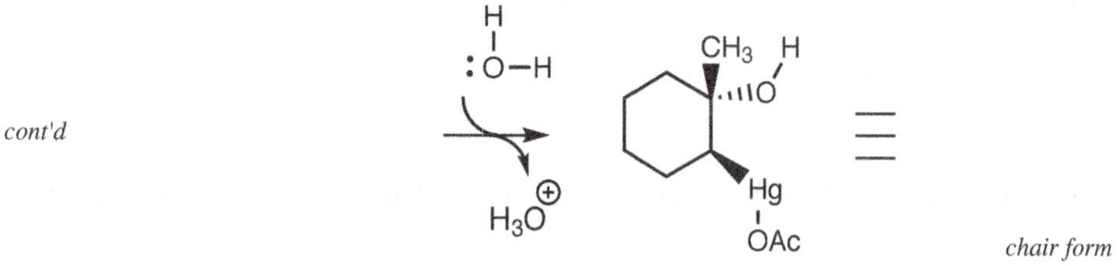

cont'd

chair form

Step 2, reductive demercuration. (See *Notes*.)

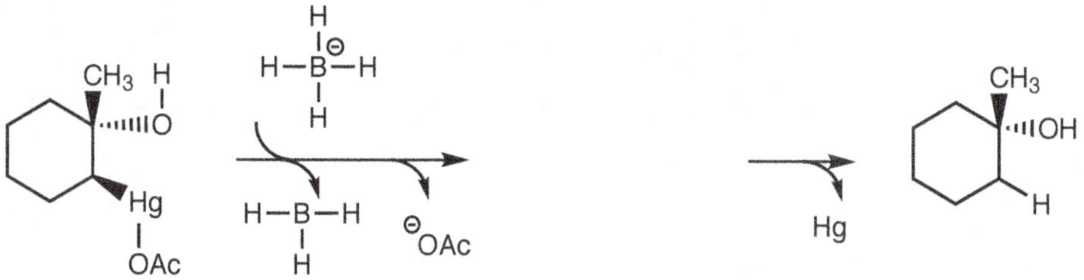

Hydroboration-Oxidation of Alkene

18. Hydroboration-oxidation of propene to give 1-propanol.

Step 1, hydroboration. Each bracket represents one of three hydroboration steps. (See *Notes*.)

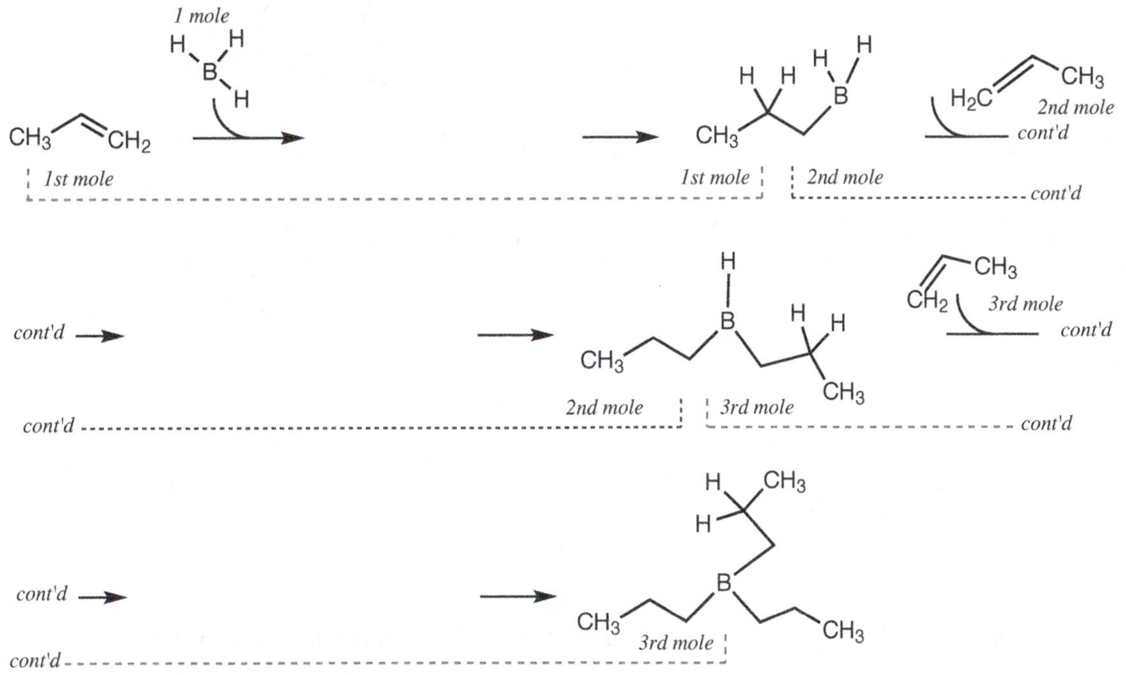

Step 2, oxidation. Each bracket represents one of three oxidation steps. (See *Notes*.)

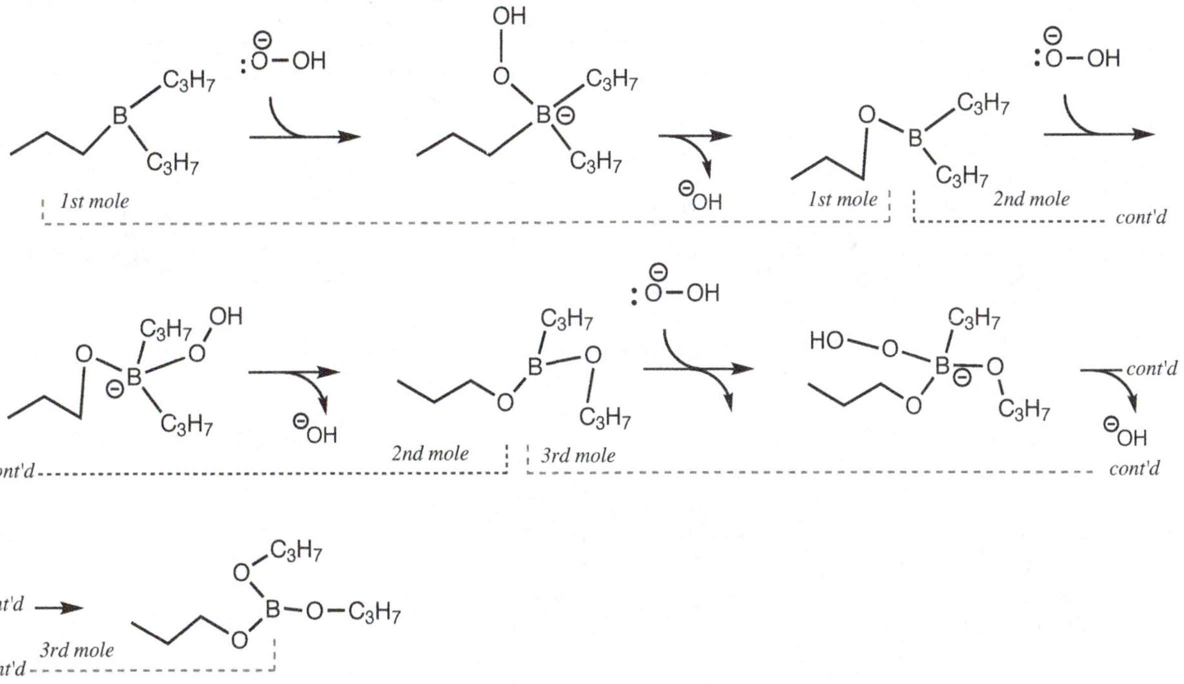

Step 3, borate ester hydrolysis to 1-propanol. Each bracket represents one of three hydrolysis steps. (See *Notes*.)

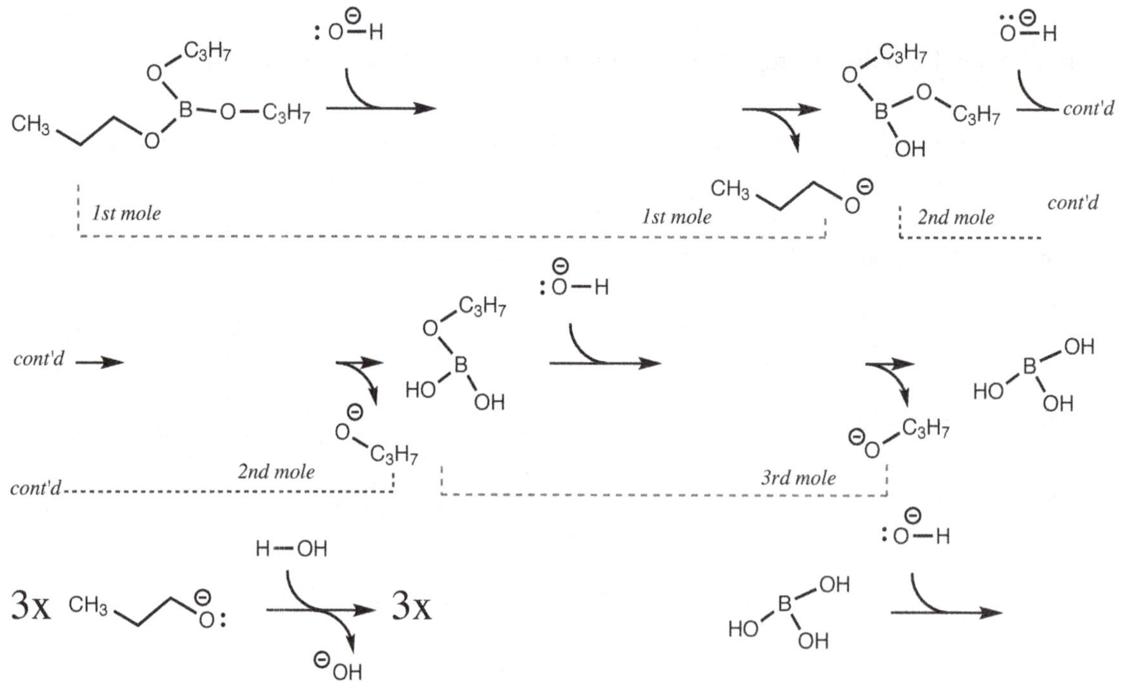

19. Hydroboration/oxidation of 1-methylcyclohexene to give *trans*-2-methylcyclohexanol. (See *Notes*.)

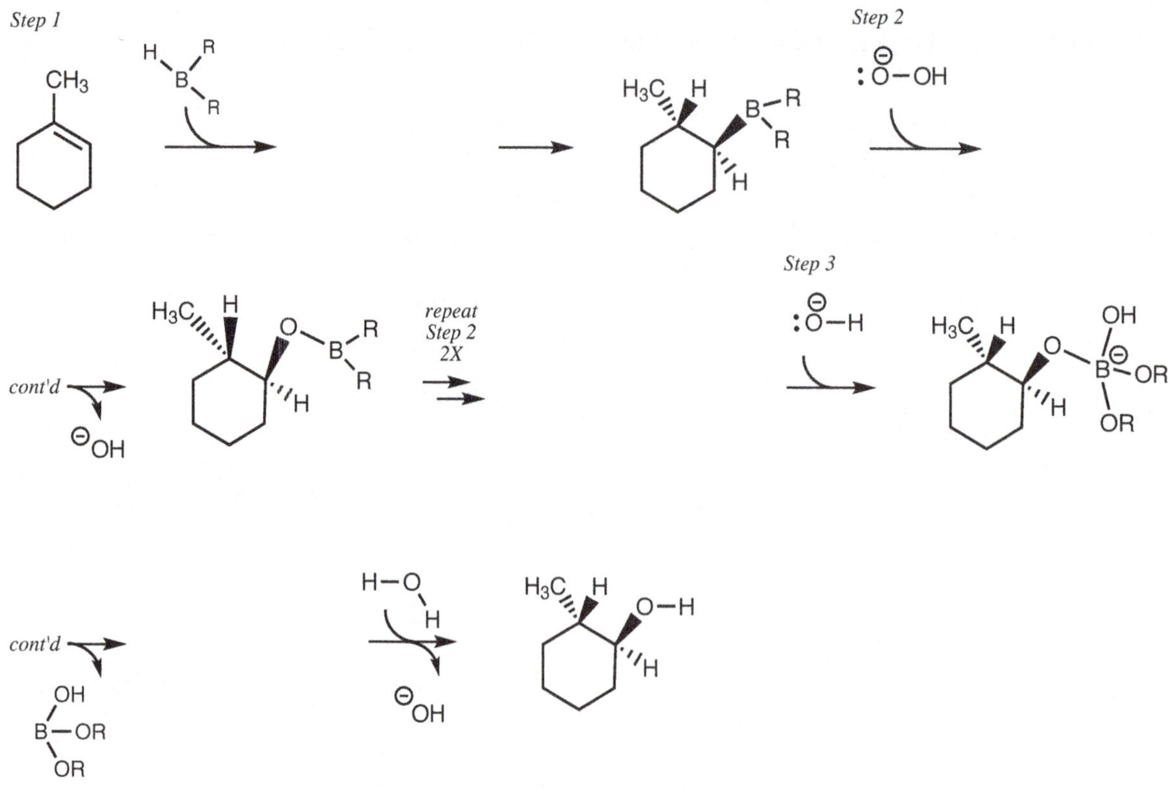

Carbon-Carbon Triple Bond Electrophilic Reactions

Addition to an Internal Acetylene

20. Addition of HCl to 2-butyne (dimethyl acetylene) to give (E)-2-chloro-2-butene and 2,2-dichlorobutane. (See *Notes*.)

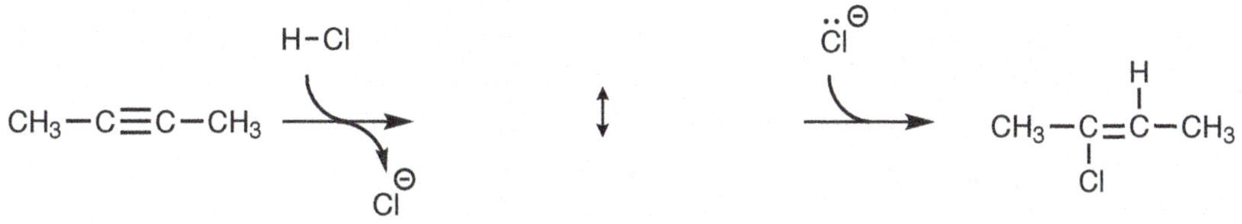

2nd Equivalent of HCl

21. Addition of bromine to ethynylcyclopentane to give *(E)*-(1,2-dibromovinyl)cyclopentane and (1,1,2,2-tetrabromoethyl)cyclopentane. (See *Notes*.)

2nd Equivalent of bromine.

22. Sulfuric acid catalyzed hydration of 1-propynylbenzene to give 1-phenyl-1-propanone. (See *Notes*.)

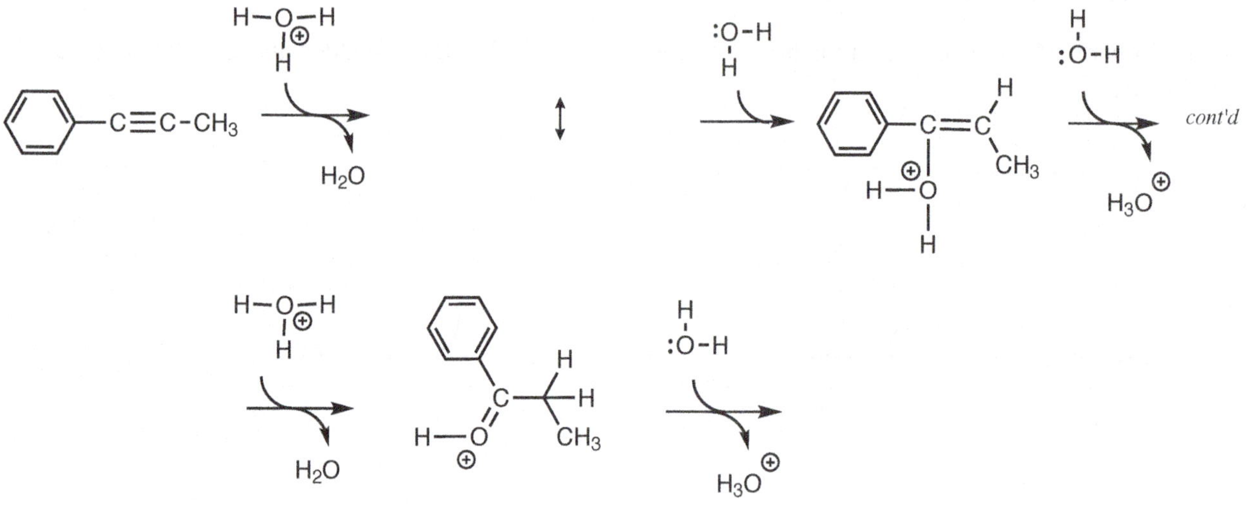

Addition to a Terminal Acetylene

23. Mercury catalyzed hydration of propyne (methyl acetylene) to give 2-propanone (acetone). (See *Notes*.)

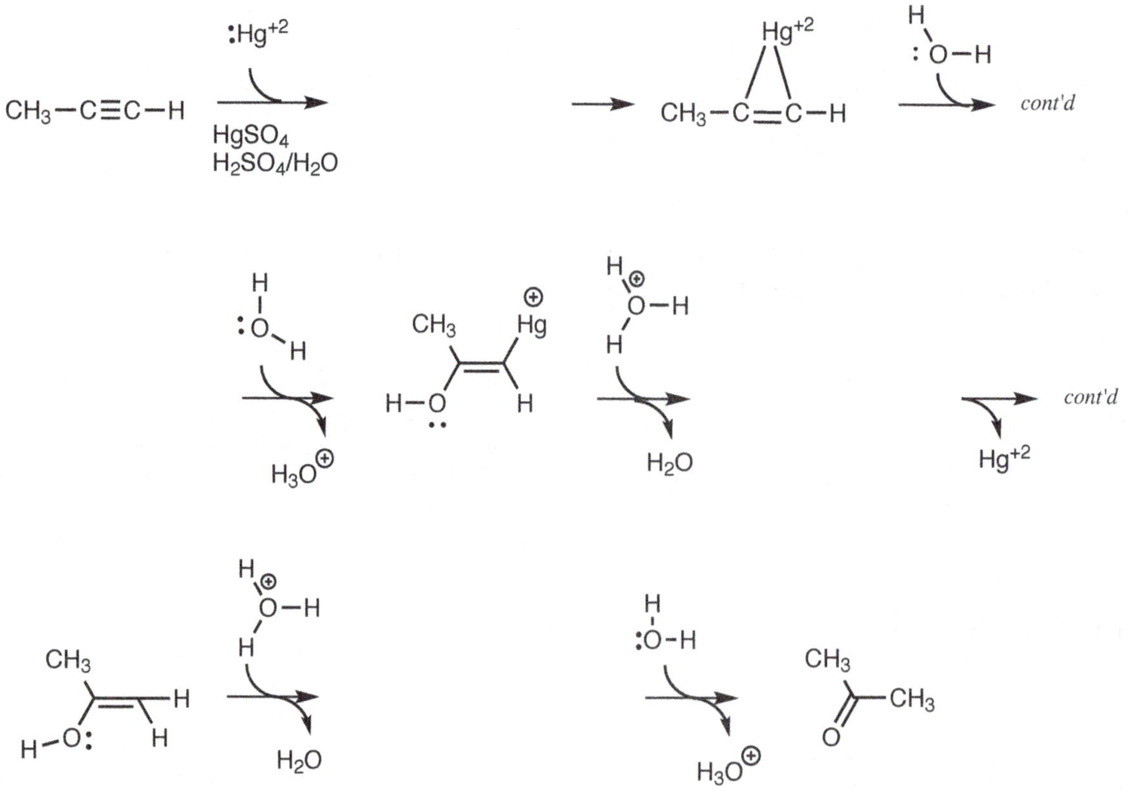

Disiamylborane Hydroboration–Oxidation of an Acetylene

24. Hydroboration-oxidation of phenylacetylene with disiamylborane to give phenylacetaldehyde. (See *Notes*.)

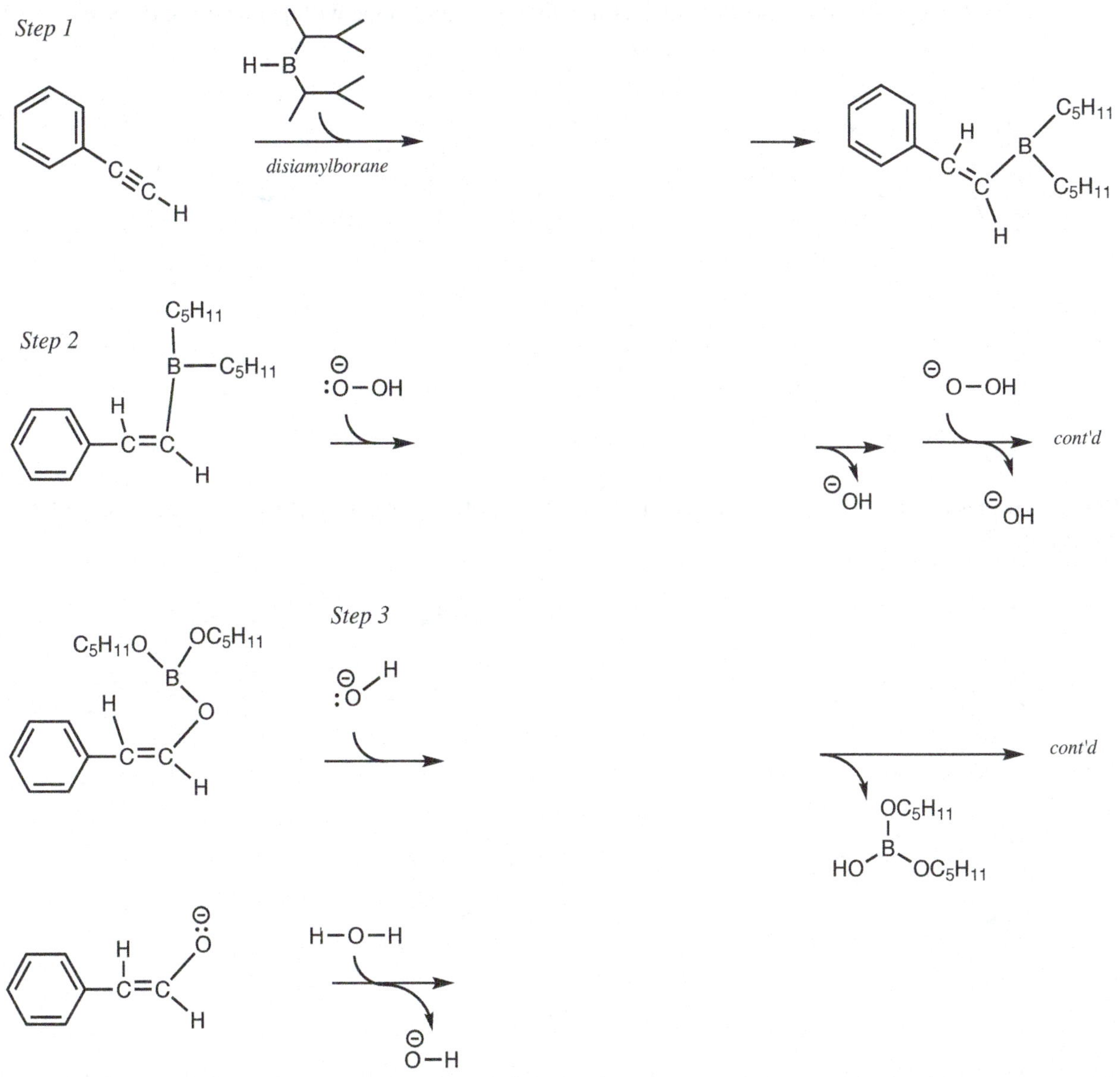

Step 1

Step 2

Step 3

6 — Rearrangement Reactions

Baeyer-Villiger Oxidation

1. Acid catalyzed Baeyer-Villiger oxidation of 2,2-dimethylcyclopentanone with peracetic acid. (See *Notes*.)

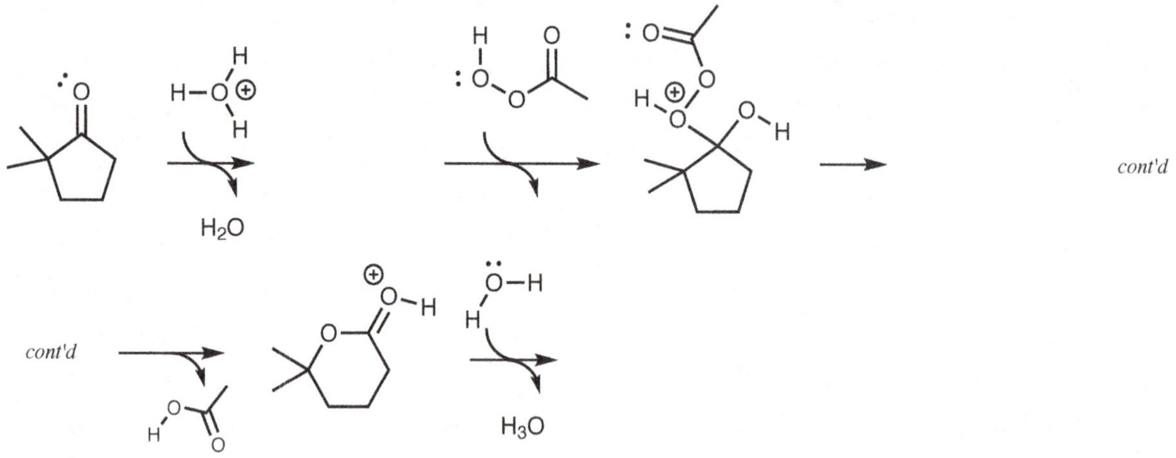

cont'd

2. Baeyer-Villiger oxidation of *o*-methoxyacetophenone to *o*-methoxyphenyl acetate with peracetic acid, acid catalyzed. (See *Notes*.)

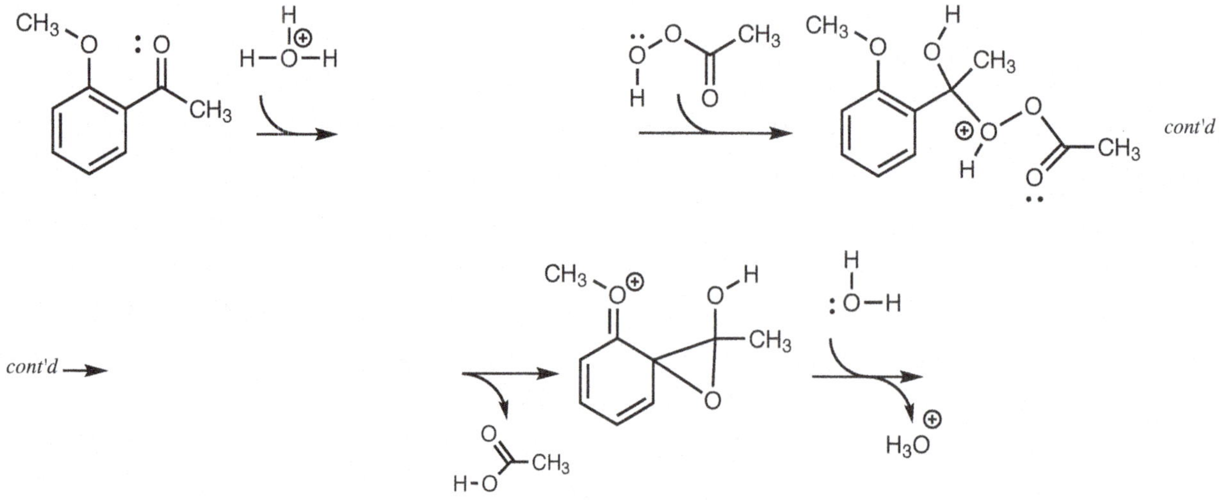

3. Acid catalyzed Baeyer-Villiger oxidation of benzaldehyde with peracetic acid to give benzoic acid. (See *Notes*.)

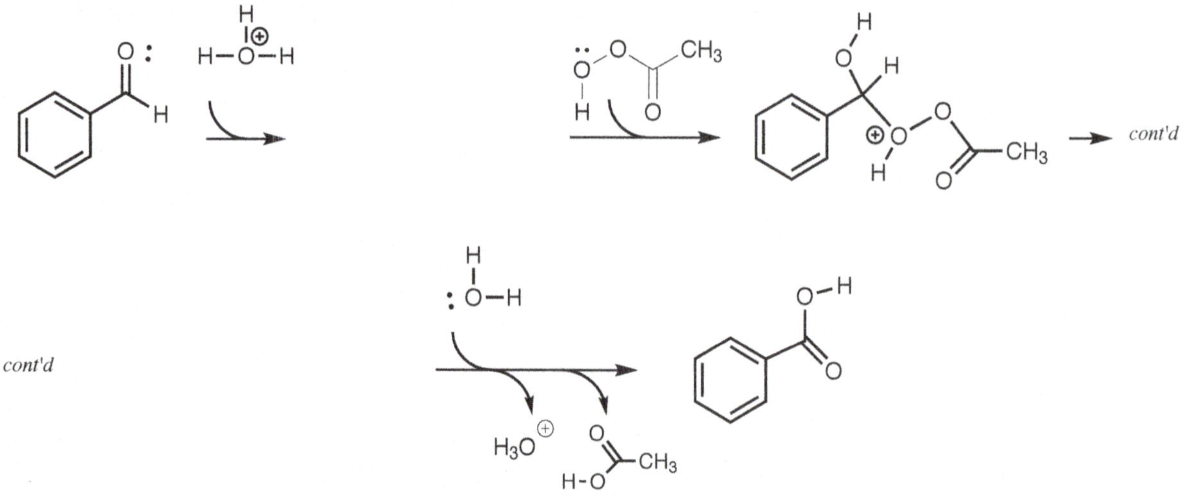

4. Baeyer-Villiger oxidation of a benzophenone with trifluoroperacetic acid. (See *Notes.*)

5. Baeyer-Villiger oxidation of *o*-methoxyacetophenone with *m*-chloroperoxybenzoic acid. (See *Notes.*)

6. Baeyer-Villiger oxidation of bicyclic phenyl ketone with trifluoroperacetic acid. (See *Notes.*)

7. Baeyer-Villiger oxidation of *p*-chlorobenzaldehyde with peracetic acid to give *p*-chlorobenzoic acid. (See *Notes.*)

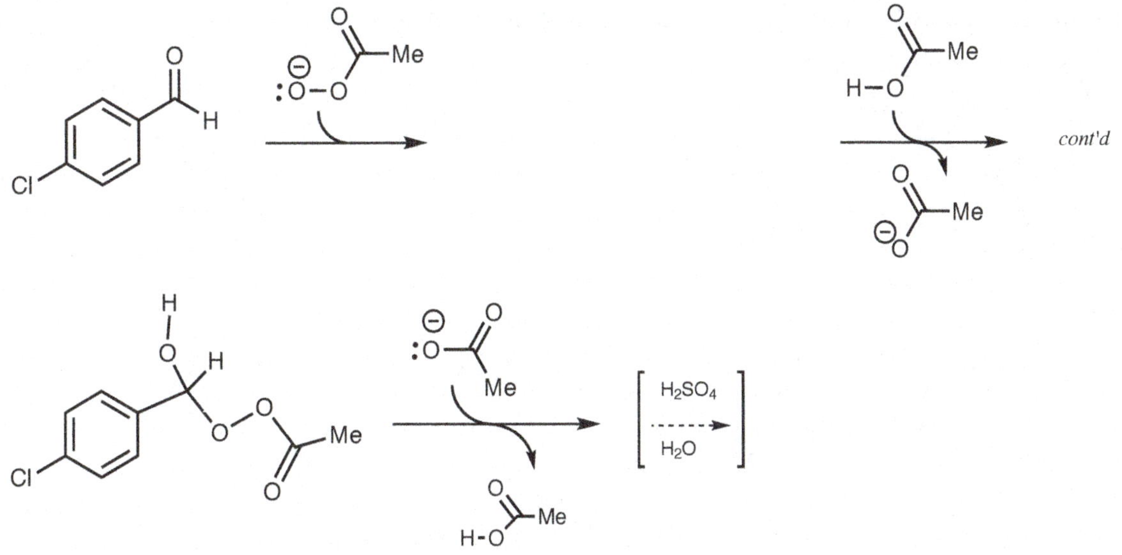

cont'd

Pinacol Rearrangement

8. Rearrangement of pinacol to pinacolone, methyl *t*-butyl ketone or 3,3-dimethyl-2-butanone. (See *Notes*.)

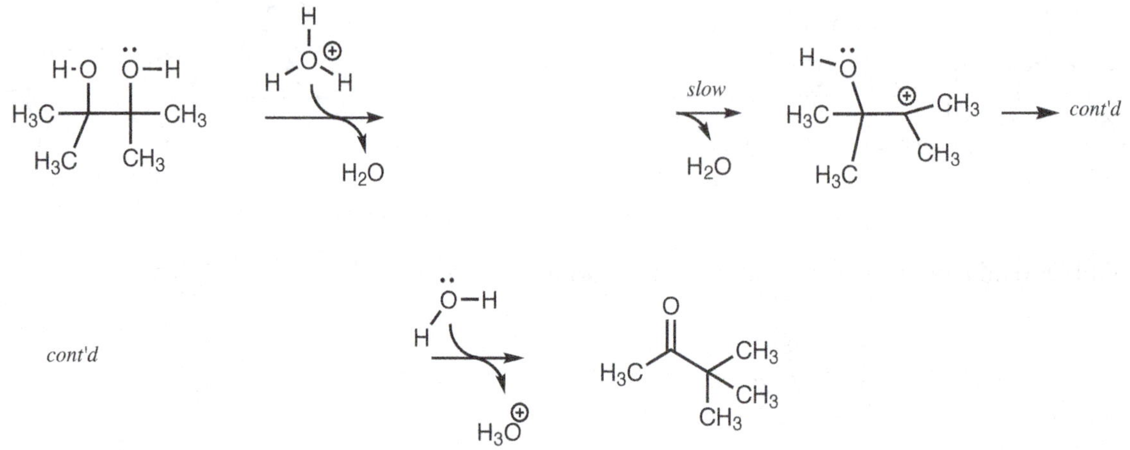

cont'd

9. Rearrangement of 1,2-dimethyl-1,2-cyclohexanediol with acid to a methyl ketone. (See *Notes*.)

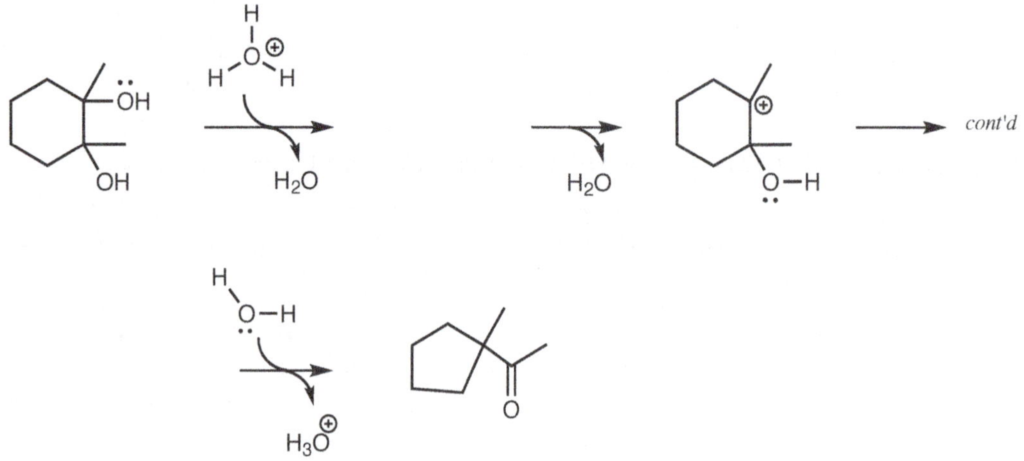

Benzilic Acid Rearrangement

10. Reaction of benzil with hydroxide to give benzilic acid after rearrangement. (See *Notes*.)

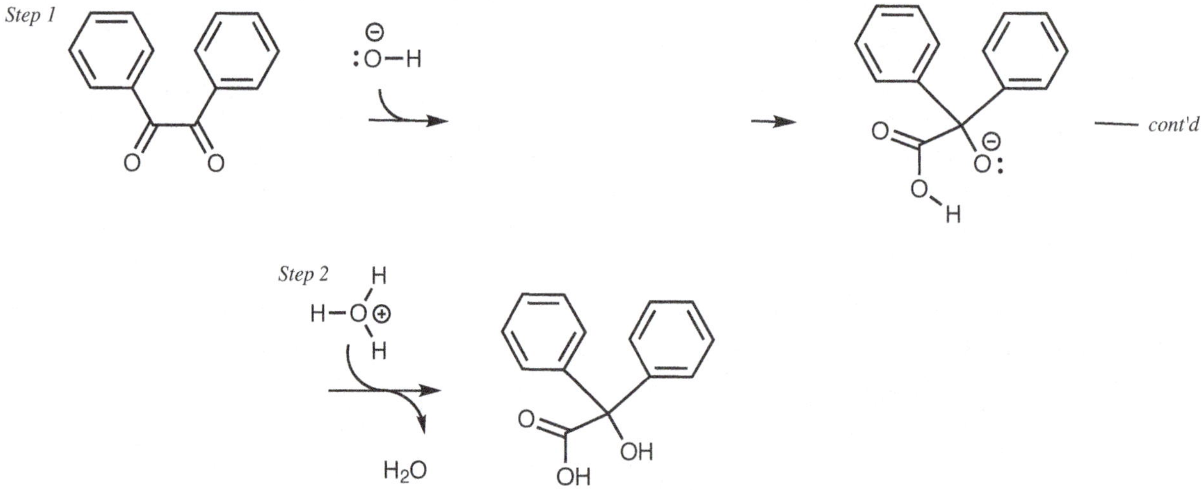

Dakin Reaction

11. Reaction of an *o*- or *p*-hydroxybenzaldehyde with basic hydrogen peroxide to give a phenol. (See *Notes*.)

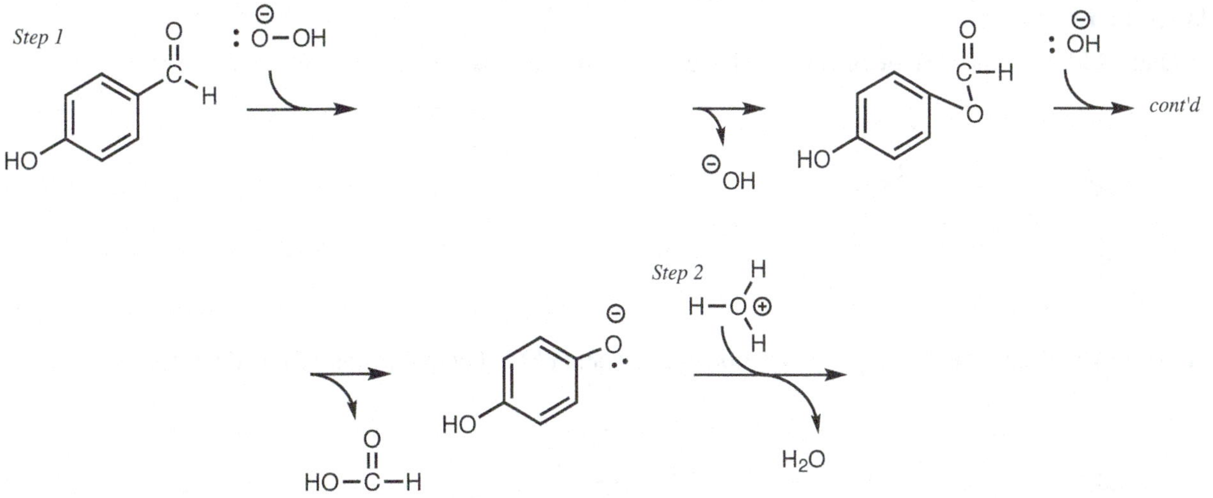

Acetone from Cumene

12. Conversion of isopropylbenzene (cumene) to acetone and phenol.

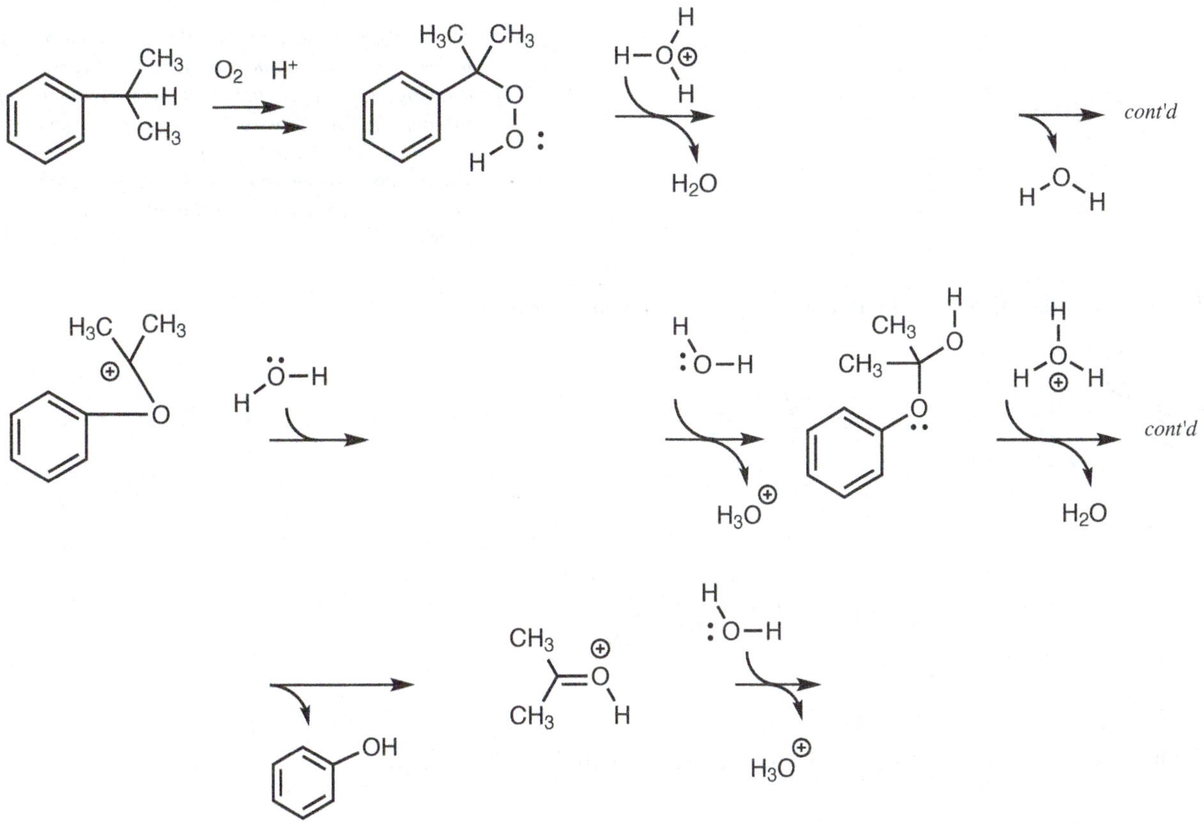

7 — Electrocyclic Reactions

Diels Alder Reactions

1. A Diels-Alder reaction between 1,3-butadiene and 2-propenal (acrolein).

2. A Diels-Alder reaction between 1,3-cyclopentadiene and (E)-2-butenal (trans-crotonaldehyde).

redraw

The preferred orientation of the dienophile is below. You should check with your text for further details.

This is the same example as above. It is more difficult to negotiate the atom movements, congestion, and drawing the final product all at once. I wrote the upper example in two steps as it is easier to visualize and convert it to the bicyclic product.

The preferred orientation has the electron-withdrawing group of the dienophile overlapping with the diene, if possible.

3. The Diels-Alder dimer of cyclopentadiene. (*Complete the structures*)

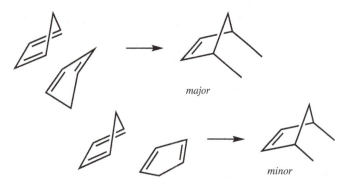

major

minor

4. A Diels-Alder reaction between 1,3-butadiene and methyl (Z)-2-butenoate.

5. A Diels-Alder reaction between 2-methyl-1,3-butadiene and *N*-methylmaleimide.

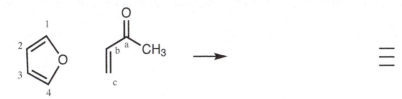

6. A Diels-Alder reaction between furan and but-3-en-2-one (methyl vinyl ketone, MVK).

stereo view

7. A Diels-Alder reaction between cyclopentadiene and dimethyl acetylenedicarboxylate.

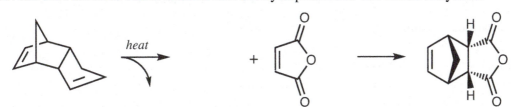

stereo view

8. A reverse-forward Diels-Alder reaction between cyclopentadiene and maleic anhydride.

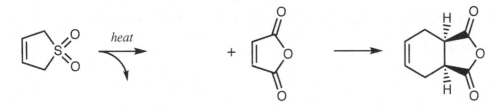

9. A reverse-forward Diels-Alder between butadiene (sulfone) and maleic anhydride.

10. A Diels-Alder reaction between (2*E*,4*E*)-hexa-2,4-diene and maleic anhydride.

11. Draw the Diels-Alder product for the reaction of 1,3-cyclohexadiene and 2-butenal (crotonaldehyde).

stereo view

12. A Diels-Alder reaction between 1-methoxy-1,3-butadiene and MVK. What are the structures of the different orientations?

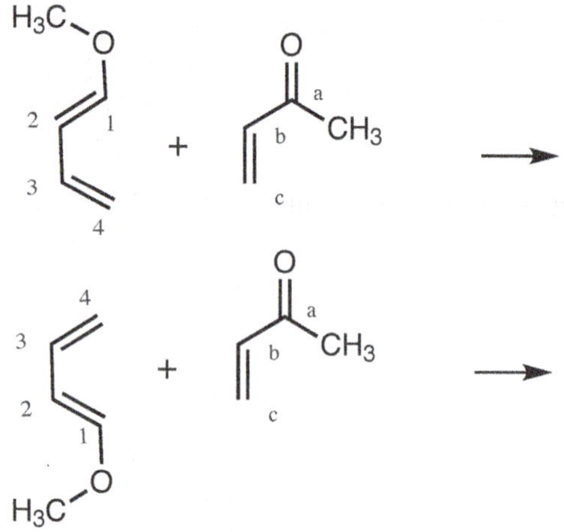

13. Draw the electrostatic charges of the starting materials by drawing the resonance structures.

    Draw the missing resonance structures

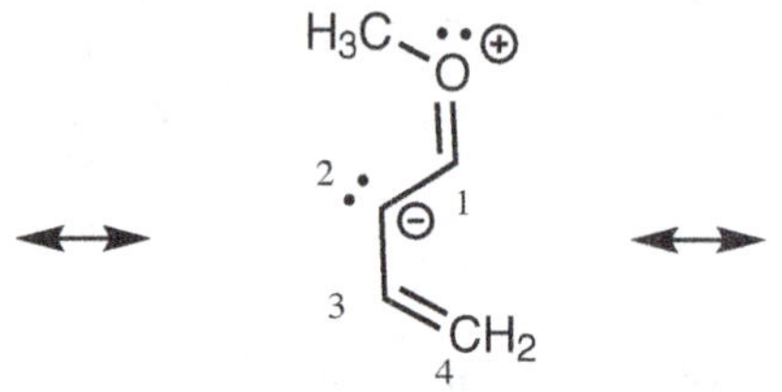

14. Examine the electrostatic charges of the starting materials by drawing the resonance structures.

    Draw the missing resonance structures

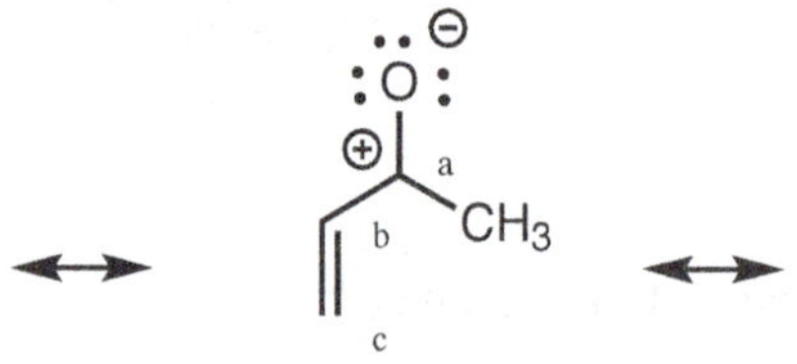

15. How do resonance structures shown in 13 and 14 predict the product for 12? The Diels-Alder reaction is thought of as a truly concerted reaction. If this reaction were ionic, which product would you predict from the resonance structures in Examples 13 and 14?

Other Electrocyclic Reactions

16. Draw the 3+2 cycloaddition product of cyclopentene and benzonitrile oxide.

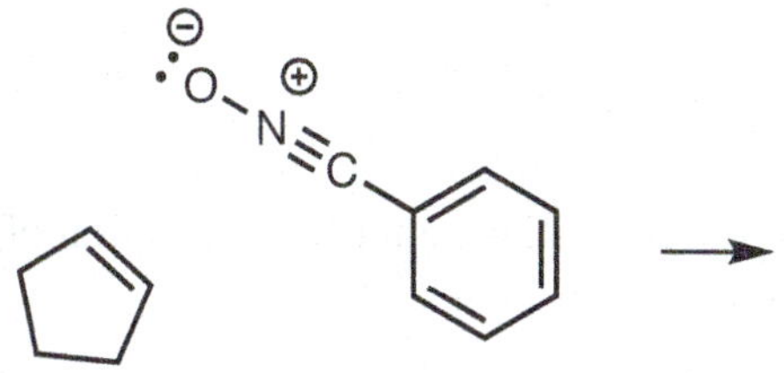

17. A Claisen rearrangement (electrocyclic) reaction to transfer a group from oxygen to carbon.

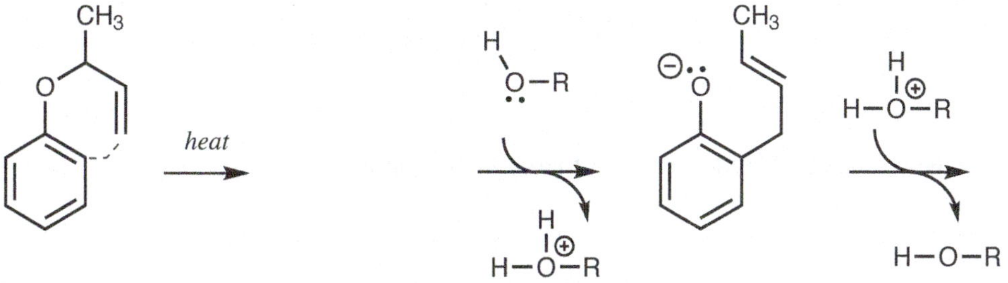

18. A double Claisen rearrangement (electrocyclic) reaction to transfer a group from oxygen to carbon to another carbon. Count the number of electrons that move in each step.

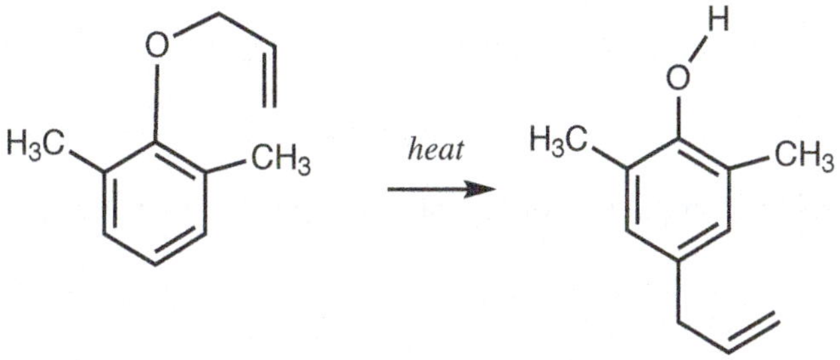

8 — Carbonyl Addition and Addition-Elimination Reactions

Additions by C, N, and O. Addition by hydrogen nucleophiles are discussed in Chapter 11, Reduction.

Grignard Addition to a Carbonyl Group

1. Addition of methyl magnesium bromide to cyclohexanone to give 1-methylcyclohexanol (for formation of Grignard reagents, see *Notes*).

2. Addition of a Grignard reagent to acetaldehyde to give 6-methyl-2-heptanol.

Alkyllithium Addition to a Carbonyl Group

3. Addition of ethyllithium to benzaldehyde to give 1-phenylpropanol (for formation of lithium reagents, see *Notes*).

4. Addition of propynyllithium to acetone to give 2-methylpent-3-yn-2-ol. (see Chapter 2.14)

Wittig Reaction

5. Wittig reaction, Step 1, formation of Wittig reagent. (See *Notes*.)

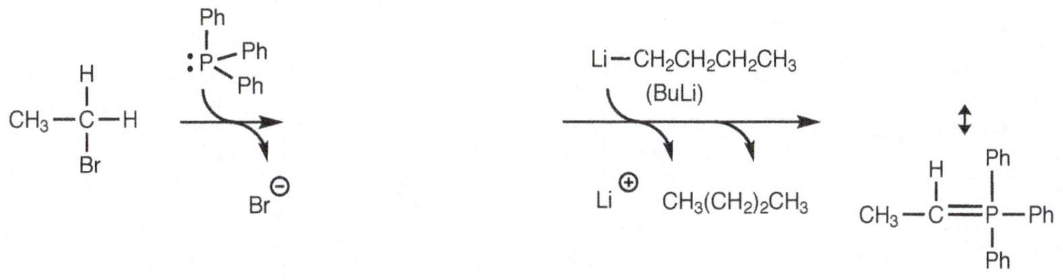

Step 2, reaction with benzaldehyde.

fast

slow

+

+

major

+

minor

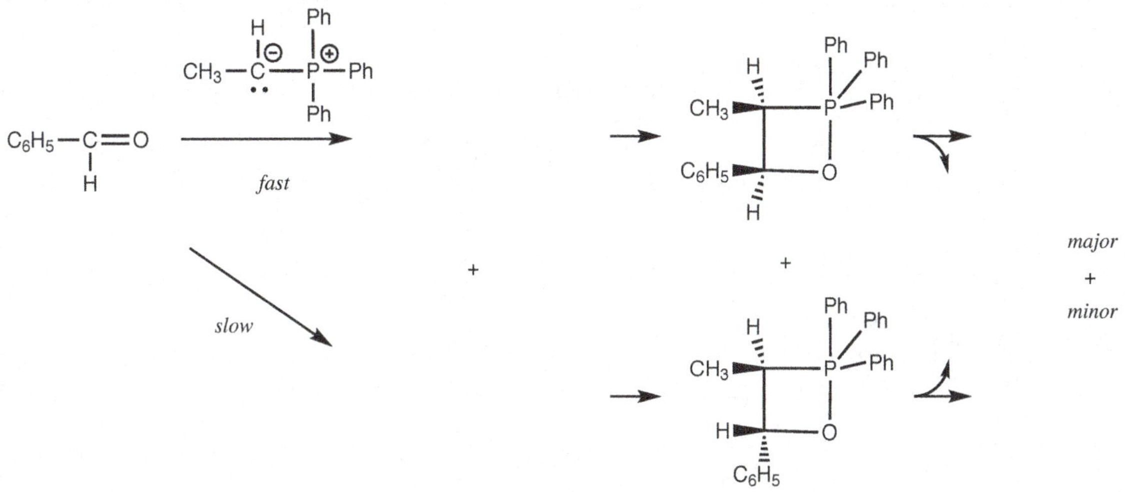

6. Wittig reaction, Horner-Wadsworth-Emmons modification, Step 1, Arbusov reaction. (See *Notes*.)

Step 2, reaction with benzaldehyde.

cont'd

major   +   *minor*

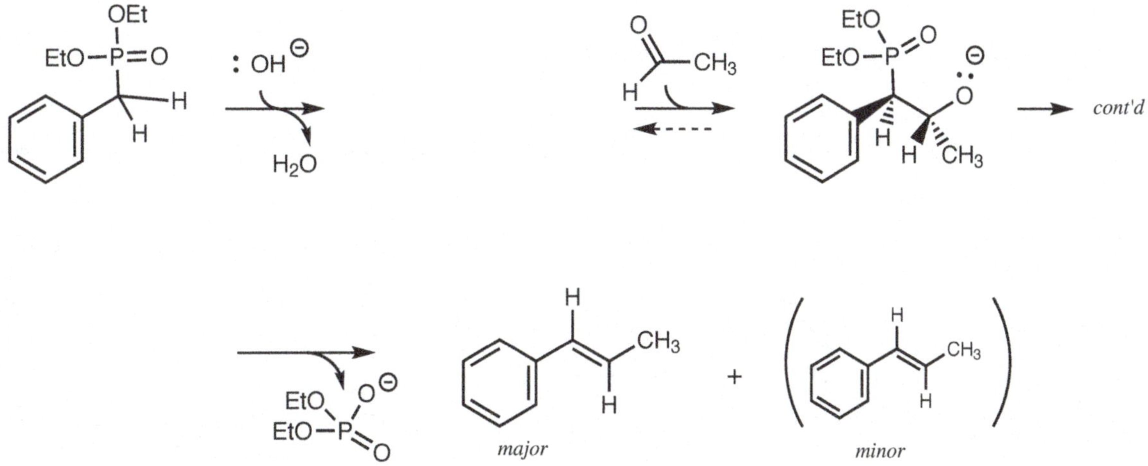

Addition-Elimination Reactions (Reversible Additions)

Ketal Formation and Hydrolysis

7. Acid catalyzed ketalization of cyclohexanone. (See *Notes*.)

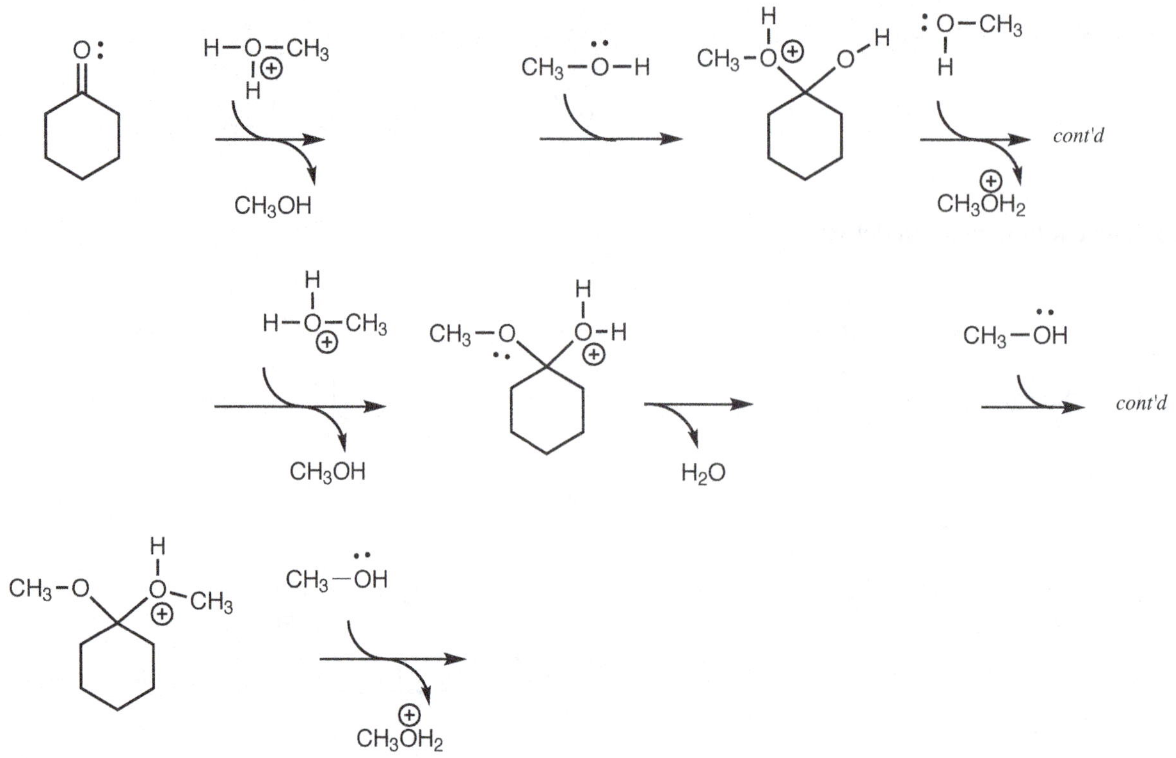

8. Acid catalyzed hydrolysis of the dioxolane acetal of benzaldehyde. (See *Notes*.)

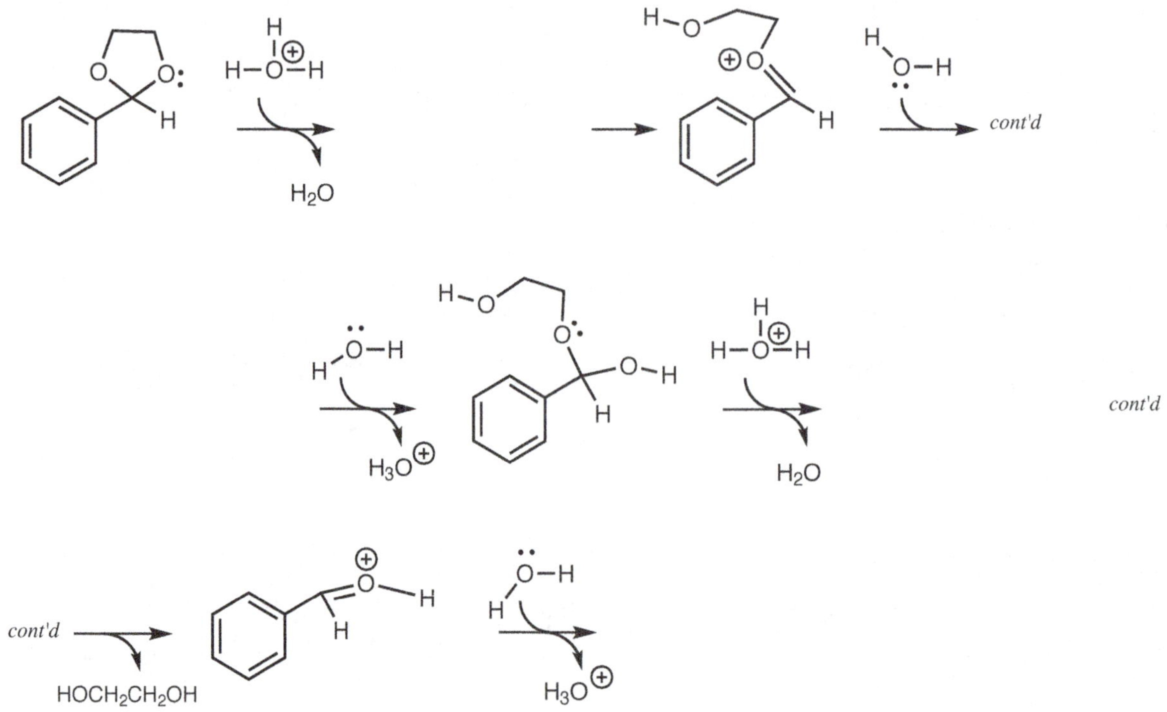

Oxime Formation

9. Formation of the oxime of cyclohexanone. (See *Notes*.)

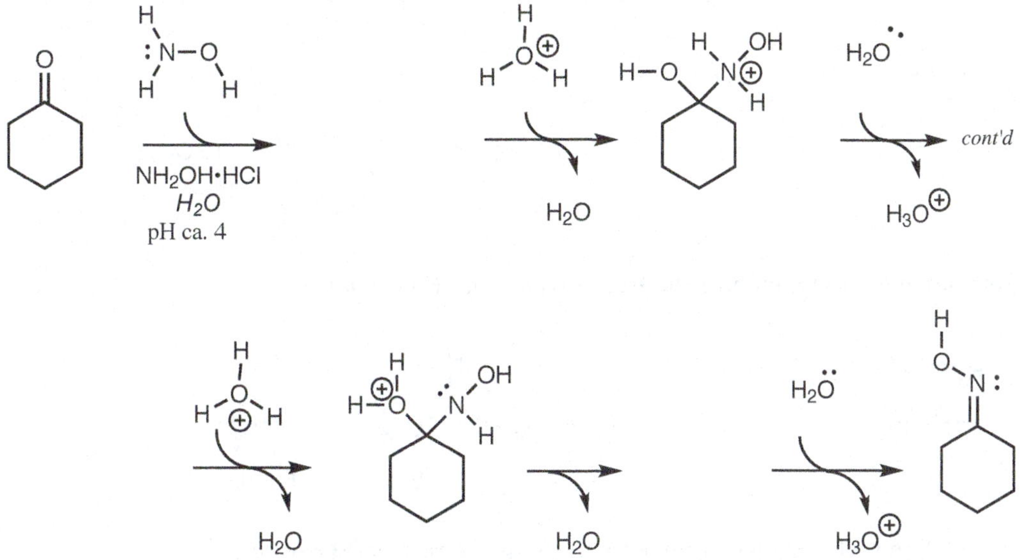

Other Additions to a Carbonyl Group

10. Formation of the cyanohydrin (2-hydroxy-2-methylpropanenitrile) from acetone. (See *Notes*.)

11. Reversion of the cyanohydrin (2-hydroxy-2-methylpropanenitrile) to form acetone. (See *Notes*.)

12. Addition of ethyl acetate enolate to 4-bromobenzaldehyde to give a benzyl alcohol.

Reactions of Acyl Chlorides, Anhydrides, Esters, and Amides

Esters from Acid Chlorides or Anhydrides

13. Reaction of benzoyl chloride with ethoxide to give ethyl benzoate. (See *Notes*.)

14. Direct reaction of ethanol with acetyl chloride to give ethyl acetate. (See *Notes*.)

15. Pyridine catalyzed acylation with benzoyl chloride to give ethyl benzoate. (See *Notes*.)

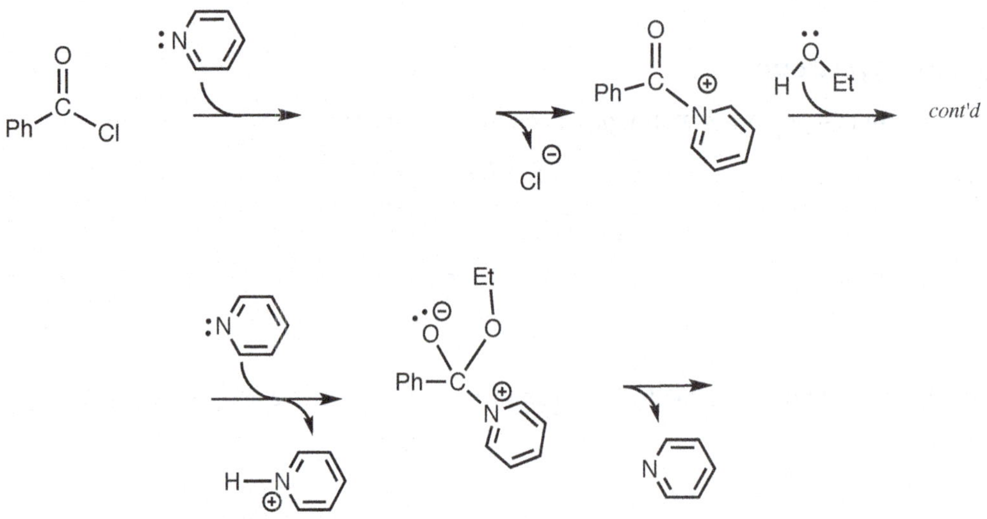

16. Reaction of acetic anhydride with ethanol catalyzed by sulfuric acid. (See *Notes*.)

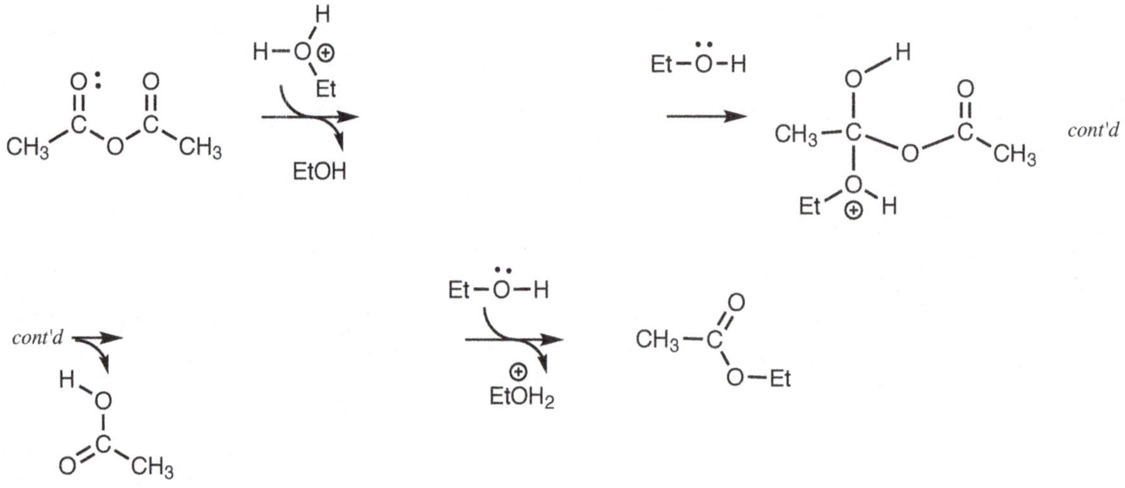

Amides from Acid Chlorides or Anhydrides

17. Reaction of benzoyl chloride with ethylamine to give *N*-ethylbenzamide. (See *Notes*.)

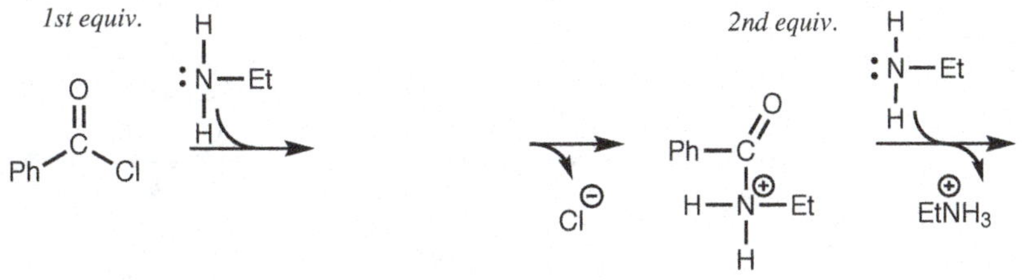

18. Reaction of acetic anhydride with aniline (and pyridine) to give acetanilide. (See *Notes*.)

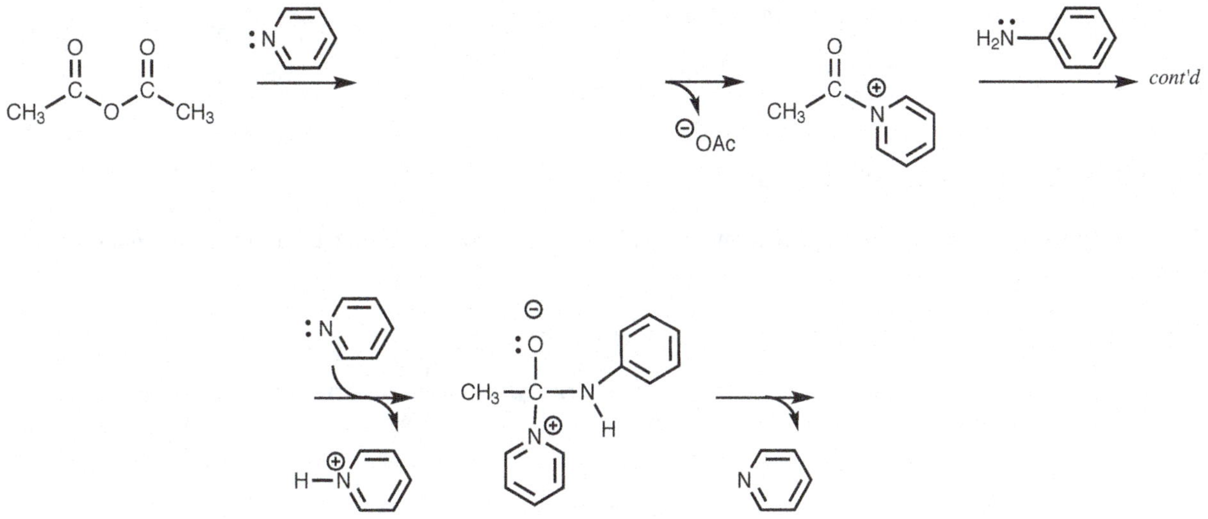

19. Reaction of acetic anhydride with aniline to give acetanilide. (See *Notes*.)

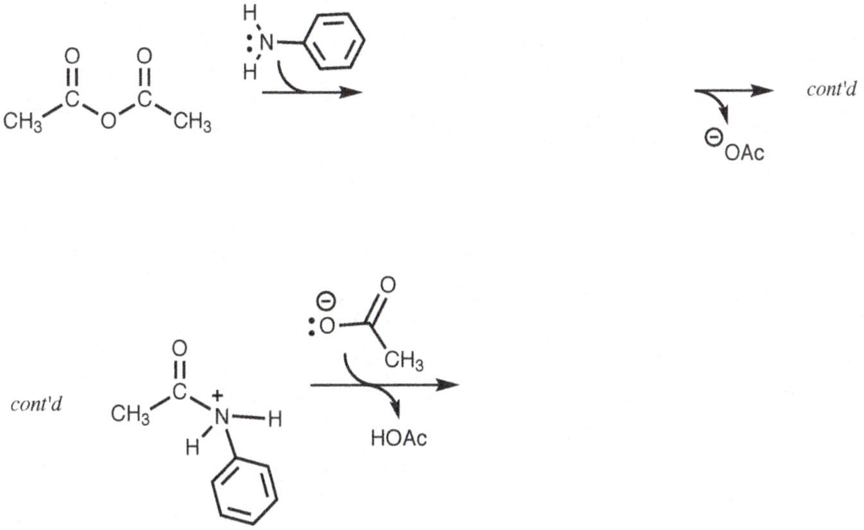

Ester from Acid with Mineral Acid Catalysis (Fischer Esterification)

20. Acid catalyzed esterfication. Formation of methyl benzoate from benzoic acid and methanol with sulfuric acid or hydrogen chloride. (See *Notes*.)

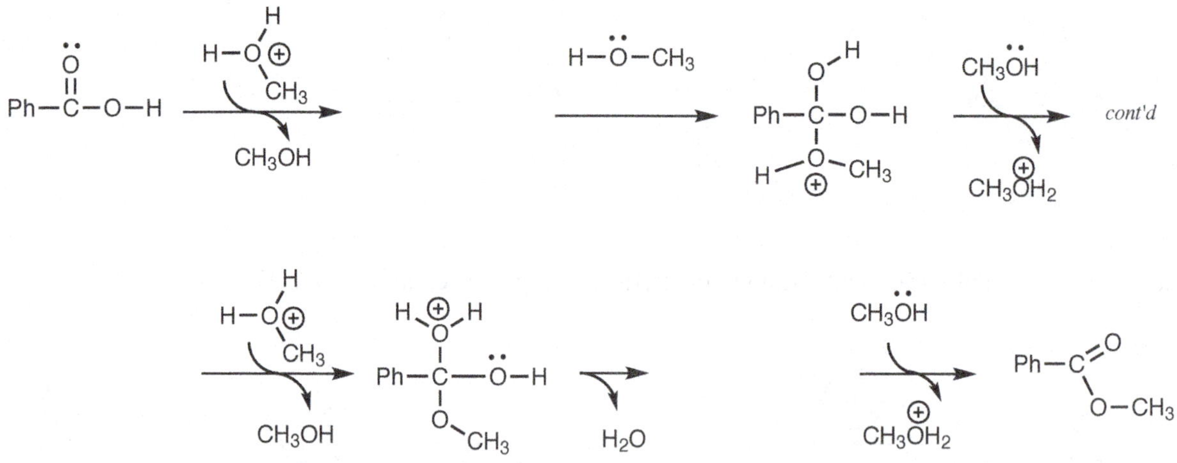

Acid Catalyzed Hydrolysis of an Ester

21. Acid catalyzed hydrolysis of methyl pentanoate (valerate) to pentanoic (valeric) acid plus methanol. (See *Notes*.)

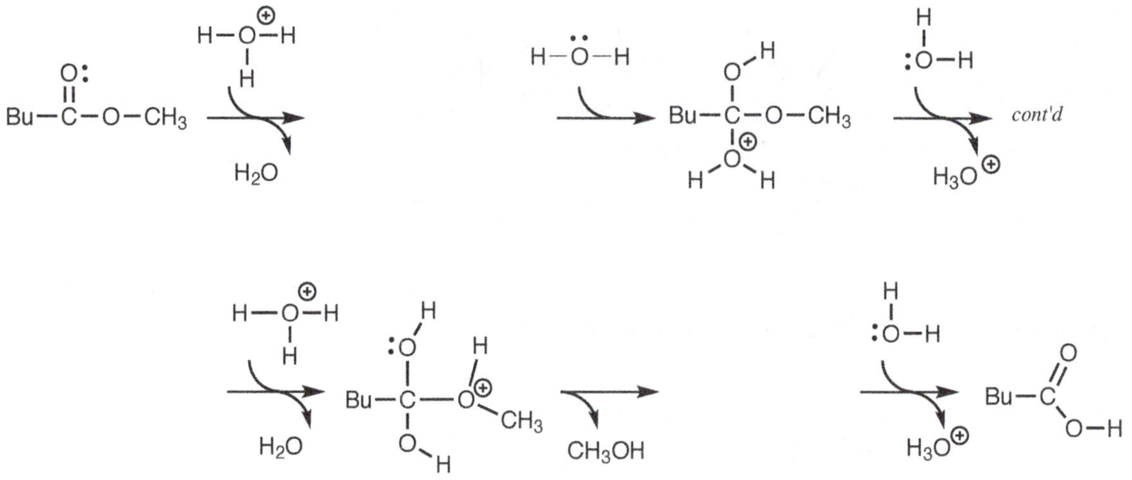

22. Acid catalyzed conversion of *t*-butyl pivalate to pivalic acid. (See *Notes*.)

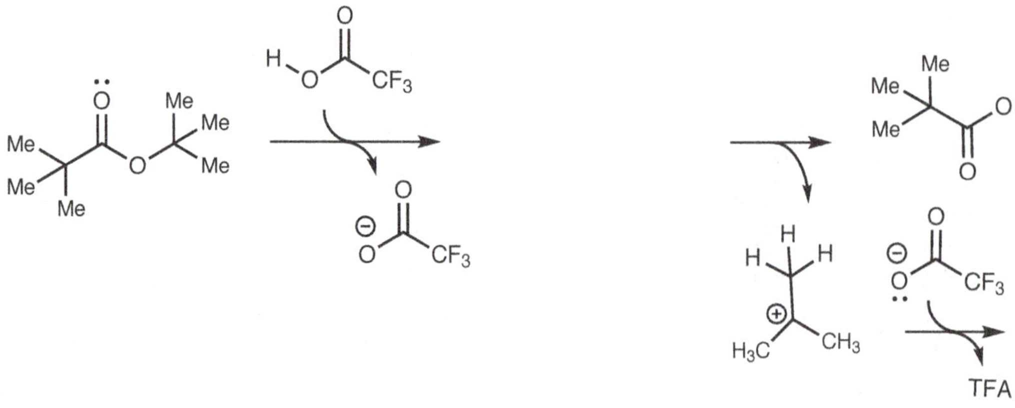

Base Hydrolysis of an Ester (Saponification)

23. Base hydrolysis of octyl isobutyrate to give octanol and isobutyric acid. Step 1, treatment with base. (See *Notes*.)

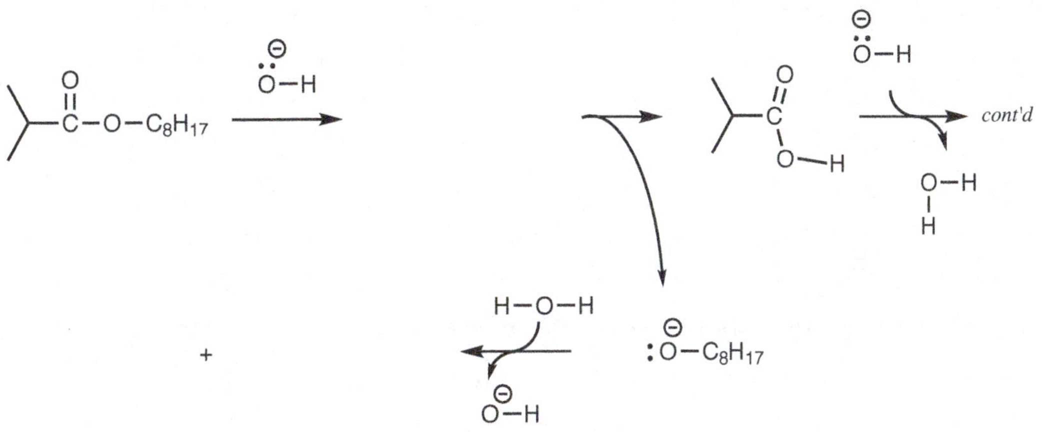

Step 2, acidification of isobutyrate and isolation of isobutyric acid.

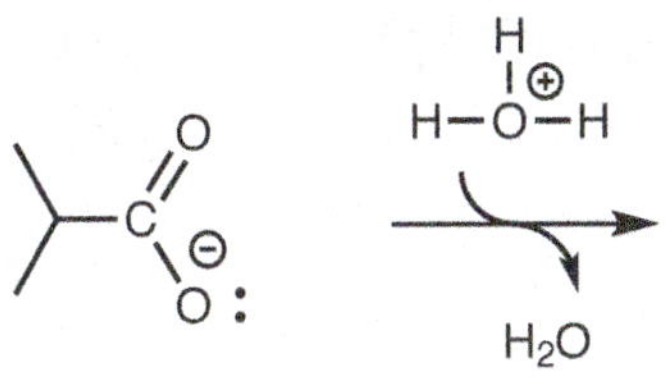

Hydrolysis of an Amide

24. Base hydrolysis of *N*-butylacetamide to give *n*-butylamine and acetic acid. Step 1, treatment with base. (See *Notes*.)

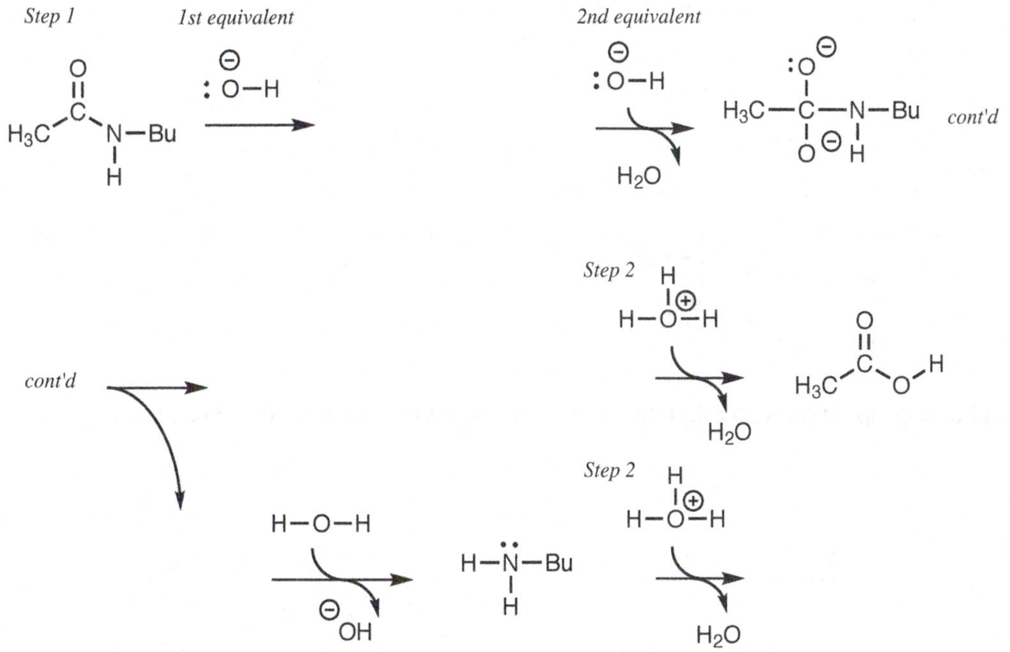

25. Acid catalyzed hydrolysis of *N,N*-dimethylacetamide to give acetic acid and dimethylammonium chloride. (See *Notes*.)

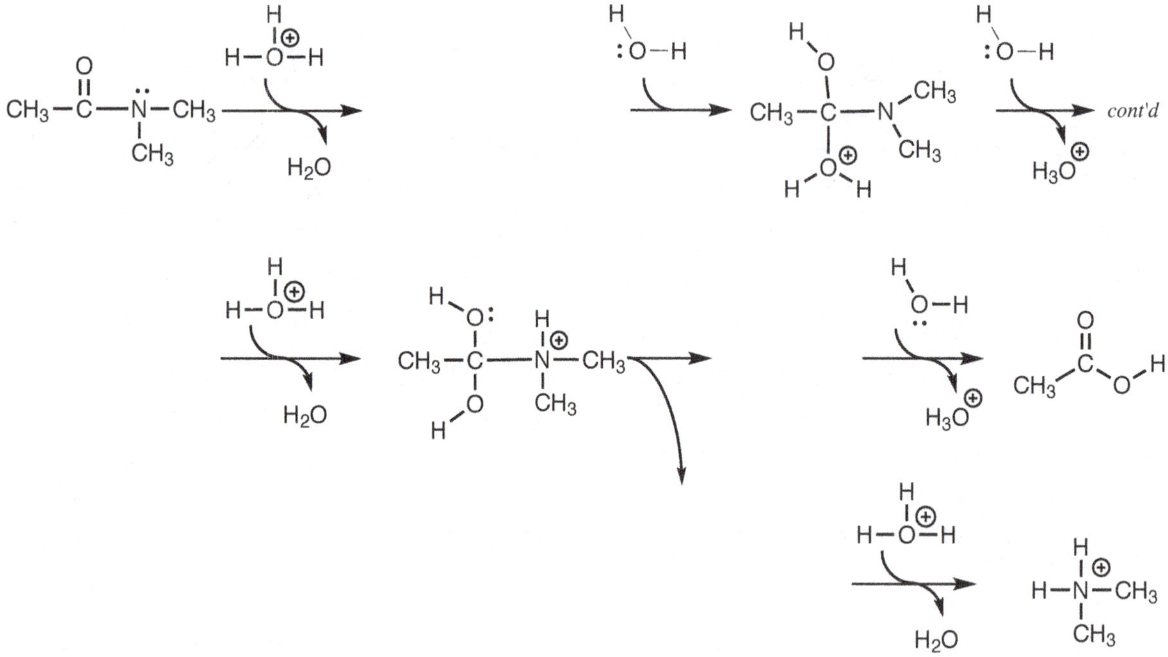

Reactions of Esters

26. Addition of methyl magnesium bromide to ethyl benzoate to give 2-phenyl-2-propanol. (See *Notes*.)

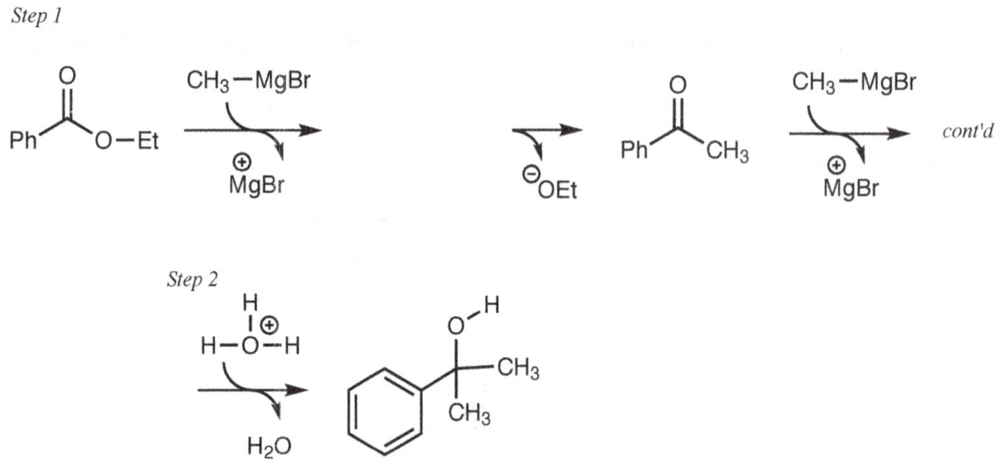

Step 1

cont'd

Step 2

27. Reaction of 4-nitrophenyl propionate with ethyl amine to give *N*-ethyl propionamide. (See *Notes*.)

Reactions of Nitriles

28. Acid catalyzed hydrolysis of benzonitrile to give benzoic acid and ammonium chloride. (See *Notes*.)

29. Base catalyzed hydrolysis of a nitrile, conversion of cyclopentanecarbonitrile to cyclopentanecarboxamide. (See *Notes*.)

30. Addition of phenyl lithium to a nitrile to give, after hydrolysis, cyclohexyl phenyl ketone. (See *Notes*.)

Step 1, addition to the nitrile.

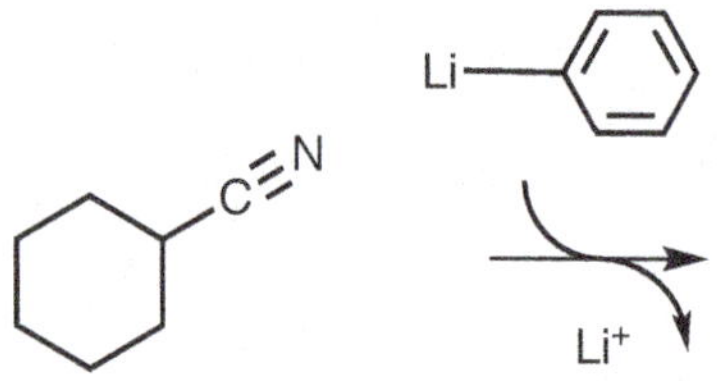

Step 2, hydrolysis of the imine.

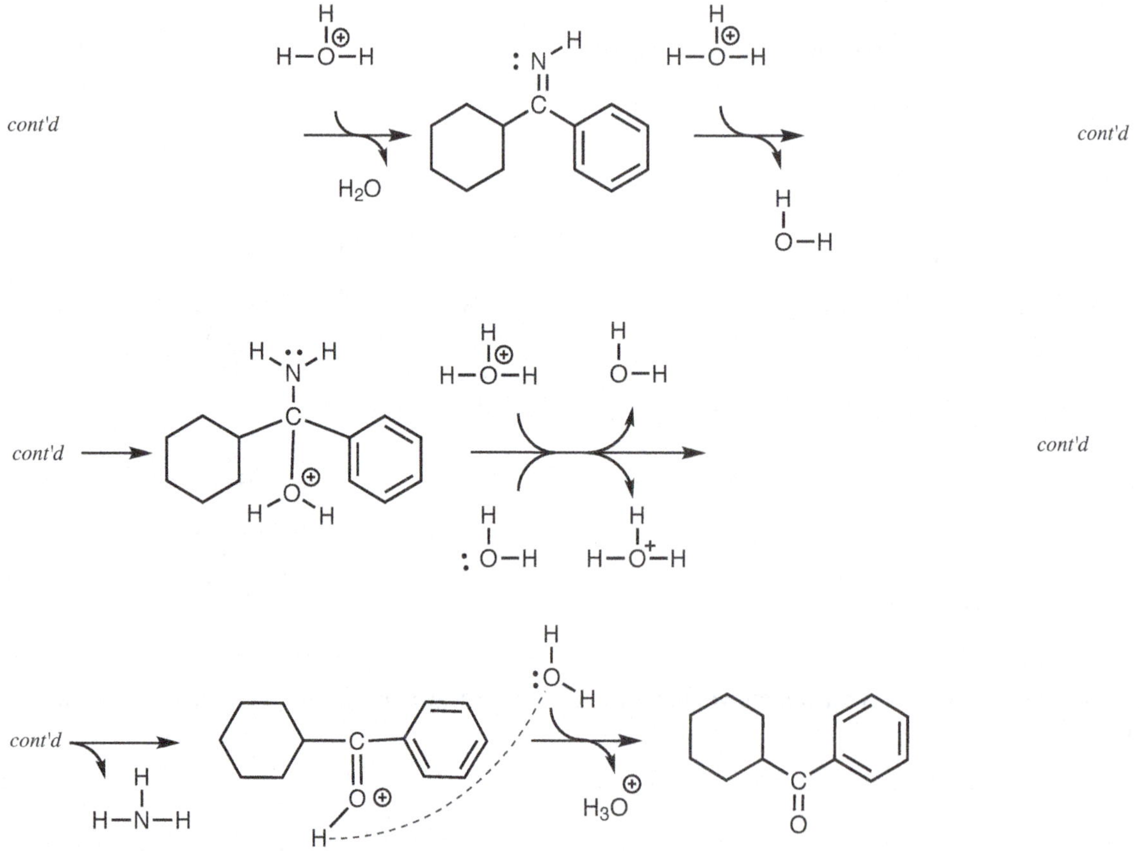

cont'd

cont'd

cont'd →

cont'd

cont'd →

Miscellaneous, Ester Formation with Diazomethane

31. Reaction of diazomethane with a carboxylic acid to form methyl 2-butenoate (methyl crotonate). (See *Notes*.)

9 - Reactions of Enols and Enolates

Aldol Reaction

1. Base catalyzed aldol condensation of butanal (butyraldehyde). (See *Notes*.)

Base catalyzed dehydration step. Under concentrated base or heating, the dehydration reaction may occur spontaneously.

2. Directed kinetic aldol condensation of 2-methylcyclohexanone with propanal (propionaldehyde). Step 1, enolate formation; step 2, reaction with propanal; step 3, neutralization. (See *Notes*.)

3.   Base catalyzed mixed or crossed aldol condensation of acetone and benzaldehyde.

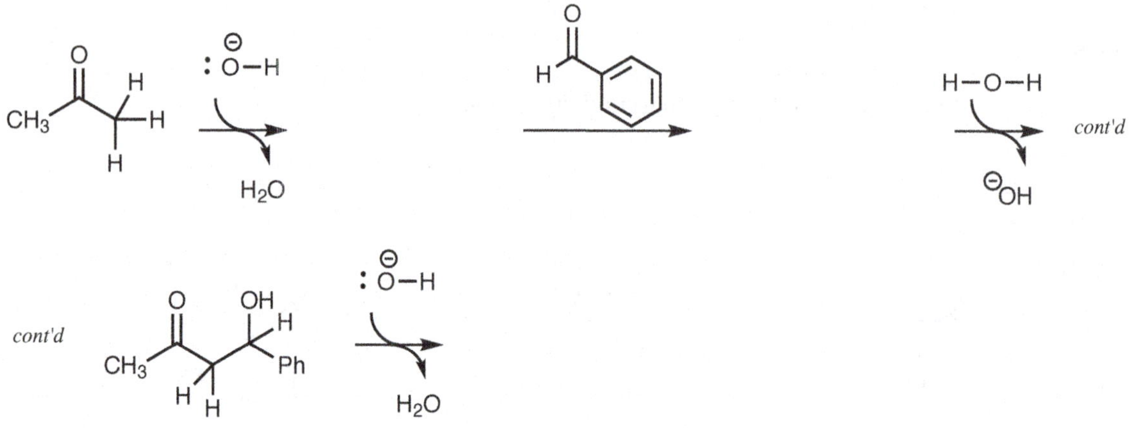

4.   Mannich reaction, acid catalyzed enolization of 2-propanone in a reaction with diethylamine, formaldehyde, and 2-propanone to give 4-(diethylamino)butan-2-one. (See *Notes*.)

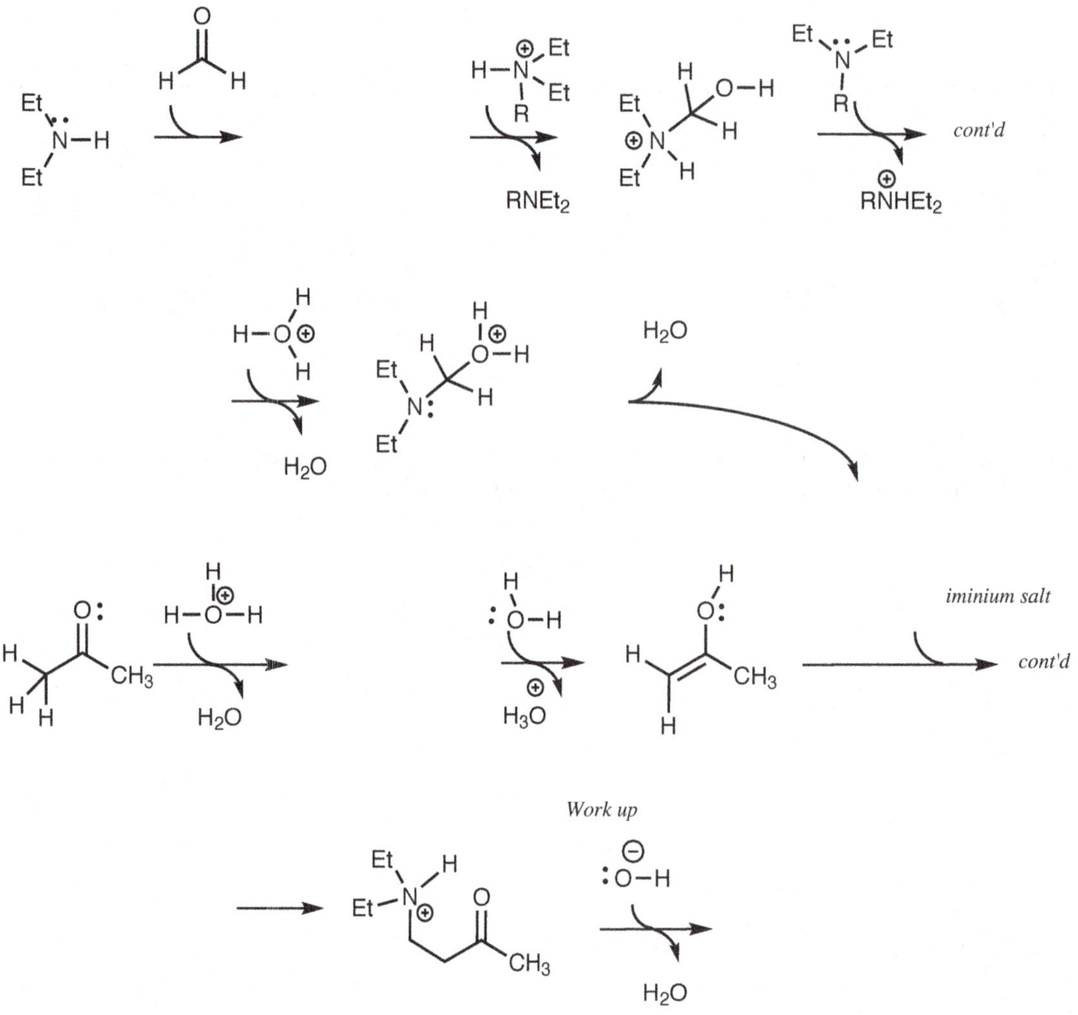

Work up

Claisen Condensation

5. Ethoxide catalyzed Claisen condensation of ethyl acetate to ethyl acetoacetate (ethyl 3-oxobutanoate). (See *Notes*.)

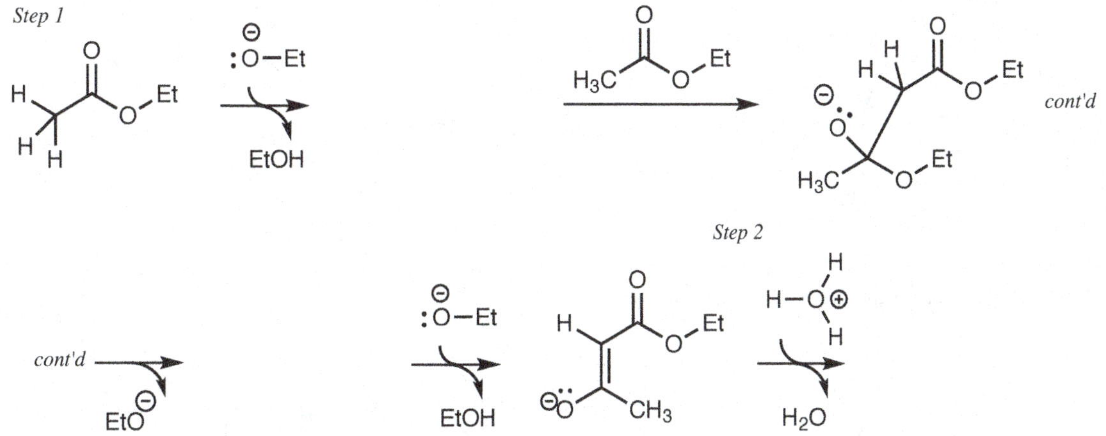

Step 1

cont'd

Step 2

cont'd

6. Ethoxide catalyzed crossed-Claisen condensation of cyclohexanone and ethyl formate. (See *Notes*.)

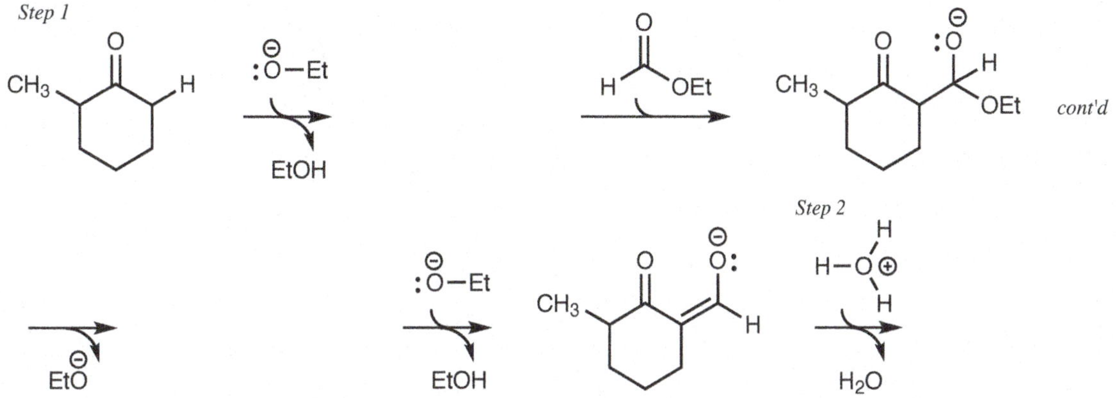

Step 1

cont'd

Step 2

7. Ethoxide catalyzed crossed-Claisen condensation of ethyl acetate and ethyl benzoate. (See *Notes*.)

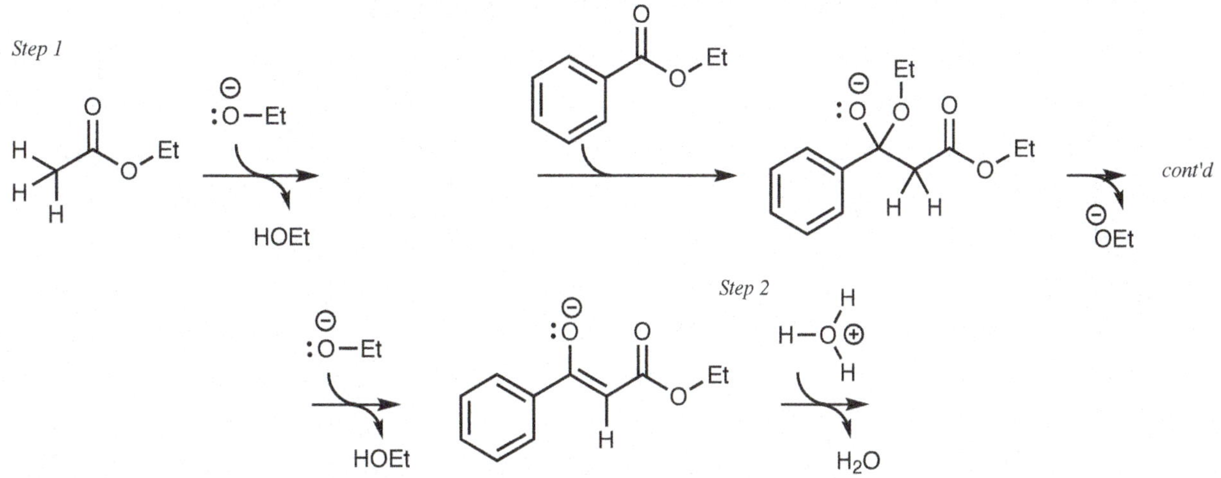

Step 1

cont'd

Step 2

Ethoxide propagation

Na—H

Na   H₂

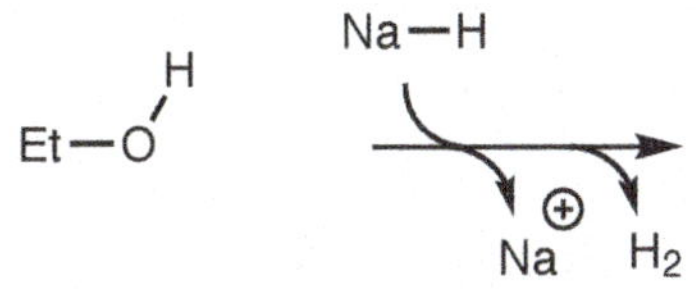

Acetoacetate Synthesis

8. Step 1, S_N2 alkylation of acetoacetate.

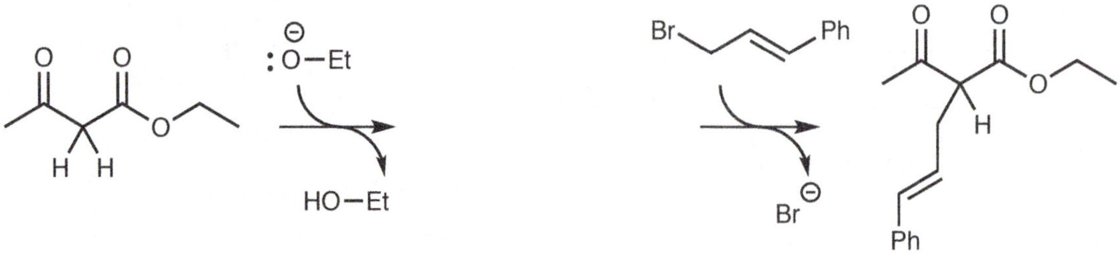

Step 2, sodium hydroxide hydrolysis of ester (saponification).

cont'd

Step 3, acidification of carboxylate, decarboxylation, and tautomerization.

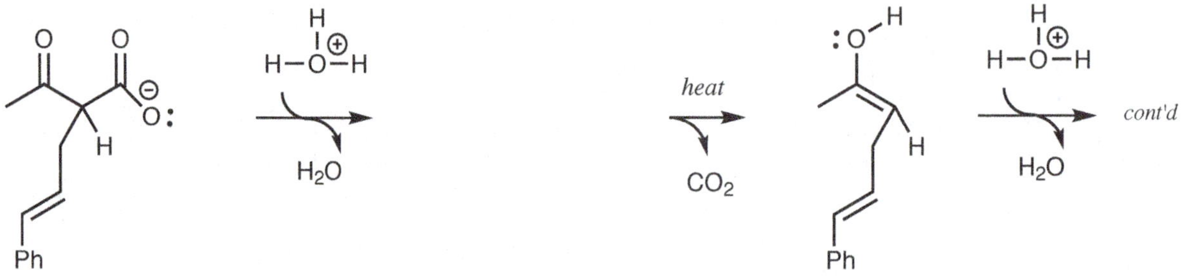

cont'd

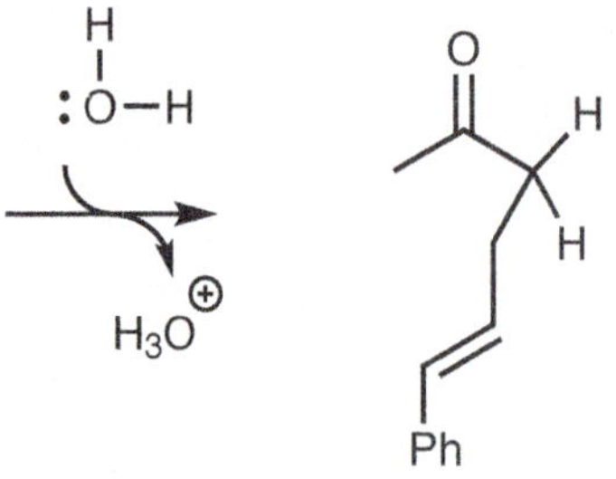

Enolate Alkylation Reactions

9. Enolization and alkylation of ethyl propionate with benzyl bromide. (See *Notes*.)

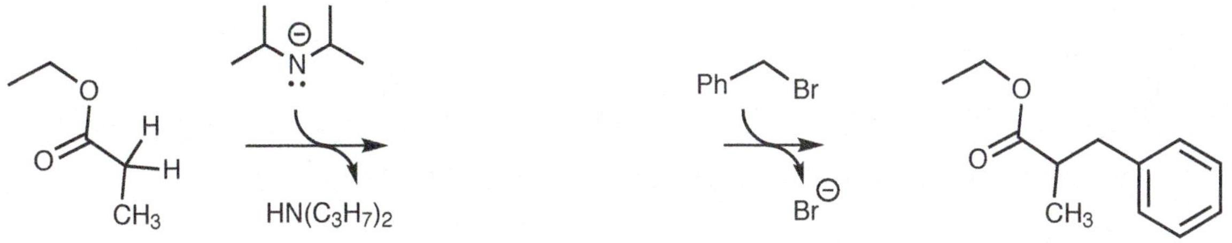

10. Sequential alkylation of a dianion of a *beta*-ketoester. (See *Notes*.)

cont'd

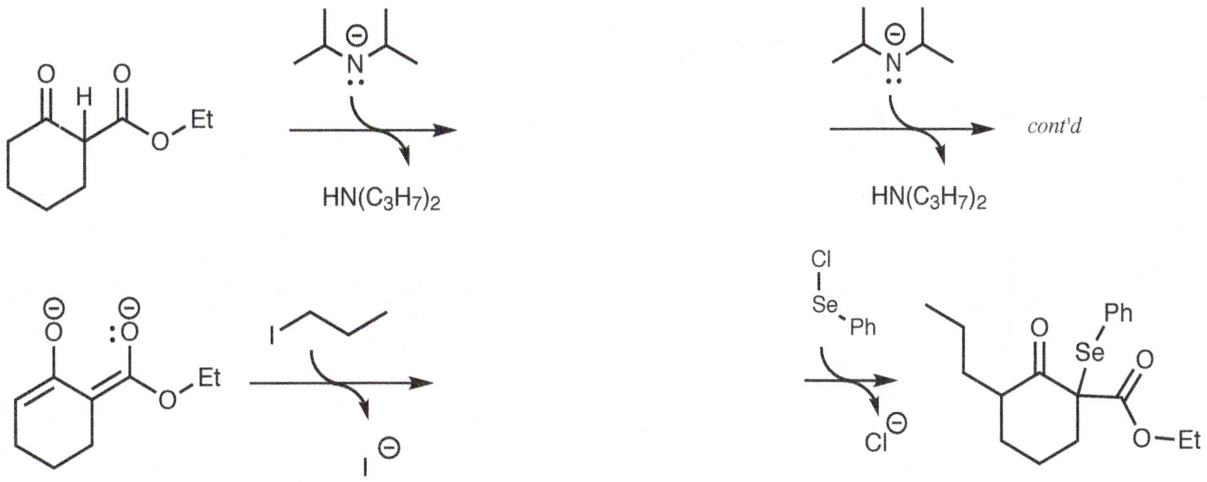

11. A retro-Claisen reaction from a *beta*-ketoester. (See *Notes*.)

cont'd

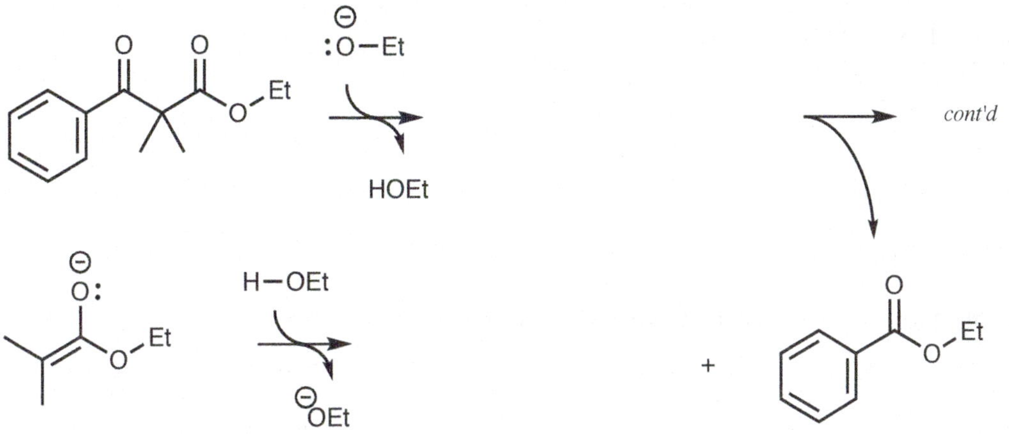

12. Enolization and alkylation of phenylacetonitrile with methyl iodide. (See *Notes*.)

Halogenation of Carbonyl Compounds

13. Basic bromination of 3-methyl-2-butanone with sodium hydroxide and bromine, bromoform reaction.

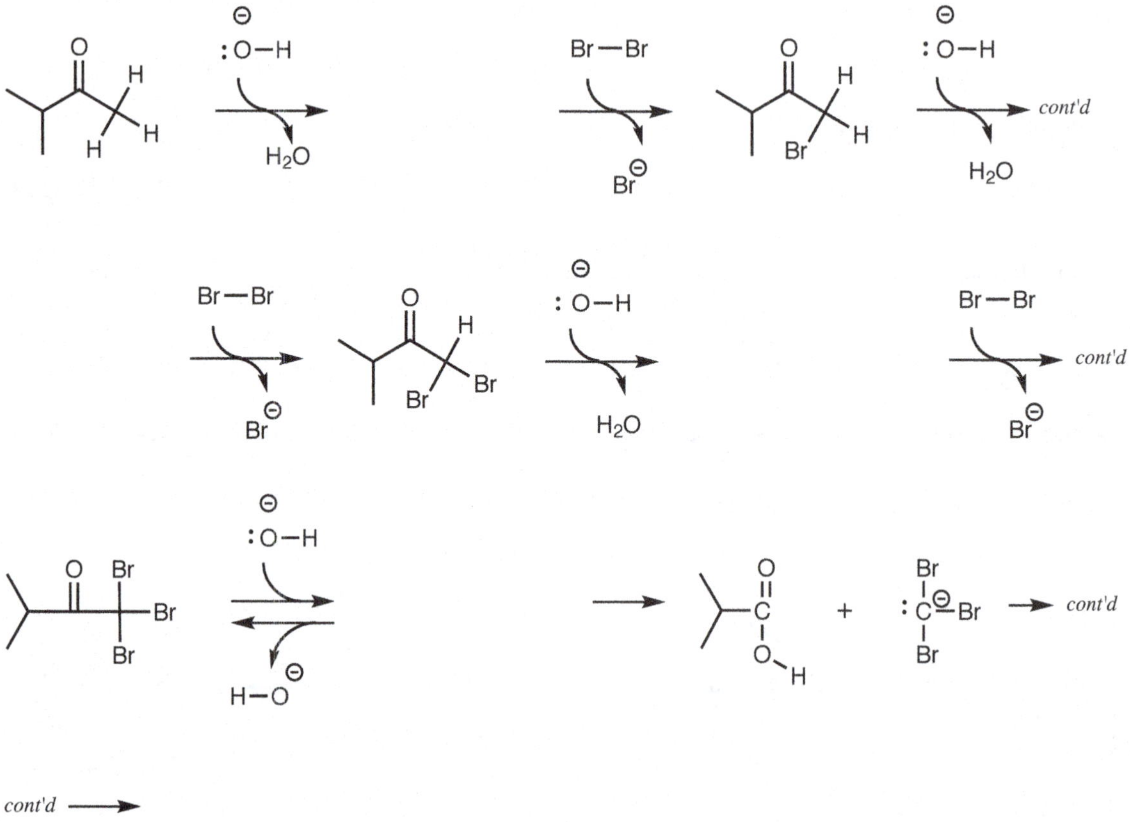

Step 2, acidification and isolation of isobutyric acid.

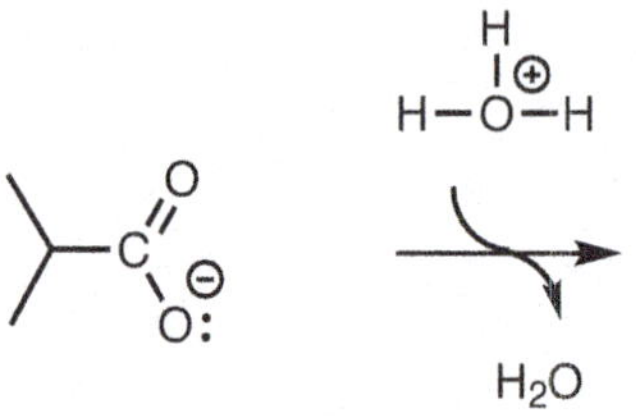

14. Acid catalyzed bromination of acetophenone to give α-bromoacetophenone

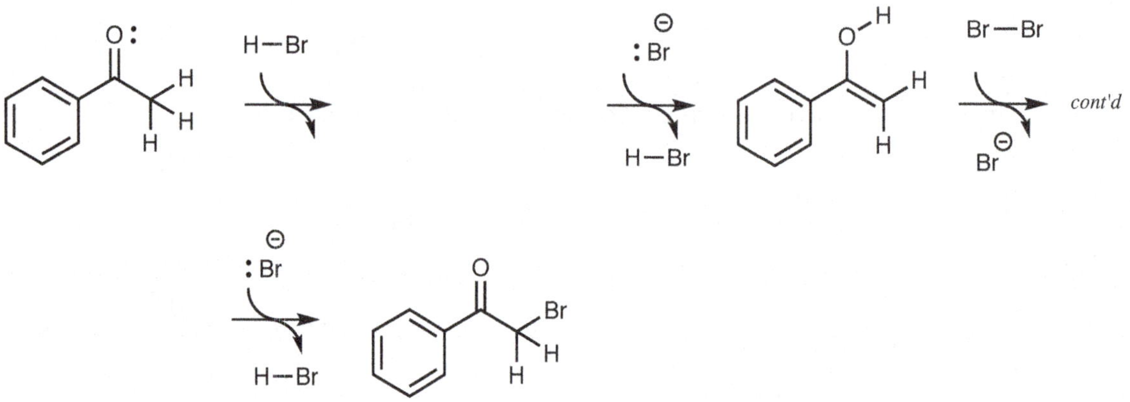

Michael Addition or 1,4-Conjugate Addition Reaction

15. Michael addition reaction of dimethylamine to methyl vinyl ketone.

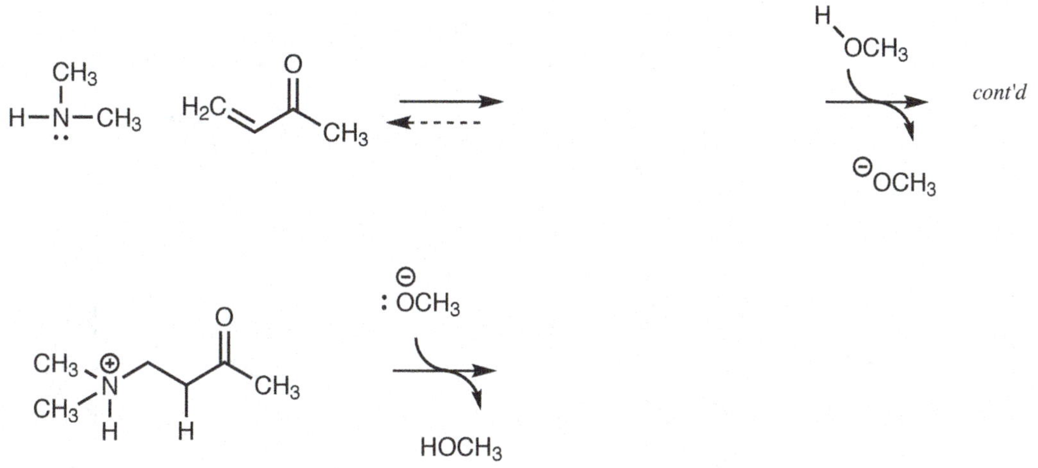

16. Michael addition reaction of ethyl acetoacetate to cyclohexenone.

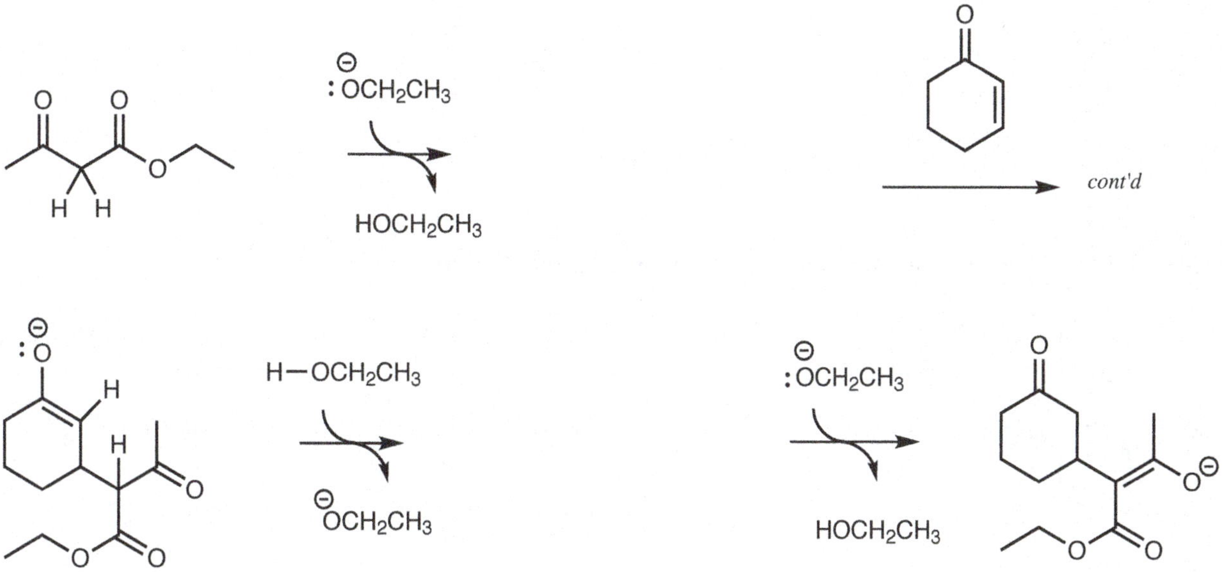

Step 2, neutralization of reaction mixture.

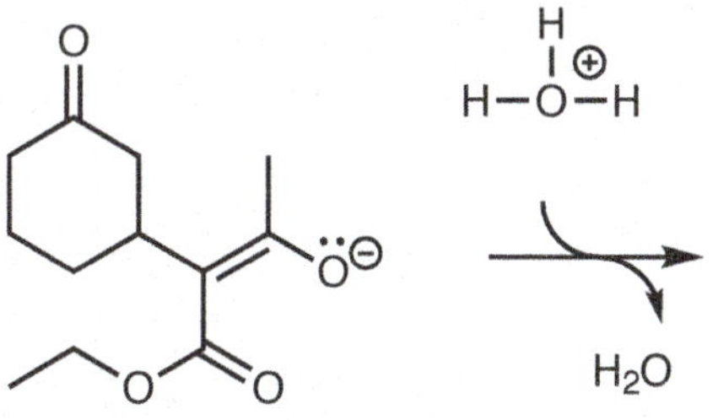

Enamine Alkylation

17. Step 1, formation of the morpholine enamine of cyclohexanone.

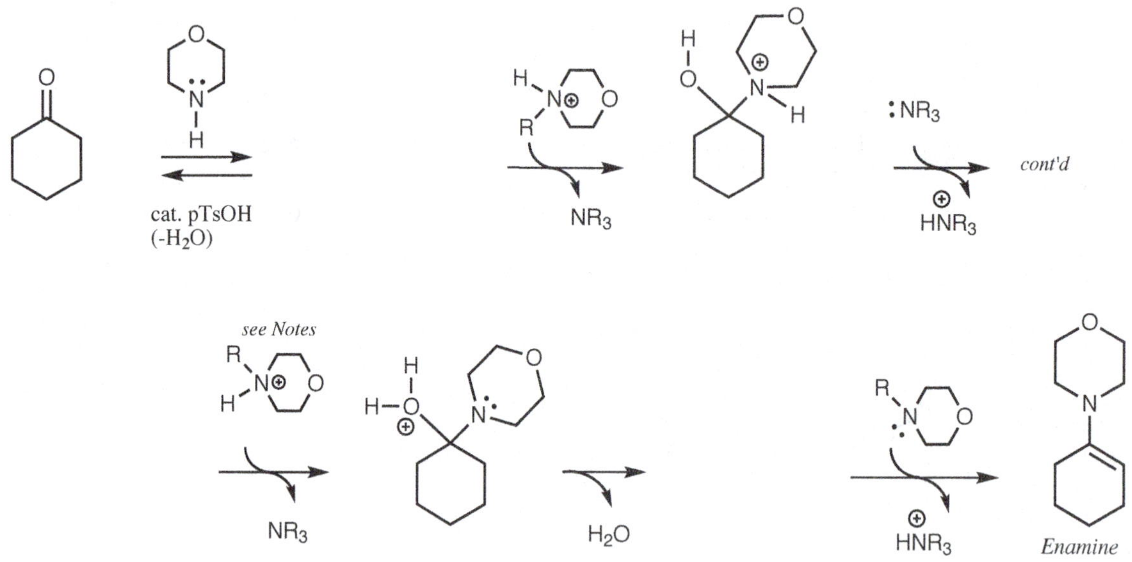

see Notes

Step 2, enamine alkylation with methyl iodide.

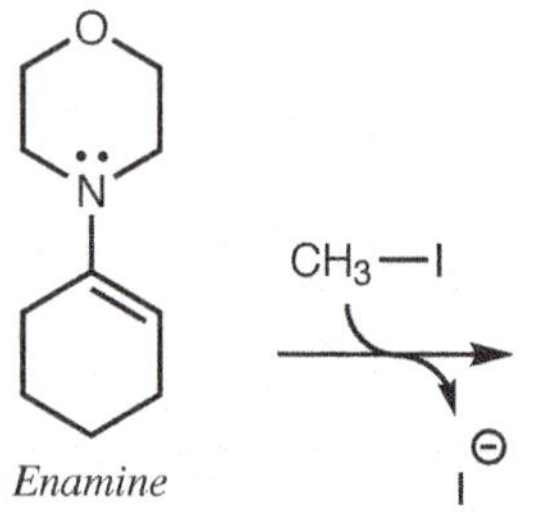

Enamine

Step 3, hydrolysis of iminium salt and isolation of substituted cyclohexanone.

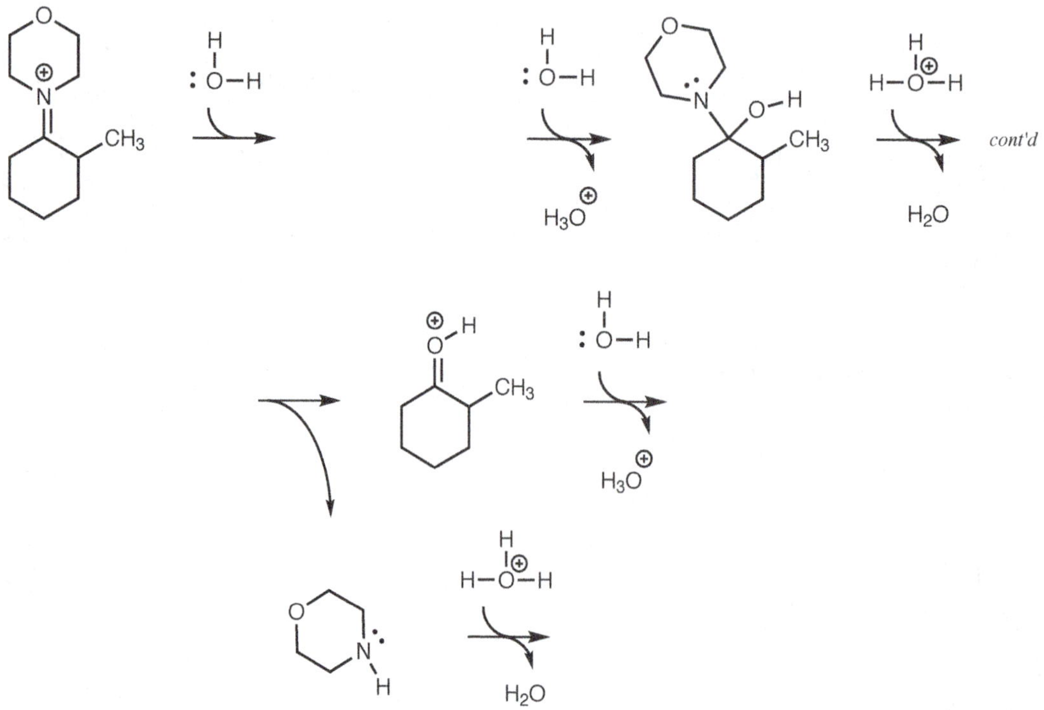

10 — Dehydration/Halogenation Agents

1. Reaction of cyclohexanecarboxamide with acetic anhydride to give cyclohexanecarbonitrile. (See *Notes.*)

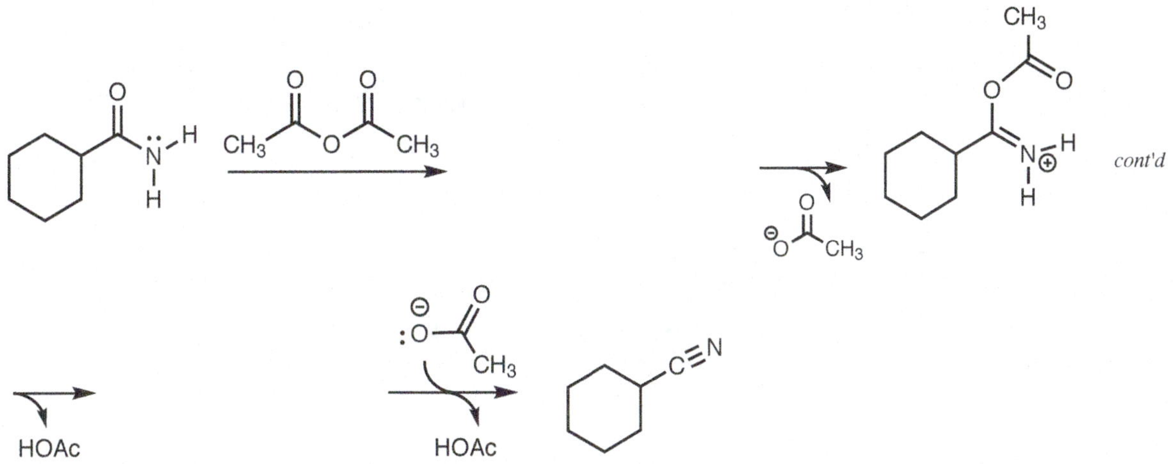

cont'd

2. S$_N$2 reaction of thionyl chloride with 1-butanol to give chlorobutane. (See *Notes.*)

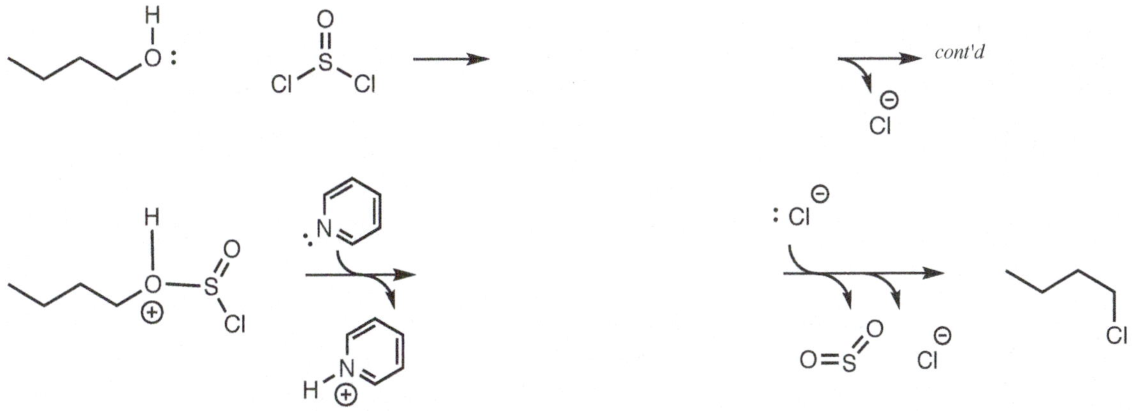

cont'd

3. Reaction of 1-butanol with tosyl chloride and pyridine to give butyl tosylate. (See *Notes.*)

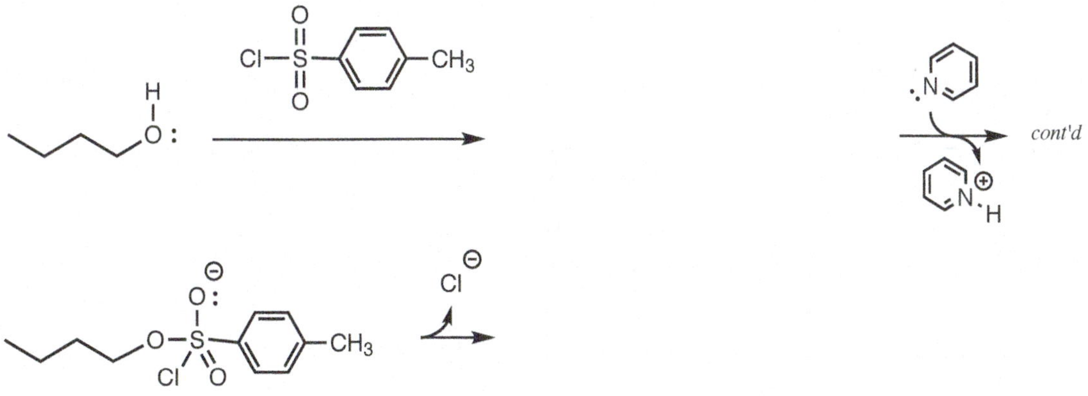

cont'd

Carboxylic Acid with Thionyl Chloride

4. Conversion of acetic acid to acetyl chloride with thionyl chloride.

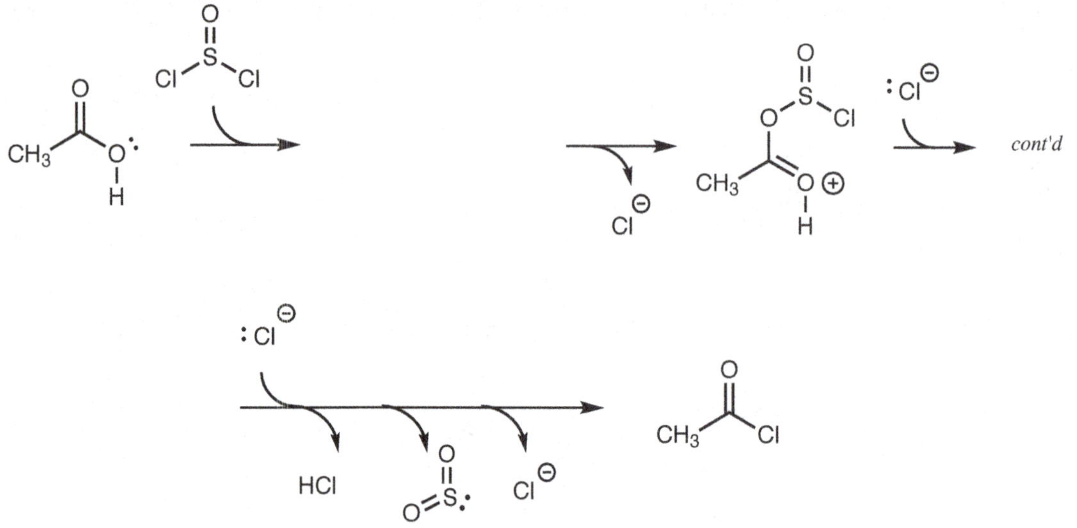

5. Conversion of acetic acid to acetyl chloride with thionyl chloride. Part 1, reaction of thionyl chloride with DMF to form iminium salt. (See Notes.)

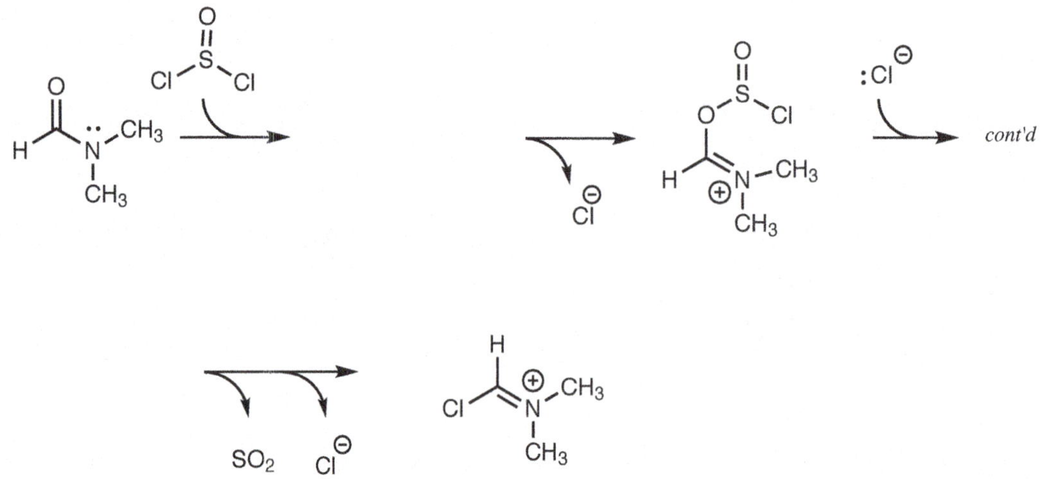

Part 2, reaction of iminium salt with carboxylic acid

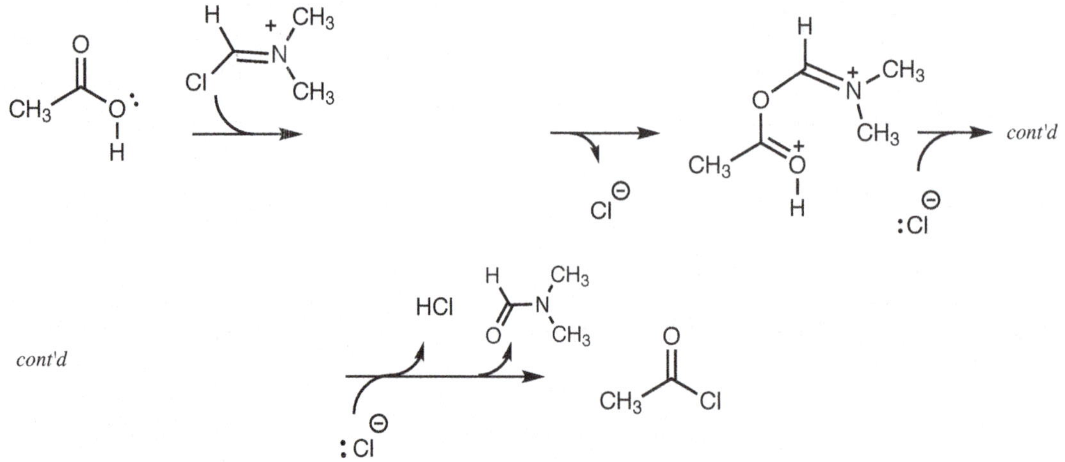

Halide from an Alcohol with a Phosphorus Reagent

6. Reaction of isobutyl alcohol with phosphorus tribromide to give isobutyl bromide. (See *Notes*.)

7. Reaction of triphenylphosphine, carbon tetrachloride and cyclopentanol to give chlorocyclopentane. (See *Notes*.)

Ester from an Alcohol and Carboxylic Acid with a Phosphorus Reagent

8. Mitsunobu reaction with triphenyl phosphine and diethyl azodicarboxylate to prepare an ester from an acid and an alcohol.

cont'd

11 — Reduction Reactions

Sodium Borohydride Reductions

1. Sodium borohydride reduction of 2-methylpropanal (isobutyraldehyde) to 2-methyl-1-propanol (isobutyl alcohol).

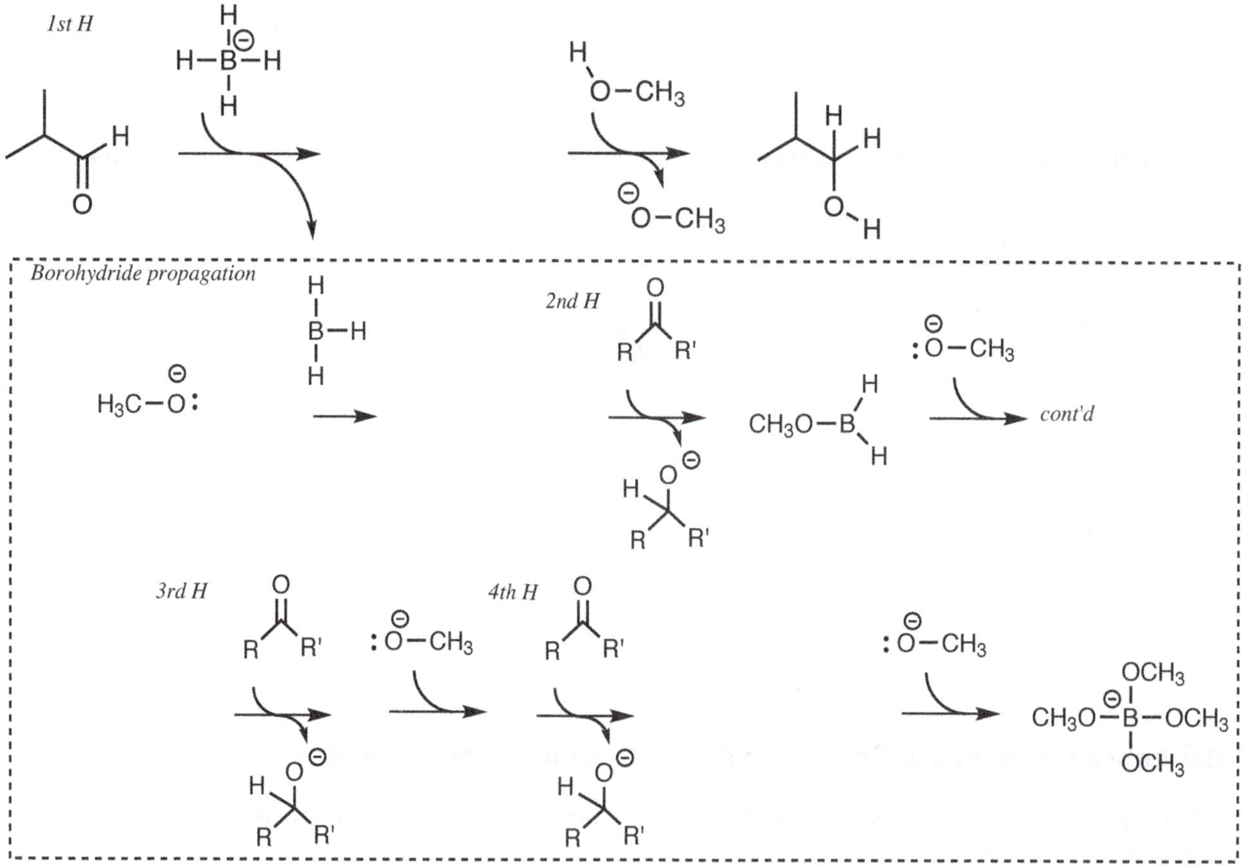

2. Sodium borohydride reduction of cyclopentanone to cyclopentanol.

3. Sodium borohydride reduction of ethyl 4-oxocyclohexanecarboxylate to give ethyl 4-hydroxycyclohexanecarboxylate. (See *Notes*.)

Lithium Aluminum Hydride Reductions

4. Lithium aluminum hydride reduction of acetophenone to 1-phenylethanol.

Step 1

Step 2   Work-up

5. Lithium aluminum hydride reduction of ethyl butanoate (butyrate) to give ethanol and butanol. (See *Notes*.)

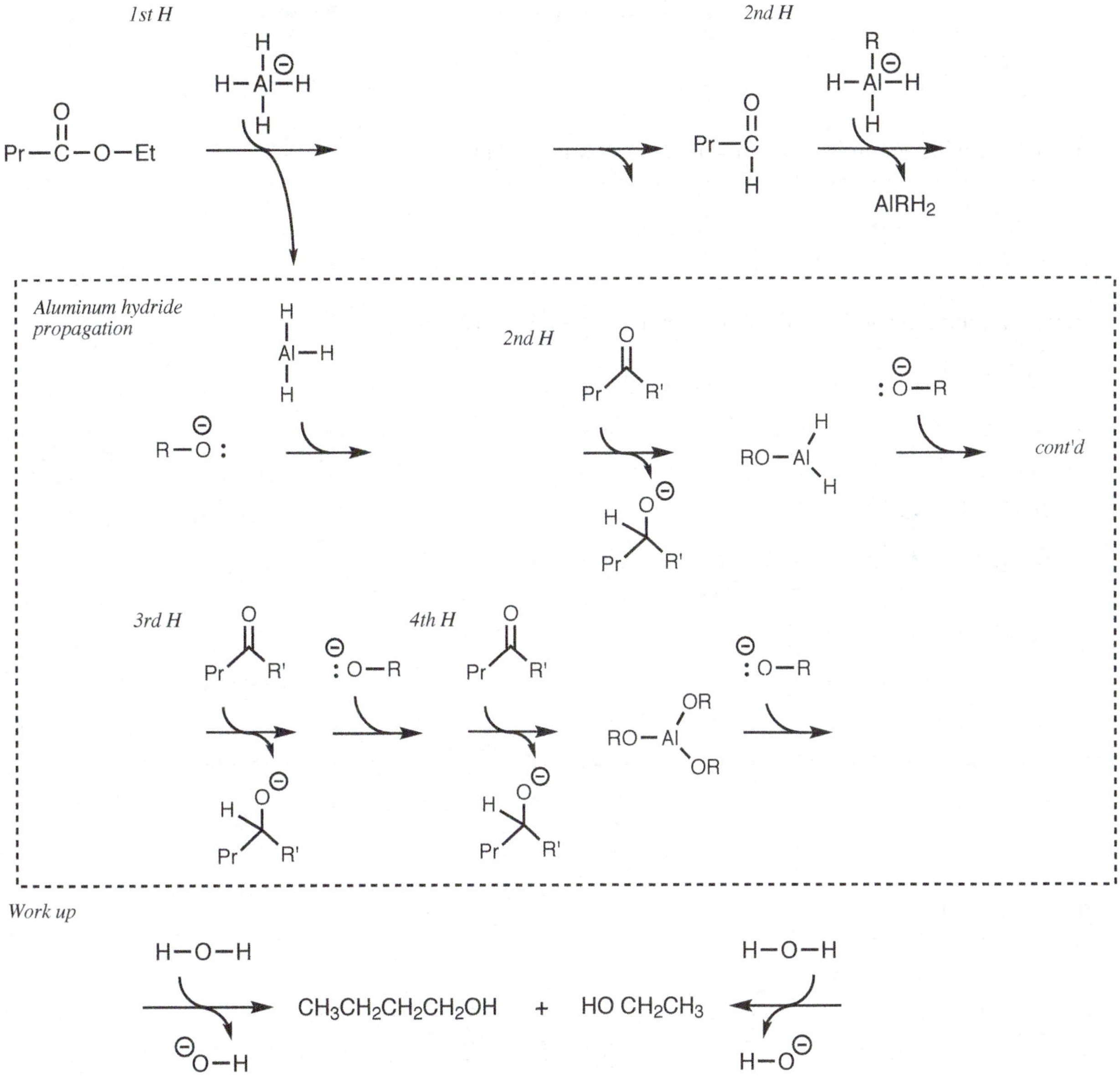

1st H   *2nd H*

Aluminum hydride propagation   *2nd H*   *cont'd*

3rd H   *4th H*

Work up

$$CH_3CH_2CH_2CH_2OH \quad + \quad HO\,CH_2CH_3$$

6. Lithium aluminum hydride reduction of a cyclic amide(1-methylpyrrolidin-2-one) to give a cyclic amine (1-methylpyrrolidine). (See *Notes*.) See Example 5 for regeneration of aluminum (IV) reductant.

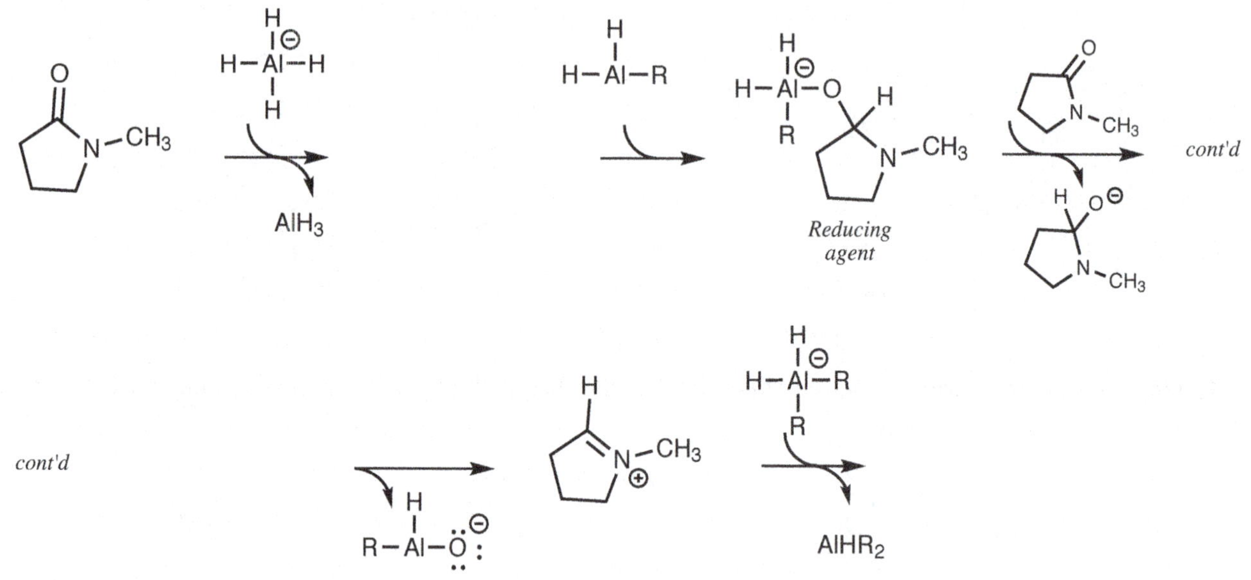

cont'd

Preferred leaving group for lithium aluminum hydride reductions.

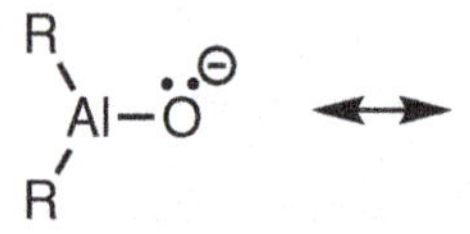

7. Lithium aluminum hydride reduction of propionyl anilide to *N*-propylaniline.

Step 1, reduction.

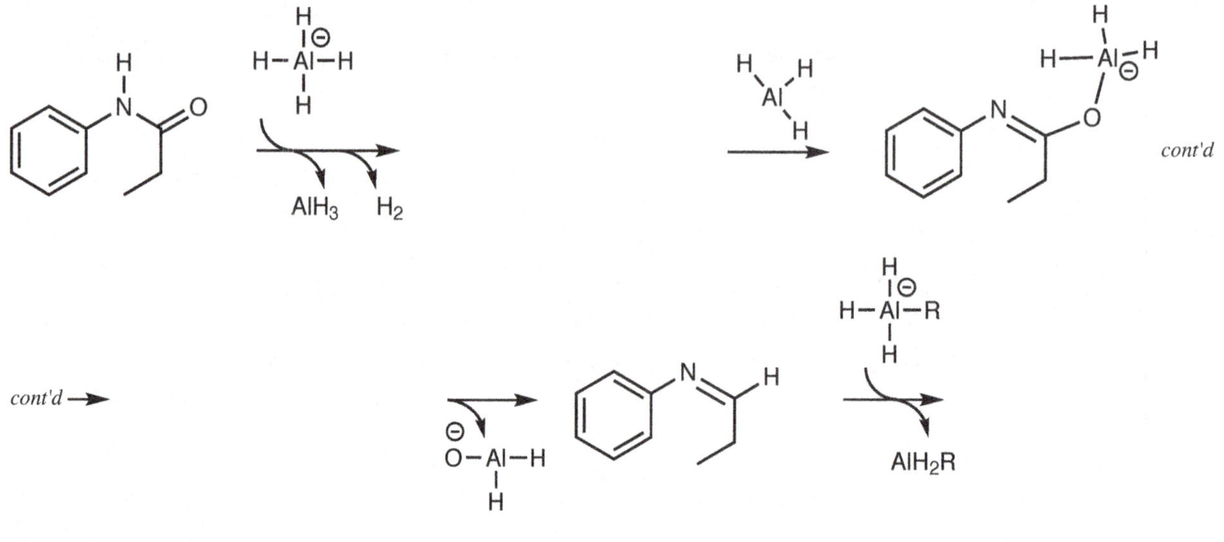

cont'd →

Step 2, work up.

cont'd

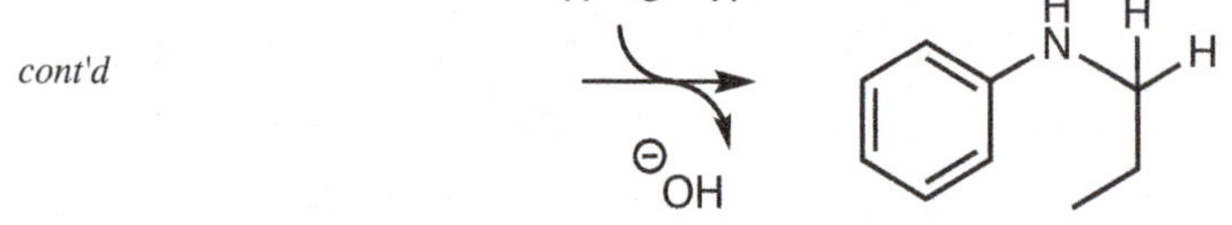

8. Lithium aluminum hydride reduction of phenylacetic acid to 2-phenylethanol. (See *Notes.*)

Step 1, reduction.

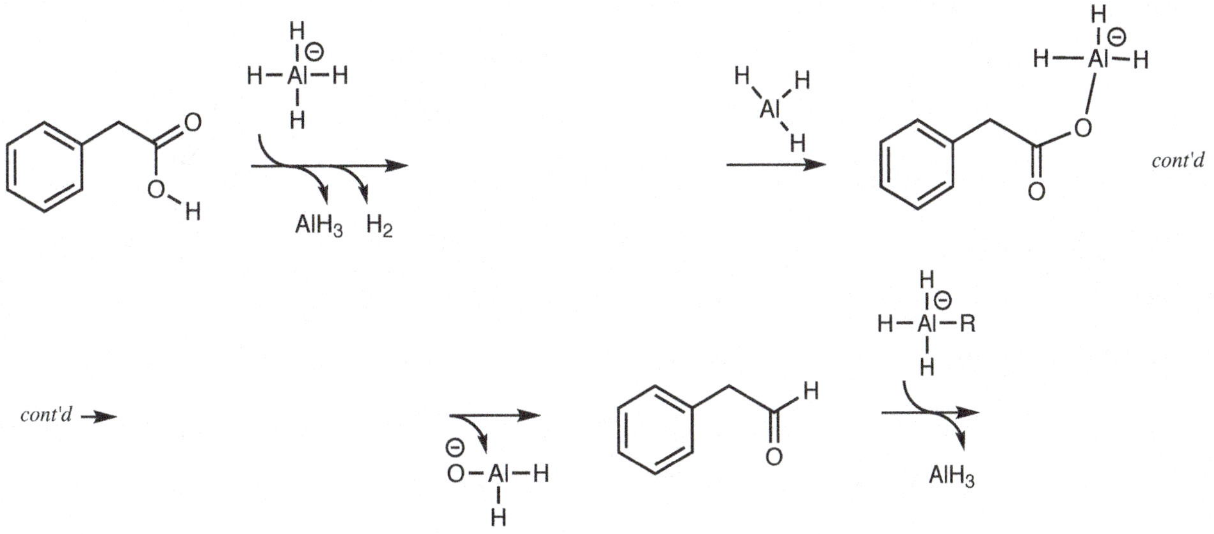

cont'd

Step 2, work up.

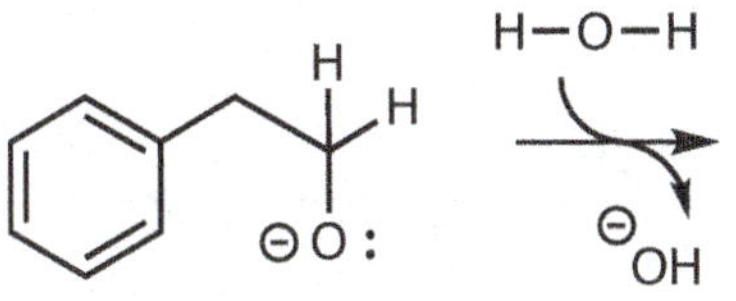

Reductive Amination

9. Step 1, formation of imine from benzaldehyde and ethylamine. (See *Notes.*)

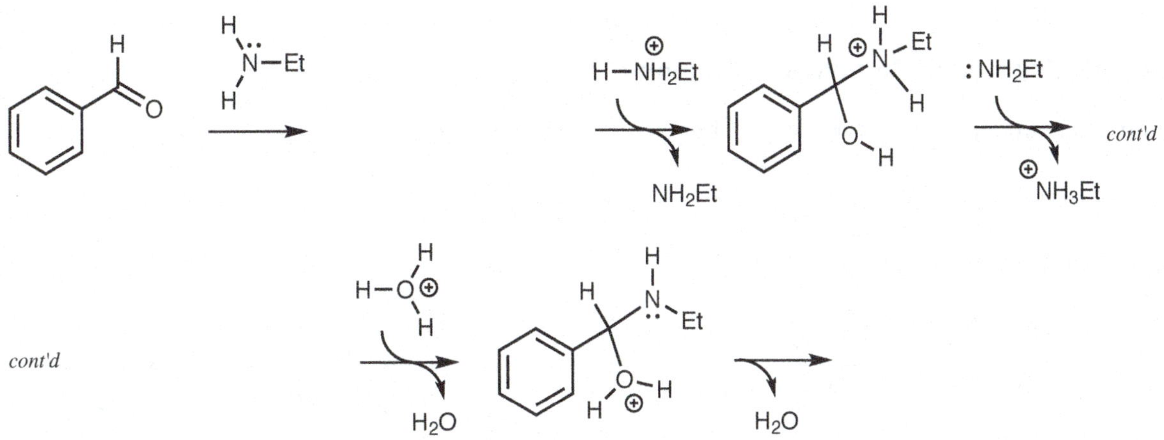

cont'd

cont'd

Step 2, reduction

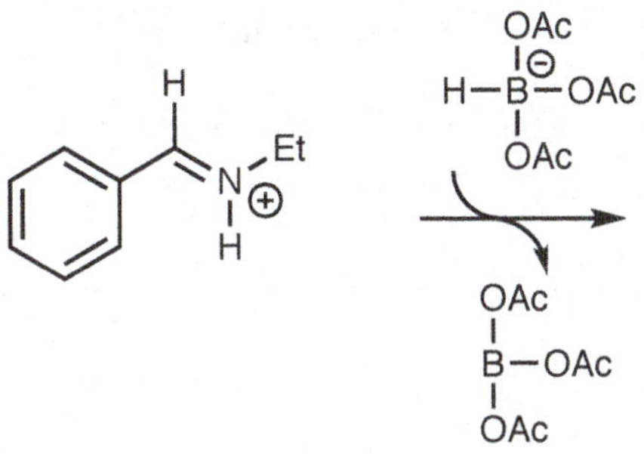

Diisobutylaluminum Hydride Reduction of an Ester

10. Step 1, addition of DIBAH or DIBAL(H) (diisobutylaluminum hydride) to ethyl isobutyrate. Reduction to give isobutyraldehyde (2-methylpropanal).

Step 2, work-up

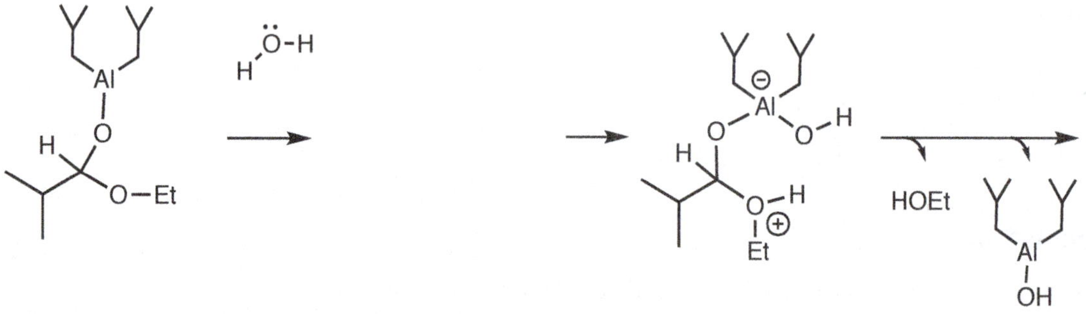

Reduction of Alkyne with Sodium and Ammonia

11. Sodium and ammonia *trans*-reduction of 2-butyne to *trans*-2-butene.

Note, use a single barbed arrow for one-electron transfers reaction.

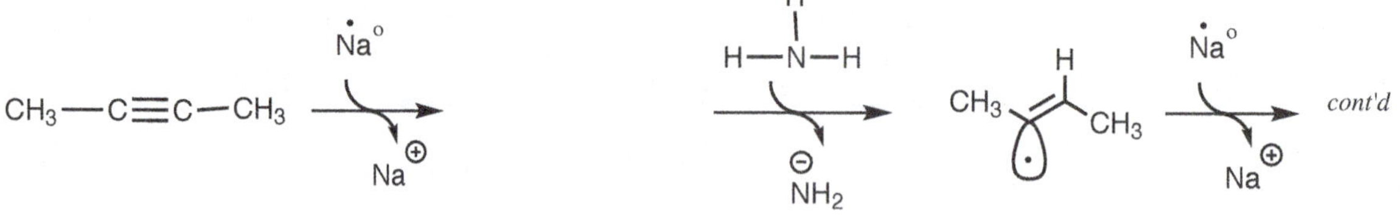

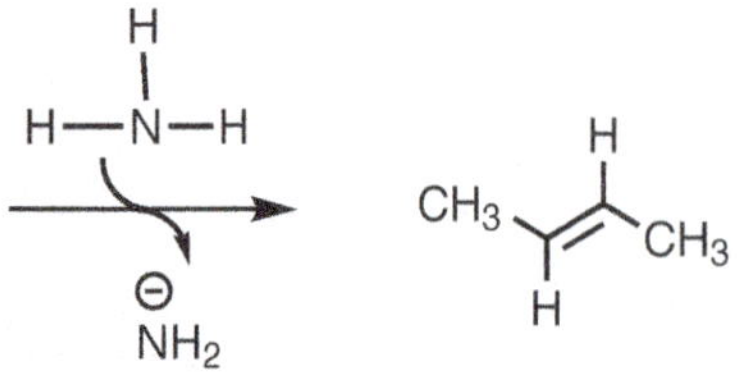

Which is the stronger base?

The pK_a of an sp^2 carbanion pK_a is 44 and NH_3 is 36.

Wolff Kischner Reduction

12. Reaction of the ketone with hydrazine under basic conditions to form the hydrazide.

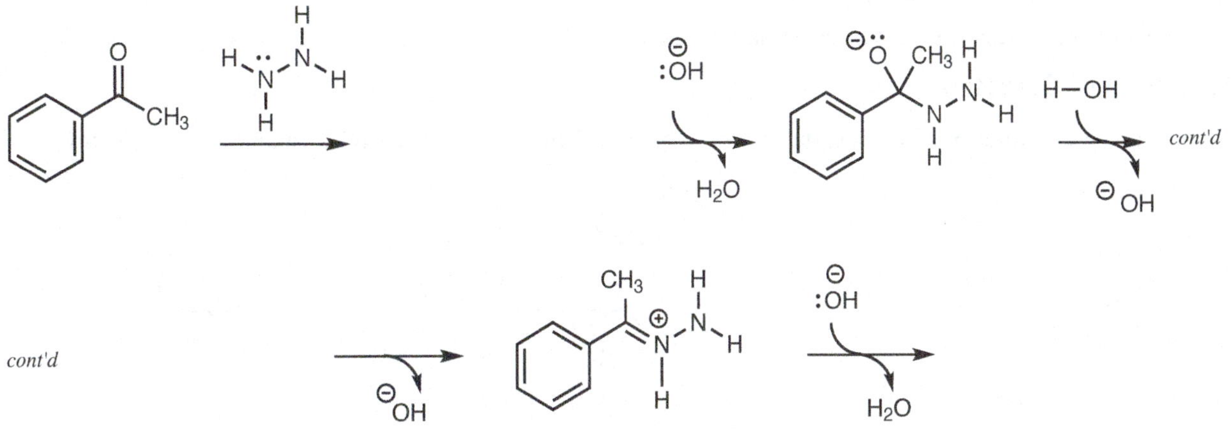

cont'd

Reaction of the *in situ* formed hydrazide with KOH to form the methylene.

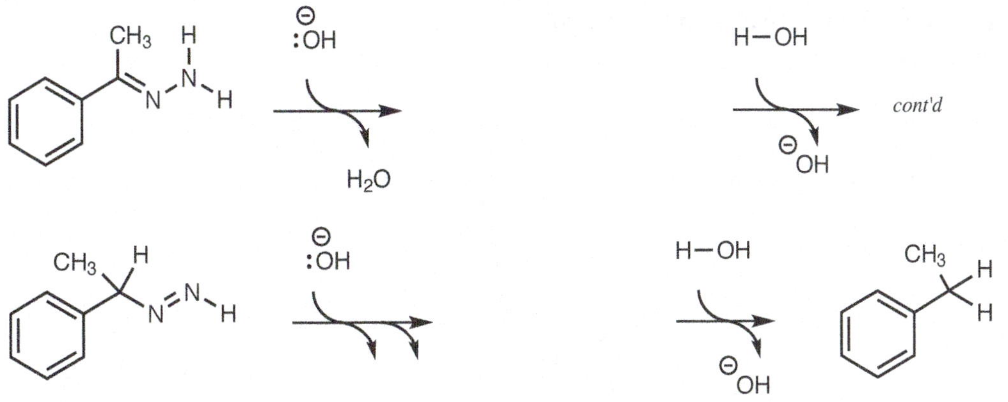

Catalytic Reduction of Nitrobenzene

Catalytic reduction of nitrobenzene to aniline. This isn't a mechanism. It shows that hydrogen can add across N-O single and double bonds. It is a very facile reduction. One can see that the oxygen atoms form water and the reduction takes three moles of hydrogen.

13. Catalytic reduction of nitrobenzene to aniline.

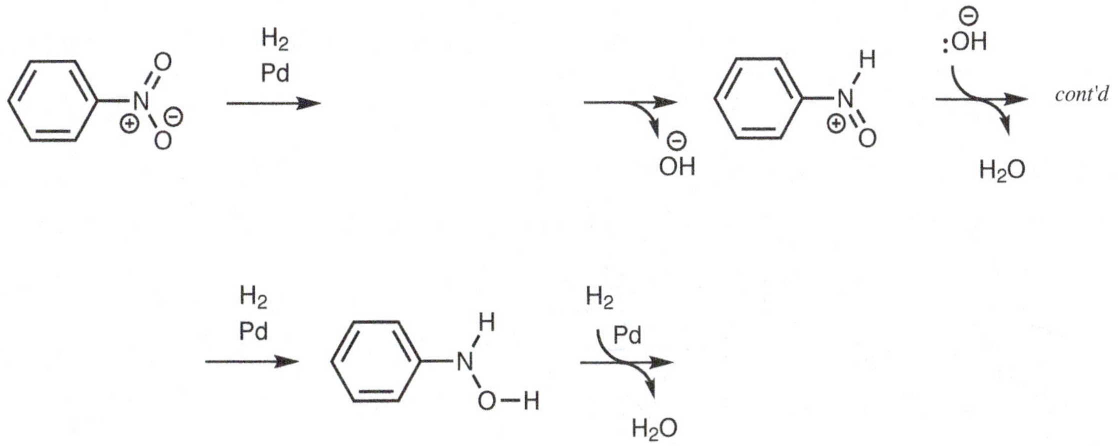

12 — Oxidation Reactions

General Form For Oxidations

See *Notes* for further discussion of oxidation reactions.

Chromic Acid Oxidation

1. Chromic acid oxidation (Jones oxidation) of 3-methyl-2-butanol to 3-methyl-2-butanone. (See *Notes*.)

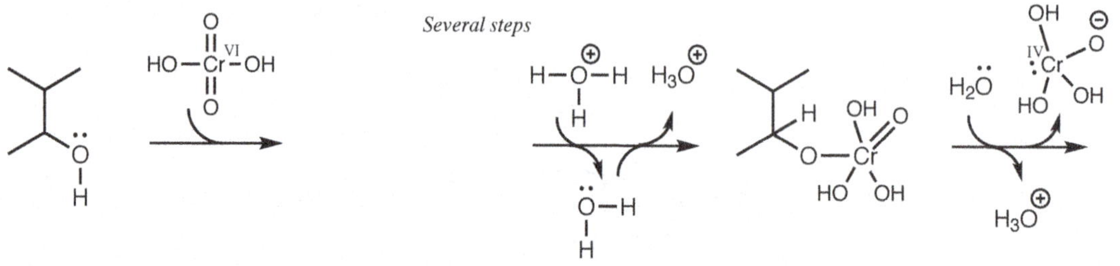

2. Chromic acid oxidation of isobutyl alcohol (2-methyl-1-propanol) to isobutyric acid (2-methylpropanoic acid). (See *Notes*.)

Step 1, chromate ester formation and oxidation to aldehyde.

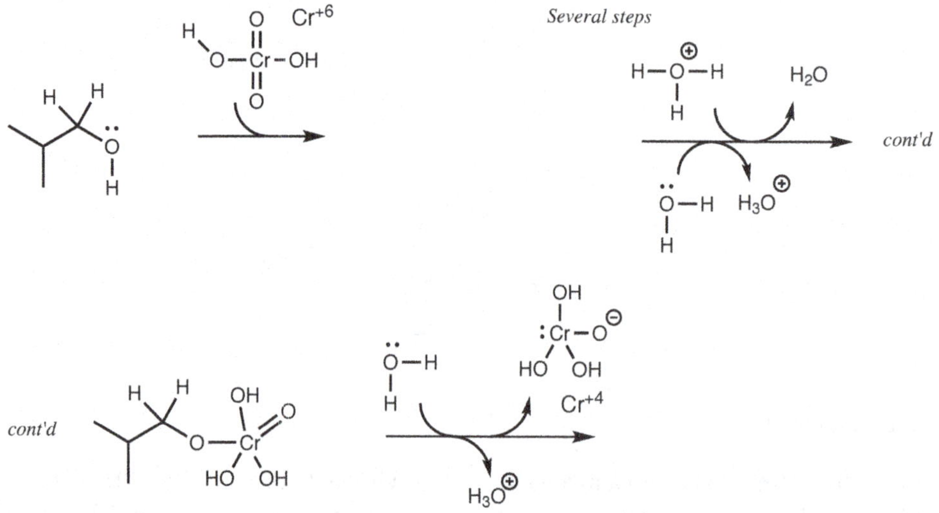

Step 2, hydration of aldehyde and oxidation of hydrate to isobutyric acid (2-methylpropanoic acid).

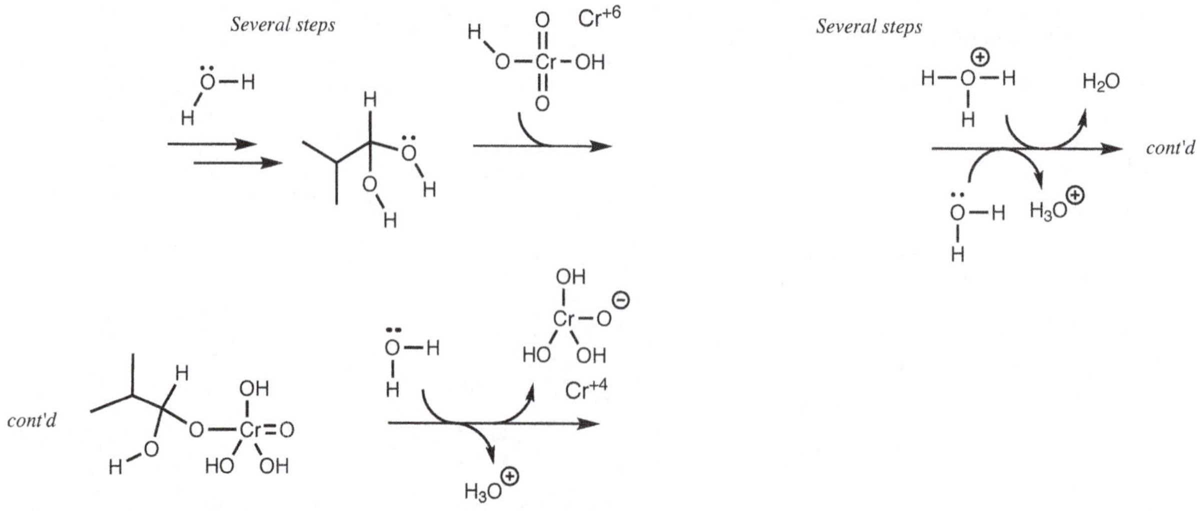

PCC, Tollens, Hypochlorite, *m*CPBA, and Sulfonium Based Oxidations

3. Pyridinium chlorochromate (PCC) oxidation of benzyl alcohol to benzaldehyde.

4. Tollens oxidation of 2-methylpropanal (isobutyraldehyde) with silver oxide (or hydroxide).

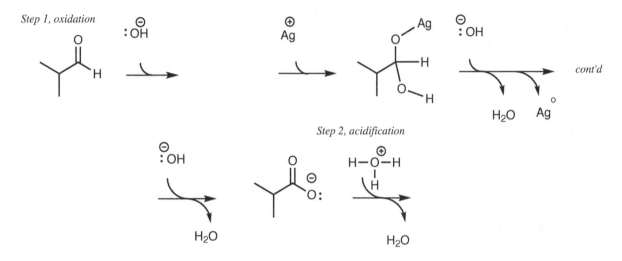

5. Oxidation of cyclohexanol to cyclohexanone with sodium hypochlorite (NaOCl, bleach).

6. Peracid epoxidation of *trans*-2-butene with *m*-chloroperoxybenzoic acid (MCPBA) to give an epoxide, (2*R*,3*R*)-2,3-dimethyloxirane. (See *Notes*.)

meta-chlorobenzoperoxoic acid
MCPBA

7. Step 1, Swern oxidation, preparation of chlorosulfonium salt.

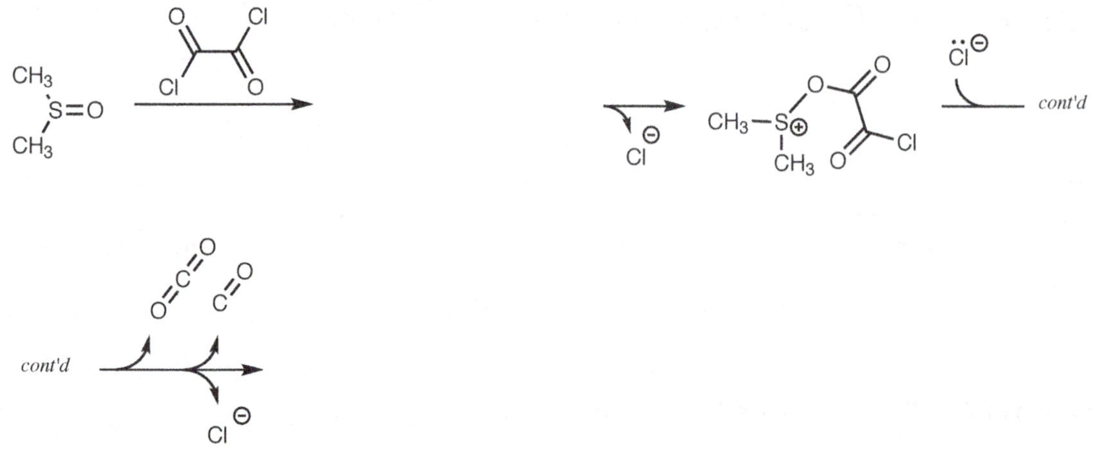

Step 2, Swern oxidation, oxidation step.

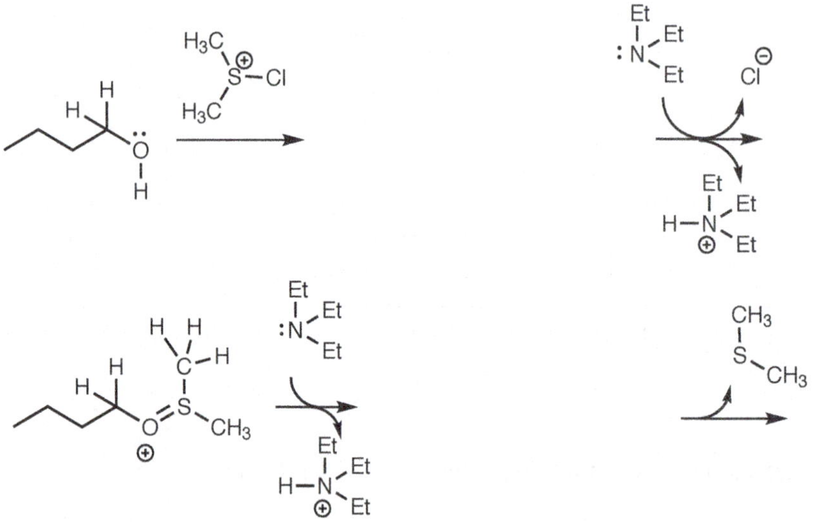

8. Corey-Kim (chlorosulfonium salt) oxidation of cyclopentanol to cyclopentanone. (See *Notes*.)

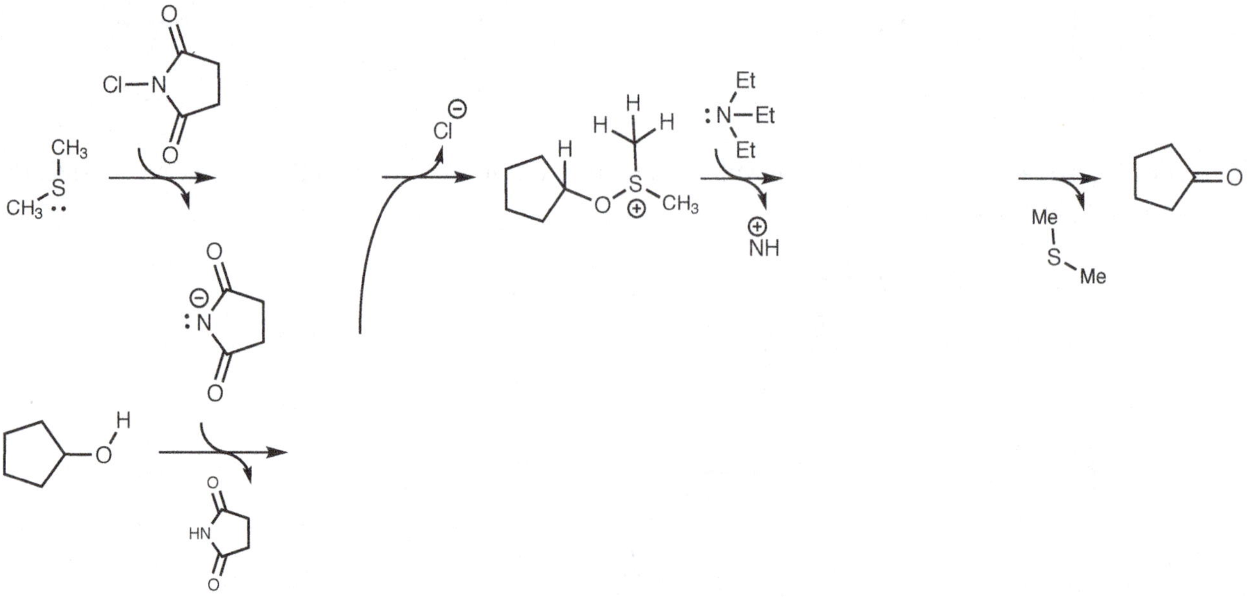

Ozone Oxidation

9. Step 1, ozone reaction with an alkene to give an ozonide.

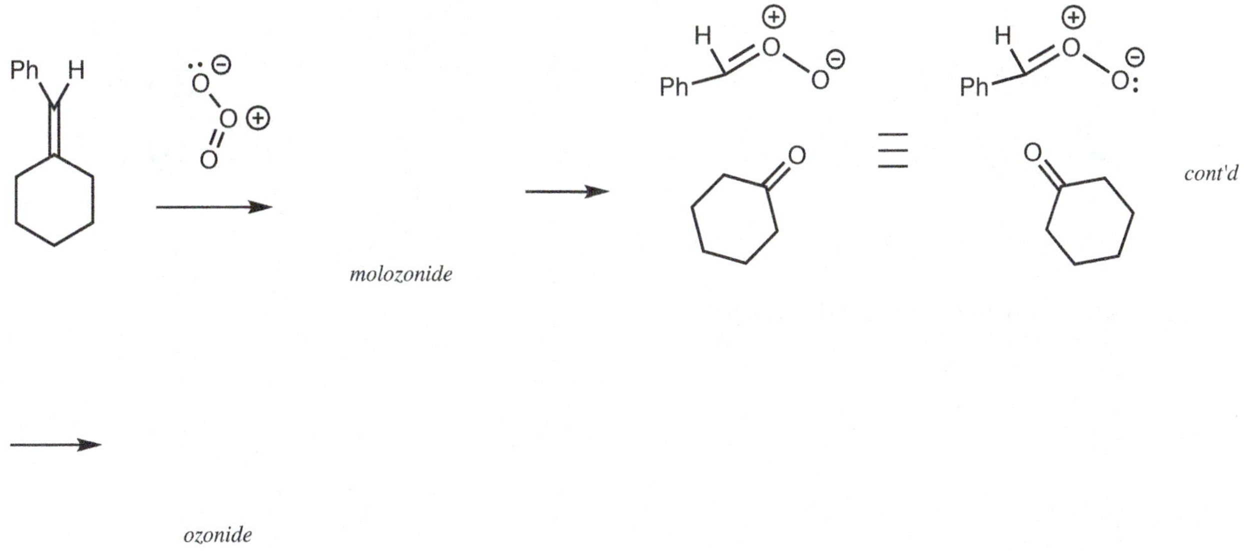

molozonide

cont'd

ozonide

Step 2, reduction of ozonide to two carbonyl compounds.

ozonide

Alternate Step 2, oxidative destruction of ozonide.

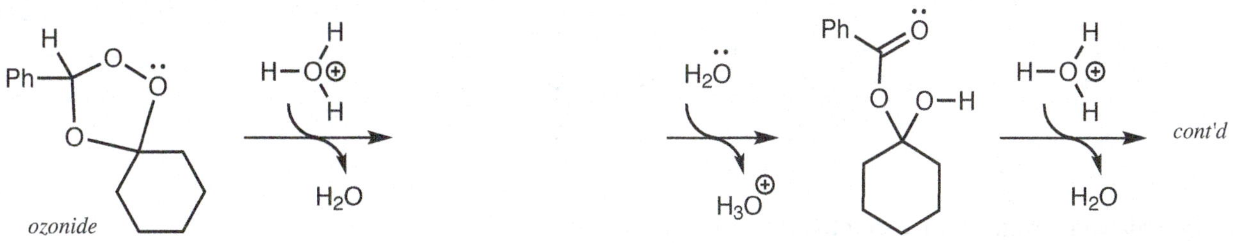

ozonide

cont'd

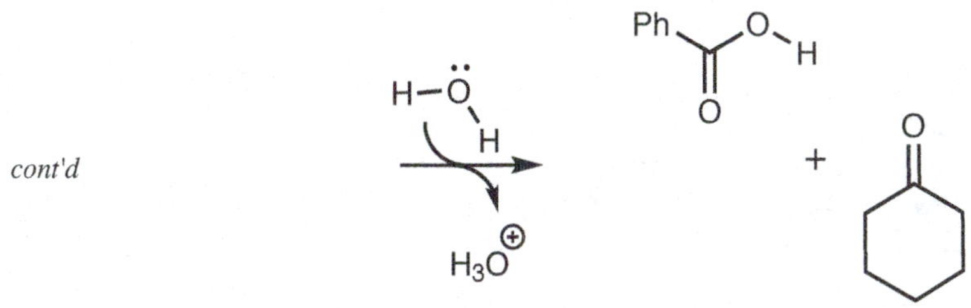

cont'd

Osmium Tetroxide, Potassium Permanganate, and Periodate Oxidations

10. Osmium tetroxide oxidation of cyclohexene to give *cis*-1,2-cyclohexanediol

Regeneration of oxidant with *tert*-butylperoxide.

11. Potassium permanganate oxidation of methylcyclohexene to 1-methylcyclohexane-1,2-diol at low temperature or cleavage at high temperature.

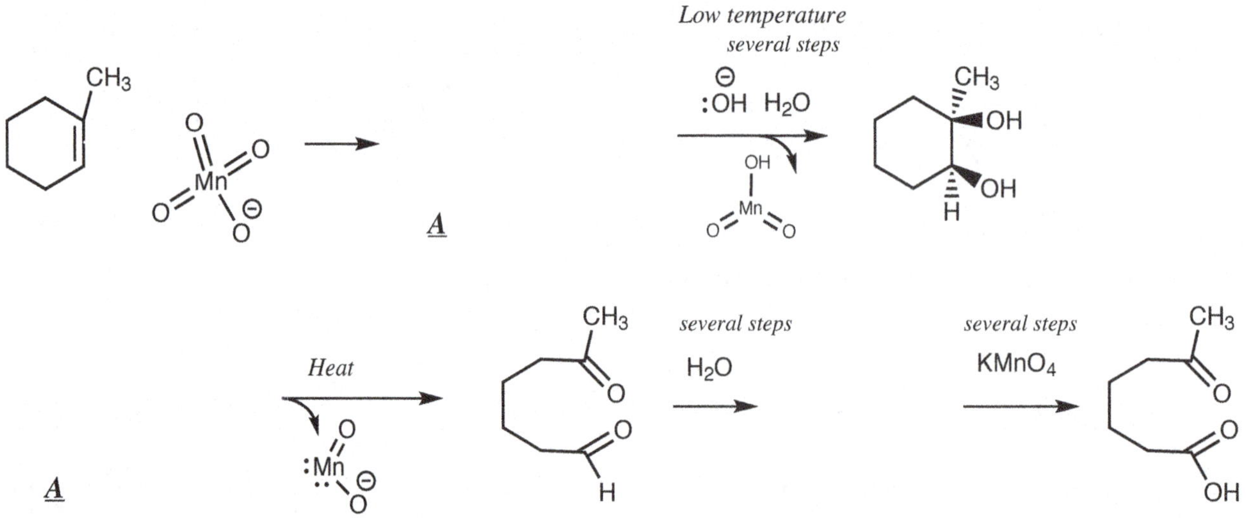

12. Periodate cleavage of a 1,2-diol to give a dicarbonyl compound.

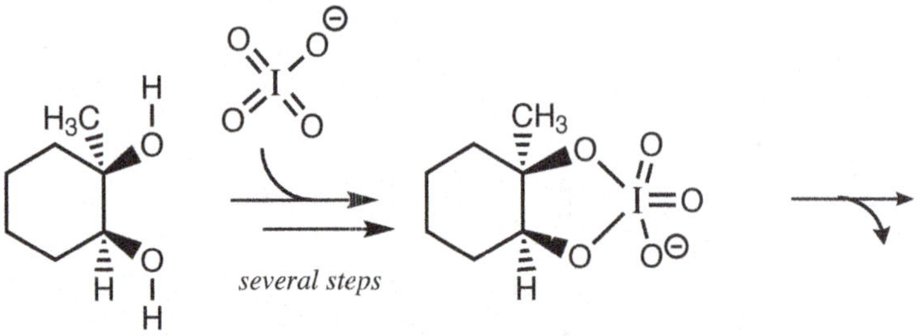

13 — Organometallic Reactions

The (transition) organometallic reactions in this chapter are of increasing importance. The reactions may recount the steps more than explain a reaction. Significantly absent is the concept of 18 valence electrons about the catalytic palladium atom.

Acyclic Heck Reaction

1. Step 1, reduction of palladium (II) to zero valent palladium with propene.

Step 2, the catalytic cycle (oxidative-addition, *syn*-addition, *syn*-elimination, and reductive-elimination) with 2-bromopropene and propene.

The alkene pi-electrons probably complex and are donated to the palladium before the syn addition takes place.

Cyclic Heck Reaction

2. Step 1, reduction of palladium (II) to zero valent palladium with cyclopentene.

AcO—Pd
         |
         OAc

syn addition →

syn elimination →

AcO

cont'd

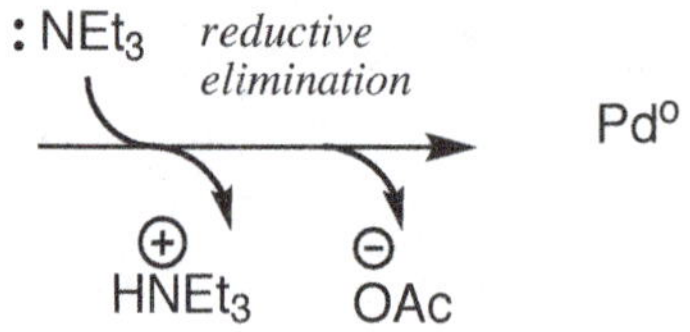

: NEt$_3$ *reductive elimination* →   Pd^o

⊕
HNEt$_3$

⊖
OAc

Step 2, the catalytic cycle (oxidative-addition, *syn*-addition, *syn*-elimination, and reductive-elimination) with iodobenzene and cyclopentene.

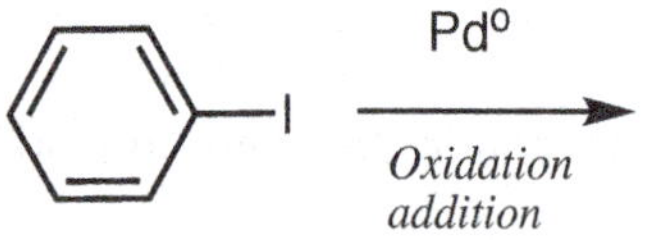

Pd^o
———→
Oxidation addition

—I

cont'd

—Pd
    |
    I

syn addition →

syn elimination →

: NEt$_3$ *reductive elimination* →   Pd^o

cont'd

⊕
HNEt$_3$

⊖
I

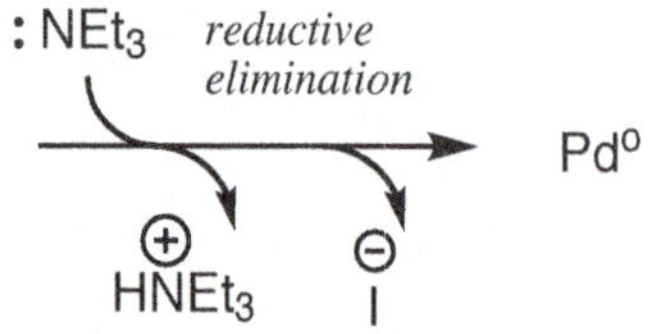

Catalytic Reduction of an Alkene

3. Catalytic hydrogenation of *cis*-3-hexene to hexane.

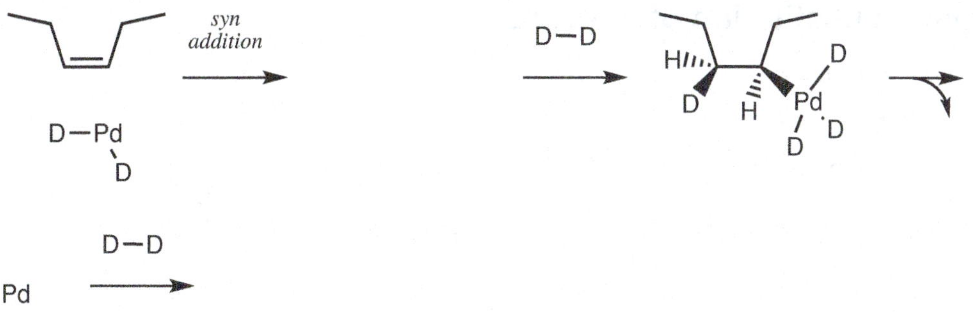

Gilman Reagent

4. Formation of Gilman reagent, lithium dimethylcuprate.

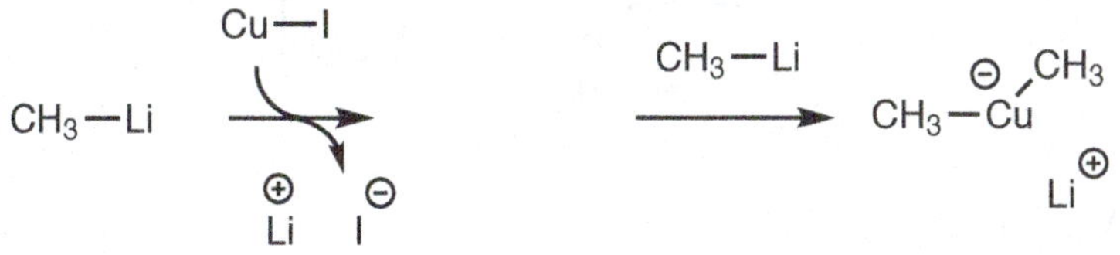

Coupling of Gilman reagent, lithium dimethylcuprate with (Z)-1-bromopent-1-ene to give *cis*-2-hexene.

(no mechanism)

5. A 1,4-conjugate addition of lithium dimethylcuprate to 2-cyclohexenone to give 3-methylcyclohexanone.

6. Reaction of lithium dimethylcuprate to benzoyl chloride to give acetophenone.

14 — Aromatic Substitution Reactions

Electrophilic Aromatic Substitution of Benzene

1. Friedel Crafts acylation of benzene.

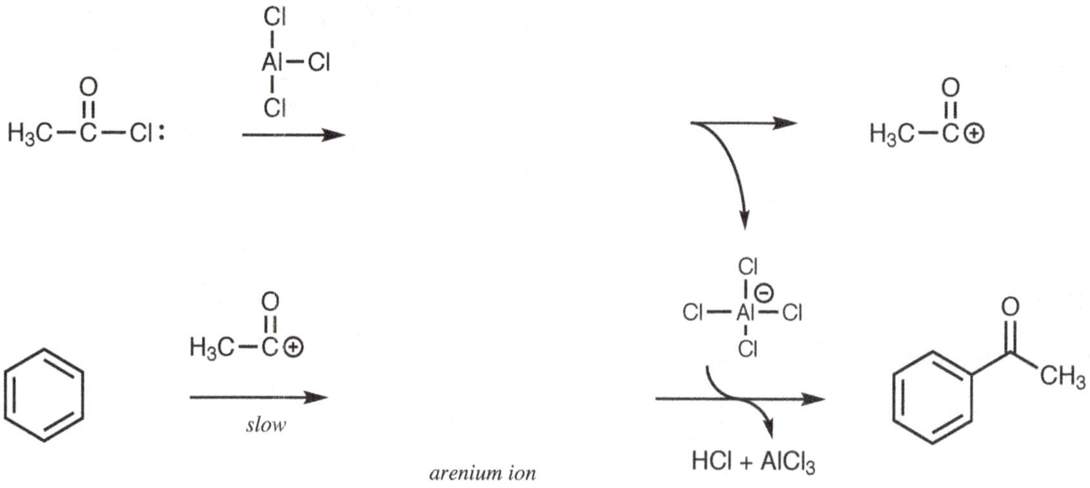

2. Friedel Crafts alkylation of benzene. (See *Notes*.)

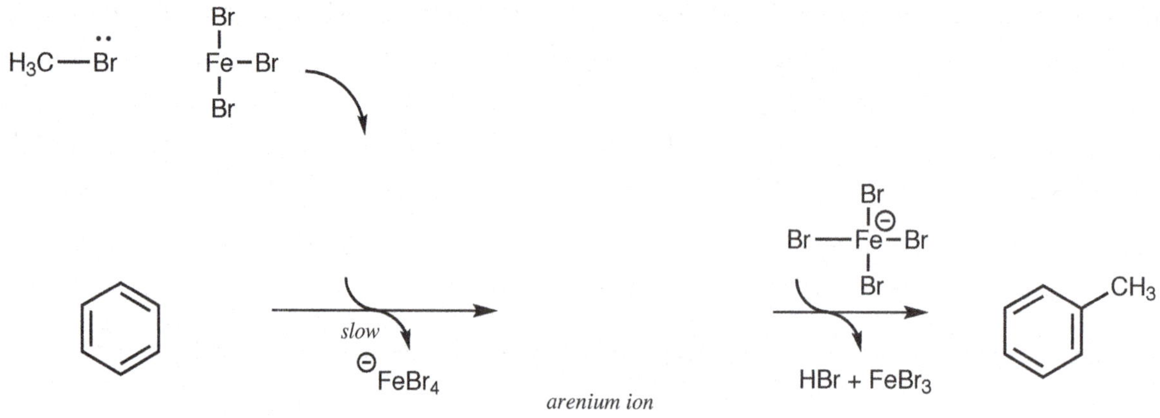

3. Ferric bromide bromination of benzene.

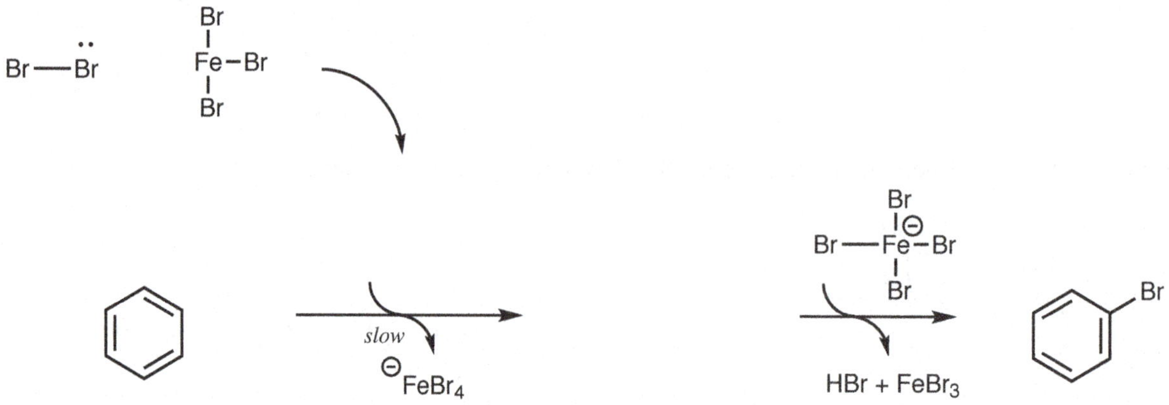

4. Nitration of benzene with nitric and sulfuric acids.

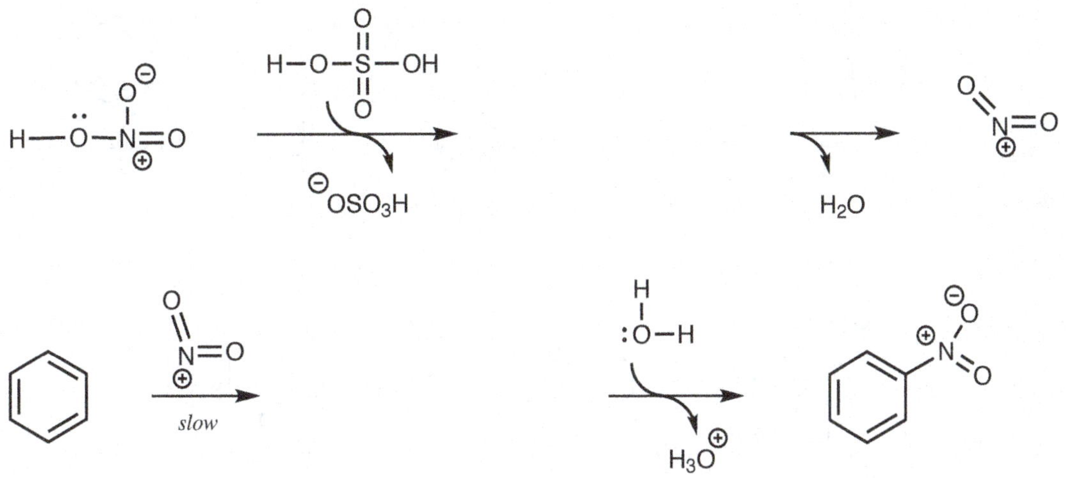

Electrophilic Substitution of Substituted Aromatic Compounds (See *Notes.*)

5. Bromination of acetophenone to give *m*-bromoacetophenone.

    For 5, 6, and 7, answer the question, "Will it react faster than benzene?"

6. Nitration of methyl *p*-chlorobenzoate to give methyl 4-chloro-3-nitrobenzoate.

7. Friedel-Crafts acylation of phenyl acetate with acetyl chloride and aluminum chloride.

plus ortho-isomer

8. Triflic acid catalyzed acetylation of toluene to give o- and p-methylacetophenone.

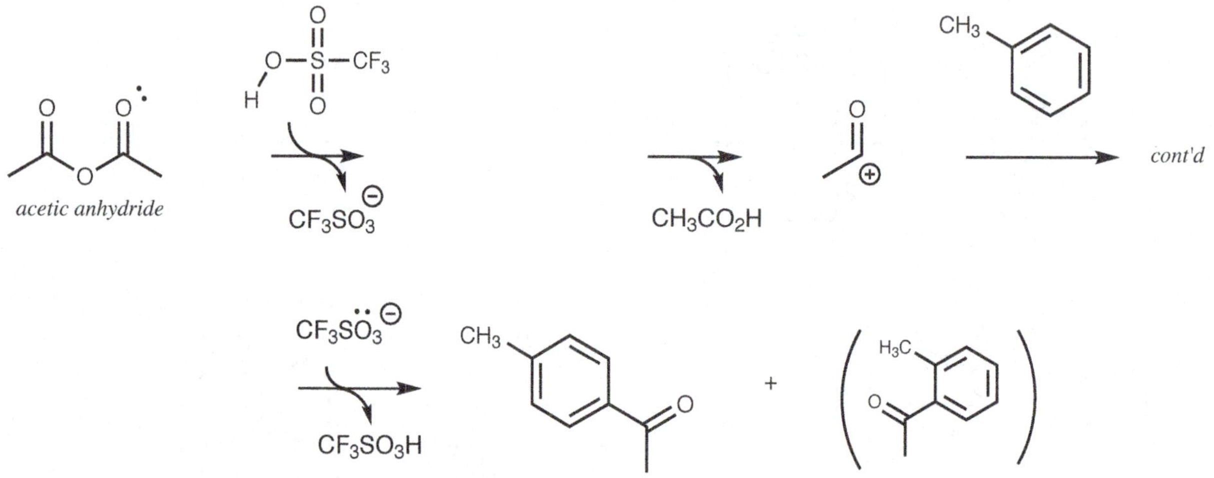

acetic anhydride

9. Aluminum chloride catalyzed chlorination of methyl 3-methoxybenzoate (methyl *m*-anisate) to give methyl 2-chloro-5-methoxybenzoate.

10. Friedel-Crafts alkylation of 4-nitro-*N*-*p*-tolylbenzamide with two equivalents of chloromethane.

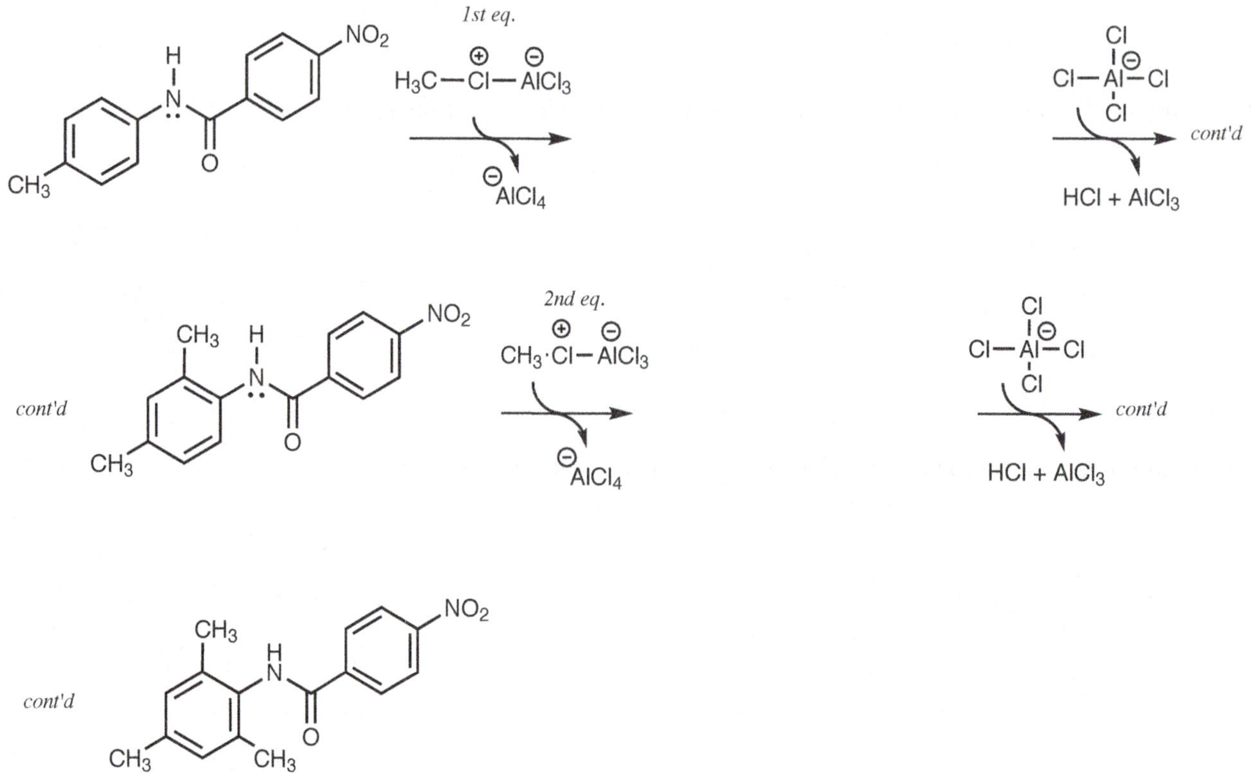

Nucleophilic Aromatic Substitution

11. Nucleophilic aromatic substitution of 1-fluoro-4-nitrobenzene with ammonia to give 4-nitroaniline.

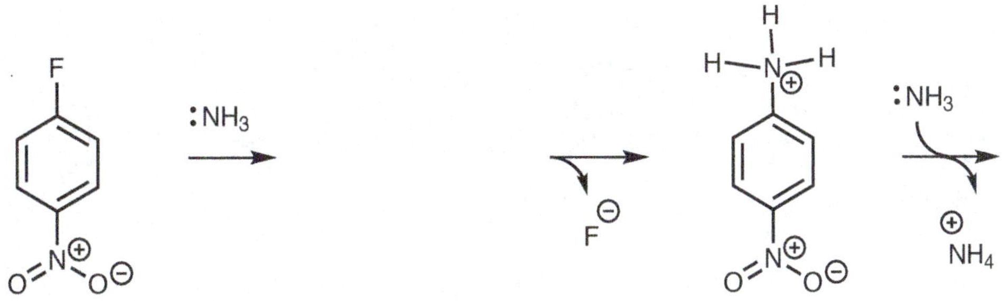

12. Nucleophilic aromatic substitution of 1,2-difluoro-4-nitrobenzene with sodium methoxide to give 2-fluoro-1-methoxy-4-nitrobenzene.

Benzyne Reaction

13. Reaction of 4-bromoanisole with sodium amide (sodamide, NaNH$_2$) to give 3- and 4-methoxyaniline via a benzyne intermediate.

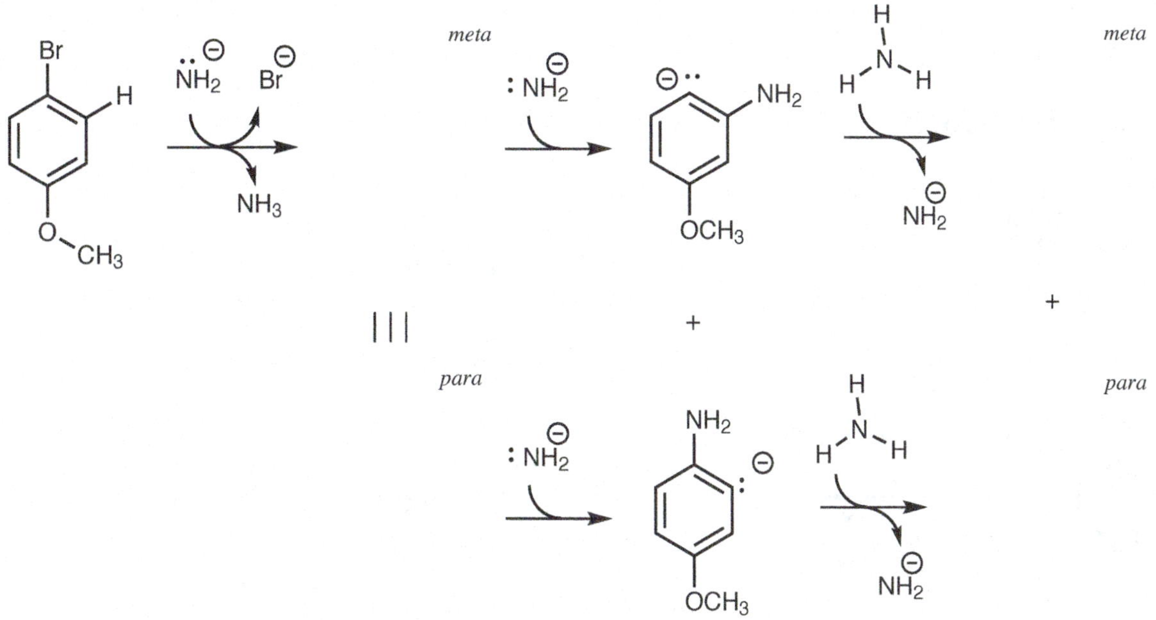

Diazonium Chemistry

14. Formation of a diazonium salt from aniline.

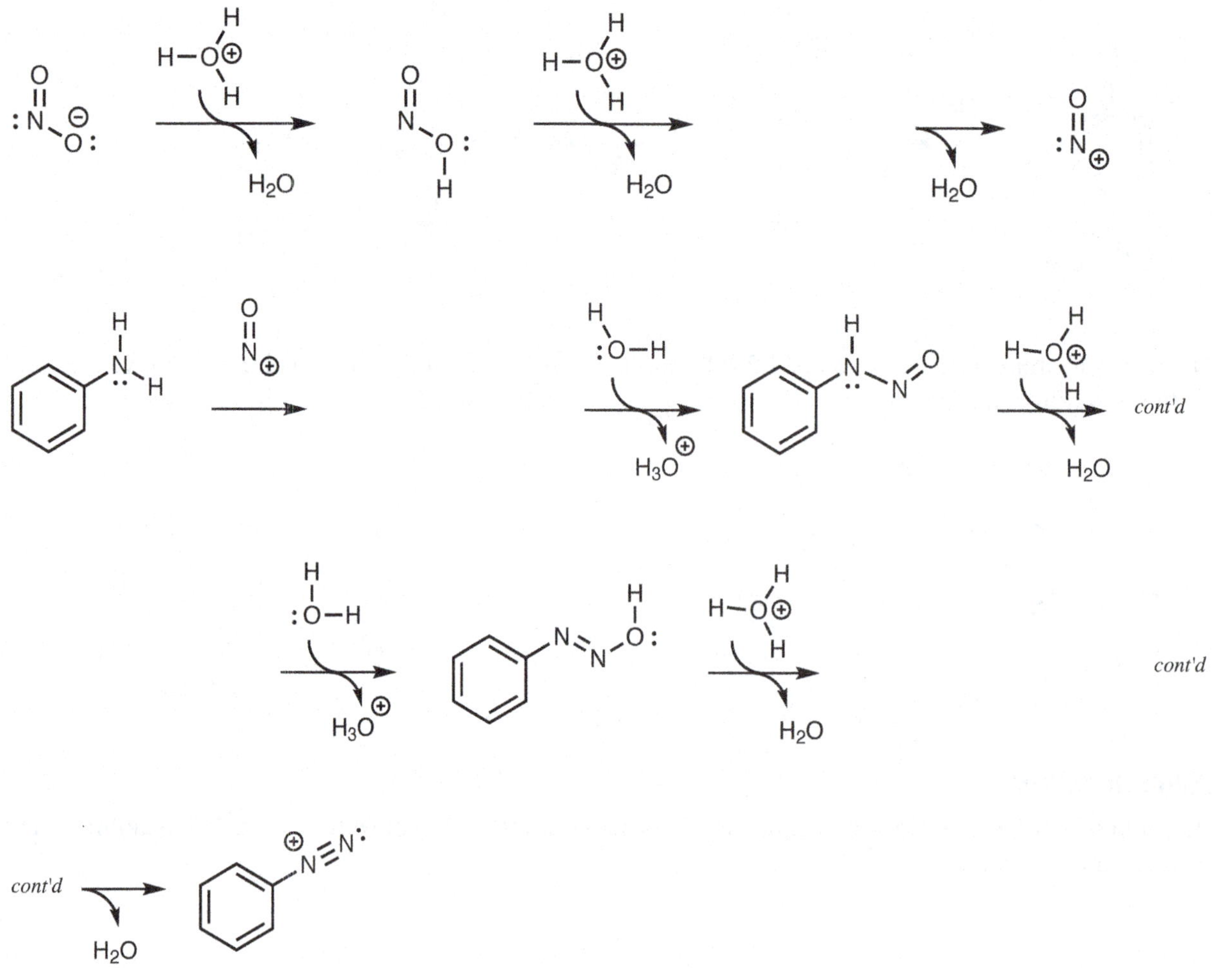

Reaction of Diazonium Salts

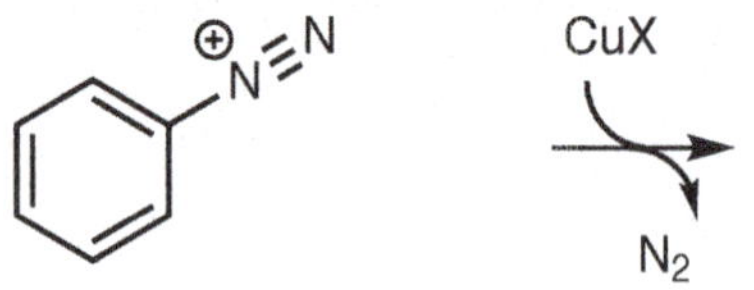

X = Cl, Br, CN, OH

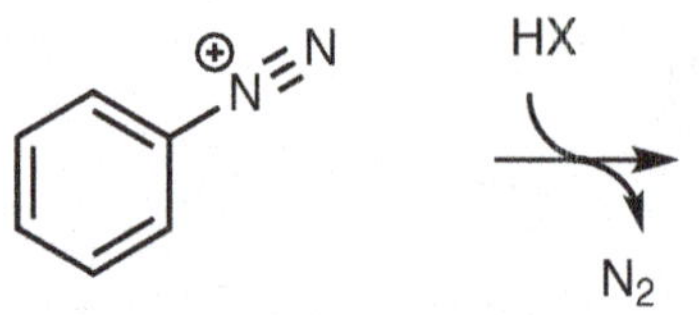

X = I, F, or OH

15 — Carbene and Nitrene Reactions

Carbene Reactions

1. Simmons-Smith carbene addition to cyclohexene to give a bicyclo[4.1.0]heptane. (See *Notes*.)

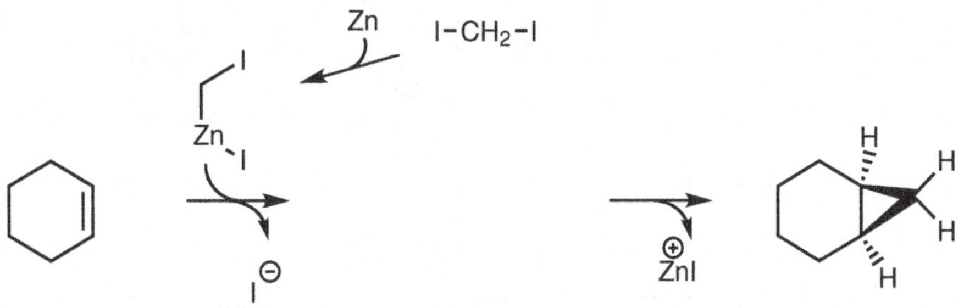

2. Dihalocarbene addition to (*E*)-1-phenylpropene to give a dibromocyclopropane.

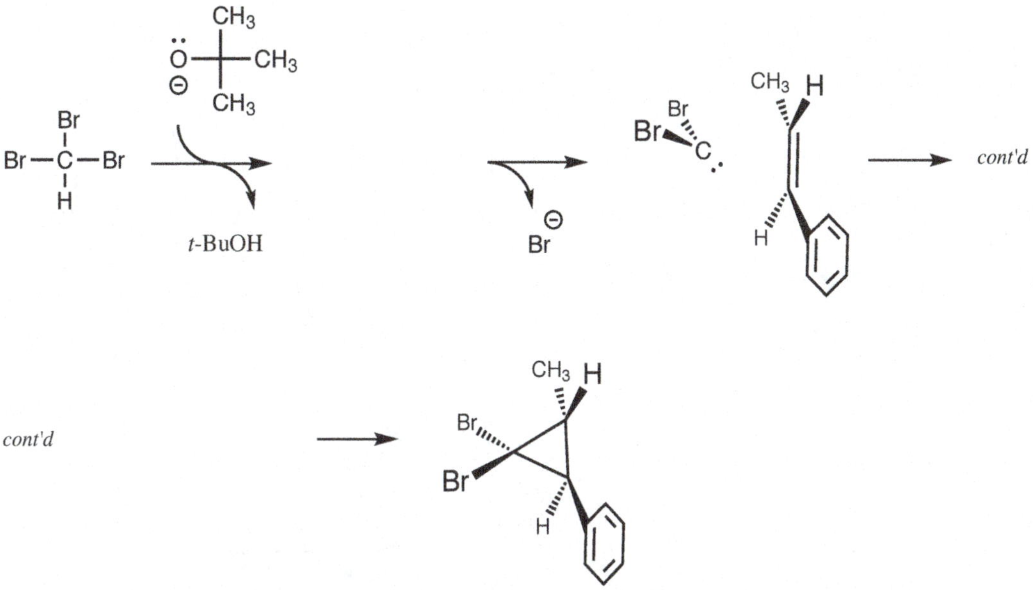

3. Dihalocarbene addition to cyclohexene to give 7,7-dichlorobicyclo[4.1.0]heptane.

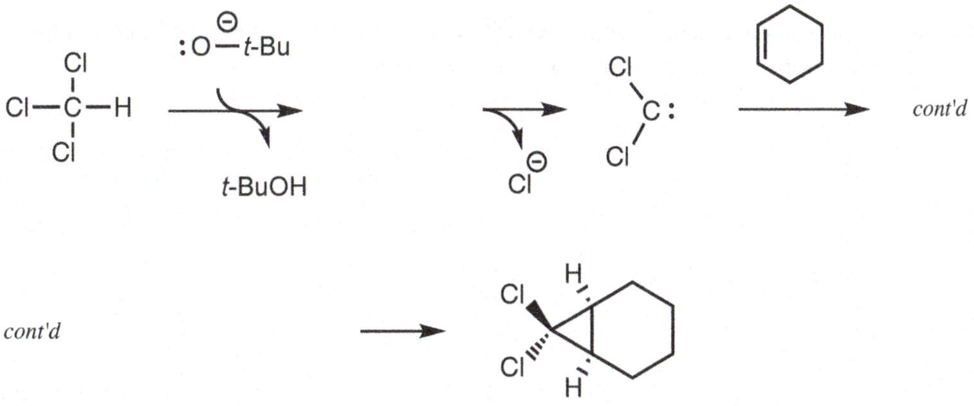

Curtius Rearrangement

4.   Step 1, Curtius rearrangement, reaction of acid chloride with azide and rearrangement via a nitrene intermediate.

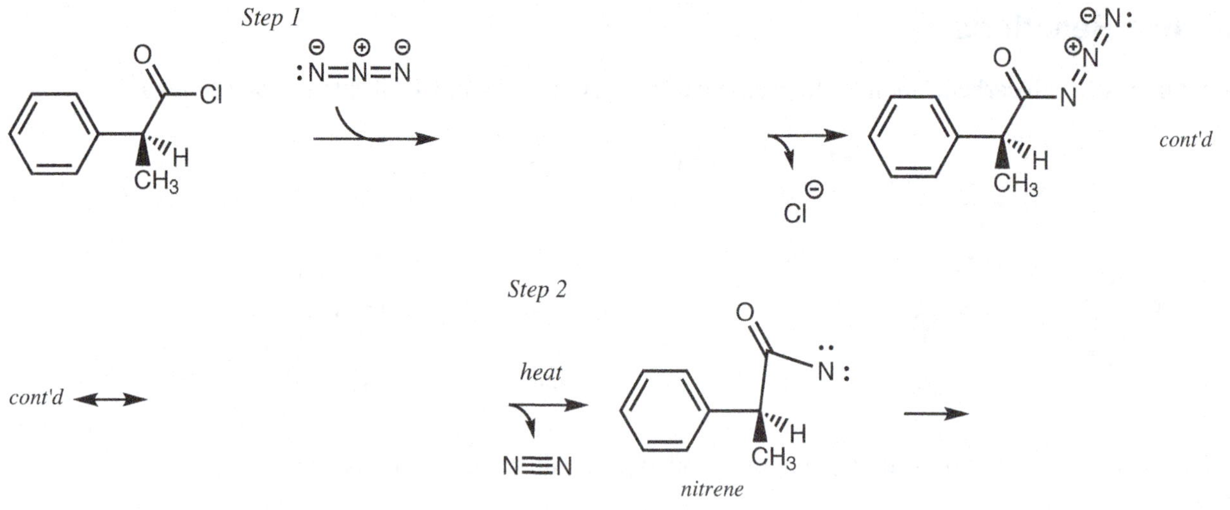

Step 1

cont'd

Step 2

heat

nitrene

cont'd ⟷

Step 2, hydrolysis of isocyanate to an amine.

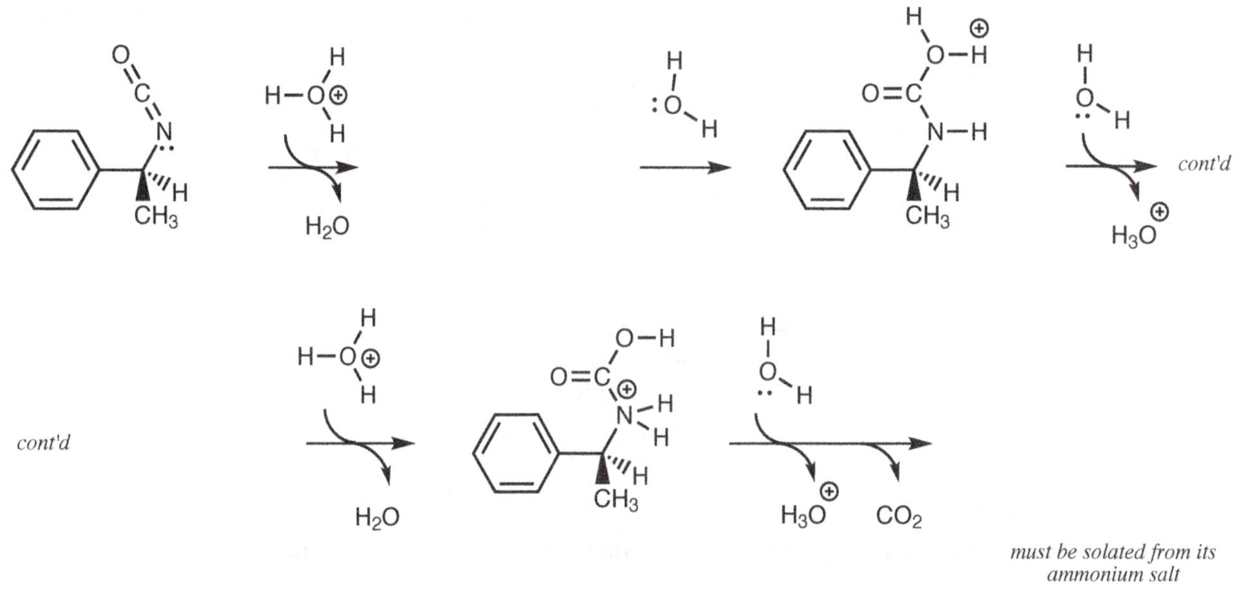

cont'd

cont'd

must be isolated from its ammonium salt

Alternate Step 2, hydrolysis of isocyanate to a carbamate. A carbamate is a nitrogen analog of a carbonate ester. The steps are similar to the above reaction with an alcohol replacing the water.

several steps

CH_3OH +   ⟶ $H-O-C-O-CH_3$

methyl hydrogen carbonate

CH_3OH +   ⟶

Ph

methyl phenylcarbamate

Hoffmann Rearrangement

5. Step 1, Hoffmann rearrangement of a primary amide to give the corresponding amine. The reaction of the amide with sodium hydroxide and bromine.

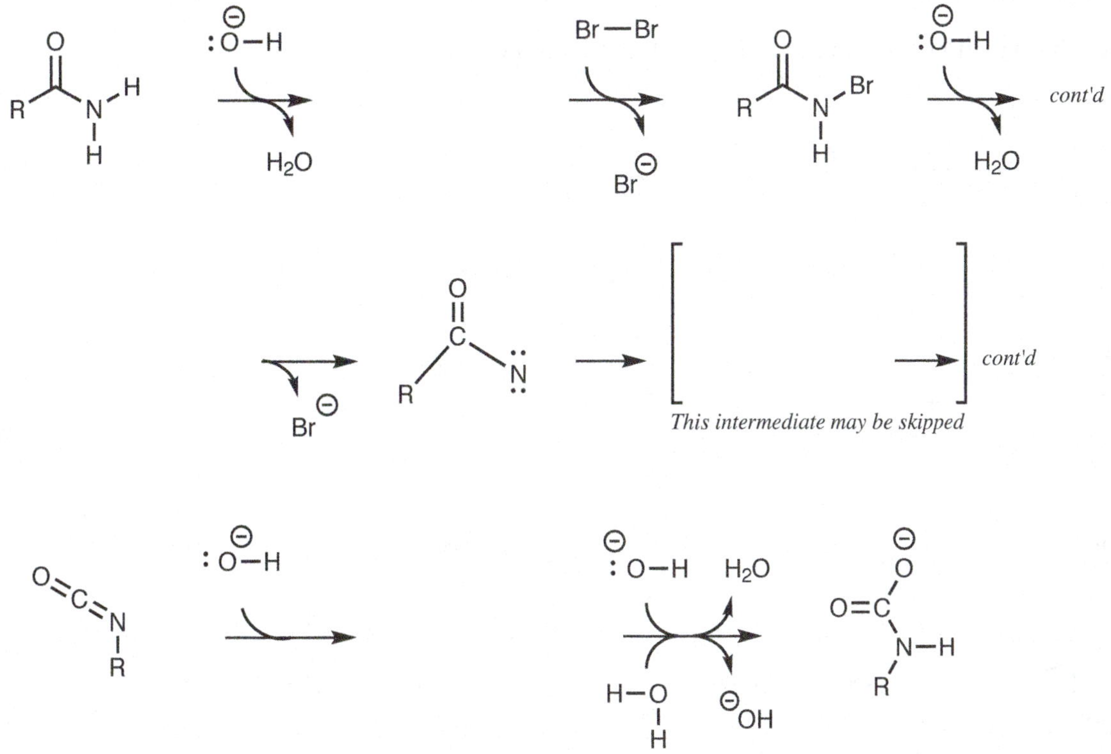
cont'd
cont'd
This intermediate may be skipped

Step 2, acidification and decarboxylation to give the amine.

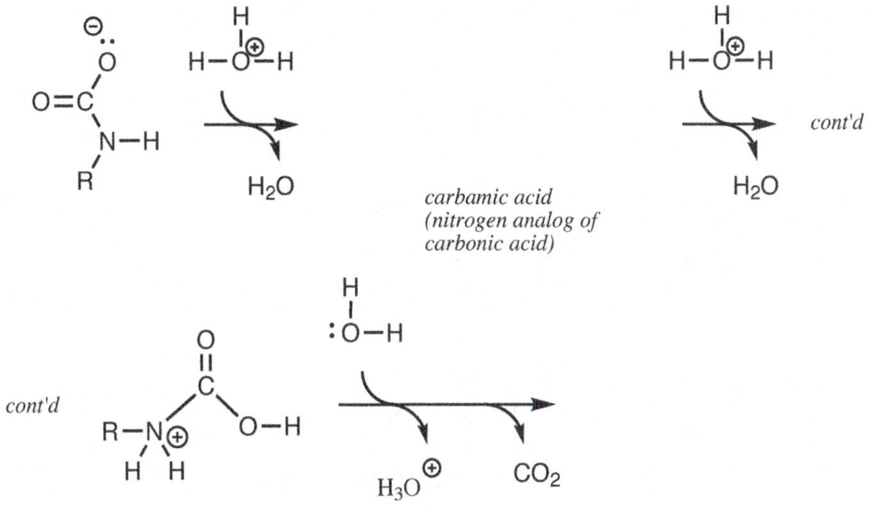
cont'd
carbamic acid
(nitrogen analog of
carbonic acid)
cont'd

16 — Radical Reactions

For discussion of radical reactions, see *Notes*.

Bromination (or Chlorination) Reactions

1. Free radical bromination of cyclohexane to give bromocyclohexane.

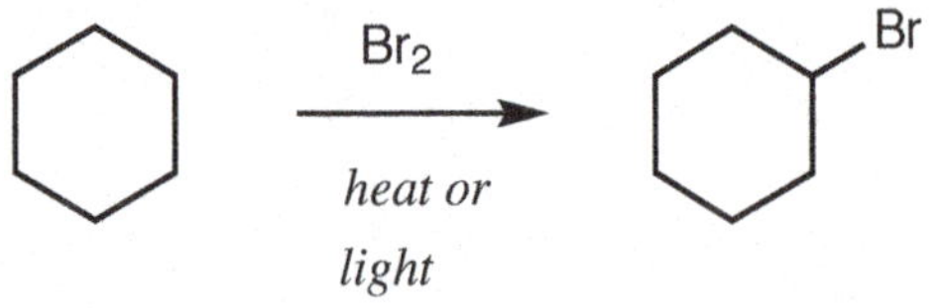

Initiation

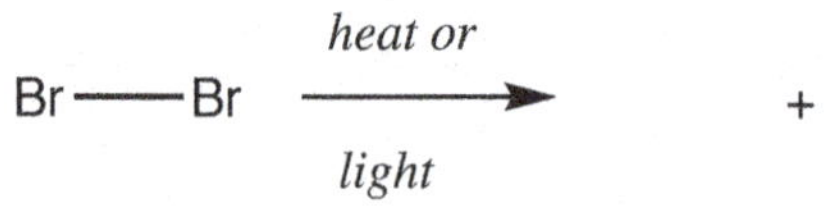

Propagation

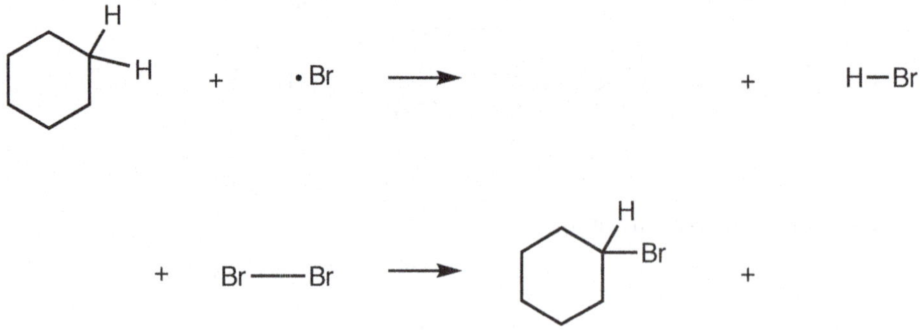

Termination

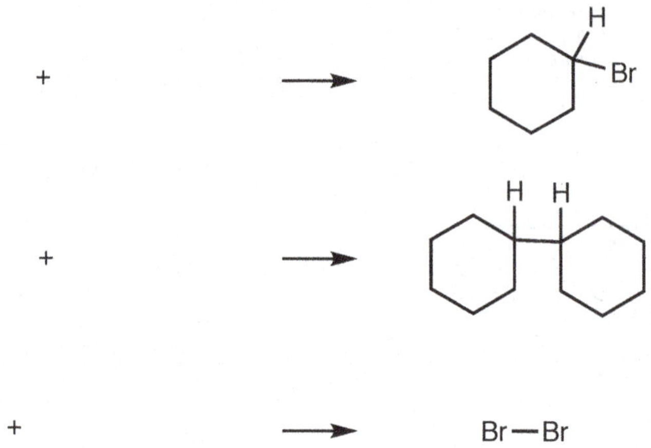

Allylic Bromination with NBS

2.   Free radical bromination of cyclohexene with *N*-bromosuccinimide, an allylic bromination.

Overall reaction

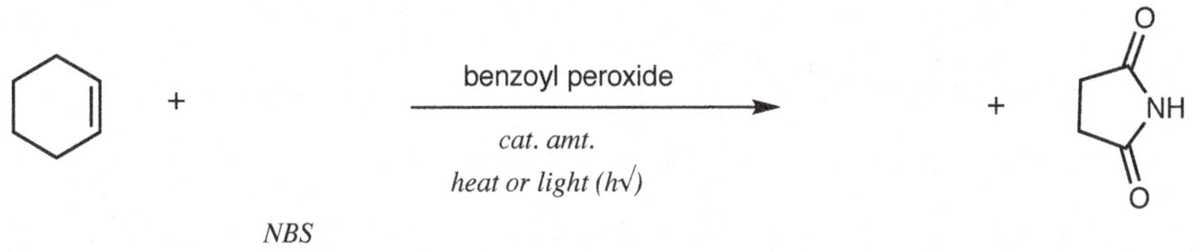

Initiation (See *Notes.*)

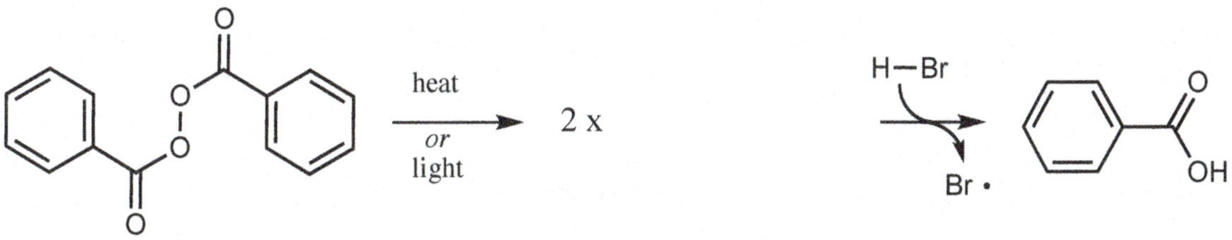

Propagation

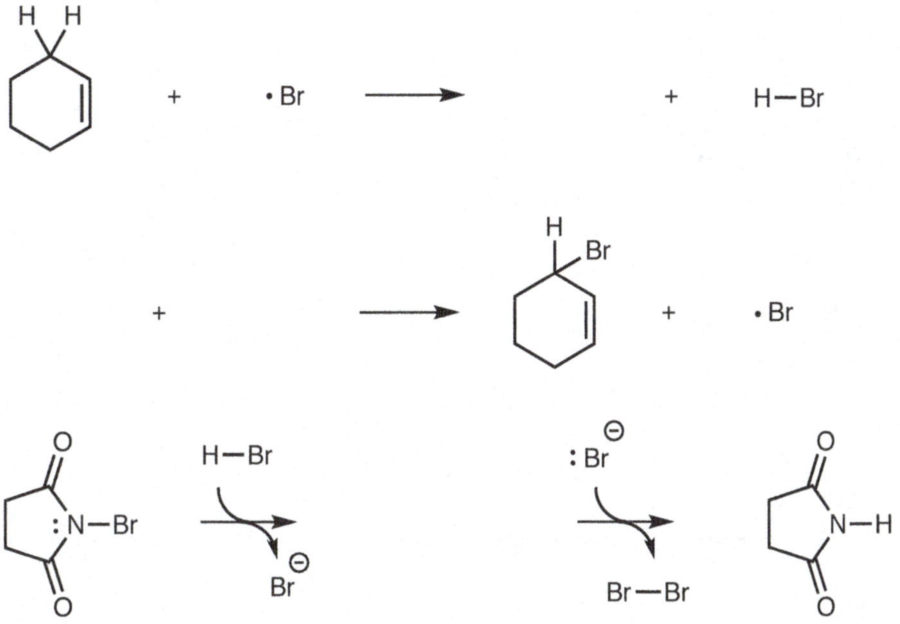

Termination

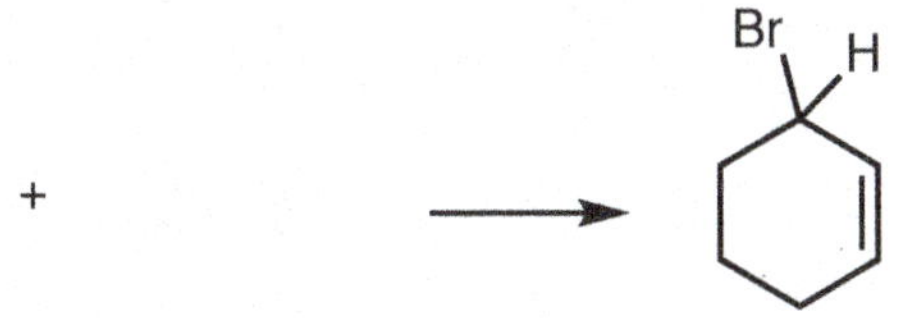

+ others

Radical Addition of Hydrogen Bromide

3. Free Radical Addition of HBr/H_2O_2 to propene to give 1-bromopropane, an *anti*-Markovnikov addition of HBr. (See *Notes.*)

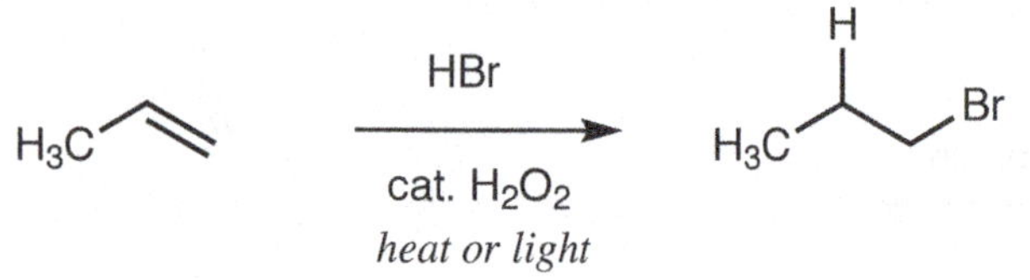

Initiation (See *Notes.*)

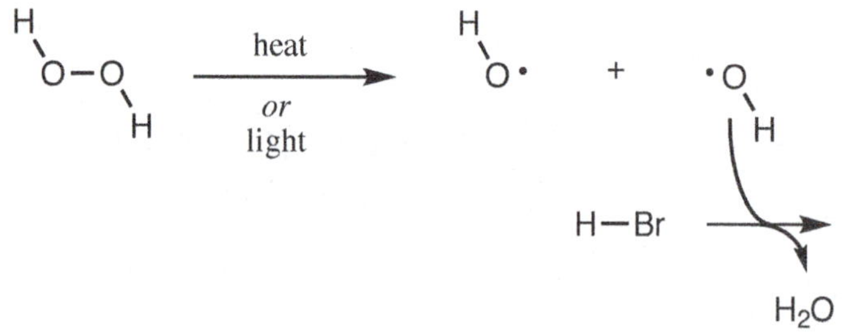

Propagation

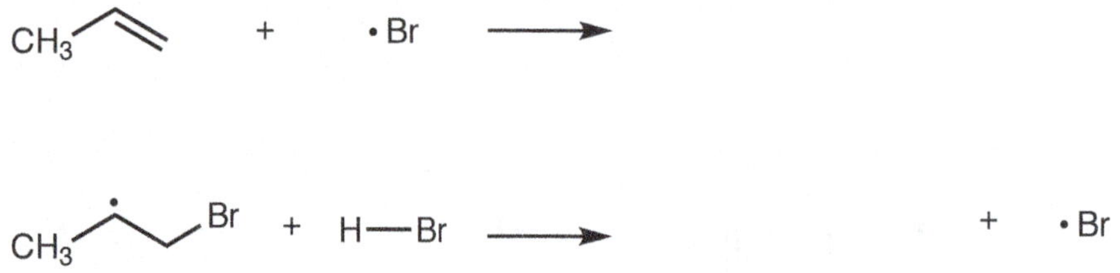

Termination

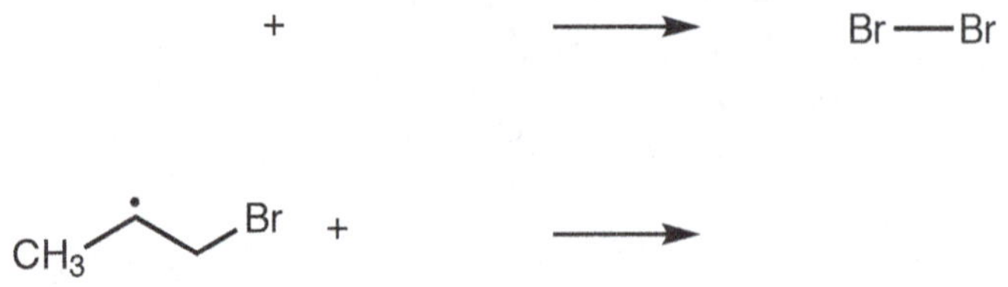

others

Benzylic Bromination with NBS

4. Benzylic bromination of ethylbenzene with NBS to give 1-bromo-1-phenylethane.

Overall reaction

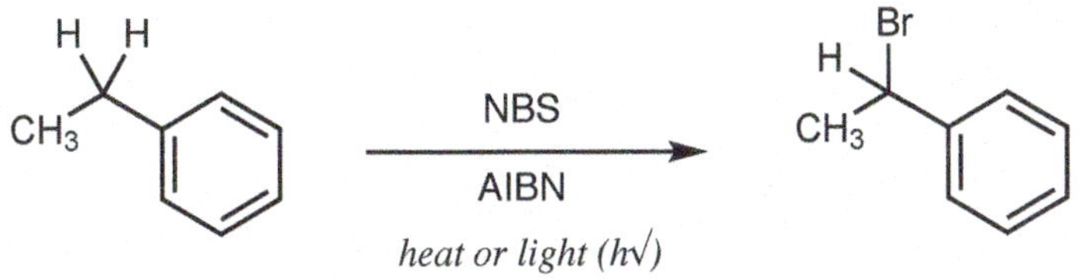

Initiation with azobisisobutyronitrile (AIBN)

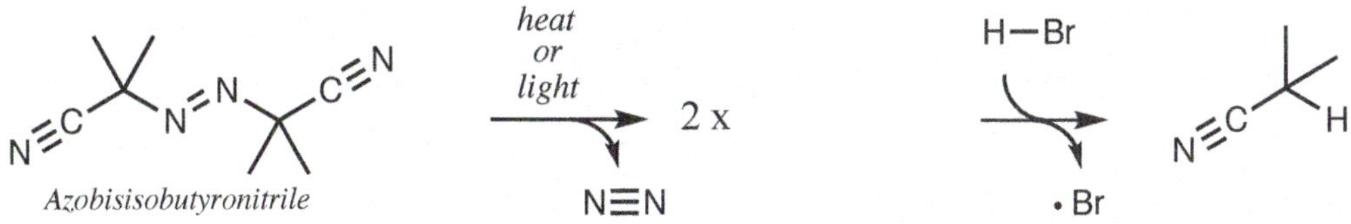

Propagation

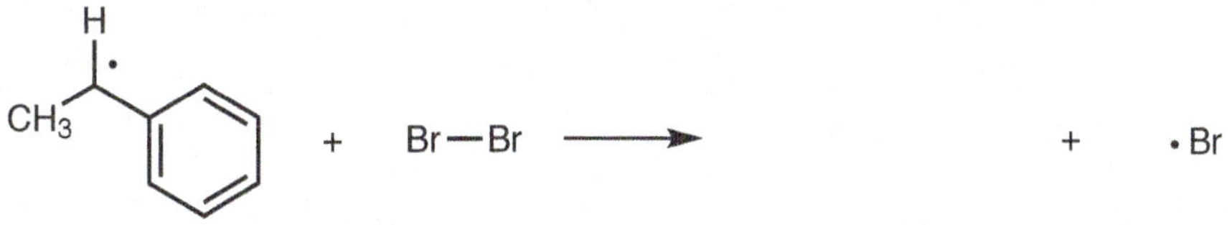

Termination

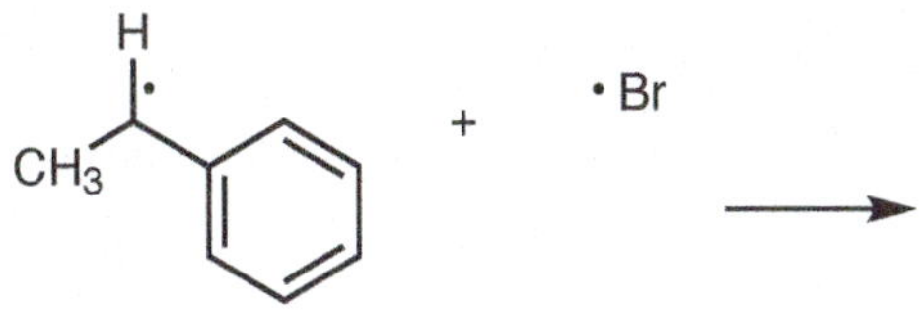

+ others

1 - Getting Ready for Reactions

Be sure to complete page 1 in *Part B* and *Part C* before proceeding with resonance structures.

About the Atom

Complete the following table by adding up the total number of protons and electrons. I hope you can learn to make some important predictions from this table. (Hint, the fastest way is to complete each row by adding or subtracting protons.)

	Charge	protons & electrons	Charge	Bond length	protons & electrons	Charge	protons & electrons
	+1		0			-1	
C	H—C—H (with H's and ⊕)	11p, 10e	H—C—H	1.10Å		H—C—H	
N	H—N—H (⊕)		H—N—H	1.01Å		H—N—H	
O	H—O—H (⊕)		H—O—H	0.96Å	10p, 10e	H—O:	
F	H—F—H (⊕)		H—F:	0.92Å		:F:	9p, 10e

1. What pattern do you note in this table, they all have the same number of _____?

2. For each column, how can the <u>compounds</u> with different elements have the same number of protons?

3. For the column with zero charge, which atom would you expect the valence electrons to be closest to the central nucleus, C, N, O, or F? Explain.

4. For the column with zero charge, why might the bond lengths decrease from CH_4 to HF? Explain.

5. Since the charge for a pair of electrons is the same, how do the electrons of oxygen and fluorine differ? Why might you expect HF to be a stronger acid than H_2O?

6. If you had two nitrogen atoms of different base strengths (ability to attract a proton), then which electrons would you predict are held closer to the nitrogen nucleus (to the more basic or less basic)?*

---

*This is the less common carbocation obtained by protonation of methane. CH_3^+ is more common, but does not fit the principle of the table because it has lost a proton and two electrons.

Guide to Drawing Resonance Structures

Resonance Structures

If you are unfamiliar with the use of the curved arrow, refer to the discussion in the *Notes* section. A good place to start pushing electrons is in drawing resonance structures. They have the elements of electron movement, but the problems will be more limited in scope.

Resonance Structures of Anions

Add the missing curved arrows for the following resonance structures. For 1-6, the first and last structures are the same. In that case, you are converting it back to the starting structure.

1.

2.

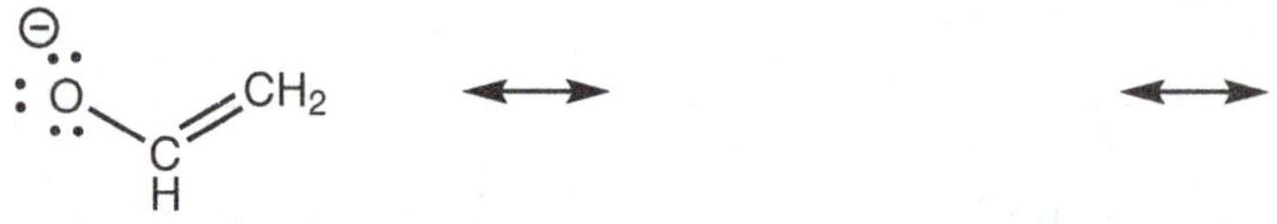

3.

4.

5.

6.

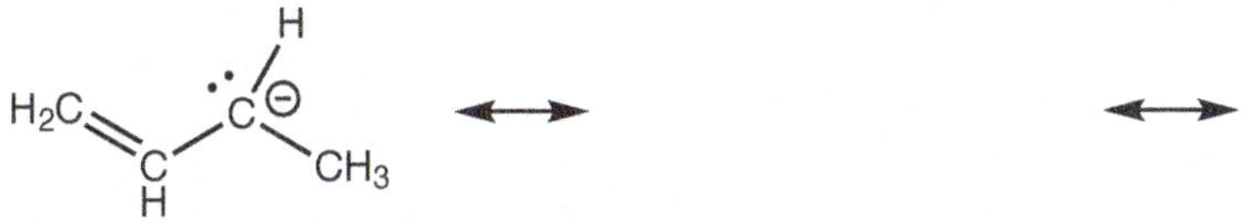

7.

8.

These examples show how the electrons can move, where the resulting charge will form, and how the charge can be distributed. It does not tell you on which atoms the greater charge density might exist nor upon which atom a reaction might next occur. You will note that since the original structure had a negative charge, the only charge that exists on any of the resulting resonance structures is a negative charge.

Resonance Structures of Cations

Conversely, if there is a positive charge, it is the positive charge that will attract (pull) electrons. Start a curved arrow from a neighboring pi or pair of non-bonded electrons and bring it toward the positive charge. You will note that only one curved arrow is necessary to create a new resonance structure. Because the original structure is a cation, completing its octet should not require further electron movement.

Add curved arrows to the following structures. For 9 and 10, the first and last structures are the same. In that case, you are converting it back to the starting structure.

9.

10.

11.

12.

13.

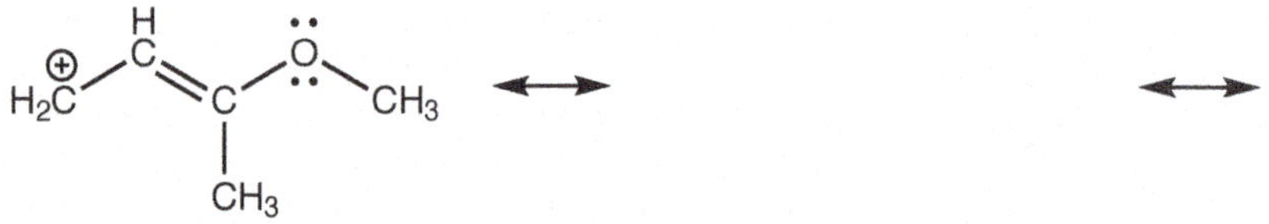

Again, these examples show how the electrons can move, where the resulting charge will form, and how the charge can be distributed. It does not tell you on which atoms the greater charge density might exist nor upon which atom a reaction might next occur. You will note that since the original structure had a positive charge, the only charges that exist on any of the resulting resonance structures are positive charges.

Resonance Structures of Neutral Compounds with Non-Bonded Electrons

If there isn't a charge and there are adjacent non-bonded electrons, then it will be the non-bonded electrons that will move (push) toward a neighboring pi bond. Start a curved arrow with the non-bonded electrons and direct them to the neighboring double bond.

Add curved arrows to the following structures. For 14-17, the first and last structures are the same. In that case, you are converting it back to the starting structure.

14.

15.

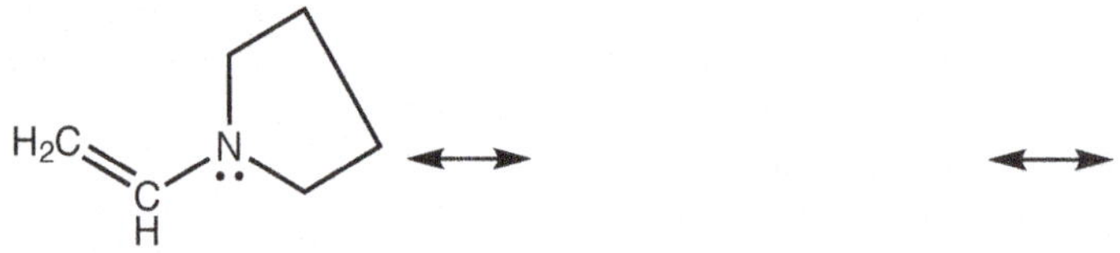

16.

17.

18.

19.

Again, these examples show how the electrons can move, where the resulting charges will form, and how the charges can be distributed. It does not tell you on which atoms the greater charge densities might exist nor upon which atoms a reaction might next occur. You will note that since the original structure was neutral, the net charges that exist on any of the resulting resonance structures are also neutral and only two atoms have a charge.

Resonance Structures of Neutral Compounds without Non-Bonded Electrons

What if there is no charge and there are no neighboring non-bonded electrons? Then we will push the pi electrons of a double bond toward the least substituted carbon or the most electron withdrawing atom, see Example 26. Start a curved arrow with the pi electrons of a double bond and direct them from the most to the least substituted carbon or for a C=O bond, toward the oxygen atom.

20.

21.

This is the same example as above, however the arrows are pointing in the opposite direction. If you are uncertain in which direction the electrons might move, a good strategy is to draw an arrow in the opposite direction and then to compare the results of the two possibilities. Compare the result below with the one above. Which is the more stable? If you do not recognize the lower arrangement as a lesser contributor, you may need to refer to your textbook for the rules of carbanion, carbocation, and resonance stability. Examples 23 and 24 are similar, which is more stable?

22.

23.

24. Draw a difference resonance structure than Example 23.

25.

26.

In these examples, we have incorporated chemical principles. The carbocations that are the most substituted are the most stable. Coinciding with this principle is that carbanions with the least substitution (or a heteroatom, last example) are the most stable.

Resonance Structures of Radicals

Radicals, compounds with unpaired electrons, are less stable than those with paired electrons. The fate of radical reactions is to form a paired-electron bond. However, sharing unpaired electrons with neighboring non-bonded electrons or pi-bonds can attain added stability. Also note the curved arrow has a single barb indicating the movement of a single electron. Two arrows are required for a pair of electrons.

27.

28.

2 — Acid-Base Chemistry

Bronsted-Lowrey Acids and Bases

Acid-base reactions are often the first intermolecular reaction you will encounter. A proton will be exchanged from the strongest acid to the strongest base.

In these examples, you must note the conjugate acid and conjugate base that result in each reaction. You should note the use and meaning of the curved arrows. If the example does not contain a curved arrow, you must supply one. The meaning of the curved arrow is important for you to understand.

- The rule for predicting the product of an acid-base reaction is simple. A reaction will generally give the product that is the weakest base (or conjugate base). The base strength of a compound is related to the acidity of the acid, the stronger the acid, the weaker the base, or the corollary, the weaker the acid, the stronger the base. In order to compare the base strengths, the acidities of the acid and conjugate acid must be determined first.

- Look at each example and write the pK_a under each acid (on the left) and conjugate acid (on the right). Be careful that you correctly identify the acid and the corresponding pK_a. Strong acids have a small pK_a and weak acids have a large pK_a. You may need to use a table to find some values.

- For each acid (or conjugate acid), write the corresponding pK_a beneath it. Write the label "B(ase)" (or "CB") under each base (or conjugate base). Look at the pK_a of the acid or conjugate acid. The strongest acid corresponds with the weakest base. Label it, "weakest base". In example 1, the pK_a of HF is 3.2. It is placed under HF. The pK_a of acetic acid is 4.75 and 4.75 is placed under it. Because HF is the stronger acid (lowest pK_a), its conjugate base will be the weakest base. The equilibrium will shift to the right.

- Label the equilibrium of each reaction, L(eft) or R(ight). Example 1, R.

1. What is the weakest base?

| Base | Acid | Conjugate Acid | Conjugate Base |

L/**R**   3.2   4.75

Write sentences to describe electron movements.

2. Label acids, bases, and conjugate acids and bases.

L/**R**   base   3.2   15.7   conj. base

Write sentences to describe electron movements.

Continue by completing the equation, adding curved arrows, pK$_a$ values, and direction of equilibrium.

3. Write a sentence(s) describing any bonds being made or broken.

    H—Ö—H   +   H—F   ⇌   +

    L/R   base   3.2   -1.7   conj. base

4. Write a sentence(s) describing any bonds being made or broken.

    H—S̈—H   +   H—N̈—H   ⇌   +
                                    |
                                    H

    L/**R**   7.0   9.2   conj. base

5. Write a sentence(s) describing any bonds being made or broken.

                        ⊖
    H—F   +   Cl   ⇌   +

    L/R   3.2   conj. Base   -8

6. Write a sentence(s) describing any bonds being made or broken.

    ⟨⟩—O—H   +   ⊖:Ö—H   ⇌   +

    L/R   10.0   conj. Base   15.7

7. Write a sentence(s) describing any bonds being made or broken.

            O
            ‖
    H₃C—C—O—H   +   H₃C—N̈—H   ⇌   +
                                    |
                                    H

    L/R   4.75   conj. Base   10.5

8. Write a sentence(s) describing any bonds being made or broken.

    CH₃-CH₂-O—H   +   CH₃—N̈—H   ⇌   +
                                    |
                                    H

    L/R   16.0   conj. Base   10.5

Continue by completing the equation, adding curved arrows, pK$_a$ values, and direction of equilibrium.

9.

$$H_3C-\overset{\overset{O}{\|}}{C}-\overset{\overset{H}{|}}{\underset{H}{C}}-H \quad + \quad \overset{\ominus}{:}\ddot{O}-\overset{\overset{CH_3}{|}}{\underset{CH_3}{C}}-CH_3 \quad \rightleftharpoons \qquad\qquad +$$

L/R   22   conj. base   19

10.

$$H-\overset{\overset{H}{|}}{\underset{H}{C}}\overset{\ominus}{:} \quad + \quad \overset{\overset{H}{|}}{\underset{H}{N}}-H \quad \rightleftharpoons \qquad\qquad +$$

L/R   38   50   conj. base

11.

$$H-\ddot{O}-H \quad + \quad H-\overset{\overset{H}{|}\oplus}{\underset{H}{N}}-H \quad \rightleftharpoons \qquad\qquad +$$

L/R   9.2   -1.7   conj. Base

12.

$$H-\overset{\overset{\cdot\cdot}{}}{\underset{H}{N}}-H \quad + \quad Ph-\overset{\overset{H}{|}\oplus}{\underset{H}{N}}-H \quad \rightleftharpoons \qquad\qquad +$$

L/R   4.6   9.2   conj. Base

13.

$$H_3C-\overset{\overset{O}{\|}}{C}-\ddot{O}-H \quad + \quad \overset{\ominus}{:}\ddot{O}-H \quad \rightleftharpoons \qquad\qquad +$$

L/R   4.75   conj. Base   15.7

14.

$$H_3C-C\equiv C-H \quad + \quad :\overset{\ominus}{C}H_3 \quad \rightleftharpoons \qquad\qquad +$$

L/R   24   conj. Base   50

Continue by completing the equation, adding curved arrows, pK$_a$ values, bases, and label the direction of equilibrium.

15.

L/R   24   16

16.

L/R   11.7   16

17. Sometimes we may be unsure how a reaction might proceed. Will H$_2$S and (CH$_3$)$_2$NH react together? What will the products be if they do? In that case, draw ALL of the possible products and analyze the results. Complete this problem as before.

From the individual equilibria, can you predict the overall result?

                        38   <-5

            7.0   10.7

18. What is the equilibrium between HOCH$_3$ and CH$_3$NHCH$_3$? First, determine the individual equilibria and then predict the overall result.

                        38   -2.2

            15.5   10.7

Lewis Acids and Bases

The prior exercise showed a hydrogen atom accepting electrons. With Lewis acids, other atoms can also accept electrons. See your text for further discussion.

Add structures, non-bonded electrons, curved arrows, and formal charges as needed to complete the following.

19. H₃C—N(CH₃)(H₃C) + H—B(H)—H   ⇌

20. Br—B(Br)(Br) + :O:(CH₃)(CH₃)   ⇌

21. Cl—Cl + Cl—Fe(Cl)(Cl)—Cl   ⇌   ⇌   +

22. Br—Br + Br—B(Br)—Br   ⇌

23.   +   ⇌   H₃C—N⁺(CH₃)(CH₃)—Al⁻(Cl)(Cl)—Cl

24. H—C(H)(H)—C(=O)—Cl + Cl—Al(Cl)—Cl   ⇌   ⇌   +

25. F—B(F)(F)—F + :O:(CH₂CH₃)(CH₂CH₃)   ⇌

26. H₃C—C(H)=C(H)—H + Hg⁺—Cl   ⇌

3 — Substitution Reactions

S_N2 Substitution Reactions

Add curved arrows to the following reactions.

1. An S_N2 reaction of 1-chlorobutane with sodium iodide. (See *Notes*.)

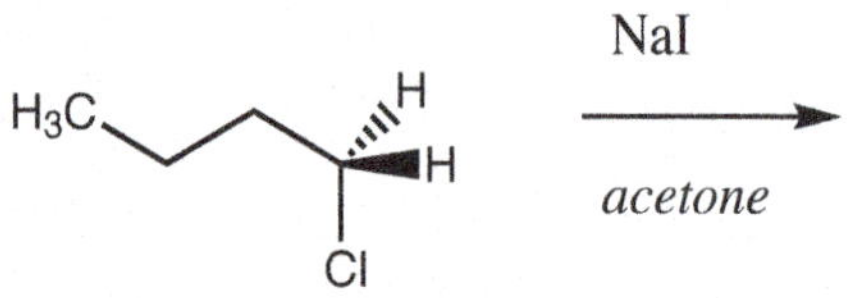

2. An S_N2 reaction of 1-bromobutane with ethoxide.

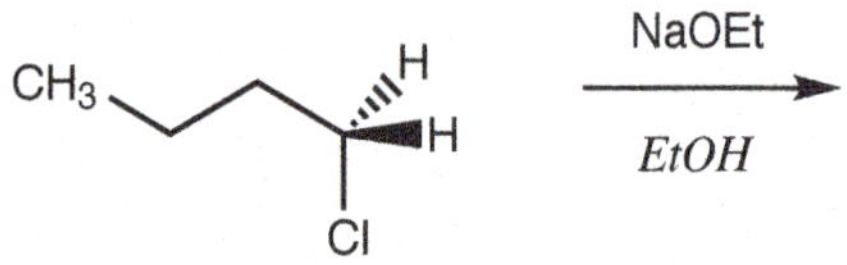

3. An S_N2 reaction of *(R)*-2-bromobutane with sodium thiocyanate. (See *Notes*.)

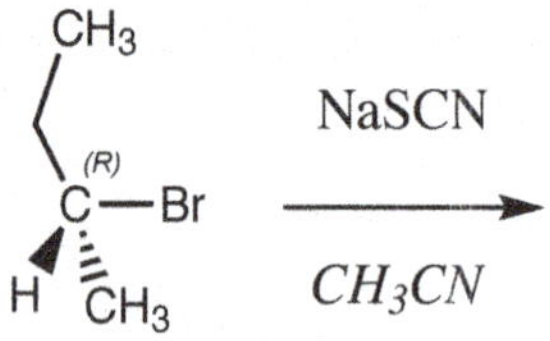

4. An S_N2 reaction of *(S)*-2-bromobutane with sodium acetate.

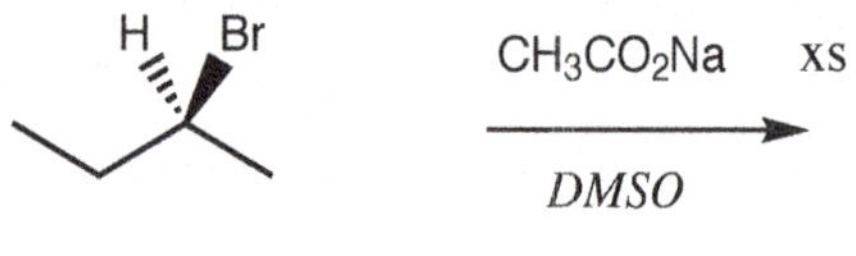

5. An S_N2 reaction of a triflate with potassium cyanide. (See *Notes*.)

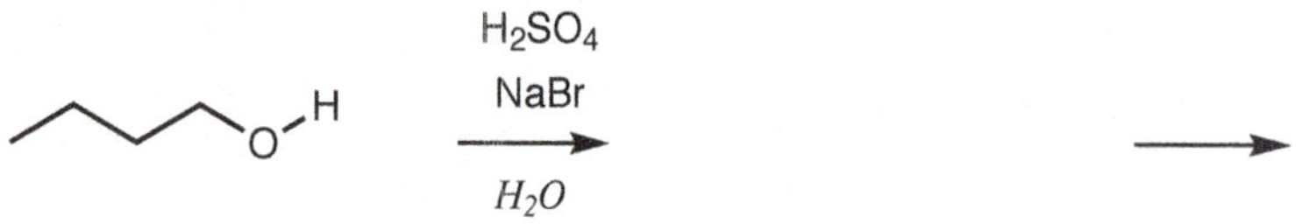

6. An S$_N$2 reaction of a triflate with sodium methanethiolate. (See *Notes*.)

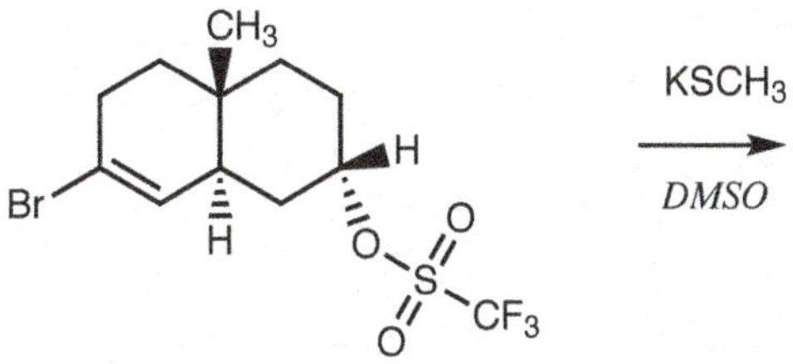

7. An S$_N$2 reaction of *cis*-1-chloro-4-methylcyclohexane with sodium azide.

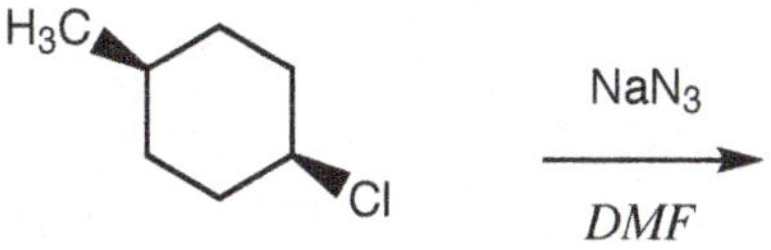

8. An S$_N$2 reaction of a ditosylate with one equivalent of potassium cyanide. (See *Notes*.)

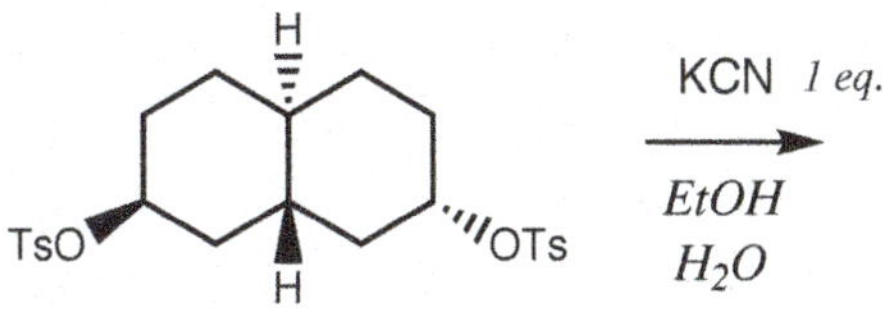

9. An S$_N$2 reaction of 1-bromo-3-chloropropane with potassium cyanide. (See *Notes*.)

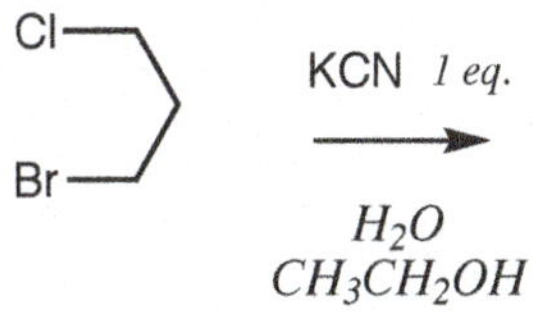

10. An S$_N$2 reaction of 1-bromobutane with acetylide anion. (See *Notes*.)

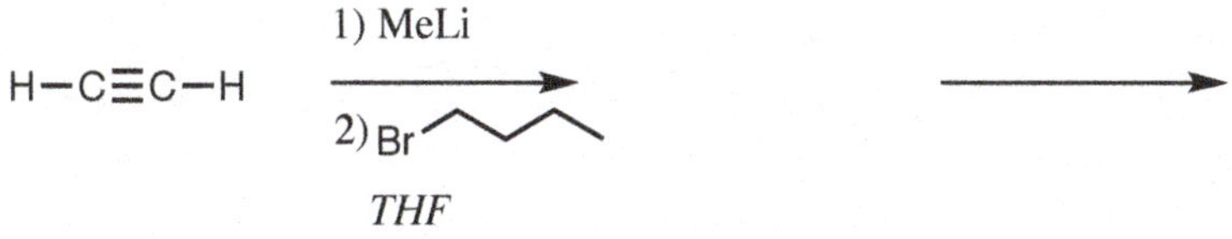

11. An S_N2 reaction of methyl phenylacetate enolate with iodopropane. (See *Notes.*)

$$CH_3CH_2CH_2\text{-}I$$

$$\xrightarrow{}$$

THF

12. A Gabriel amine synthesis with formation of phthalimide anion (potassium carbonate) and alkylation with 1-bromo-2-butene. (See *Notes.*)

$$K_2CO_3$$
DMF
$$\xrightarrow{}$$
$$CH_2CH=CHCH_2Br$$

$$\xrightarrow{}$$

13. An S_N2 reaction of bromomethane with phenoxide. (See *Notes.*)

1. $NaOCH_3$
   CH₃OH
$$\xrightarrow{}$$
2. CH_3Br
   CH₃OH

$$\xrightarrow{}$$

14. An S_N2 reaction of benzyl bromide with sodium ethoxide.

$$NaOCH_2CH_3$$
$$\xrightarrow{}$$
EtOH

15. An S_N2 reaction of *p*-nitroanisole with hydrogen iodide. (See *Notes.*)

$$HI$$
$$\xrightarrow{}$$
HOAc

$$\xrightarrow{}$$

16. An S$_N$2 reaction of 1-bromo-2-butanol with sodium hydroxide.

$$\xrightarrow[\text{H}_2\text{O}]{\text{NaOH}}$$

17. An S$_N$2 reaction of the anion of 1-butyne with *(R)*-2-ethyloxirane. (See *Notes.*)

$$\text{CH}_3\text{CH}_2-\text{C}\equiv\text{C}-\text{H} \xrightarrow[\text{THF}]{\text{CH}_3-\text{Li}}$$

+   CH$_3$ ⋯⋯ ⟶ *cont'd*

Work up

$$\xrightarrow[\text{H}_2\text{O}]{\text{HCl}}$$

18. An acid catalyzed opening of 2-ethyloxirane with methanol. (See *Notes.*)

$$\xrightarrow[\substack{\text{CH}_3\text{OH} \\ \text{CH}_3\text{OH}}]{\text{HCl}}$$

⟶   ⟶ *cont'd*

cont'd

19. An S$_N$2 reaction of the enolate of acetophenone with *(R)*-2-methyloxirane (propylene oxide). (See *Notes.*)

Step one   *Step two*

1) [epoxide with CH$_3$ and H]

2) HCl, H$_2$O   ⟶

S_N1 Substitution Reactions

20. An S_N1 solvolysis reaction of *t*-butyl iodide. (See *Notes*.)

$$CH_3-\underset{\underset{CH_3}{|}}{\overset{\overset{CH_3}{|}}{C}}-I \quad \xrightarrow{H_2O} \qquad \longrightarrow \qquad \longrightarrow$$

21. An S_N1 solvolysis reaction of *(R)*-(1-chloroethyl)benzene.

$$\xrightarrow{\substack{dioxane-\\H_2O}} \qquad \longrightarrow \qquad \longrightarrow$$

22. An S_N1 reaction of 1-methylcyclohexanol with hydrogen chloride. (See *Notes*.)

$$\xrightarrow{conc.\ HCl} \qquad \longrightarrow \qquad \longrightarrow$$

23. An S_N1 solvolysis reaction of *(1S,3R)*-1-bromo-1,3-dimethylcyclohexane. (See *Notes*.)

Top-face attack. -

$$\xrightarrow{H_2O} \qquad \longrightarrow \qquad \longrightarrow$$

Bottom-face attack. -

$$\longrightarrow \qquad \longrightarrow$$

24. An S_N1 solvolysis reaction of 2-bromo-3-methylbutane. (See *Notes*.)

$$\xrightarrow{H_2O} \qquad \longrightarrow \qquad \longrightarrow \qquad cont'd$$

$$\longrightarrow$$

25. An S_N1 reaction of 2-methylbut-3-en-2-ol with hydrogen bromide. (See *Notes*.)

major

CH3
H–O
CH3
CH3

HBr
H₂O

+

26. An S_N1 solvolysis reaction of 1-bromo-3-methyl-2-butene. (See *Notes*.)

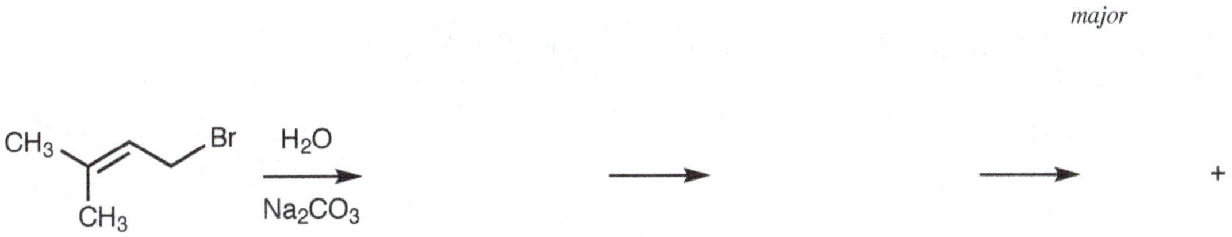

major

CH3
CH3
Br   H₂O

Na₂CO₃

+

27. An S_N1 reaction of 2-methylcyclopentanol with hydrobromic acid.

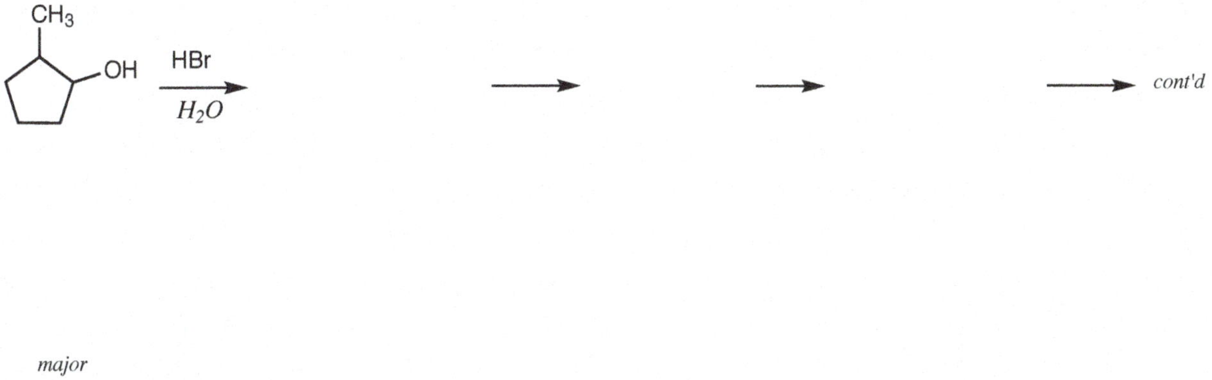

CH3
OH   HBr

H₂O

cont'd

major

28. An S_N1 solvolysis reaction of 3-chlorocyclopentene. (See *Notes*.)

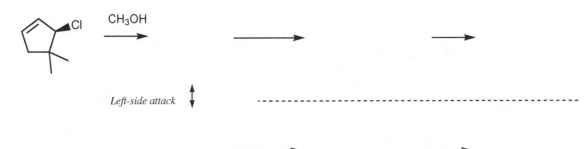

Right-side attack

Cl   CH₃OH

Left-side attack

4 — Elimination Reactions

Alkene Formation

1. An elimination reaction of 1-chlorooctadecane with potassium *t*-butoxide. (See *Notes*.)

$C_{15}H_{31}$

$\xrightarrow[\text{toluene}]{\text{KO-}t\text{-Bu}}$

2. An elimination reaction of 2-bromobutane with sodium ethoxide. (See *Notes*.)

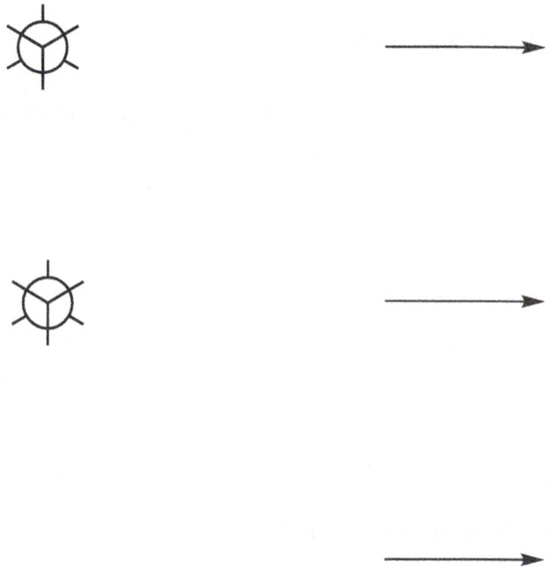

$\xrightarrow[\text{EtOH}]{\text{NaOEt}}$

racemic

+   +   (   )

55%   19%   16%   10%

$\longrightarrow$

$\longrightarrow$

$\longrightarrow$

3. An elimination reaction of 2-bromo-2-methylbutane.

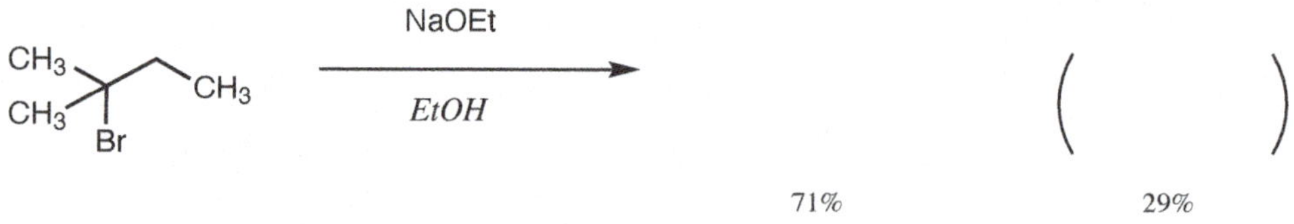

$\xrightarrow[\text{EtOH}]{\text{NaOEt}}$

(   )

71%   29%

4. An elimination reaction of (*1R,2R*)- or (*1S,2S*)-1-bromo-1,2-diphenylpropane. (See *Notes*.)

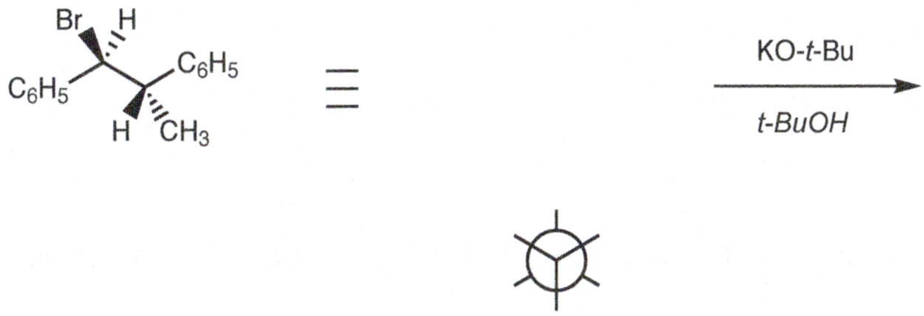

5. An elimination reaction of *trans*-1-chloro-2-isopropylcyclohexene. (See *Notes*.)

stereoview

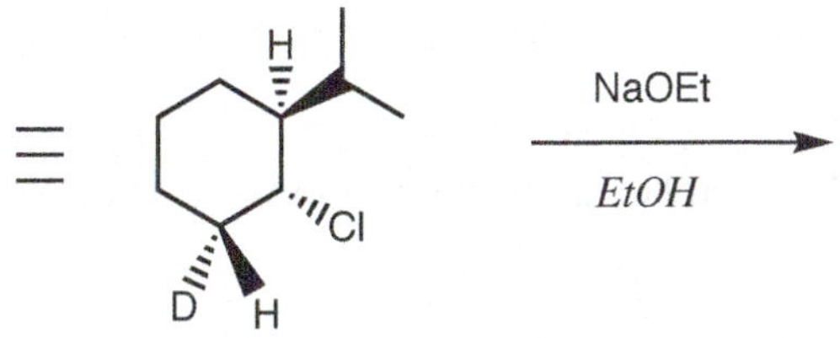

6. An E2 elimination reaction of *cis*-1-chloro-2-isopropylcyclohexene. (See *Notes*.)

stereoview   *major*

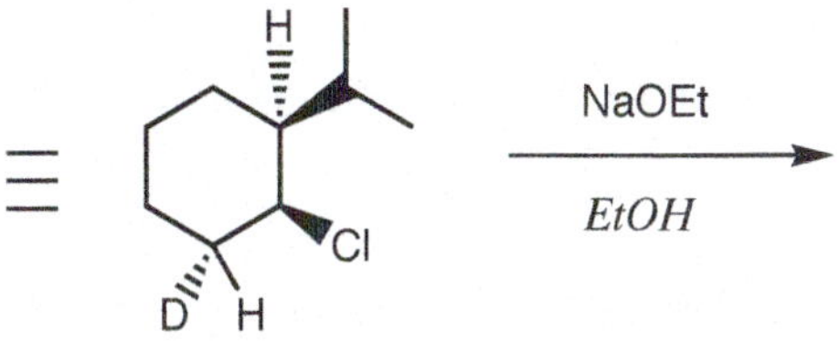

7. A competitive elimination reaction of *cis*- and *trans*-1-bromo-4-*t*-butylcyclohexane and one equivalent of *t*-butoxide. (See *Notes*.)

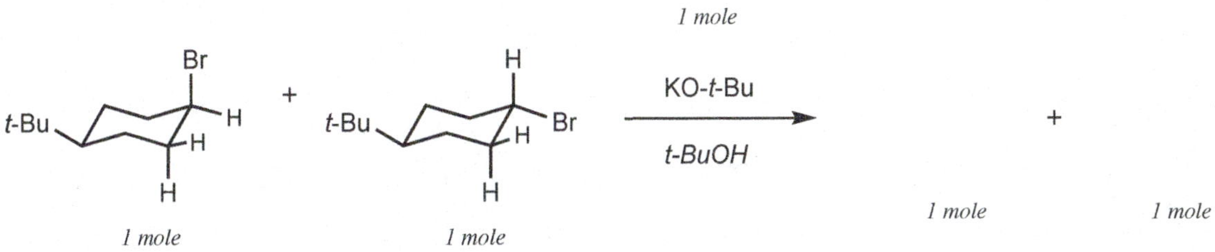

1 mole   *1 mole*   *1 mole*   *1 mole*

8. An elimination reaction of 1-chloro-1-methylcyclohexane with ethoxide. (See *Notes*.)

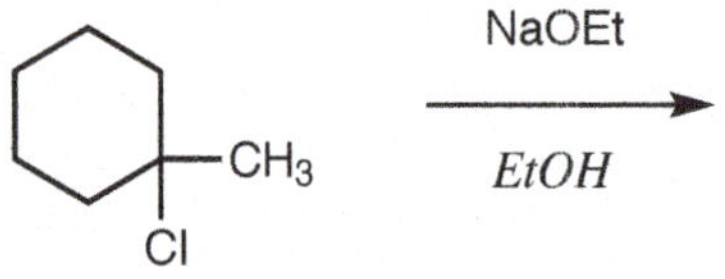

CH3
Cl
NaOEt
EtOH

9. An elimination reaction of 1-chloro-1-methylcyclohexane with *t*-butoxide to give methylenecyclohexane.

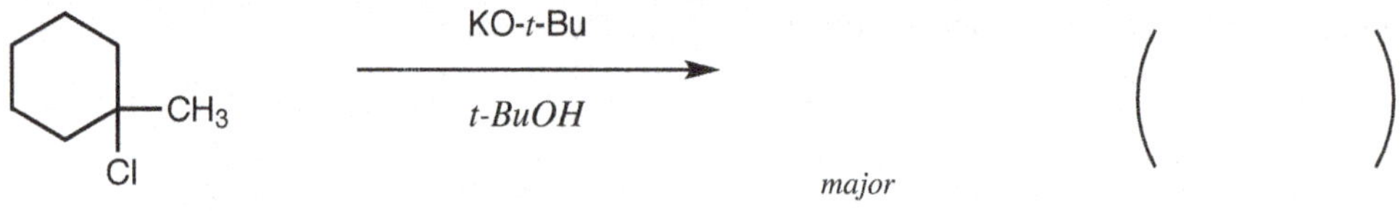

CH3
Cl
KO-t-Bu
t-BuOH
major

10. An elimination reaction of 3-chloro-3-methylcyclohexanone with *t*-butoxide. (See *Notes*.)

O
CH3
Cl
KO-t-Bu
t-BuOH

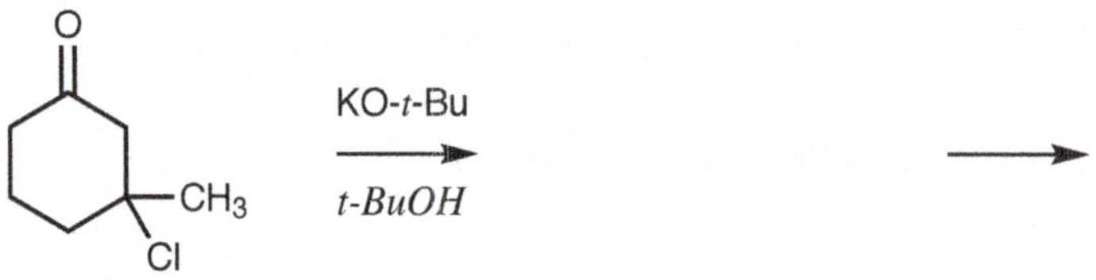

11. An elimination reaction of 4-bromo-cyclohexene with *t*-butoxide. (See *Notes*.)

Br
D
D
KO-t-Bu
t-BuOH
major

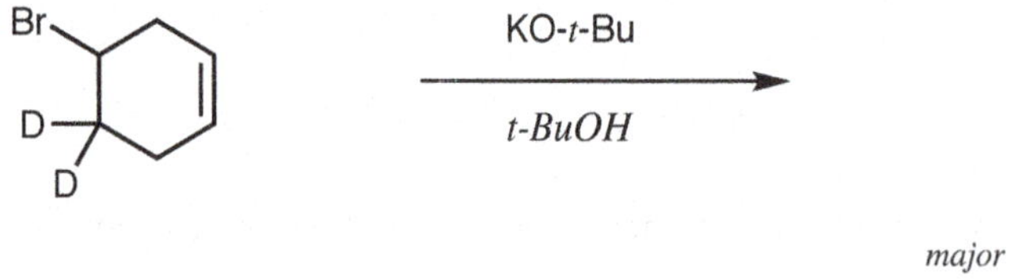

12. An elimination reaction of 3-bromo-1-cyclohexene with *t*-butoxide. (See *Notes*.)

Br
H
H
D D
KO-t-Bu
t-BuOH
major

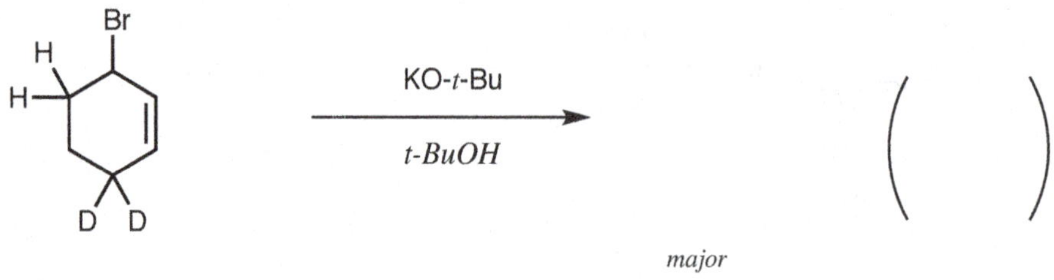

13. An elimination reaction of 2-methylcyclopentanol by treatment with sulfuric acid. (See *Notes.*)

CH$_3$

$\xrightarrow[\substack{H_2O \\ heat}]{H_2SO_4}$

$\longrightarrow$

$\longrightarrow$

14. An elimination reaction of 3-methyl-3-pentanol by treatment with sulfuric acid. (See *Notes.*)

$\xrightarrow[\substack{H_2O \\ heat}]{H_2SO_4}$

$\longrightarrow$

$\longrightarrow$

conformer A

10

$\longrightarrow$

conformer B   50

conformer C   40

15. A solvolysis-elimination reaction of a benzyl bromide in aqueous base. (See *Notes.*)

conformer A

$\xrightarrow[\substack{H_2O}]{\substack{NaOH \\ (dilute)}}$

$\longrightarrow$

conformer B

$\longrightarrow$

16. An elimination reaction of 2-cyclobutyl-2-propanol and sulfuric acid. (See *Notes.*)

CH$_3$

CH$_3$

$\xrightarrow[\substack{H_2O \\ heat}]{H_2SO_4}$

$\longrightarrow$

$\longrightarrow$

cont'd

$\longrightarrow$

$\longrightarrow$

$\equiv$

17. A Hofmann elimination reaction. (See *Notes*.)

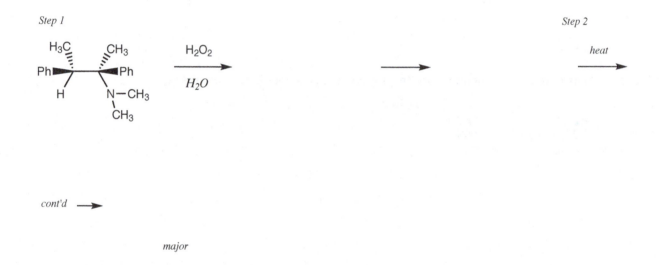

18. A Cope elimination reaction. (See *Notes*.)

19. A selenoxide oxidation and elimination reaction. (See *Notes*.)

Acetylene Formation

20. A bromination and elimination reaction of a *trans* alkene. (See *Notes*.)

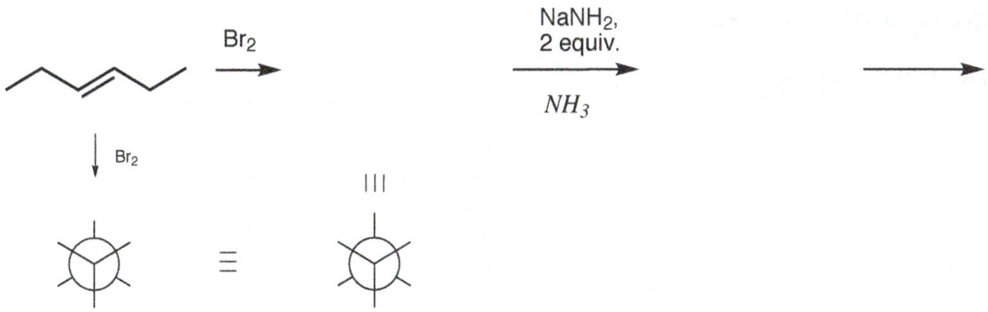

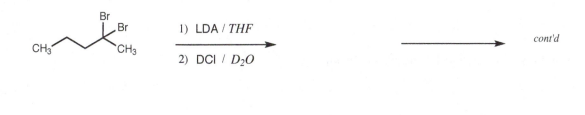

21. A bromination and elimination reaction of a *cis* alkene. (See *Notes*.)

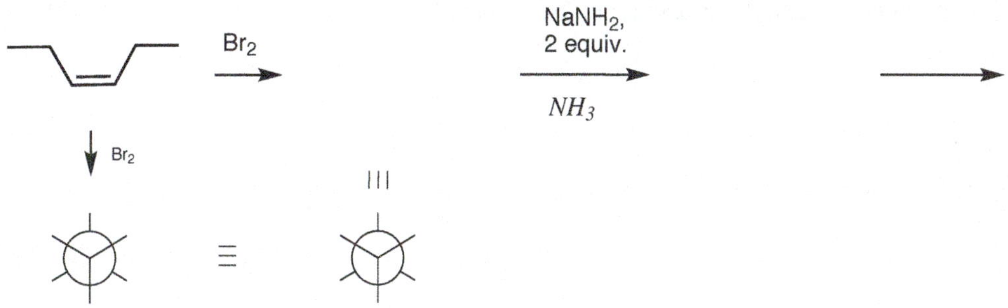

22. An elimination reaction of 1,1-dibromopentane with *t*-butoxide. (See *Notes*.)

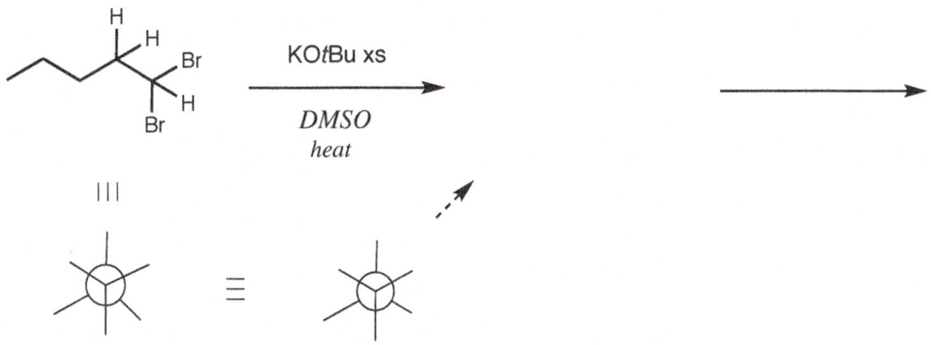

23. An elimination reaction of 2,2-dibromopentane with LDA. (See *Notes*.)

1) LDA / THF

2) DCl / D₂O

cont'd

5 — Electrophilic Addition to Alkenes and Alkynes

Addition of HX and H₂O to Alkenes

1. Addition of hydrogen bromide to propene. (See *Notes*.)

2. Acid catalyzed addition of water to methylcyclopentene. (See *Notes*.)

3. Addition of acetic acid to propene catalyzed by sulfuric acid. (See *Notes*.)

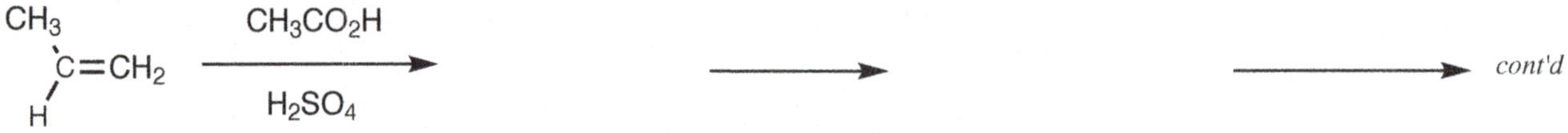

cont'd

4. Addition of hydrogen bromide to 1-methylcyclohexene.

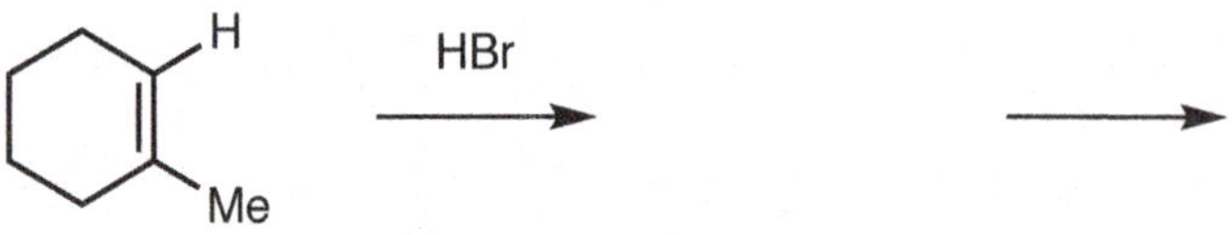

5. Addition of hydrogen chloride to *(E)*-3-hexene.

HCl

$\longrightarrow$ +

6. Addition of hydrogen chloride to *(Z)*-3-hexene.

HCl

$\longrightarrow$ +

7. Addition of hydrogen bromide to 3-methyl-1-butene (See *Notes.*)

HBr

$\longrightarrow$ $\longrightarrow$

8. Addition of HBr to 2-cyclobutylpropene. (See *Notes.*)

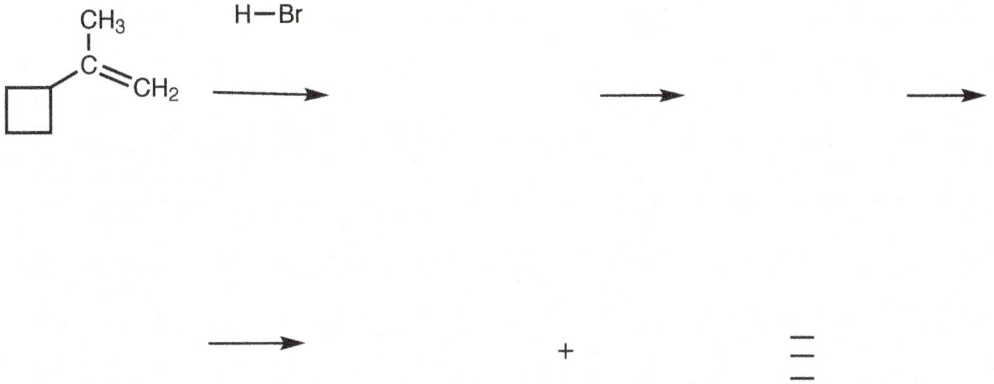

H—Br

$\longrightarrow$ $\longrightarrow$ $\longrightarrow$ *cont'd*

$\longrightarrow$ + $\equiv$

9. Addition of hydrogen chloride to allylbenzene. (See *Notes*.)

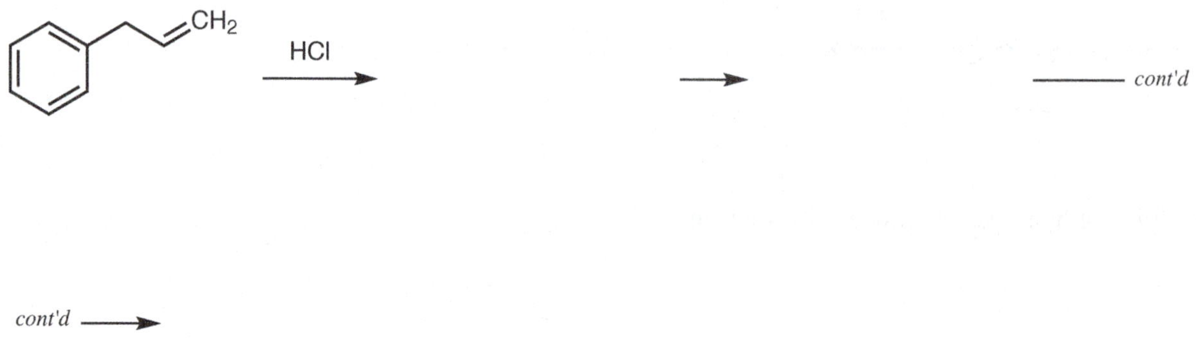
CH2
HCl
cont'd
cont'd

10. Addition of hydrogen chloride to 2-methyl-1,3-butadiene (isoprene). (See *Notes*.)

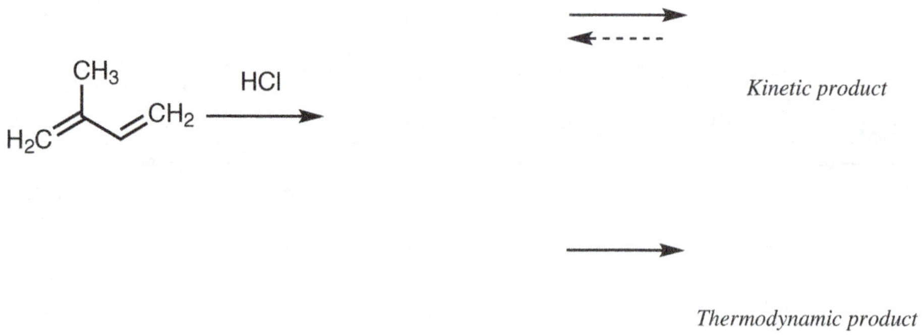
CH3
H2C
CH2
HCl
Kinetic product
Thermodynamic product

11. Addition of bromine to 2-methyl-1,3-butadiene. (See *Notes*.)

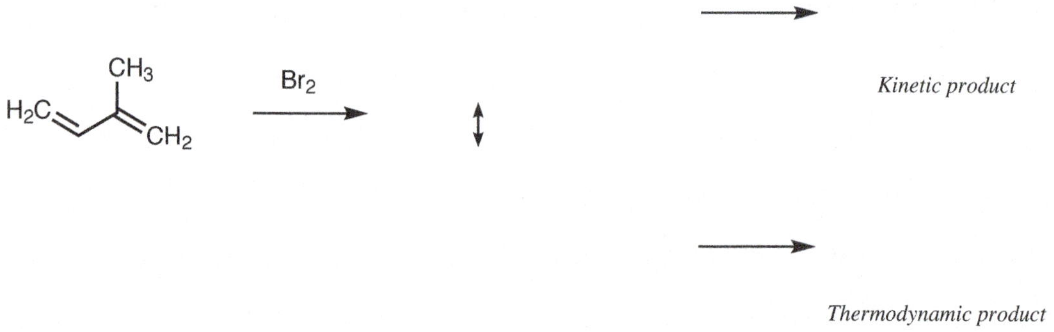
CH3
H2C
CH2
Br2
Kinetic product
Thermodynamic product

Bromination

12. Bromination of cyclohexene. (See *Notes*.)

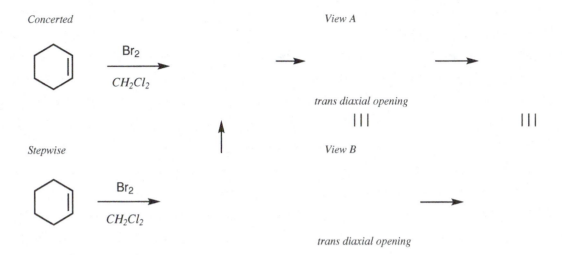

Concerted

View A

trans diaxial opening

| | |

| | |

Stepwise

View B

trans diaxial opening

13. Bromination of *trans*-2-butene.

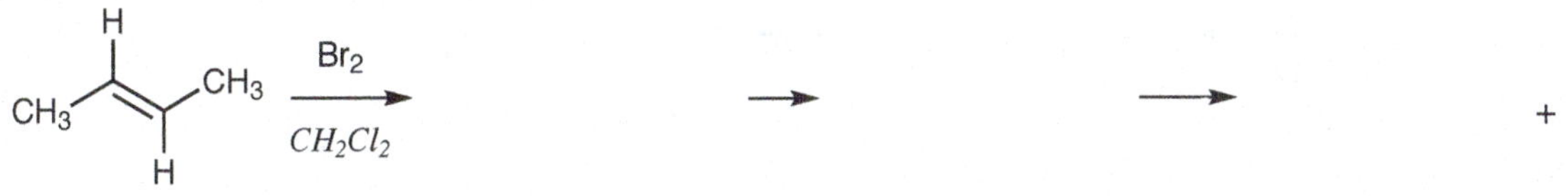

+

14. Bromination of *cis*-2-butene.

+

15. Bromination of methylcyclohexene. (See *Notes*.)

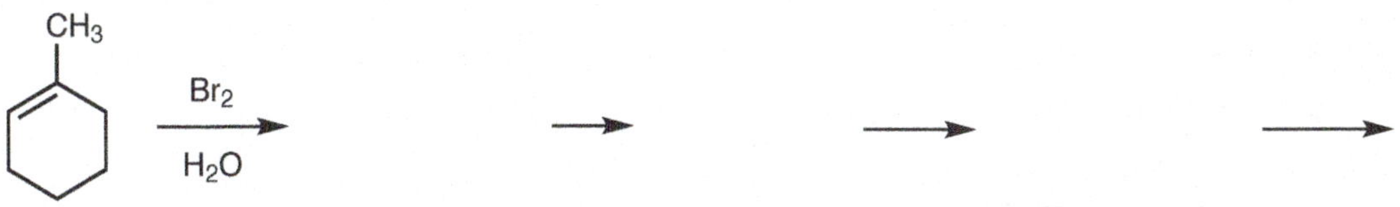

Oxymercuration

16. Step 1, oxymercuration of 3-methyl-1-butene. (See *Notes*.)

CH$_3$

H$_3$C

CH$_2$ Hg(OAc)$_2$

H$_2$O

THF

cont'd

cont'd

Step 2, reductive demercuration.

NaBH$_4$

H$_2$O

THF

17. Step 1, oxymercuration of 1-methylcyclohexene. (See *Notes*.)

CH$_3$ Hg(OAc)$_2$

H$_2$O

THF

cont'd

cont'd

chair form

Step 2, reductive demercuration.

NaBH$_4$

H$_2$O

THF

Hydroboration-Oxidation of Alkene

18. Hydroboration-oxidation of propene.

Step 1, hydroboration. Each bracket represents one of three hydroboration steps. (See *Notes*.)

Step 2, oxidation. Each bracket represents one of three oxidation steps. (See *Notes*.)

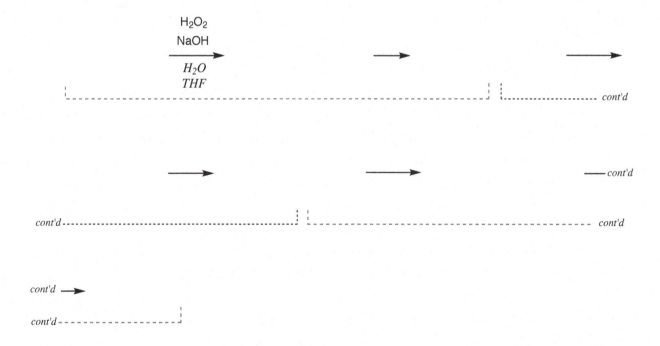

Step 3, borate ester hydrolysis. Each bracket represents one of three hydrolysis steps.

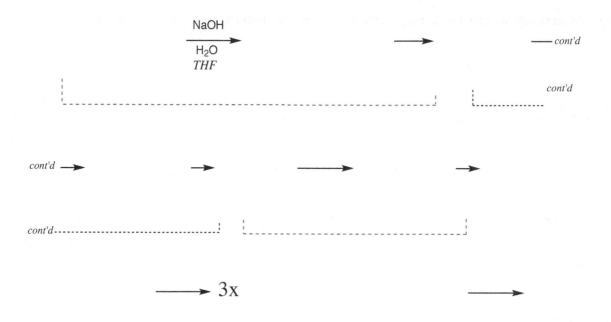

19. Hydroboration/oxidation of 1-methylcyclohexene. (See *Notes.*)

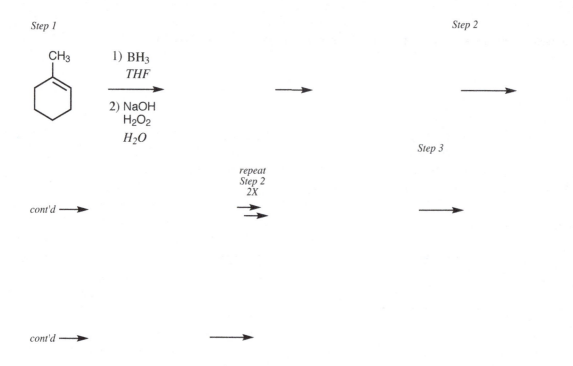

Carbon-Carbon Triple Bond Electrophilic Reactions

Addition to an Internal Acetylene

20. Addition of HCl to 2-butyne (dimethyl acetylene).

$$CH_3-C\equiv C-CH_3 \xrightarrow{\text{HCl}} \qquad \updownarrow \qquad \longrightarrow$$

2nd Equivalent of HCl

$$\xrightarrow{\text{H-Cl}} \qquad \longrightarrow$$

21. Addition of bromine to ethynylcyclopentane. (See *Notes*.)

$$\text{—C}\equiv\text{C-H} \xrightarrow{\text{Br-Br}} \qquad \longrightarrow \qquad \longrightarrow$$

2nd Equivalent of bromine.

$$\xrightarrow{\text{Br-Br}} \qquad \longrightarrow$$

22. Sulfuric acid catalyzed hydration of 1-propynylbenzene.

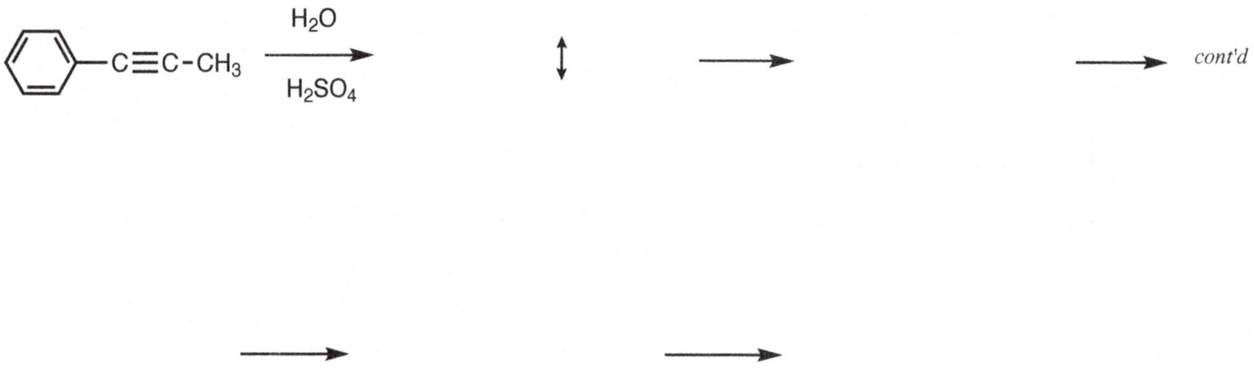

Addition to a Terminal Acetylene

23. Mercury catalyzed hydration of propyne (methyl acetylene).

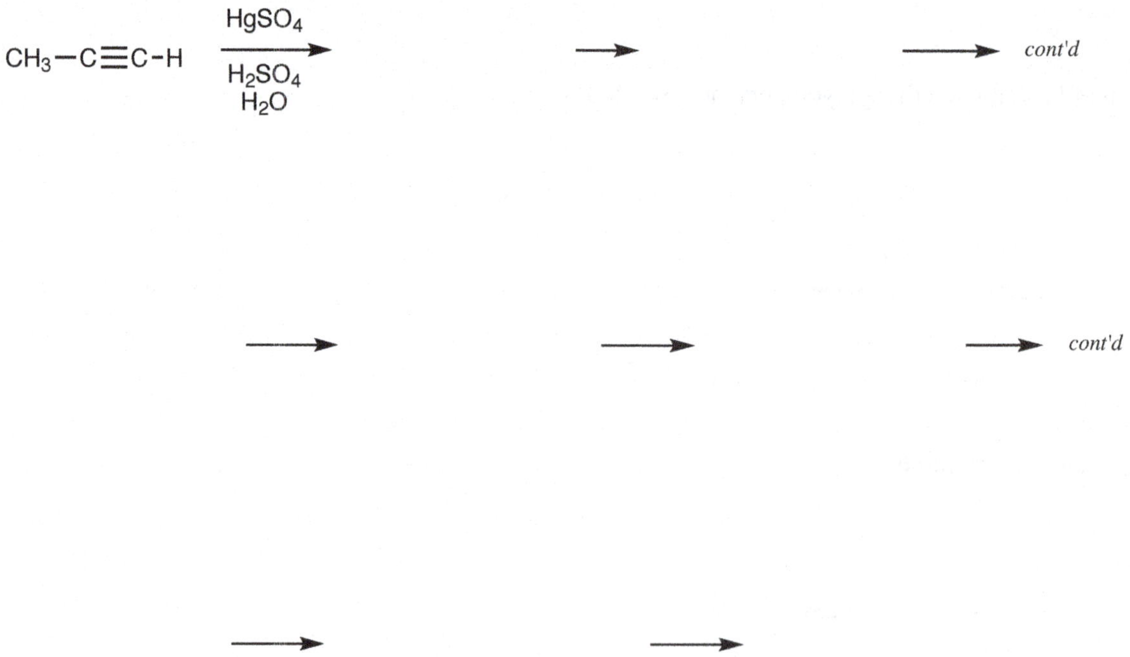

Disiamylborane, Hydroboration–Oxidation of an Acetylene

24. Hydroboration-oxidation of phenylacetylene with disiamylborane.

Step 1

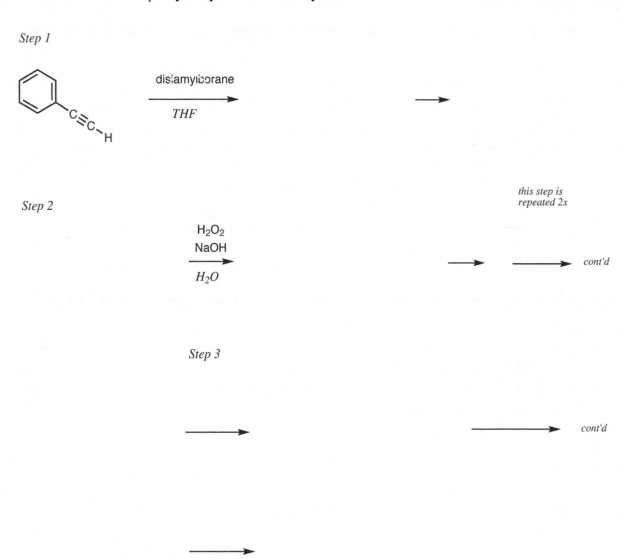

disiamylborane

THF

this step is repeated 2x

Step 2

H₂O₂
NaOH

H₂O

cont'd

Step 3

cont'd

6 — Rearrangement Reactions

Baeyer-Villiger Oxidation

1. Acid catalyzed Baeyer-Villiger oxidation of 2,2-dimethylcyclopentanone with peracetic acid. (See *Notes.*)

cont'd

cont'd

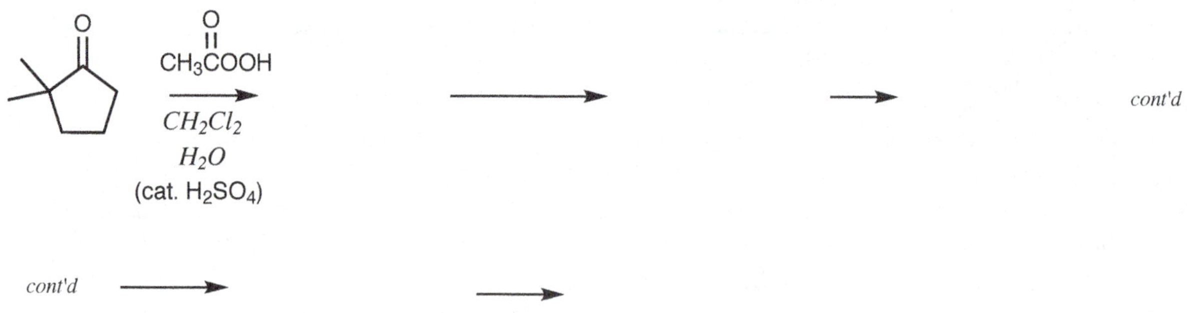

2. Baeyer-Villiger oxidation of *o*-methoxyacetophenone with peracetic acid, acid catalyzed. (See *Notes.*)

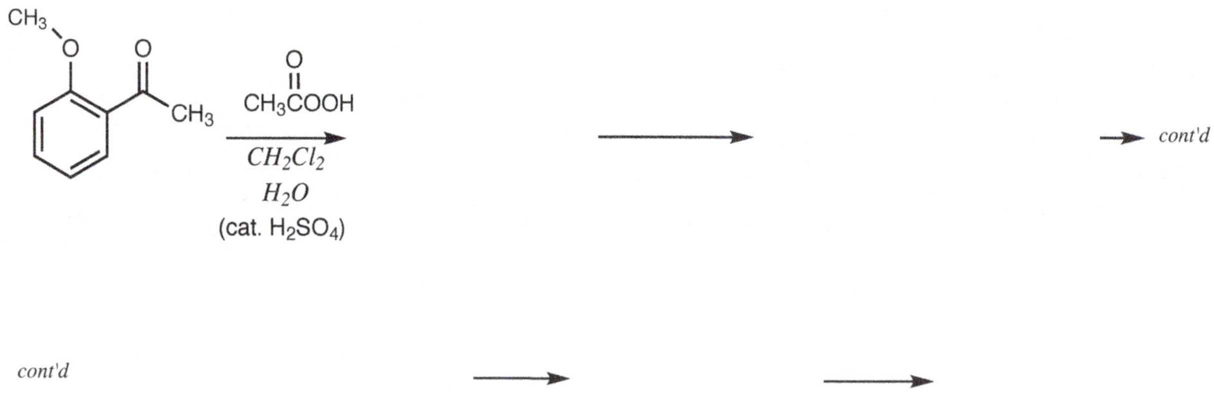

cont'd

cont'd

3. Acid catalyzed Baeyer-Villiger oxidation of benzaldehyde with peracetic acid. (See *Notes.*)

cont'd

cont'd

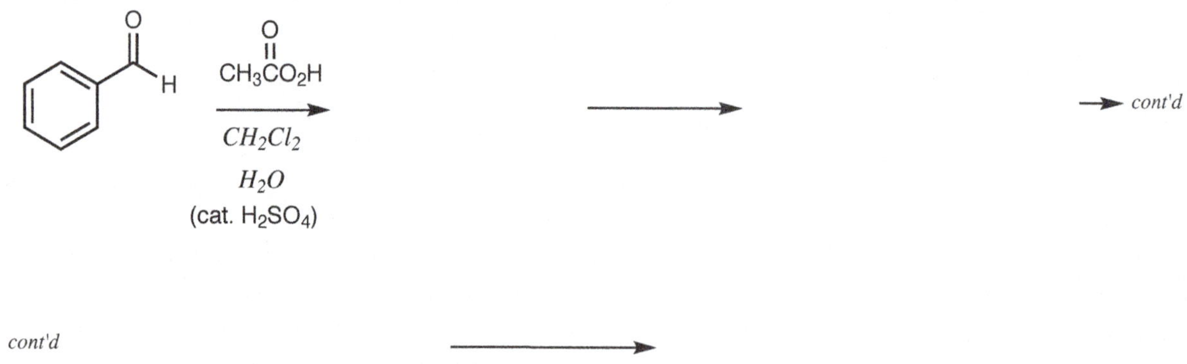

4. Baeyer-Villiger oxidation of a benzophenone with trifluoroperacetic acid. (See *Notes*.)

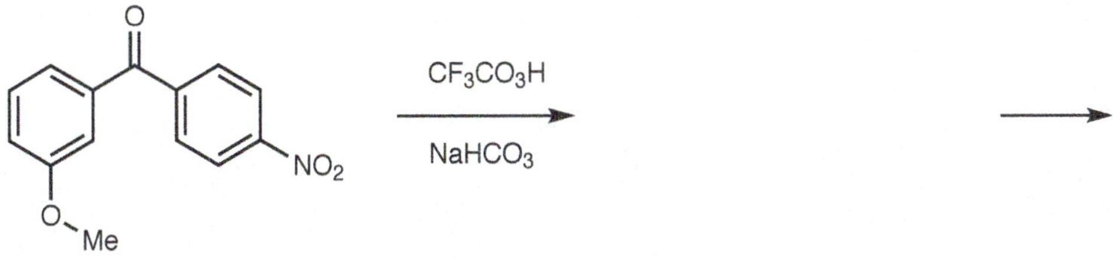

5. Baeyer-Villiger oxidation of *o*-methoxyacetophenone with *m*-chloroperoxybenzoic acid. (See *Notes*.)

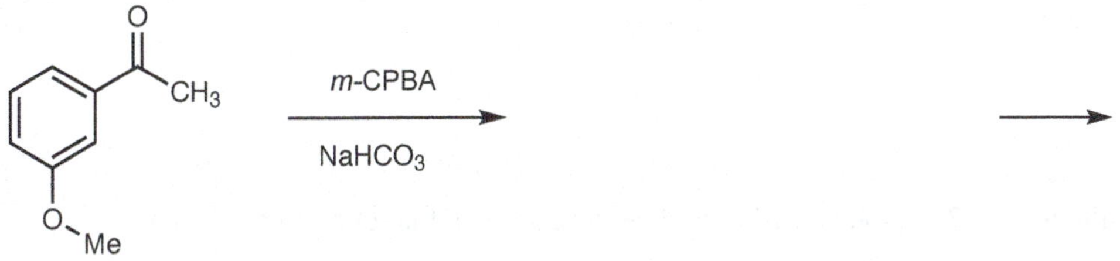

6. Baeyer-Villiger oxidation of bicyclic phenyl ketone with trifluoroperacetic acid. (See *Notes*.)

7. Baeyer-Villiger oxidation of *p*-chlorobenzaldehyde with peracetic acid to give *p*-chlorobenzoic acid. (See *Notes*.)

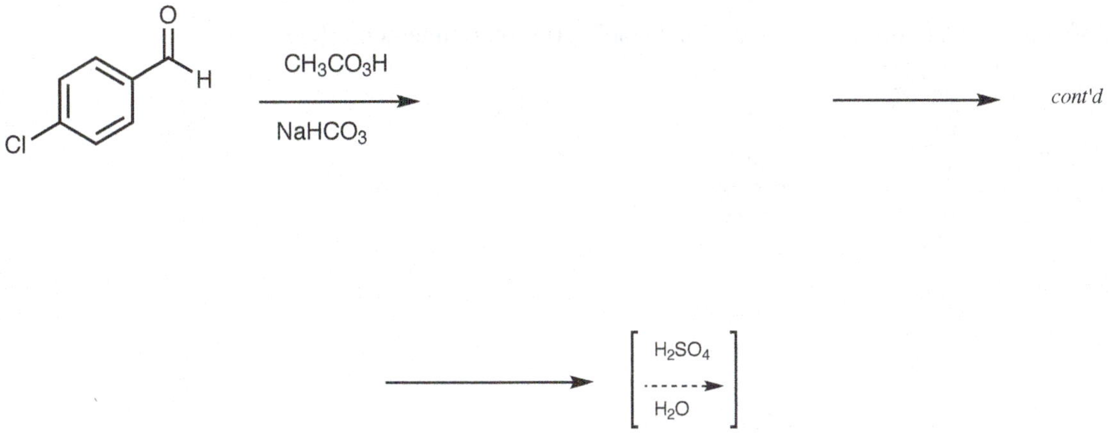

cont'd

Pinacol Rearrangement

8. Rearrangement of pinacol to pinacolone, methyl *t*-butyl ketone or 3,3-dimethyl-2-butanone. (See *Notes*.)

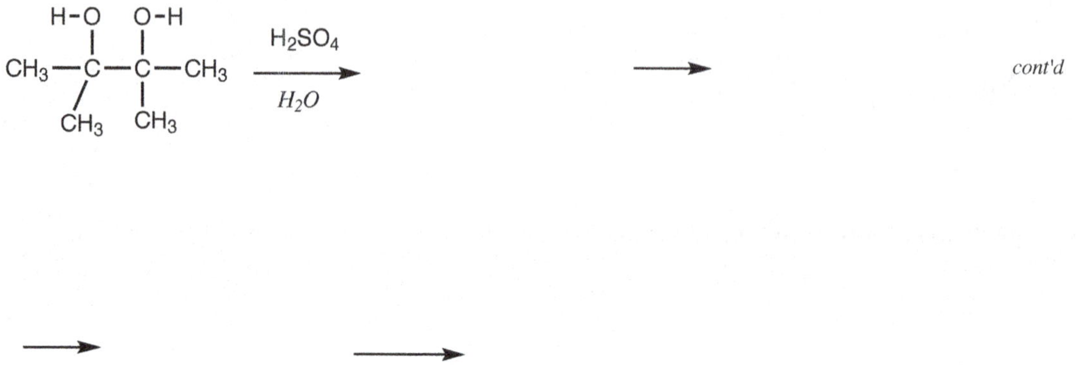

9. Rearrangement of 1,2-dimethyl-1,2-cyclohexanediol with acid to a methyl ketone. (See *Notes*.)

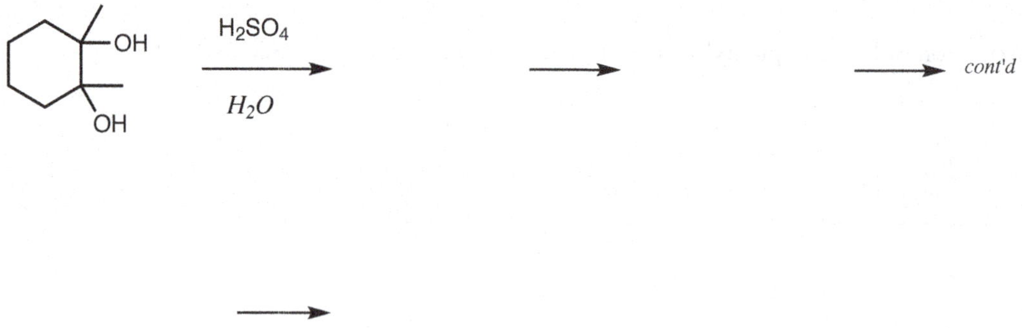

Benzilic Acid Rearrangement

10. Reaction of benzil with hydroxide to give benzilic acid after rearrangement. (See *Notes*.)

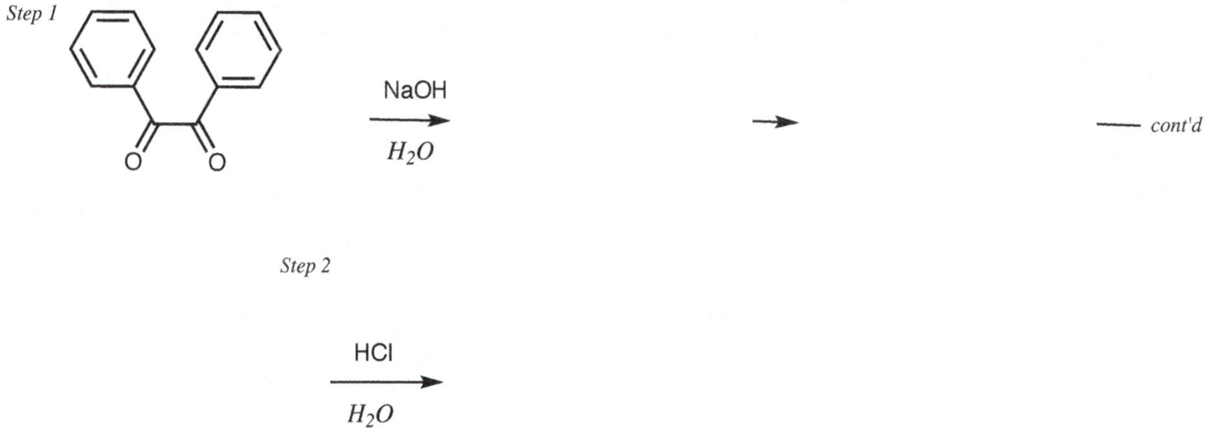

Dakin Reaction

11. Reaction of an *o*- or *p*-hydroxybenzaldehyde with basic hydrogen peroxide to give a phenol. (See *Notes*.)

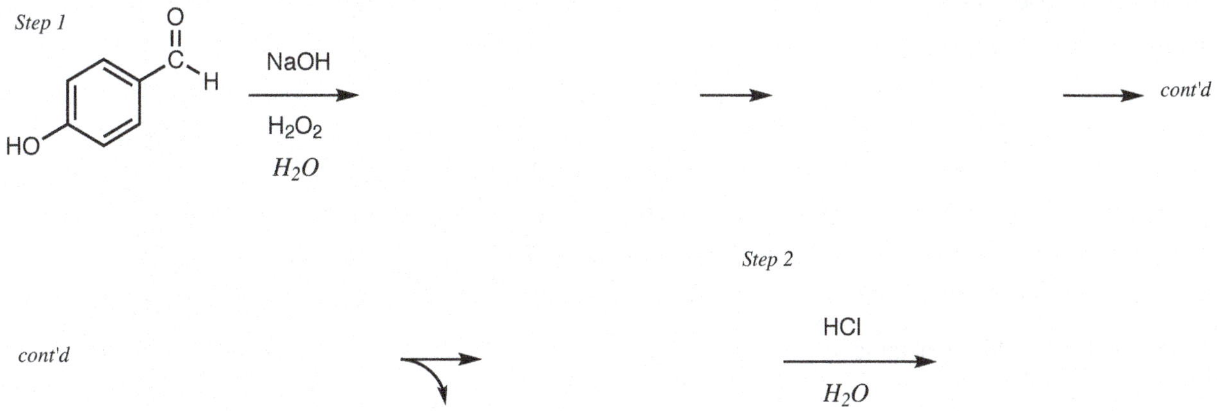

Step 1

NaOH

H$_2$O$_2$

H$_2$O

⟶ *cont'd*

cont'd

Step 2

HCl

H$_2$O

Acetone from Cumene

12. Conversion of isopropylbenzene (cumene) to acetone and phenol.

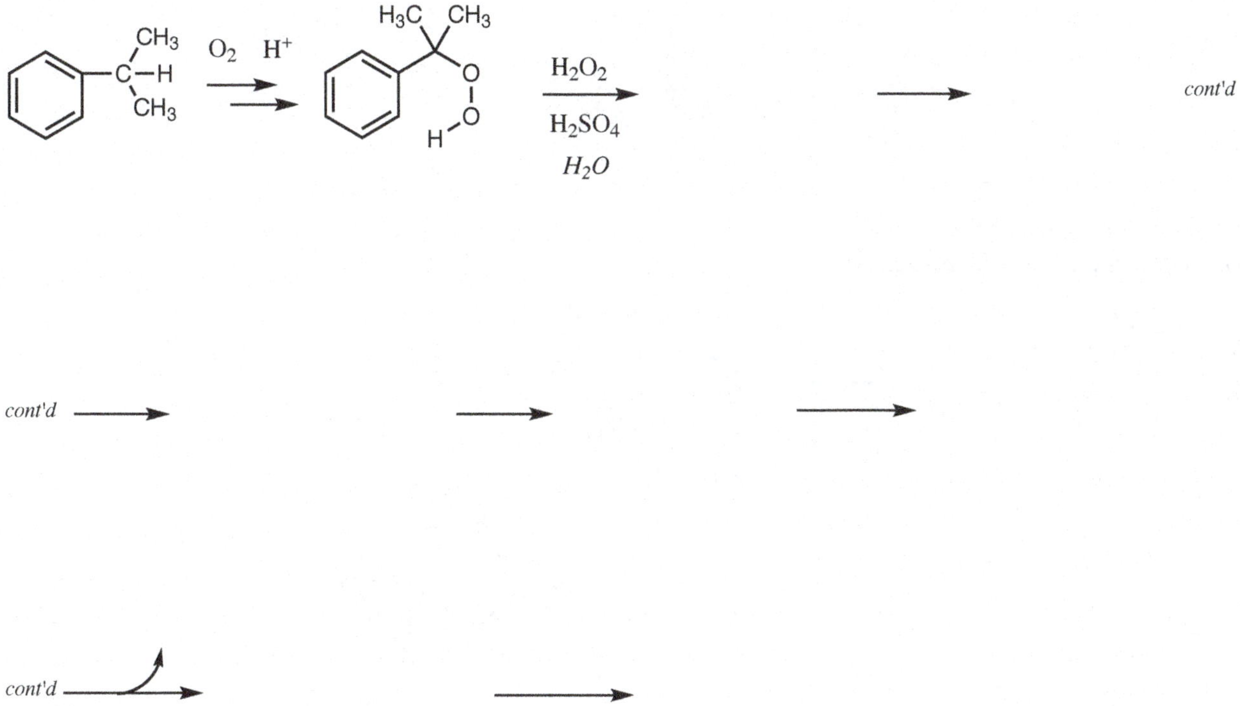

O$_2$ H$^+$

H$_2$O$_2$

H$_2$SO$_4$

H$_2$O

cont'd

cont'd ⟶

⟶

⟶

cont'd ⟶

⟶

7 — Electrocyclic Reactions

Diels Alder Reactions

1. Draw the Diels-Alder reactants.

    +   ⟶

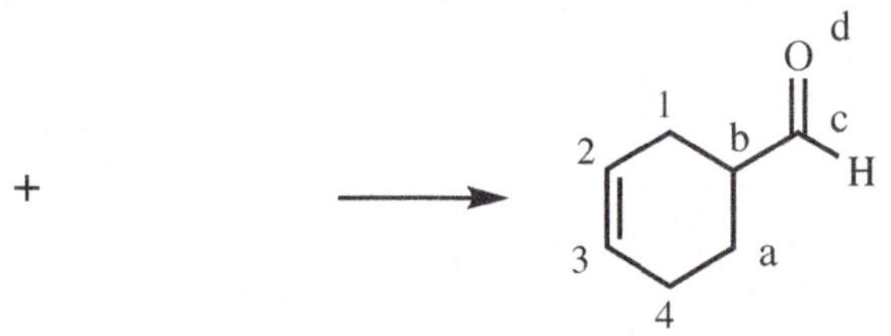

2. Draw the Diels-Alder reactants.

    +   ⟶   ≡

    redraw

    The preferred orientation.

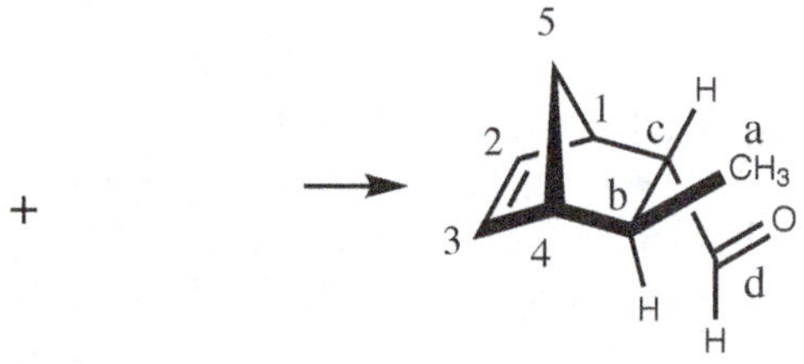

    +   ⟶

3. Draw the Diels-Alder reactants.

    +   ⟶

    major

    +   ⟶

    minor

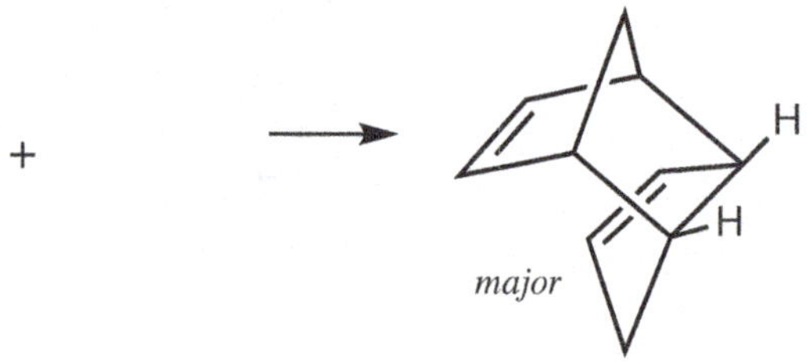

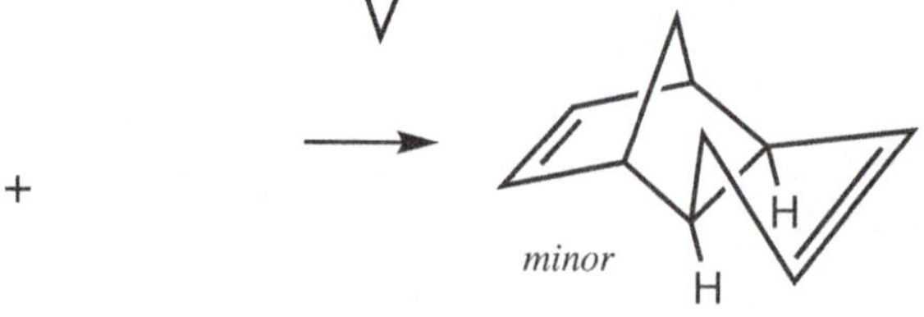

4. Draw the Diels-Alder reactants.

    +   ⟶

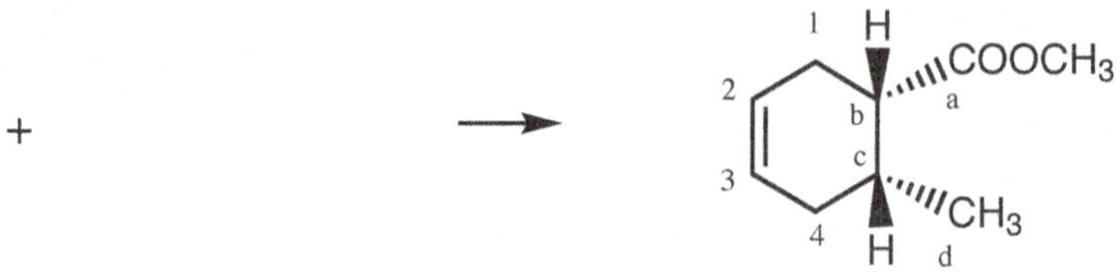

5. Draw the Diels-Alder reactants.

+ ⟶

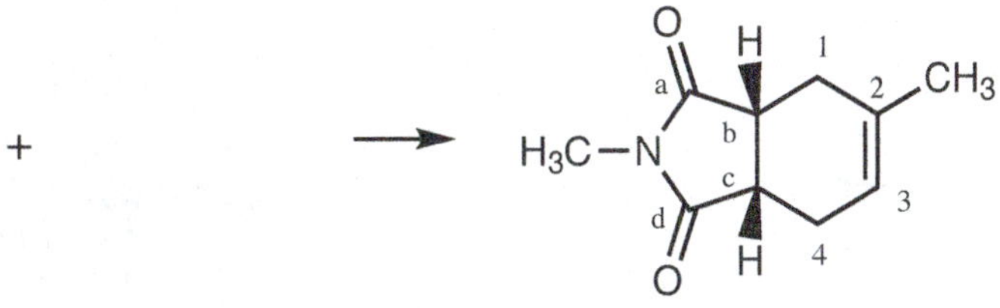

6. Draw the Diels-Alder reactants.

+ ⟶ ≡

stereo view

7. Draw the Diels-Alder reactants.

+ ⟶ ≡

stereo view

8. Draw the Diels-Alder reactants for a reverse-forward Diels-Alder reaction.

heat ⟶ + ⟶

9. Draw the Diels-Alder reactants for a reverse-forward Diels-Alder reaction.

heat ⟶ + ⟶

SO₂

10. Draw the Diels-Alder reactants.

+ ⟶ ≡

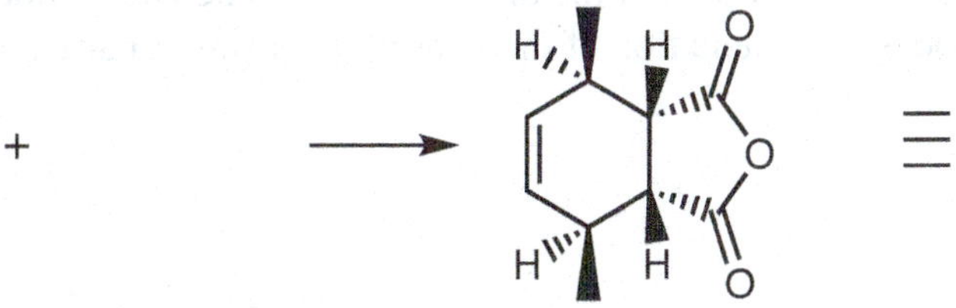

11. Draw the Diels-Alder reactants.

+   ⟶   ≡   *stereo view*

12. Draw the starting materials (and their correct orientations) that give the following Diels-Alder product. Use Examples 13 and 14 to predict the structure of the major product?

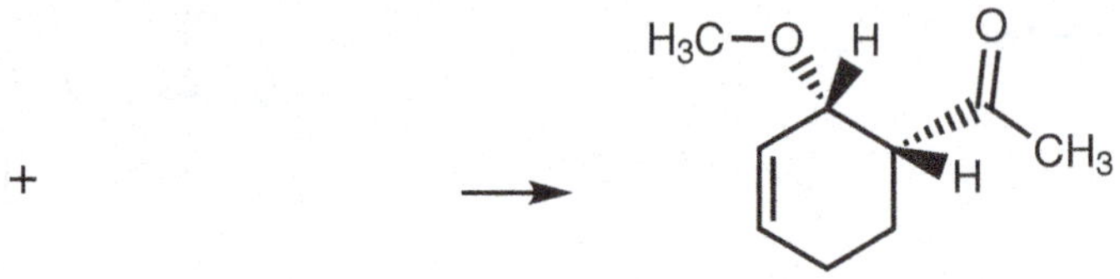

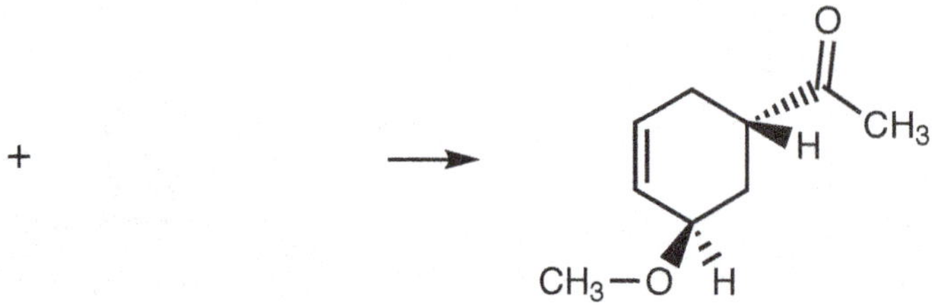

+   ⟶

+   ⟶

13. Draw the electrostatic charges of the starting materials by drawing the resonance structures.

   Draw the resonance structures.

   ⊕   ⊕

   ⟷   ⟷

14. Examine the electrostatic charges of the starting materials by drawing the resonance structures.

   Draw the resonance structures.

   ⊖   ⊖

   ⟷   ⟷

15. How do the resonance structures shown in 13 and 14 predict the product for 12? The Diels-Alder reaction is thought of as a truly concerted reaction. Does the orientation of the final product suggest any non-concerted character leading to the product?

Other Electrocyclic Reactions

16. Draw the Diels-Alder reactants.

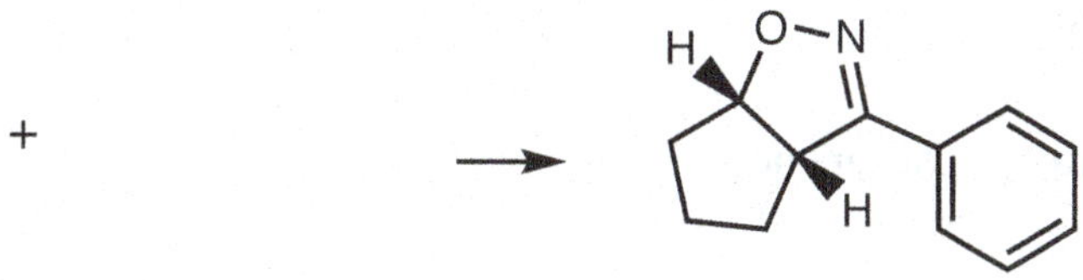

17. A Claisen rearrangement (electrocyclic) reaction to transfer a group from oxygen to carbon.

18. A double Claisen rearrangement (electrocyclic) reaction to transfer a group from oxygen to carbon to another carbon. Count the number of electrons that move in each step.

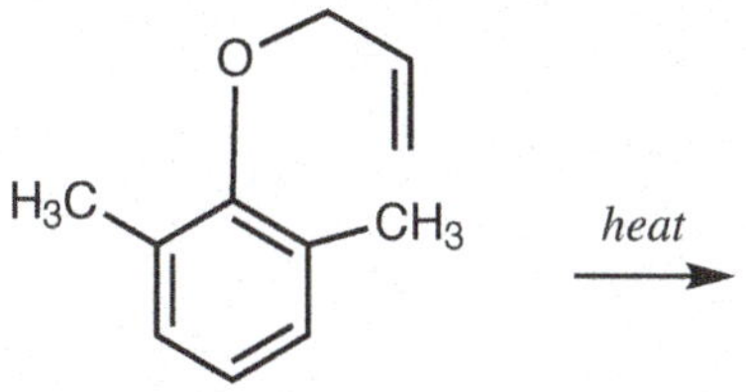

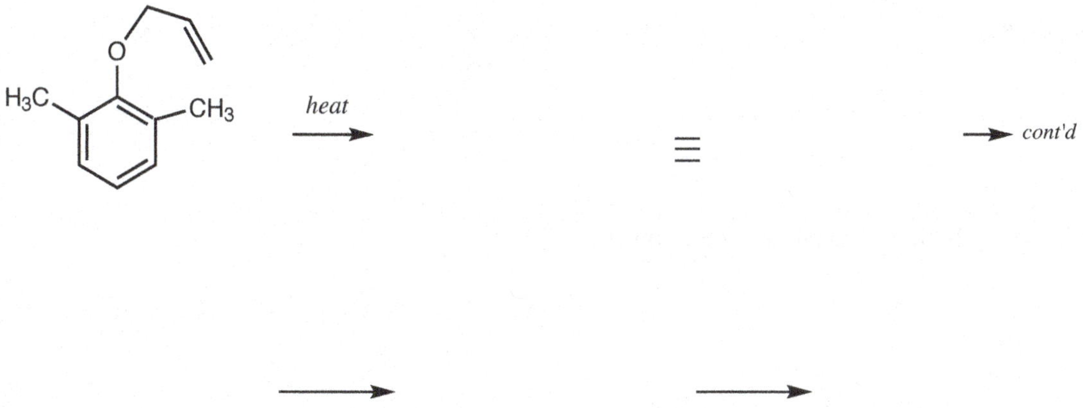

cont'd

8 — Carbonyl Addition and Addition-Elimination Reactions

Additions by C, N, and O. Addition by hydrogen nucleophiles are discussed in Chapter 11, Reduction.

Grignard Addition to a Carbonyl Group

1. Addition of methyl magnesium bromide to cyclohexanone.

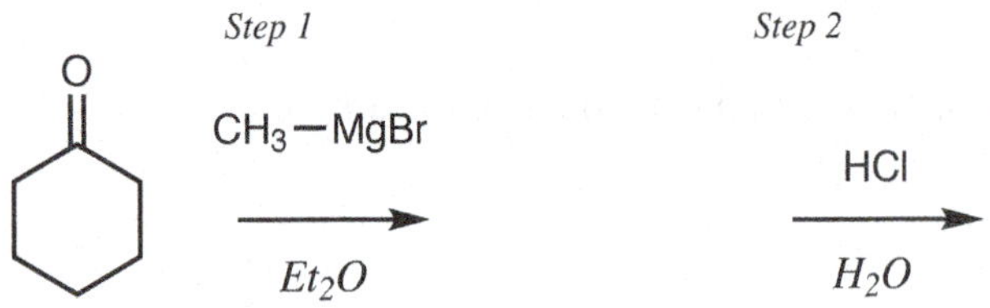

2. Addition of a Grignard reagent to acetaldehyde.

Alkyllithium Addition to a Carbonyl Group

3. Addition of ethyllithium to benzaldehyde.

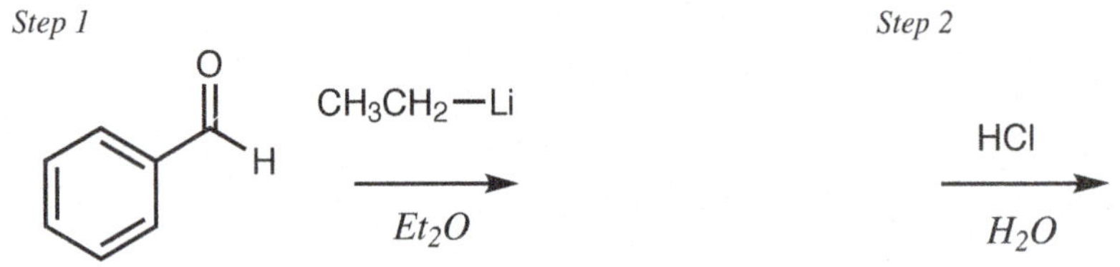

4. Addition of propynyllithium to acetone. (see Chapter 2.14)

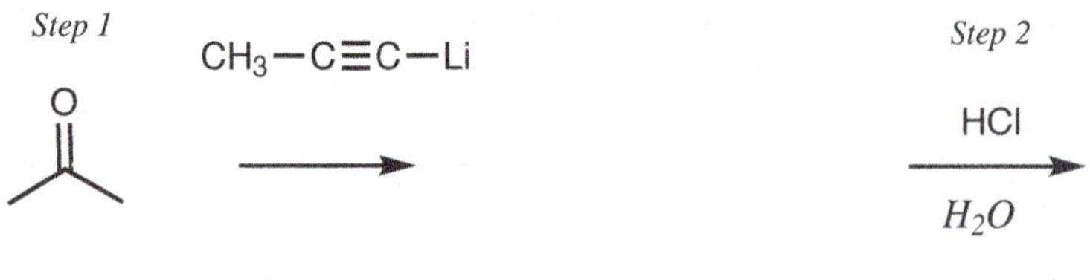

Wittig Reaction

5.   Wittig reaction, Step 1, formation of Wittig reagent.

$$
\begin{array}{c}
\text{H} \\
\text{CH}_3-\overset{\displaystyle |}{\underset{\displaystyle |}{\text{C}}}-\text{H} \\
\text{Br}
\end{array}
\quad
\begin{array}{c}
\text{1) P(C}_6\text{H}_5)_3 \\
\textit{THF} \\
\longrightarrow \\
\text{2) BuLi} \\
\textit{THF}
\end{array}
\qquad \longrightarrow \qquad \updownarrow
$$

Step 2, reaction with benzaldehyde

6.   Wittig reaction, Horner-Wadsworth-Emmons modification, Step 1, Arbusov reaction. (See Notes.)

Step 2, reaction with benzaldehyde

major   *minor*

Addition-Elimination Reactions (Reversible Additions)

Ketal Formation and Hydrolysis

7. Acid catalyzed ketalization of cyclohexanone.

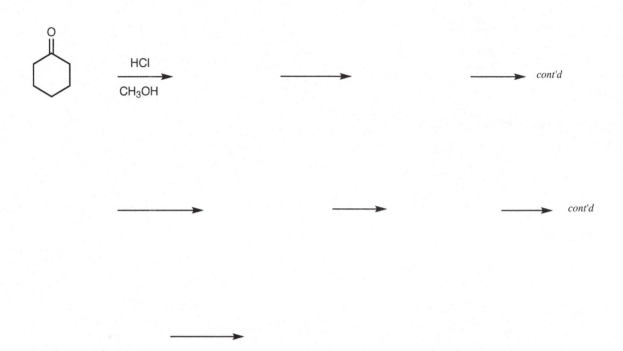

8. Acid catalyzed hydrolysis of the dioxolane acetal of benzaldehyde.

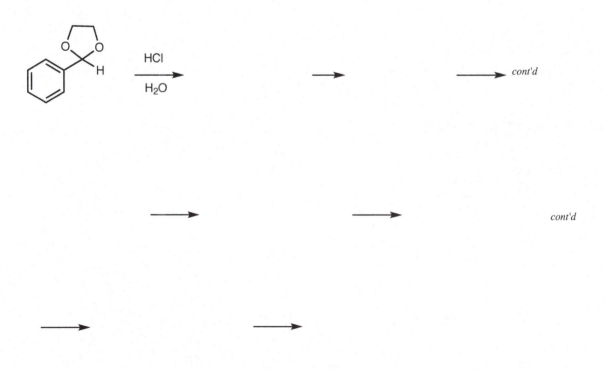

Oxime Formation

9. Formation of the oxime of cyclohexanone.

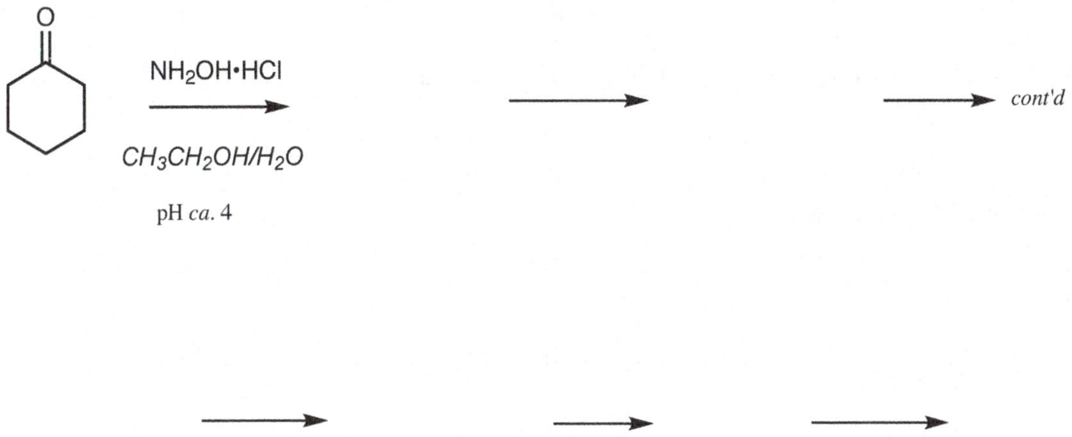

NH₂OH·HCl

CH₃CH₂OH/H₂O

pH *ca.* 4

cont'd

Addition of Cyanide to a Carbonyl Group

10. Reaction of cyanide with acetone.

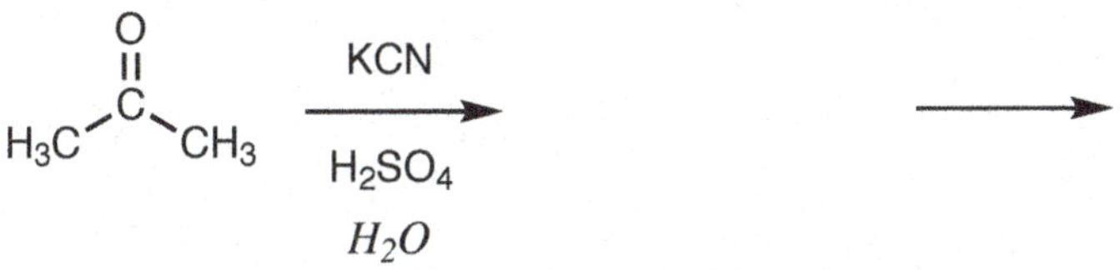

KCN

H₂SO₄

H_2O

11. Reaction of 2-hydroxy-2-methylpropanenitrile with base.

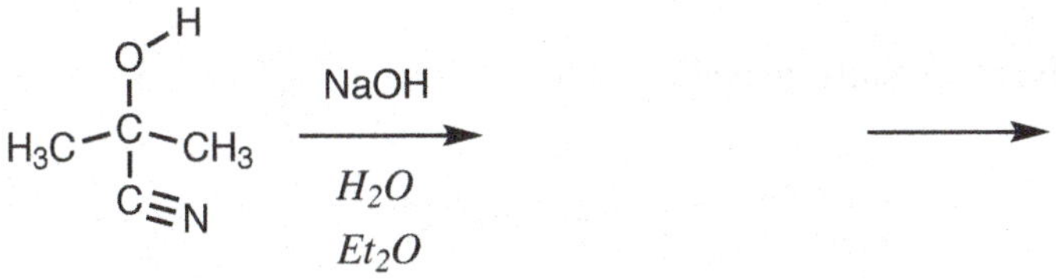

NaOH

H_2O

Et_2O

12. Reaction of ethyl acetate enolate and 4-bromobenzaldehyde

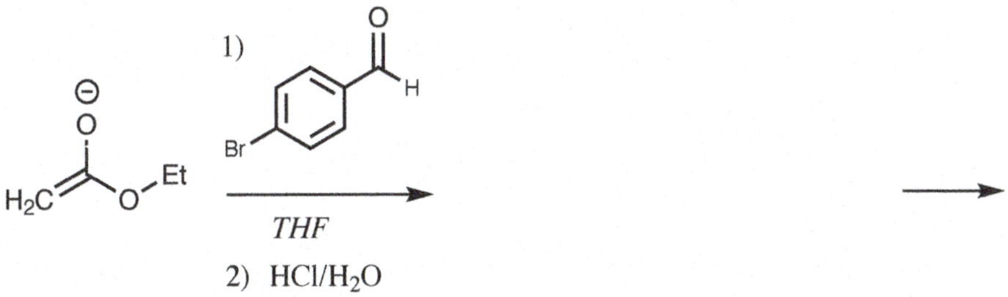

1)

THF

2) HCl/H₂O

Reactions of Acyl Chlorides, Anhydrides, Esters, and Amides

Esters from Acid Chlorides or Anhydrides

13. Reaction of benzoyl chloride with ethoxide.

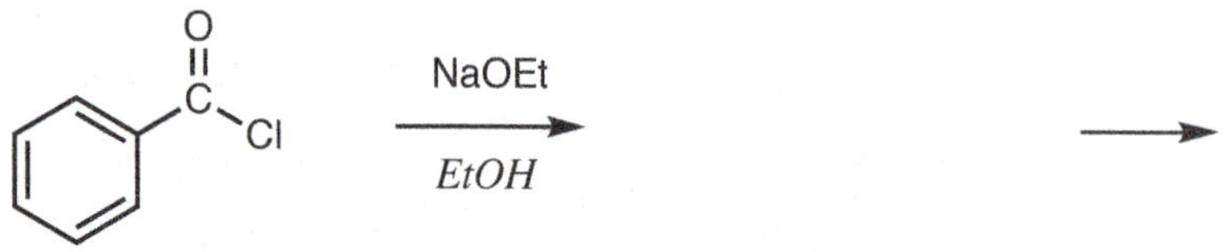

14. Reaction of ethanol with acetyl chloride.

15. Pyridine catalyzed acylation ethanol with benzoyl chloride.

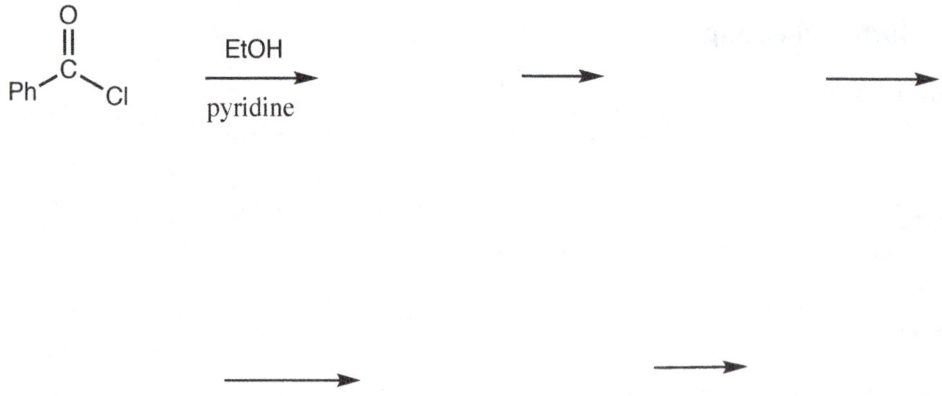

cont'd

16. Reaction of acetic anhydride with ethanol catalyzed by sulfuric acid.

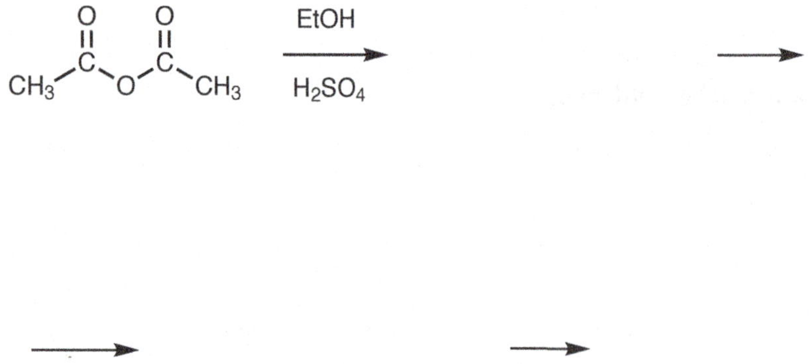

cont'd

Amides from Acid Chlorides or Anhydrides

17. Reaction of benzoyl chloride and ethylamine.

18. Reaction of acetic anhydride, aniline, and pyridine.

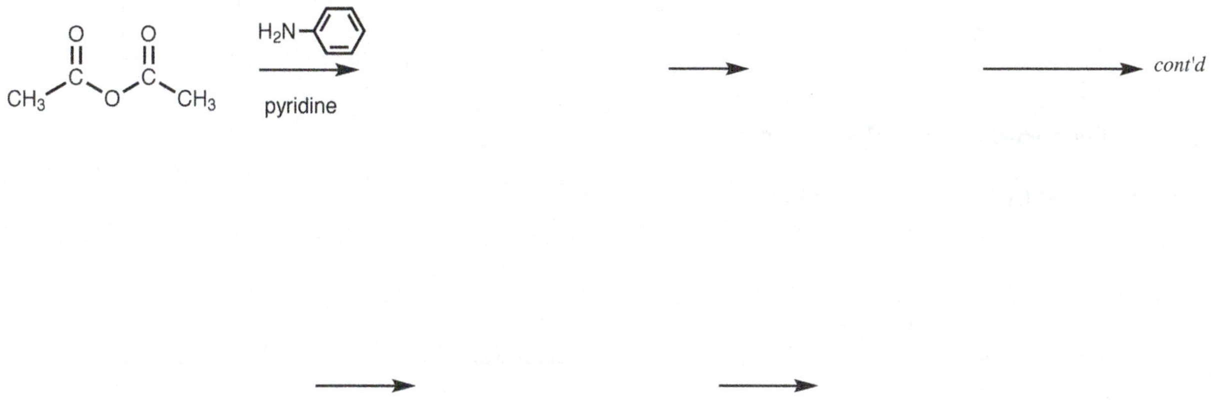

cont'd

19. Reaction of acetic anhydride with aniline.

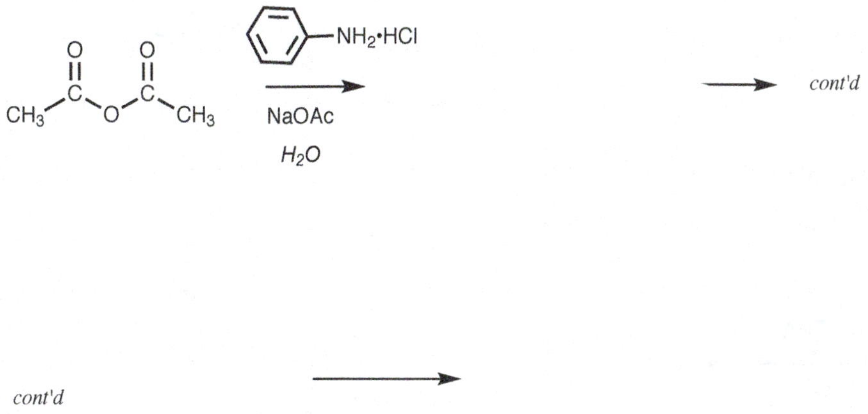

cont'd

cont'd

Ester from Acid with Mineral Acid Catalysis (Fischer Esterification)

20. Reaction of benzoic acid and methanol with sulfuric acid or hydrogen chloride.

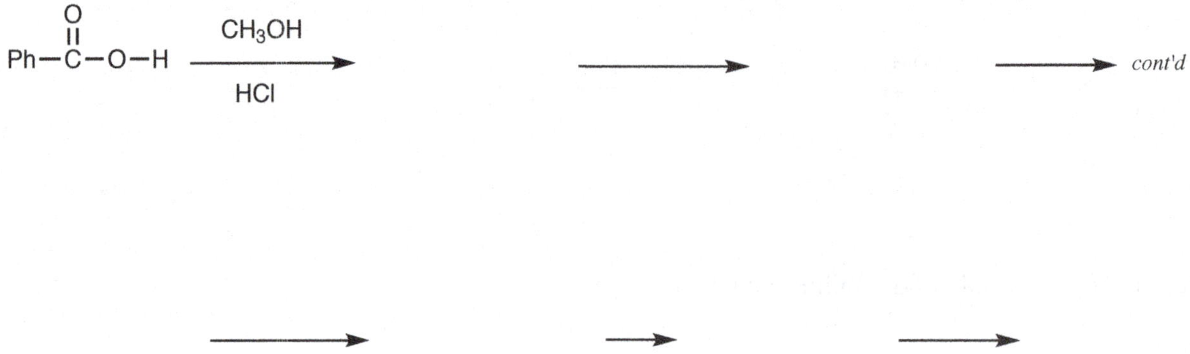

Acid Catalyzed Hydrolysis of an Ester

21. Acid catalyzed hydrolysis of methyl pentanoate (valerate).

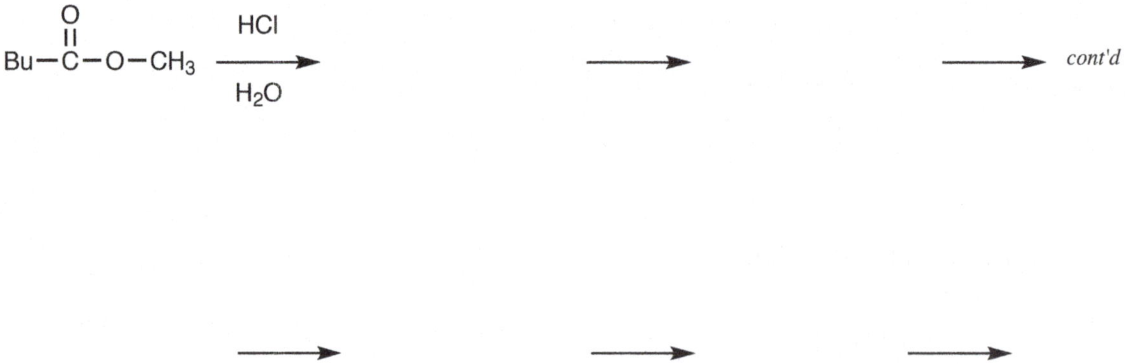

22. Treatment of *t*-butyl pivalate with acid.

Base Hydrolysis of an Ester (Saponification)

23. Base hydrolysis of octyl isobutyrate to give octanol and isobutyric acid. Step 1, treatment with base.

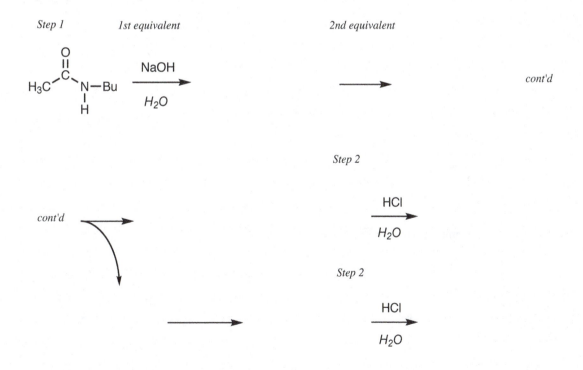

Step 2, acidification of isobutyrate and isolation of isobutyric acid.

Hydrolysis of an Amide

24. Base hydrolysis of *N*-butylacetamide.

Step 1   *1st equivalent*   *2nd equivalent*

$$H_3C-\overset{\overset{O}{\|}}{C}-\underset{\underset{H}{|}}{N}-Bu \xrightarrow[H_2O]{NaOH}$$

cont'd

Step 2

$$\xrightarrow[H_2O]{HCl}$$

cont'd

Step 2

$$\xrightarrow[H_2O]{HCl}$$

25. Acid catalyzed hydrolysis of *N,N*-dimethylacetamide.

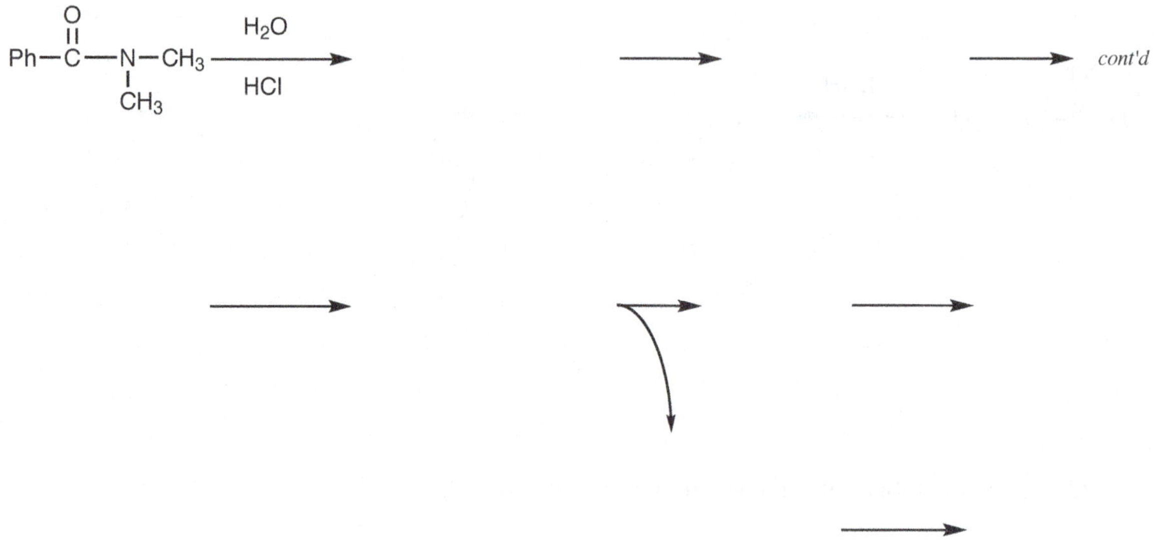

Reactions of Esters

26. Reaction of methyl magnesium bromide with ethyl benzoate.

Step 1

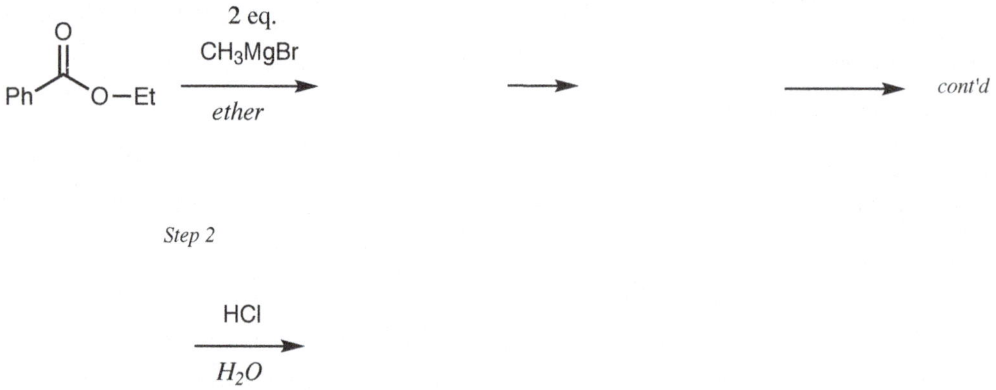

Step 2

27. Reaction of 4-nitrophenyl propionate with ethylamine.

Reactions of Nitriles

28. Acid catalyzed hydrolysis of benzonitrile. (See *Notes*.)

Ph—C≡N : $\xrightarrow[\substack{H_2O \\ \textit{heat}}]{HCl}$ ⟶ ⟶ *cont'd*

29. Base catalyzed hydrolysis of cyclopentanecarbonitrile.

—C≡N $\xrightarrow[\substack{H_2O \\ \textit{mild conditions}}]{NaOH}$ ⟶ ⟶ *cont'd*

30. Reaction of phenyl lithium with cyclohexanecarbonitrile.

   Step 1, reaction with nitrile.

   Step 2, hydrolysis of the imine.

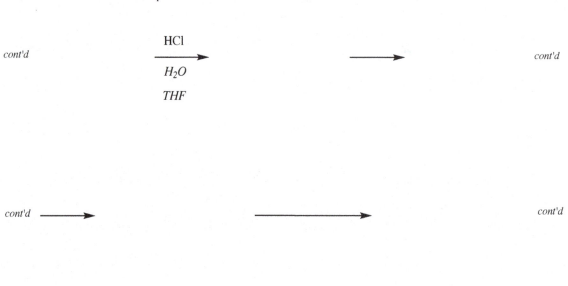

Step 2

cont'd   HCl
                    ──────►   ──────►   cont'd
                    H₂O
                    THF

cont'd ──────►   ──────────►   cont'd

cont'd ──────►   ──────►

Miscellaneous

31. Reaction of diazomethane with a carboxylic acid. (See Notes.)

9 — Reactions of Enols and Enolates

Aldol Reaction

1. Base catalyzed aldol condensation of butanal (butyraldehyde). (See *Notes*.)

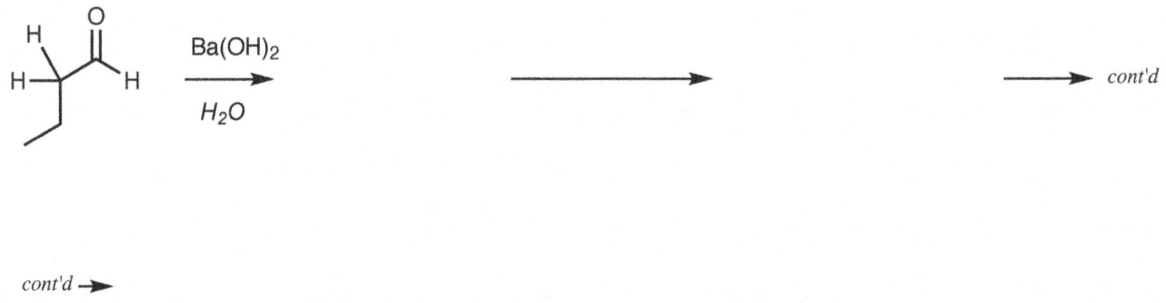

Base catalyzed dehydration step. Under concentrated base or heating, the dehydration reaction may occur spontaneously.

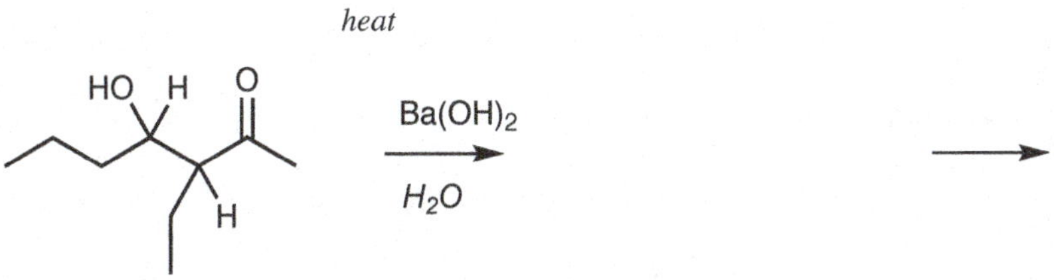

2. Directed kinetic aldol condensation of 2-methylcyclohexanone with propanal (propionaldehyde). Step 1, enolate formation; step 2, reaction with propanal; step 3, neutralization.

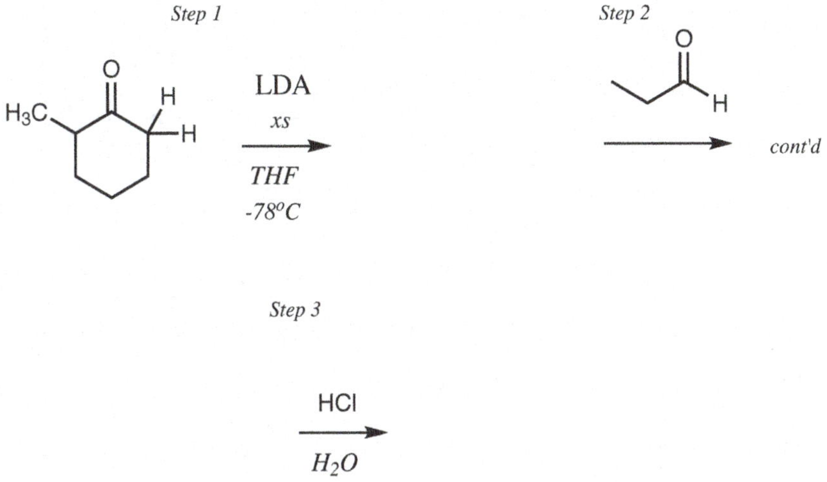

3.   Base catalyzed mixed or crossed aldol condensation of acetone and benzaldehyde.

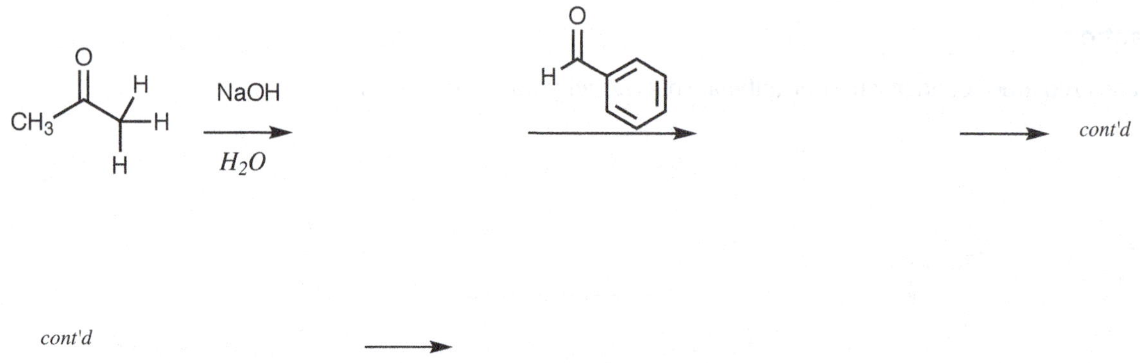

cont'd

4.   Mannich reaction, acid catalyzed enolization of 2-propanone in a reaction with diethylamine, formaldehyde, and 2-propanone to give 4-(diethylamino)butan-2-one. (See *Notes*.)

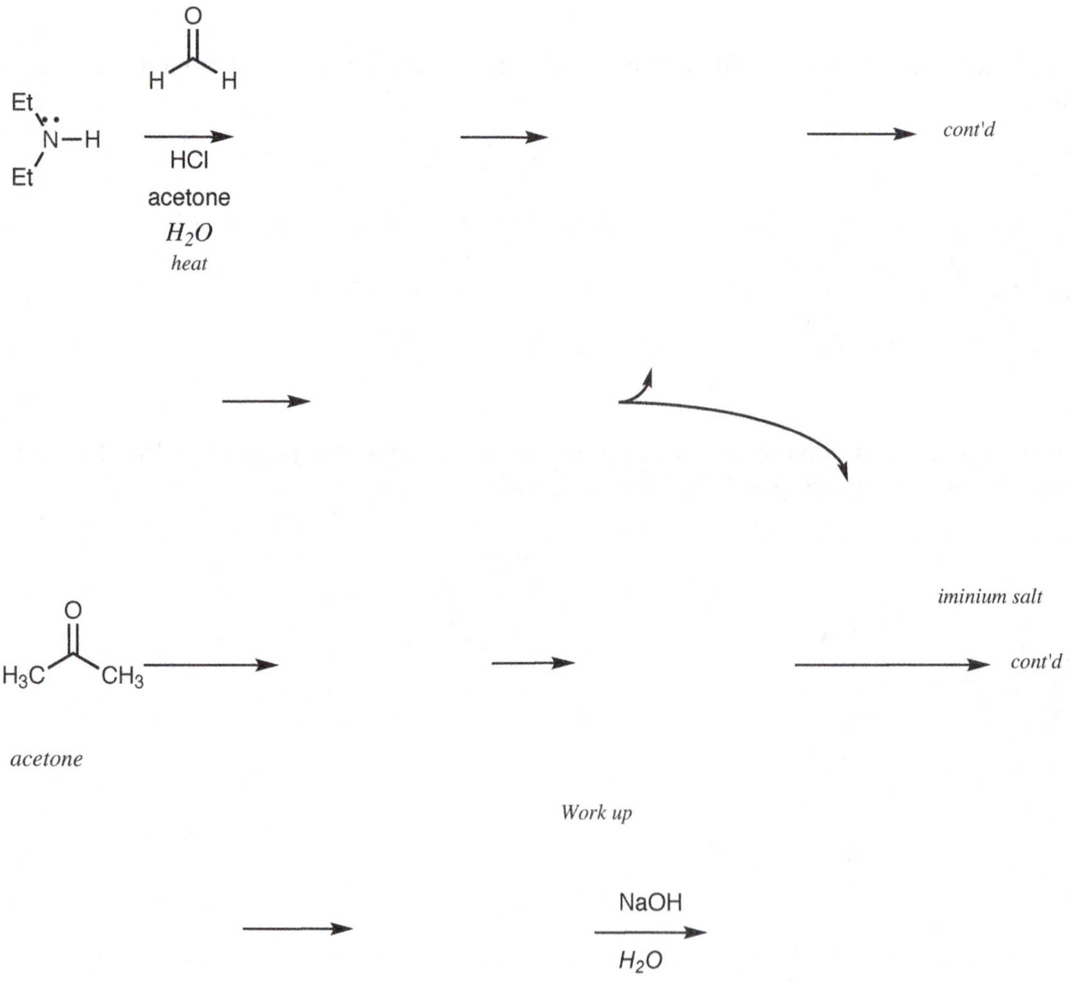

iminium salt

acetone

Work up

Claisen Condensation

5. Ethoxide catalyzed Claisen condensation of ethyl acetate. (See *Notes.*)

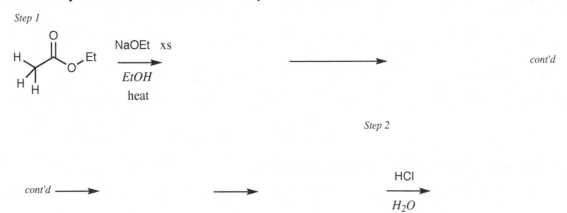

Step 1

NaOEt xs

EtOH
heat

cont'd

Step 2

cont'd ⟶

HCl

H_2O

6. Ethoxide catalyzed crossed-Claisen condensation of cyclohexanone and ethyl formate. (See *Notes.*)

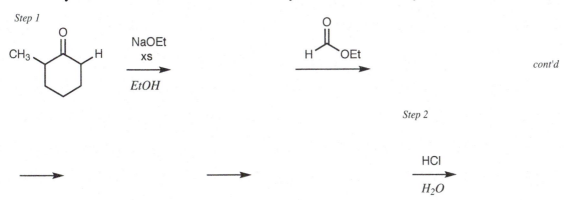

Step 1

NaOEt
xs

EtOH

cont'd

Step 2

HCl

H_2O

7. Ethoxide catalyzed crossed-Claisen condensation of ethyl acetate and ethyl benzoate. (See *Notes.*)

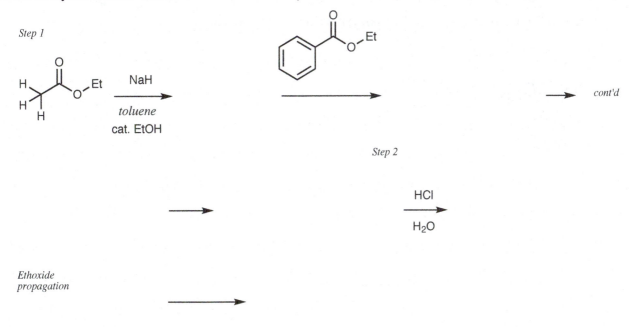

Step 1

NaH

toluene
cat. EtOH

cont'd

Step 2

HCl

H_2O

*Ethoxide
propagation*

Acetoacetate Synthesis

8. Step 1, S_N2 alkylation of acetoacetate.

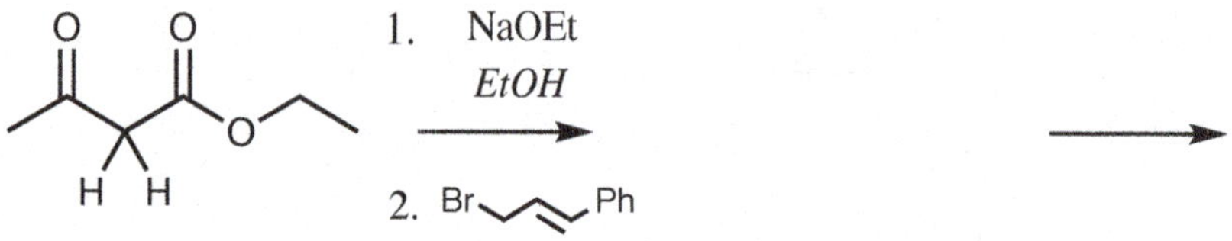
O
O
O
H H
1. NaOEt
EtOH
2. Br
Ph

Step 2, sodium hydroxide hydrolysis of ester (saponification).

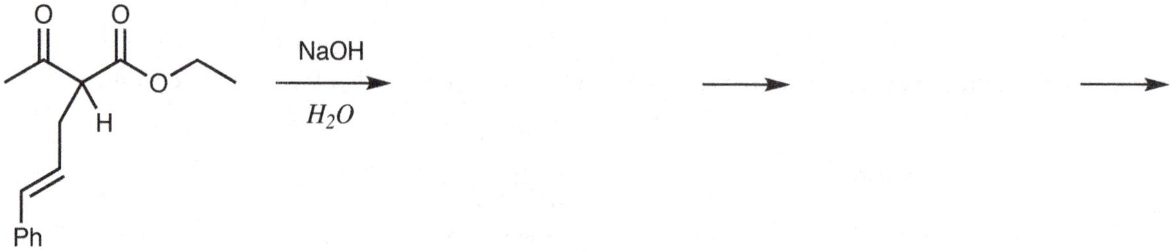
O
O
O
H
Ph
NaOH
H2O

cont'd

cont'd →

Step 3, acidification of carboxylate, decarboxylation, and tautomerization.

O
O
O
H
Ph
1. HCl / H2O
2. heat

cont'd

cont'd →

Enolate Alkylation Reactions

9. Enolization and alkylation of ethyl propionate with benzyl bromide. (See *Notes*.)

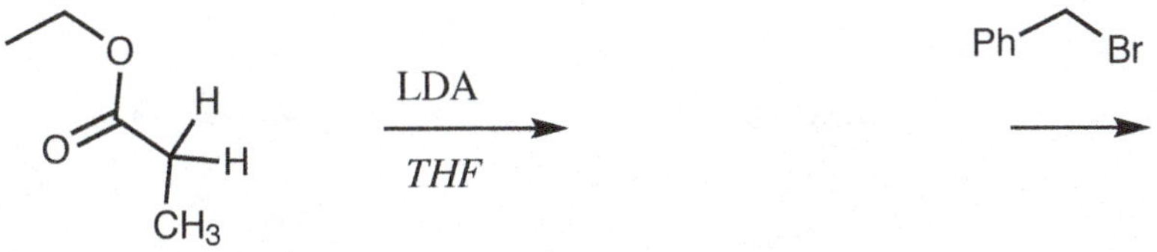

10. Sequential alkylation of a dianion of a *beta*-ketoester. (See *Notes*.)

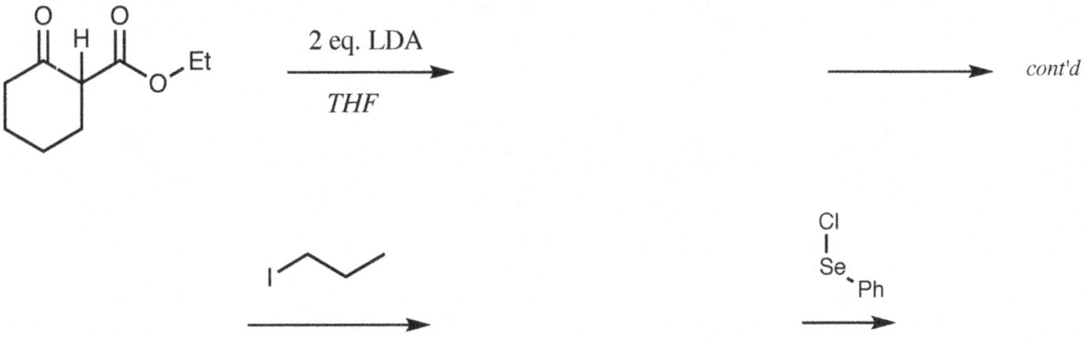

11. A retro-Claisen reaction from a *beta*-ketoester. (See *Notes*.)

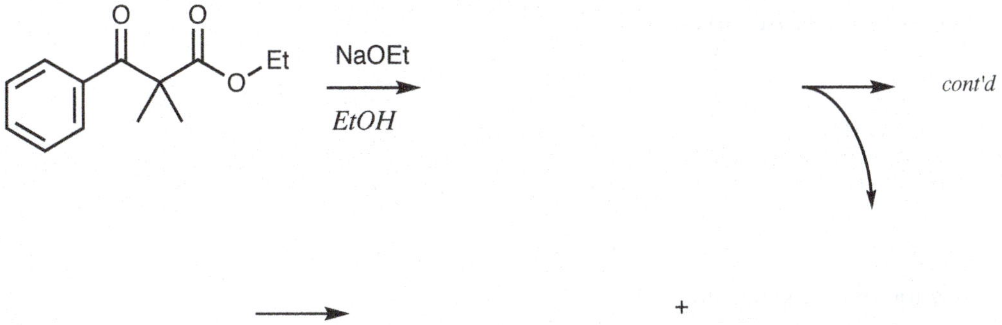

12. Enolization and alkylation of phenylacetonitrile with methyl iodide. (See *Notes*.)

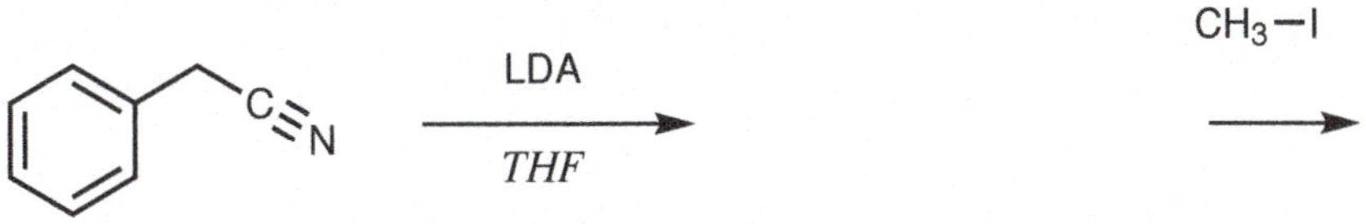

Halogenation of Carbonyl Compounds

13. Basic bromination of 3-methyl-2-butanone with sodium hydroxide and bromine, bromoform reaction.

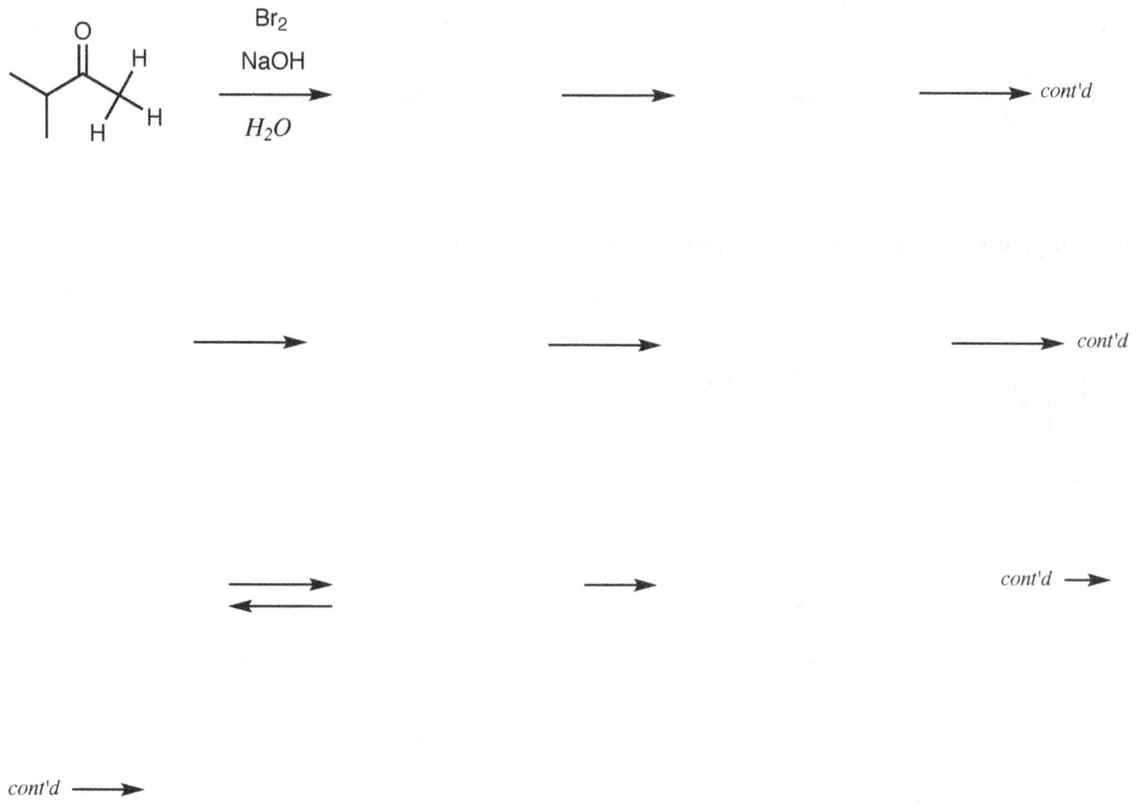

Br$_2$

NaOH

H$_2$O

cont'd

cont'd

cont'd

cont'd

Step 2, acidification and isolation of isobutyric acid.

cont'd

HCl

H$_2$O

14. Acid catalyzed bromination of acetophenone.

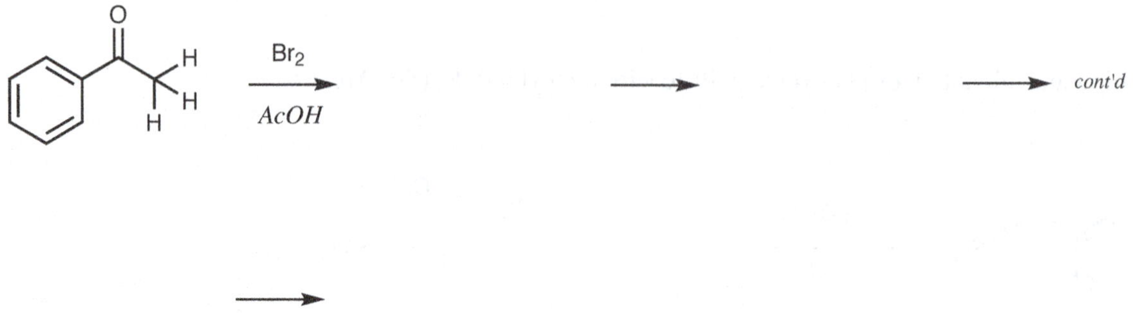

Br$_2$

AcOH

cont'd

Michael Addition or 1,4-Conjugate Addition Reaction

15. Michael addition reaction of dimethylamine to methyl vinyl ketone.

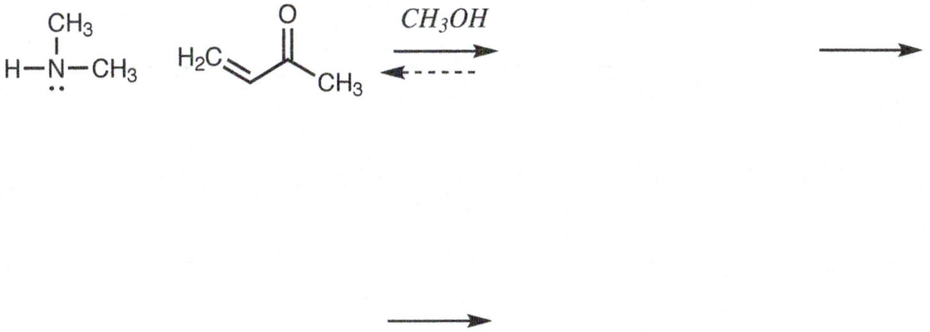

16. Michael addition reaction of ethyl acetoacetate to cyclohexenone.

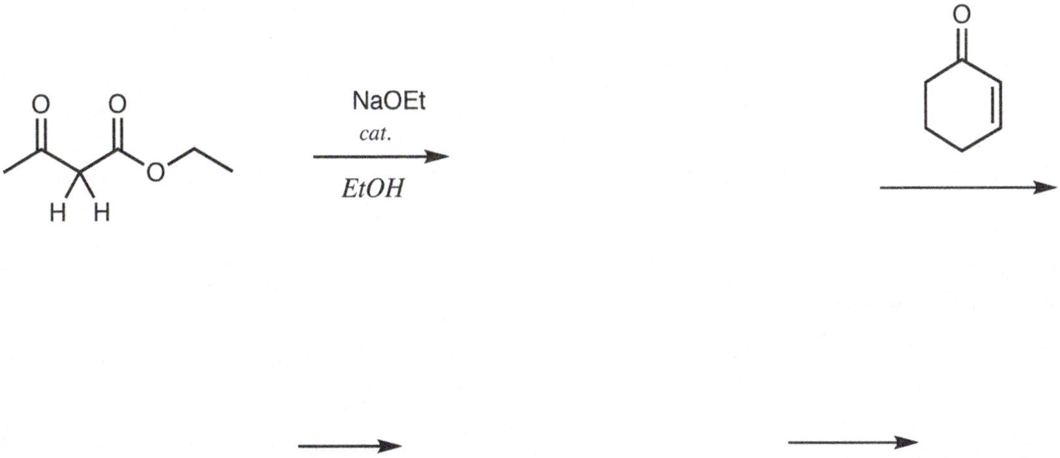

Step 2, neutralization of reaction mixture.

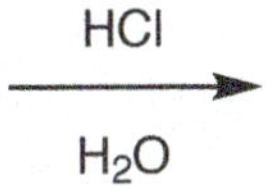

Enamine Alkylation

17. Step 1, formation of the morpholine enamine of cyclohexanone.

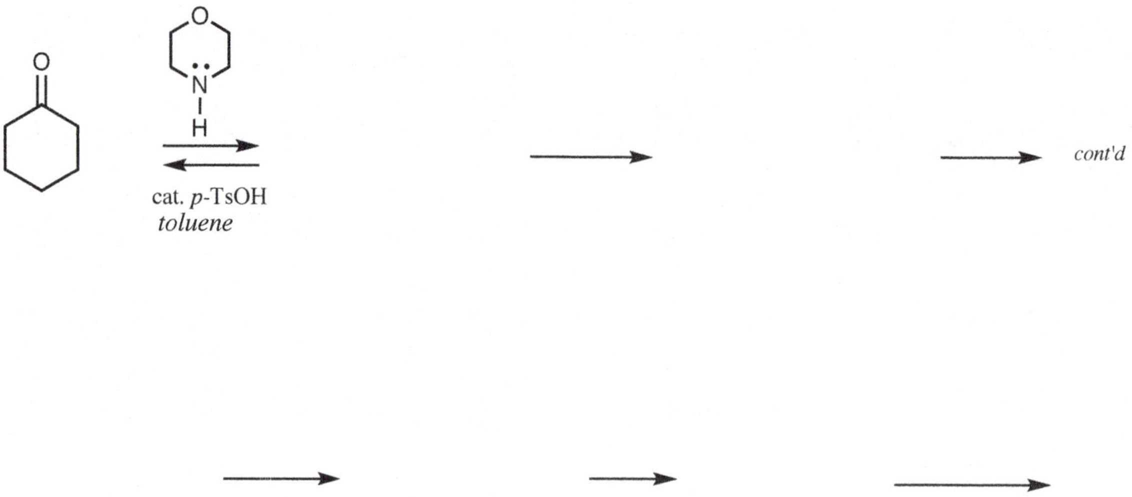

cat. *p*-TsOH
toluene

cont'd

Step 2, enamine alkylation with methyl iodide.

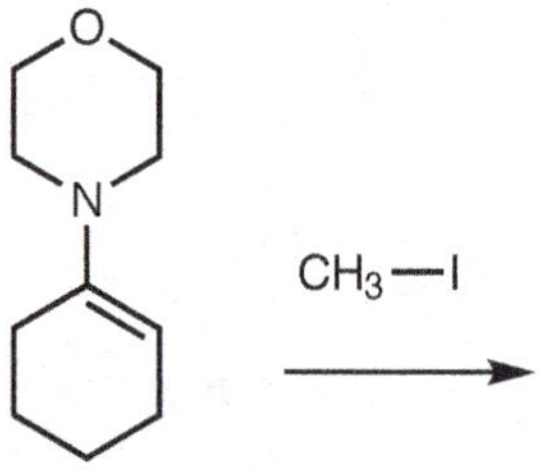

Enamine

CH₃ — I

Step 3, hydrolysis of iminium salt.

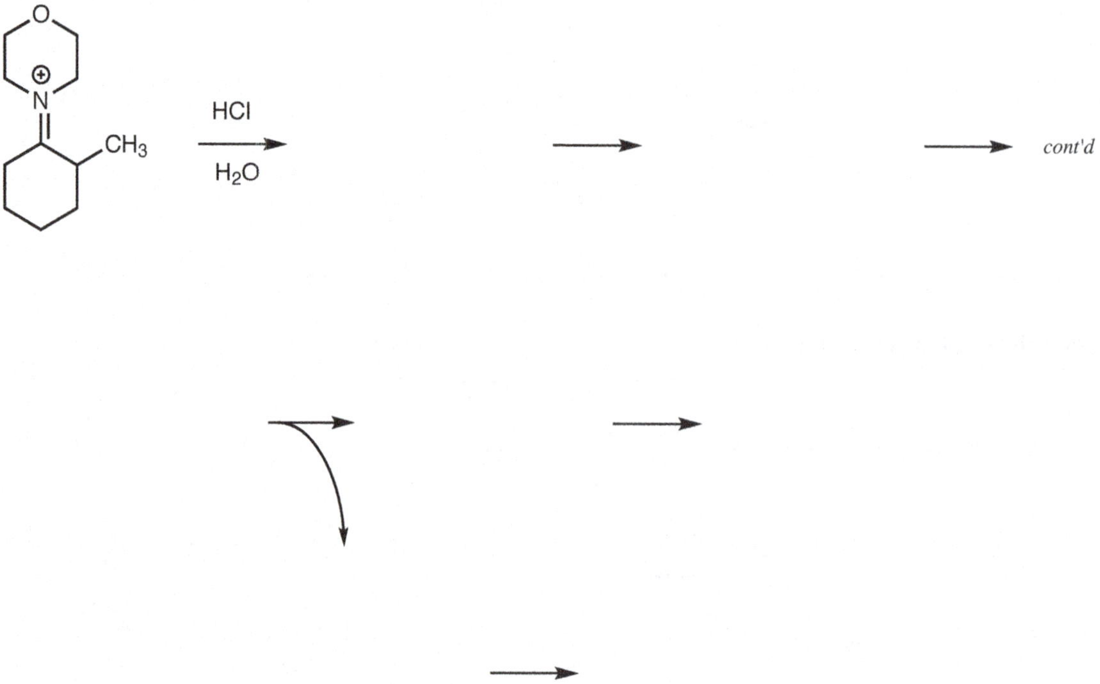

HCl

H₂O

cont'd

10 — Dehydration/Halogenation Agents

1. Reaction of cyclohexanecarboxamide with acetic anhydride to give cyclohexanecarbonitrile. (See *Notes*.)

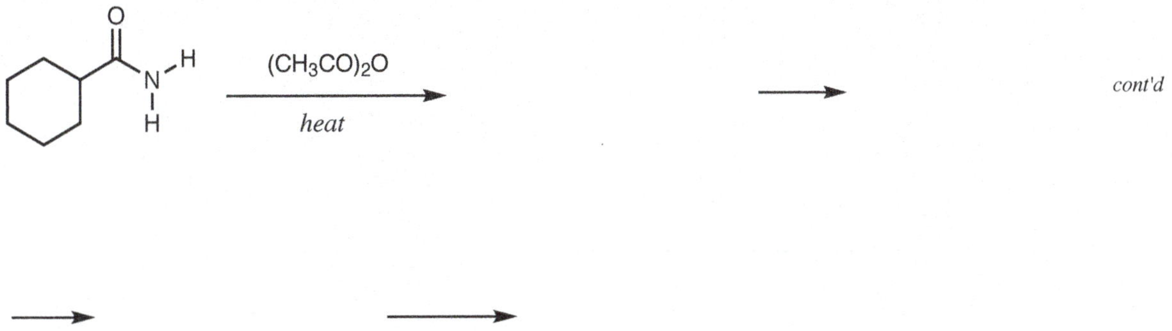

2. S$_N$2 reaction of thionyl chloride with 1-butanol to give chlorobutane. (See *Notes*.)

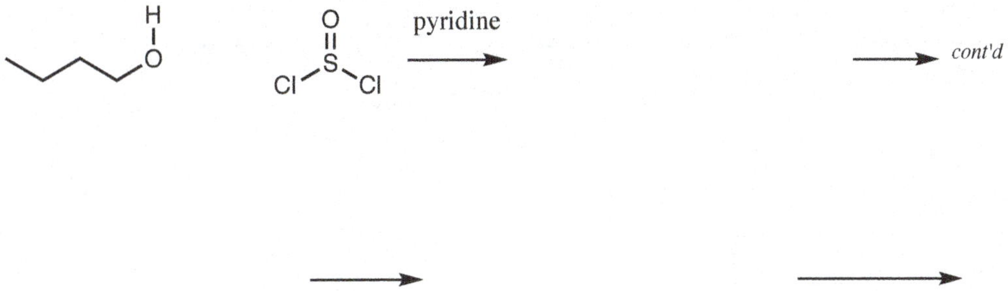

3. Reaction of 1-butanol with tosyl chloride and pyridine to give butyl tosylate. (See *Notes*.)

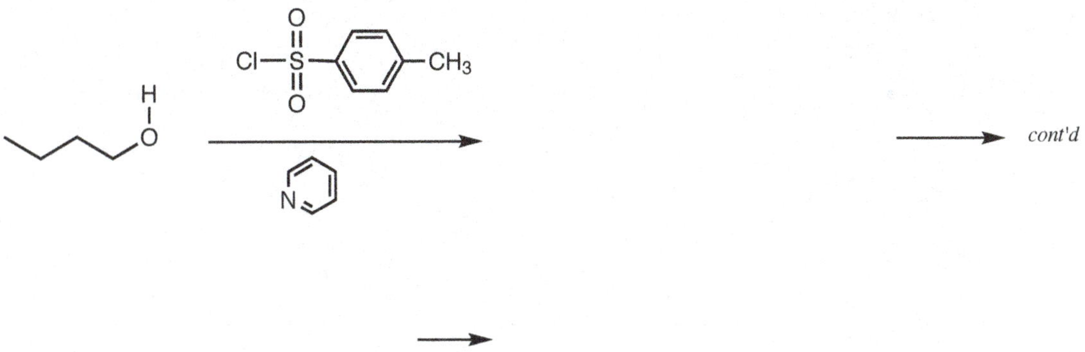

Carboxylic Acid with Thionyl Chloride

4. Conversion of acetic acid to acetyl chloride with thionyl chloride.

cont'd   ⟶

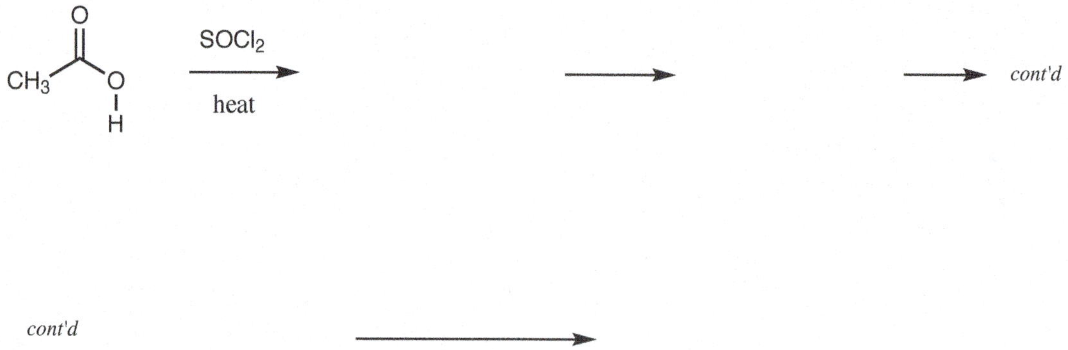

5. Conversion of acetic acid to acetyl chloride with thionyl chloride. Part 1, reaction of thionyl chloride with DMF to form iminium salt. (See Notes.)

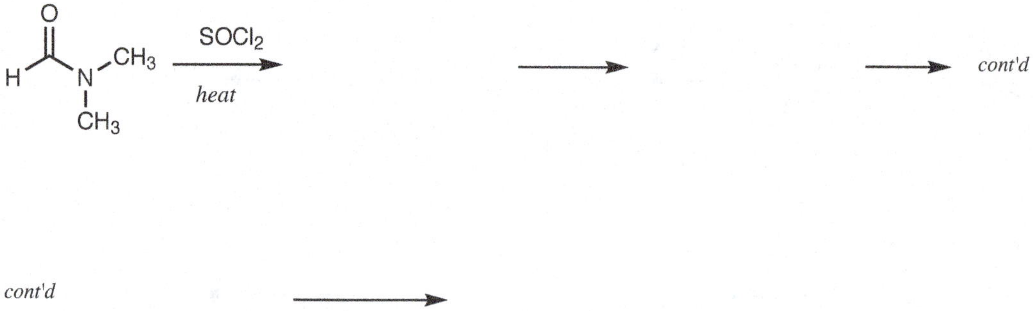

cont'd   ⟶

Part 2, reaction of iminium salt with carboxylic acid

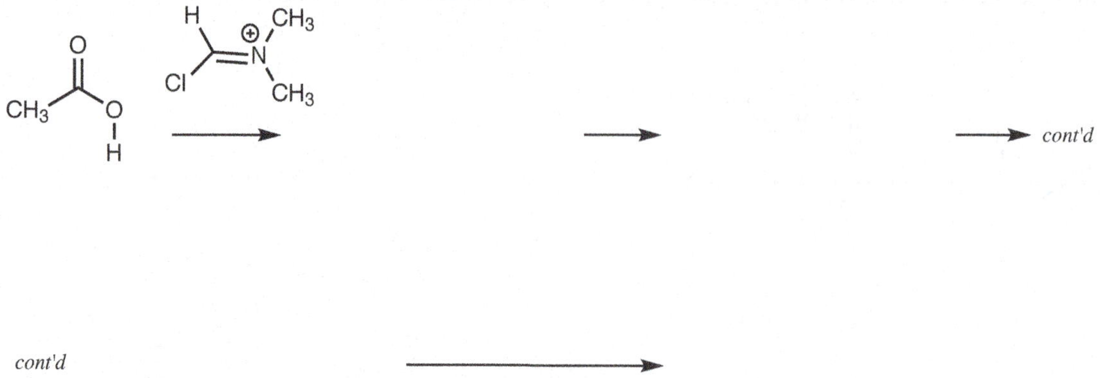

cont'd   ⟶

Halide from an Alcohol with a Phosphorus Reagent

6. Reaction of isobutyl alcohol with phosphorus tribromide to give isobutyl bromide. (See *Notes.*)

7. Reaction of triphenylphosphine, carbon tetrachloride and cyclopentanol to give chlorocyclopentane. (See *Notes.*)

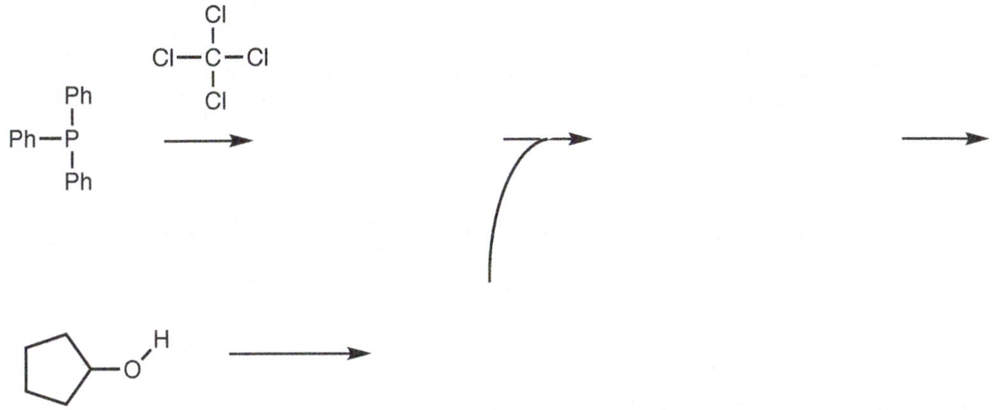

Ester from an Alcohol and Carboxylic Acid with a Phosphorus Reagent

8. Mitsunobu reaction with triphenyl phosphine and diethyl azodicarboxylate to prepare an ester from an acid and an alcohol.

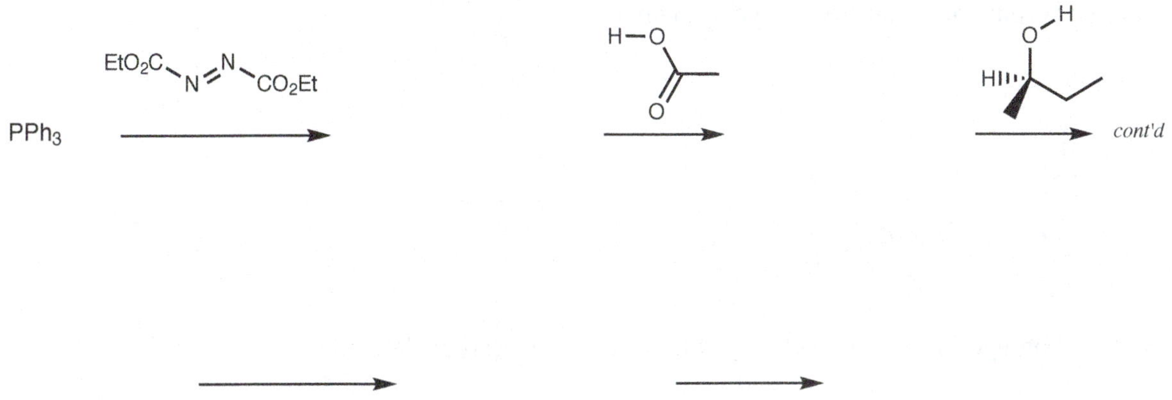

11 — Reduction Reactions

Sodium Borohydride Reductions

1. Sodium borohydride reduction of 2-methylpropanal (isobutyraldehyde).

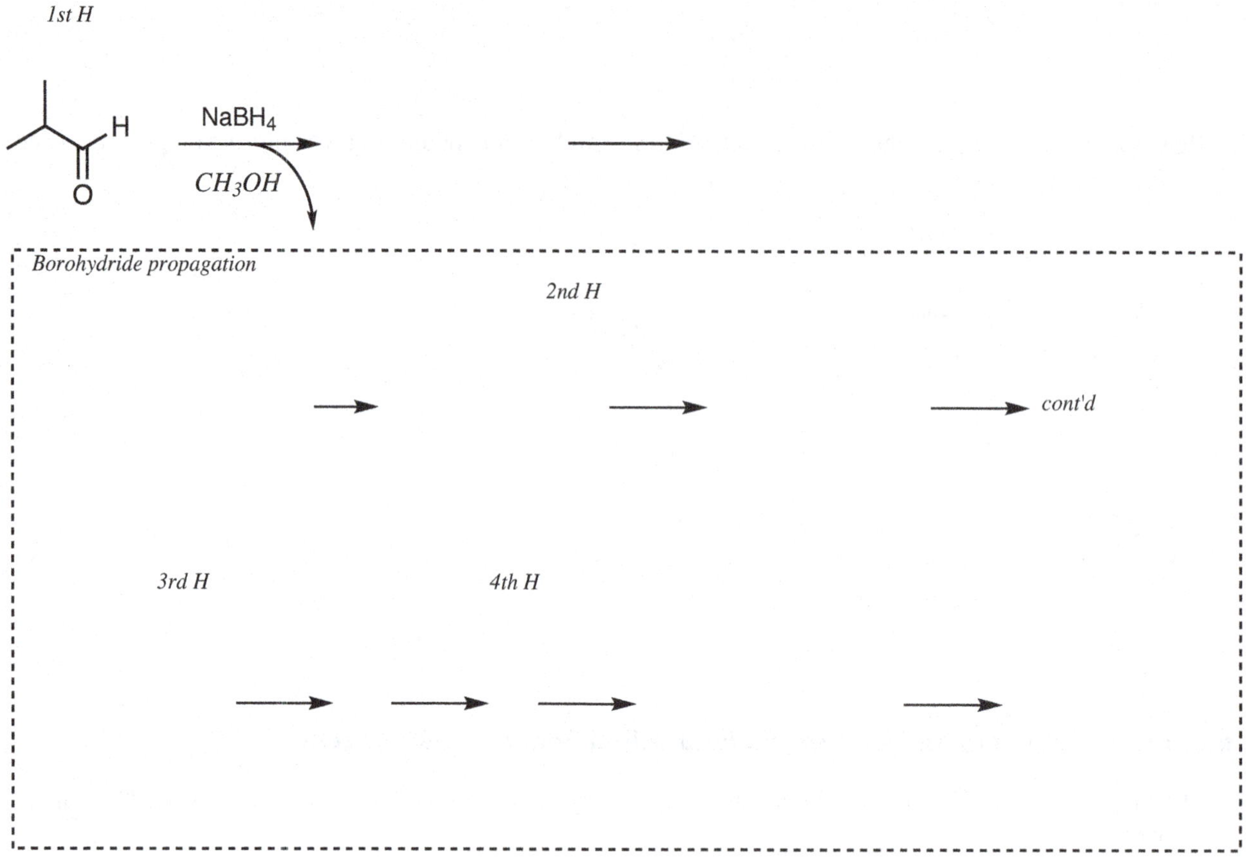

1st H

2. Sodium borohydride reduction of cyclopentanone.

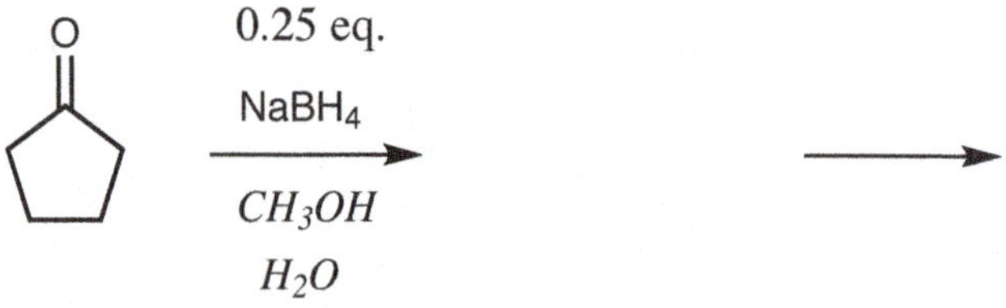

3. Sodium borohydride reduction of ethyl 4-oxocyclohexanecarboxylate. (See *Notes*.)

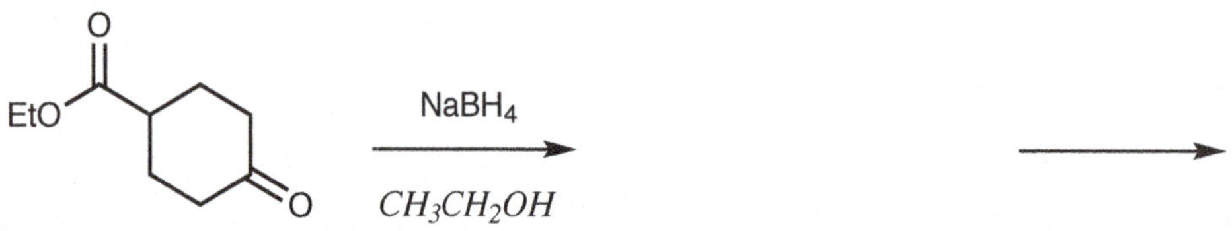

Lithium Aluminum Hydride Reductions

4. Lithium aluminum hydride reduction of acetophenone.

Step 1

0.25 eq.

LiAlH$_4$

Et$_2$O

Step 2   Work-up

H$_2$O

5. Lithium aluminum hydride reduction of ethyl butanoate (butyrate). (See *Notes*.)

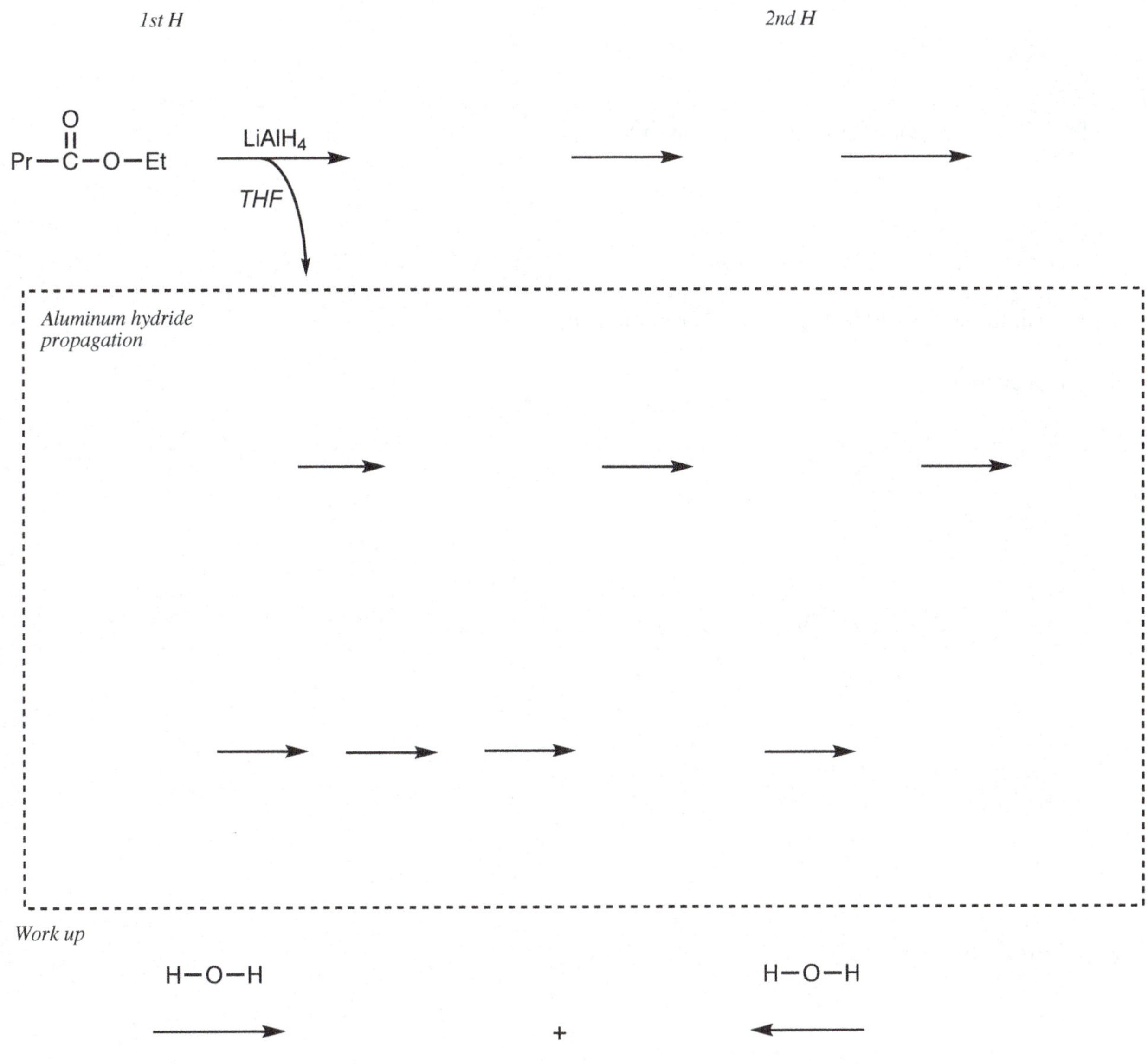

1st H   *2nd H*

Pr—C—O—Et

LiAlH$_4$

THF

Aluminum hydride propagation

Work up

H—O—H   H—O—H

+

6.   Lithium aluminum hydride reduction of a cyclic amide(1-methylpyrrolidin-2-one). (See *Notes*.)

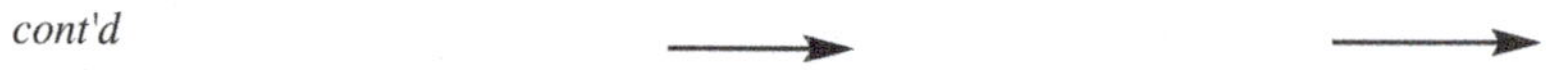

cont'd

cont'd

Preferred leaving group for lithium aluminum hydride reductions.

7.   Lithium aluminum hydride reduction of propionyl anilide.

Step 1, reduction.

cont'd

Step 2, work up.

cont'd   H₂O

8. Lithium aluminum hydride reduction of phenylacetic acid. (See *Notes*.)

   Step 1, reduction.

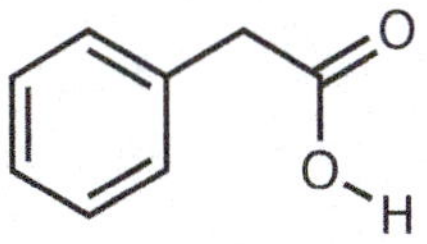

LiAlH₄

THF

heat

cont'd

Step 2, work up.

cont'd

H₂O

Reductive Amination

9. Step 1, formation of imine from benzaldehyde and ethylamine. (See *Notes*.)

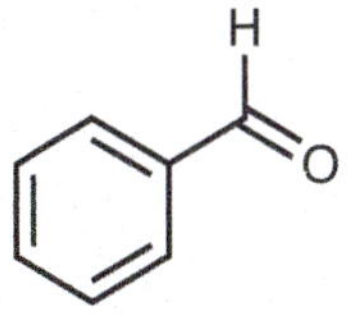

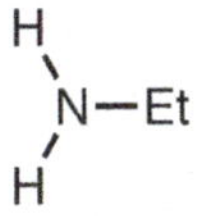

HCl

EtOH / H₂O

cont'd

cont'd

Step 2, reduction

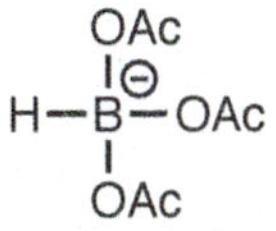

EtOH / H₂O

cont'd

Diisobutylaluminum Hydride Reduction of an Ester

10. Step 1, addition of DIBAH or DIBAL(H) (diisobutylaluminum hydride) to ethyl isobutyrate.

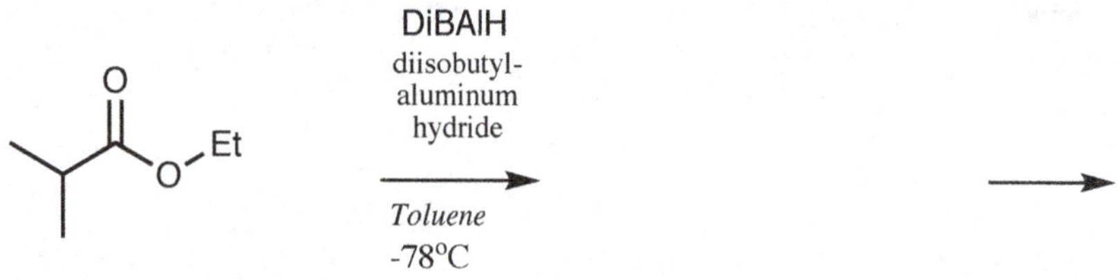

DiBAlH
diisobutyl-
aluminum
hydride

Toluene
-78°C

Step 2, work-up

H_2O

Reduction of Alkyne with Sodium and Ammonia

11. Sodium and ammonia reaction of 2-butyne.

Note, use a single barbed arrow for
one-electron transfers reaction.

CH_3—$C{\equiv}C$—CH_3 Na
                            NH_3 *cont'd*

cont'd

Wolff Kischner Reduction

12. Reaction of the ketone with hydrazine under basic conditions.

cont'd

cont'd

Reaction of the *in situ* formed hydrazide with KOH.

cont'd

cont'd

Catalytic Reduction of Nitrobenzene

Catalytic reduction of nitrobenzene. This isn't a mechanism. It shows that hydrogen can add across N-O single and double bonds. It is a very facile reduction. One can see that the oxygen atoms form water and the reduction takes three moles of hydrogen.

13. Catalytic reduction of nitrobenzene.

cont'd

12 — Oxidation Reactions

General Form For Oxidations

See *Notes* for further discussion of oxidation reactions.

Chromic Acid Oxidation

1. Chromic acid oxidation (Jones oxidation) of 3-methyl-2-butanol. (See *Notes*.)

Several steps

2. Chromic acid oxidation of isobutyl alcohol. (See *Notes*.)

   Step 1, chromate ester formation and oxidation.

Several steps

CrO$_3$ or
H$_2$CrO$_4$
⟶
H$_2$SO$_4$ / *H$_2$O*
acetone   ⟶   *cont'd*

cont'd   ⟶

   Step 2, hydration and second oxidation.

Several steps   *Several steps*
H$_2$O
CrO$_3$ or
H$_2$CrO$_4$
⟶
⟶   ⟶   ⟶   *cont'd*
H$_2$SO$_4$ / *H$_2$O*
acetone

cont'd   ⟶

PCC, Tollens, Hypochlorite, *m*CPBA, and Sulfonium Based Oxidations

3. Pyridinium chlorochromate (PCC) oxidation of benzyl alcohol.

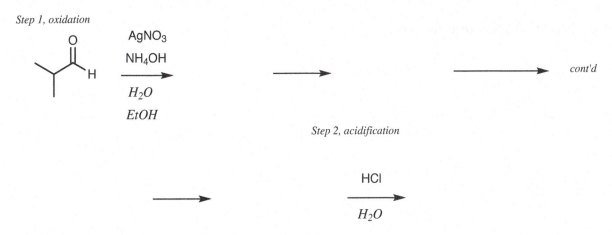

Several steps

Ph—CH₂OH →[PCC / CH₂Cl₂] ——————→

4. Tollens oxidation of 2-methylpropanal (isobutyraldehyde) with silver oxide (or hydroxide).

Step 1, oxidation

→[AgNO₃ / NH₄OH / H₂O / EtOH] ——————→ ——————→ *cont'd*

Step 2, acidification

——————→ →[HCl / H₂O]

5. Oxidation of cyclohexanol with sodium hypochlorite (NaOCl, bleach).

→[NaOCl / HCl / H₂O / HOAc] ——————→ ——————→

6. Peracid epoxidation of *trans*-2-butene with *m*-chloroperoxybenzoic acid (MCPBA). (See *Notes*.)

→[MCPBA] ——————→

7. Step 1, Swern oxidation, preparation of chlorosulfonium salt.

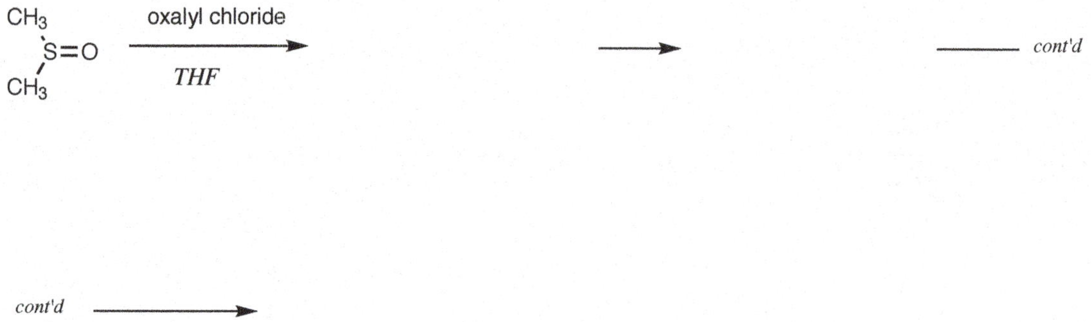

cont'd

cont'd

Step 2, Swern oxidation, oxidation step.

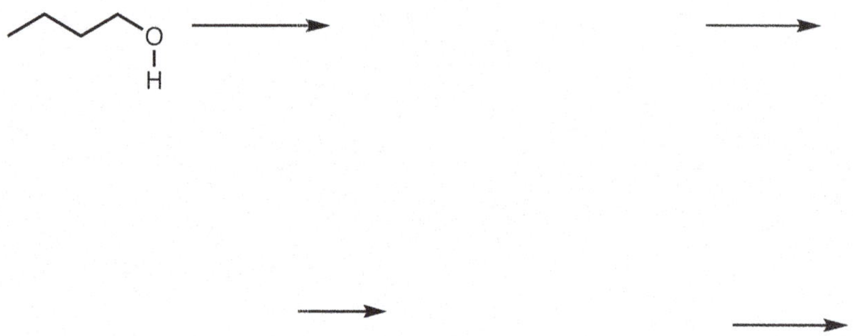

cont'd

8. Corey-Kim (chlorosulfonium salt) oxidation of cyclopentanol to cyclopentanone. (See *Notes*.)

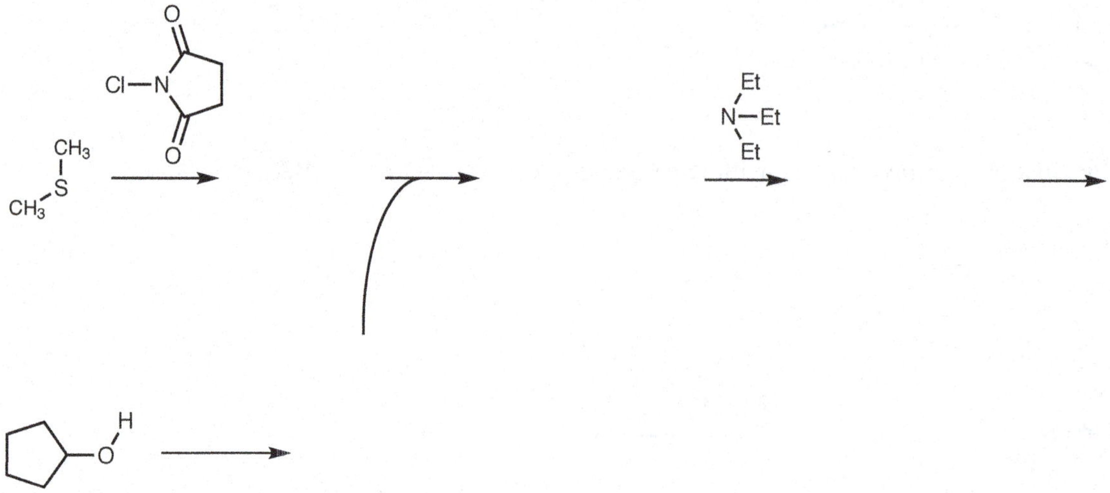

Ozone Oxidation

9. Step 1, ozone reaction with an alkene to give an ozonide.

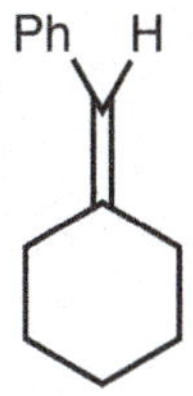

cont'd

molozonide

cont'd ⟶

ozonide

Step 2, reduction of ozonide to two carbonyl compounds.

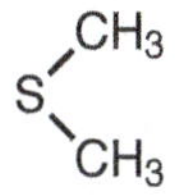

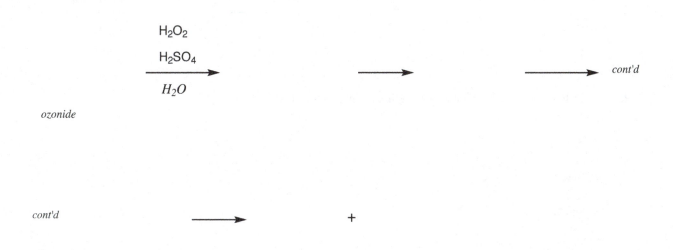

ozonide

Alternate step 2, oxidative destruction of ozonide.

H_2O_2

H_2SO_4

H_2O

cont'd

ozonide

cont'd   +

Osmium Tetroxide, Potassium Permanganate, and Periodate Oxidations

10. Osmium tetroxide oxidation of cyclohexene to give *cis*-1,2-cyclohexanediol

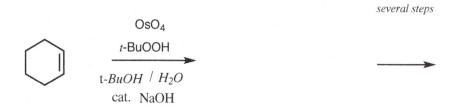

several steps

OsO₄

t-BuOOH

t-*BuOH* / *H₂O*

cat. NaOH

Regeneration of oxidant with *tert*-butylperoxide.

$$:\overset{\overset{O}{\|}}{\underset{\underset{O}{\|}}{Os}}=O$$

t-BuOOH
————————→
cat. NaOH
t-*BuOH* / *H₂O*

————→

11. Potassium permanganate oxidation of methylcyclohexene to 1-methylcyclohexane-1,2-diol at low temperature or cleavage at high temperature.

CH₃

several steps

KMnO₄
————→
H₂O

A

NaOH, H₂O
————————→
0°C

several steps   *several steps*

Heat
————————→

1. KMnO₄
————————→
2. HCl

A   H₂O

12. Periodate cleavage of a 1,2-diol to give a dicarbonyl compound.

H₃C   H
     |
     O

NaIO₄
NaOH
———————⇒
H₂O
several steps

O
|
H   O
|
H

————→

13 — Organometallic Reactions

The (transition) organometallic reactions in this chapter are of increasing importance. The reactions may recount the steps more than explain a reaction. Significantly absent is the concept of 18 valence electrons about the catalytic palladium atom.

Acyclic Heck Reaction

1. Step 1, reduction of palladium (II) to zero valent palladium with propene.

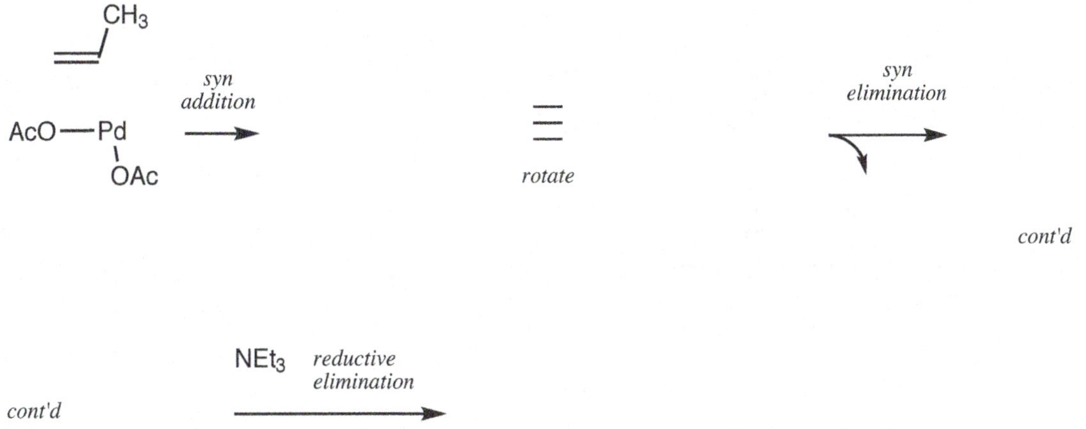

Step 2, the catalytic cycle (oxidative-addition, *syn*-addition, *syn*-elimination, and reductive-elimination) with 2-bromopropene and propene.

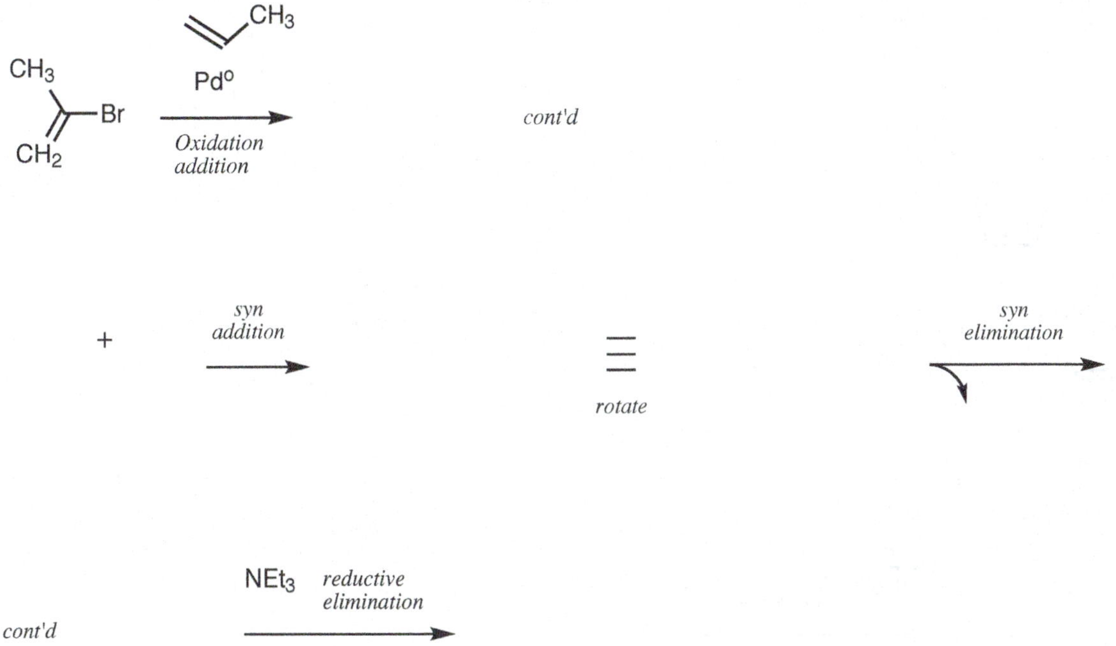

Cyclic Heck Reaction

2. Step 1, reduction of palladium (II) to zero valent palladium with cyclopentene.

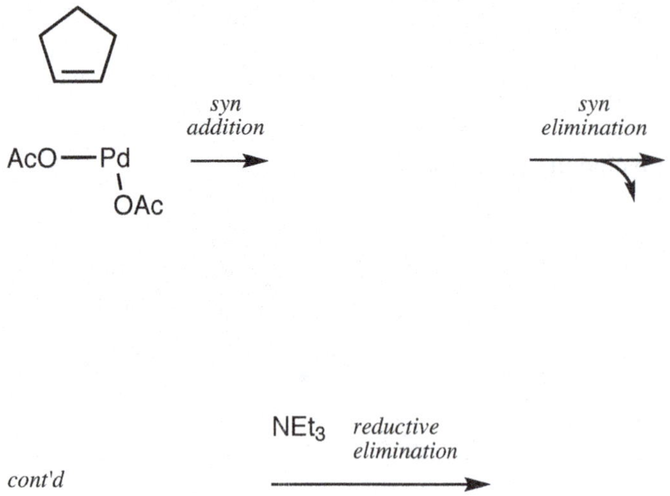

AcO—Pd
      |
     OAc

*syn
addition*

*syn
elimination*

cont'd

NEt₃   *reductive
        elimination*

Step 2, the catalytic cycle (oxidative-addition, *syn*-addition, *syn*-elimination, and reductive-elimination) with iodobenzene and cyclopentene.

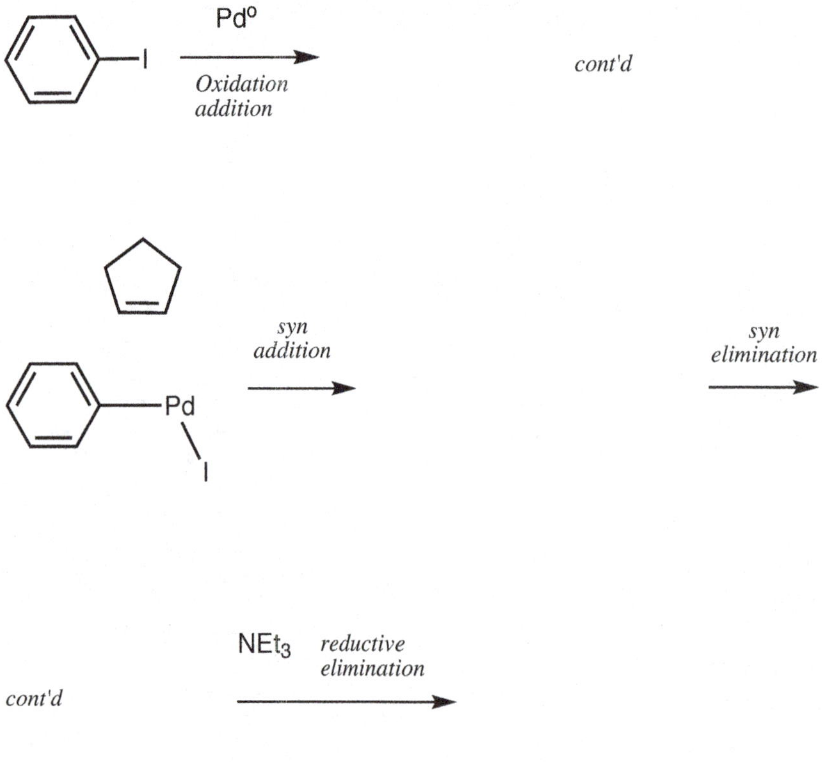

Pd⁰

*Oxidation
addition*

cont'd

—Pd
   \
    I

*syn
addition*

*syn
elimination*

cont'd

NEt₃   *reductive
        elimination*

Catalytic Reduction of an Alkene

3. Catalytic hydrogenation of *cis*-3-hexene to hexane.

Gilman Reagent

4. Formation of Gilman reagent, lithium dimethylcuprate.

$$CH_3-Li \quad \xrightarrow[\textit{THF}]{\substack{0.5 \text{ eq.} \\ CuI}} \quad \longrightarrow$$

Coupling of Gilman reagent, lithium dimethylcuprate with (Z)-1-bromopent-1-ene to give *cis*-2-hexene.

(no mechanism)

5. A 1,4-conjugate addition of lithium dimethylcuprate to 2-cyclohexenone to give 3-methylcyclohexanone.

6. Reaction of lithium dimethylcuprate to benzoyl chloride to give acetophenone.

14 — Aromatic Substitution Reactions

Electrophilic Aromatic Substitution of Benzene

1. Friedel Crafts acylation of benzene.

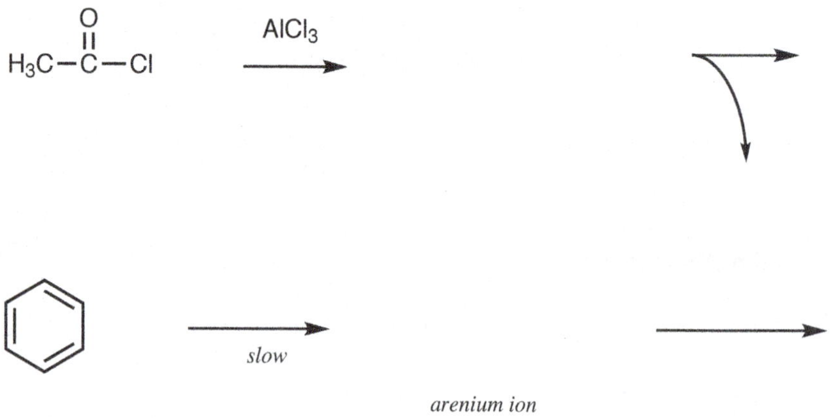

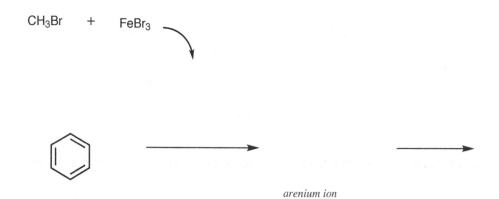

arenium ion

2. Friedel Crafts alkylation of benzene.

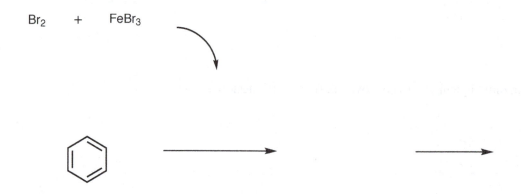

arenium ion

3. Ferric bromide bromination of benzene.

4. Nitration of benzene with nitric and sulfuric acids.

$$HNO_3 \quad \xrightarrow{\quad H_2SO_4 \quad} \qquad \longrightarrow$$

$$\xrightarrow[\substack{H_2SO_4 \\ H_2O}]{HNO_3} \qquad \longrightarrow$$

Electrophilic Substitution of Substituted Aromatic Compounds

5. Bromination of acetophenone to give *m*-bromoacetophenone.

   (For 5, 6, and 7, answer the question, "Will it react faster than benzene?")

$$\xrightarrow[FeBr_3]{Br_2} \qquad \longrightarrow$$

6. Nitration of methyl *p*-chlorobenzoate to give methyl 4-chloro-3-nitrobenzoate.

$$\xrightarrow[H_2O]{\substack{HNO_3 \\ H_2SO_4}} \qquad \longrightarrow$$

7. Friedel-Crafts acylation of phenyl acetate with acetyl chloride and aluminum chloride.

$$\xrightarrow[2.\ HCl\ /\ H_2O]{\substack{1.\ CH_3COCl \\ AlCl_3}} \qquad \longrightarrow$$

8. Triflic acid catalyzed acetylation of toluene to give *o*- and *p*-methylacetophenone.

CF_3SO_3H

Ac_2O $\longrightarrow$ $\longrightarrow$ $\longrightarrow$ *cont'd*

$\longrightarrow$ +

9. Aluminum chloride catalyzed chlorination of methyl 3-methoxybenzoate (methyl *m*-anisate) to give methyl 2-chloro-5-methoxybenzoate.

Cl_2 $\longrightarrow$ $\longrightarrow$

$FeCl_3$

10. Friedel-Crafts alkylation of 4-nitro-*N*-*p*-tolylbenzamide with two equivalents of chloromethane.

2 eq. CH_3Cl $\longrightarrow$ $\longrightarrow$ *cont'd*

2 eq. $AlCl_3$

cont'd $\longrightarrow$ $\longrightarrow$ *cont'd*

cont'd

Nucleophilic Aromatic Substitution

11. Nucleophilic aromatic substitution of 1-fluoro-4-nitrobenzene with ammonia to give 4-nitroaniline.

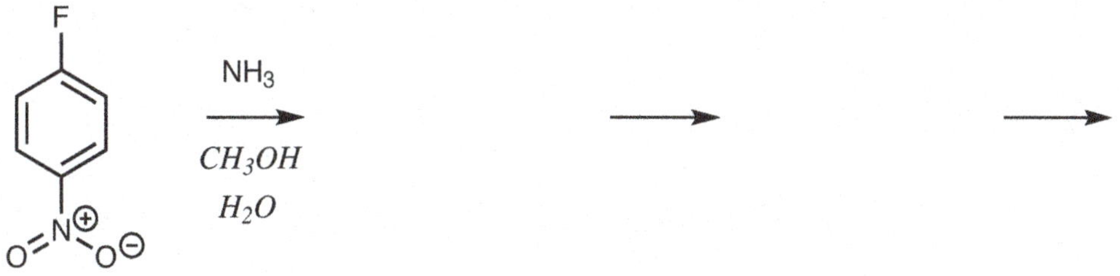

12. Nucleophilic aromatic substitution of 1,2-difluoro-4-nitrobenzene with sodium methoxide to give 2-fluoro-1-methoxy-4-nitrobenzene.

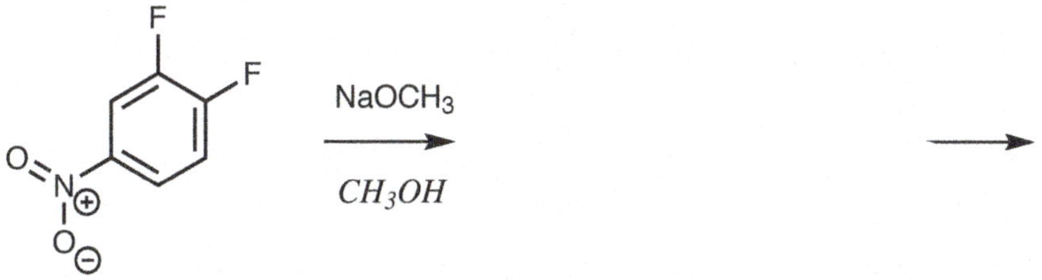

Benzyne Reaction

13. Reaction of 4-bromoanisole with sodium amide (sodamide, NaNH$_2$).

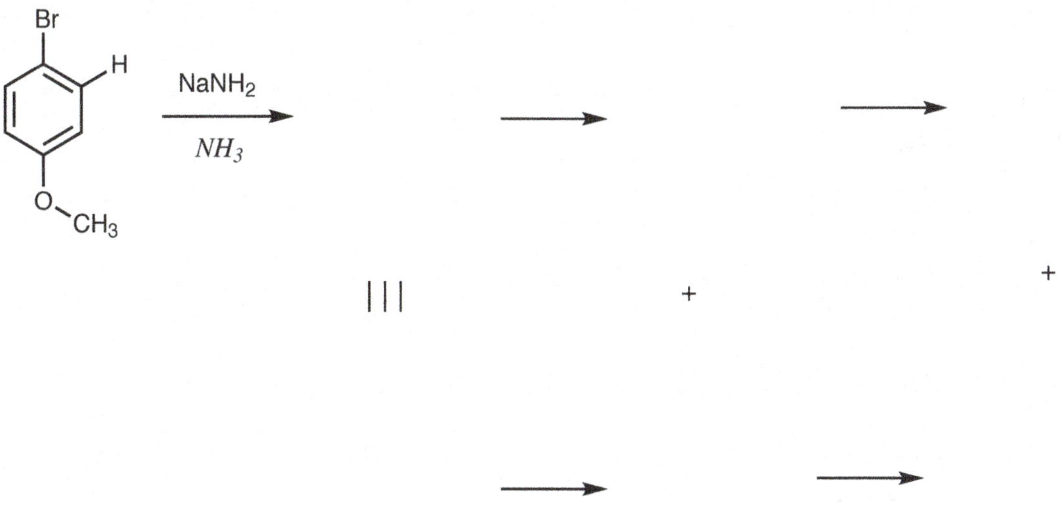

Diazonium Chemistry

14. Formation of a diazonium salt from aniline.

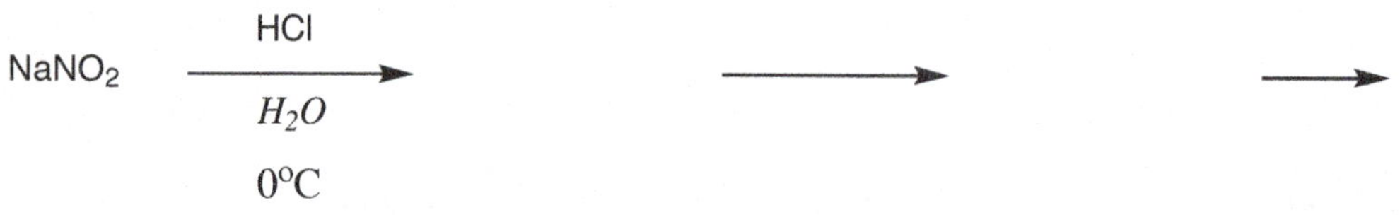

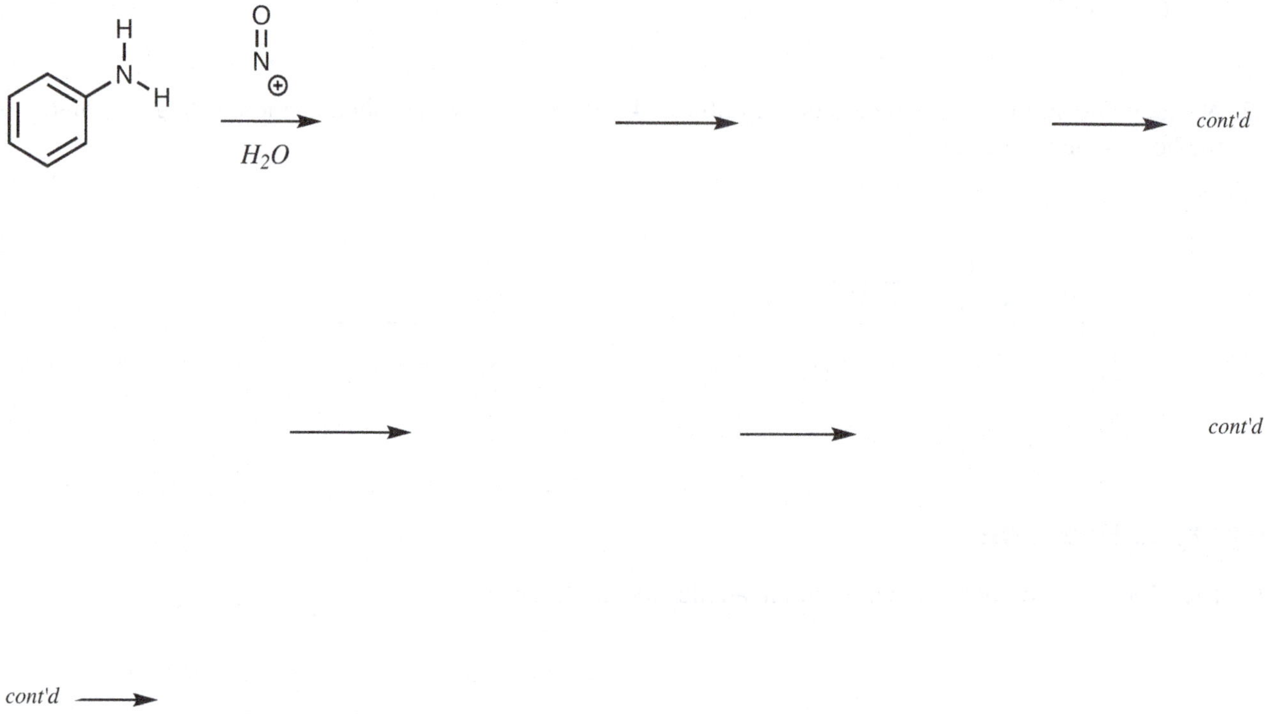

NaNO$_2$ $\xrightarrow[\text{0°C}]{\text{HCl}\atop H_2O}$ $\longrightarrow$ $\longrightarrow$

$\xrightarrow{H_2O}$ $\longrightarrow$ $\longrightarrow$ *cont'd*

$\longrightarrow$ $\longrightarrow$ *cont'd*

cont'd $\longrightarrow$

Reaction of Diazonium Salts

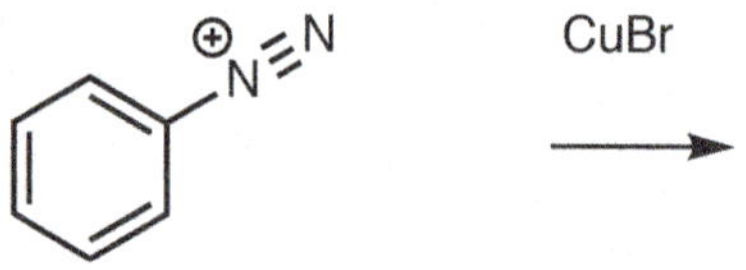

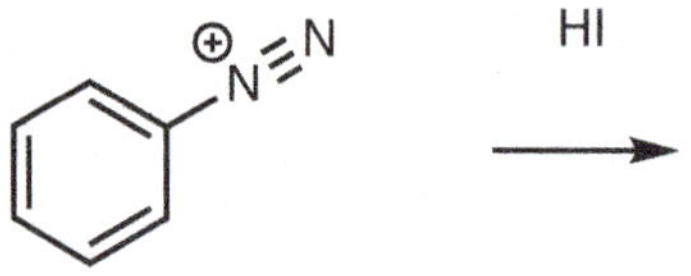

CuBr $\longrightarrow$

HI $\longrightarrow$

15 — Carbene and Nitrene Reactions

Carbene Reactions

1. Simmons-Smith carbene addition to cyclohexene to give a bicyclo[4.1.0]heptane. (See *Notes*.)

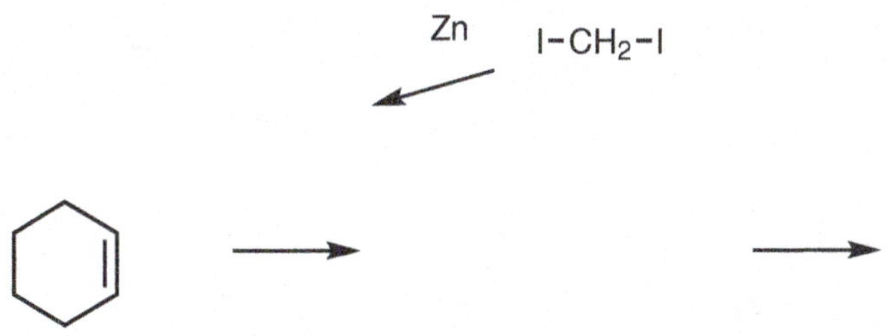

2. Dihalocarbene addition to (*E*)-1-phenylpropene to give a dibromocyclopropane.

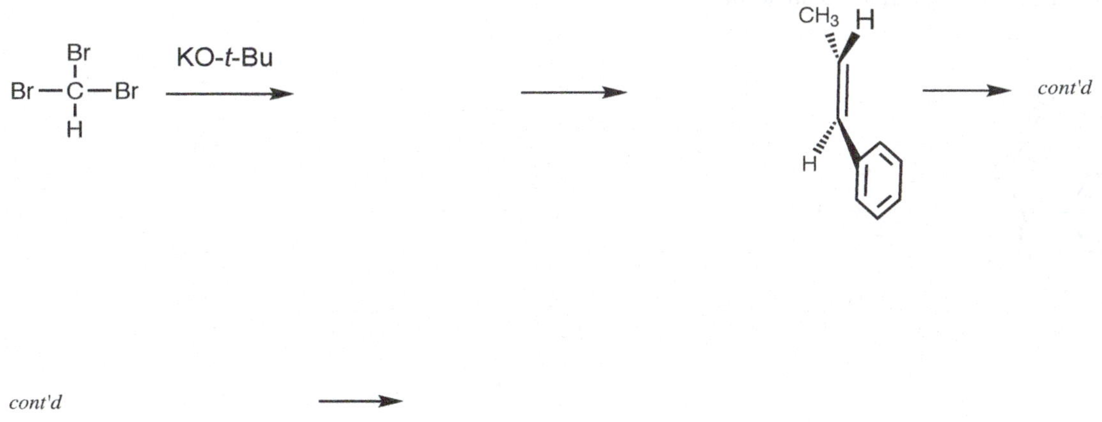

cont'd ⟶

3. Dihalocarbene addition to cyclohexene to give 7,7-dichlorobicyclo[4.1.0]heptane.

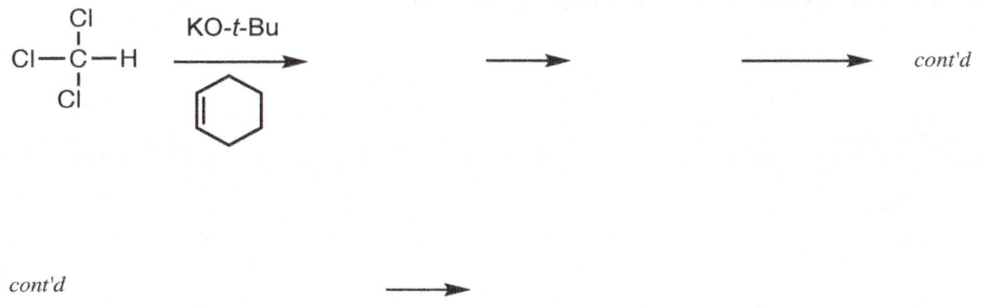

cont'd ⟶

Curtius Rearrangement

4. Step 1, reaction of acid chloride with azide and rearrangement via a nitrene intermediate.

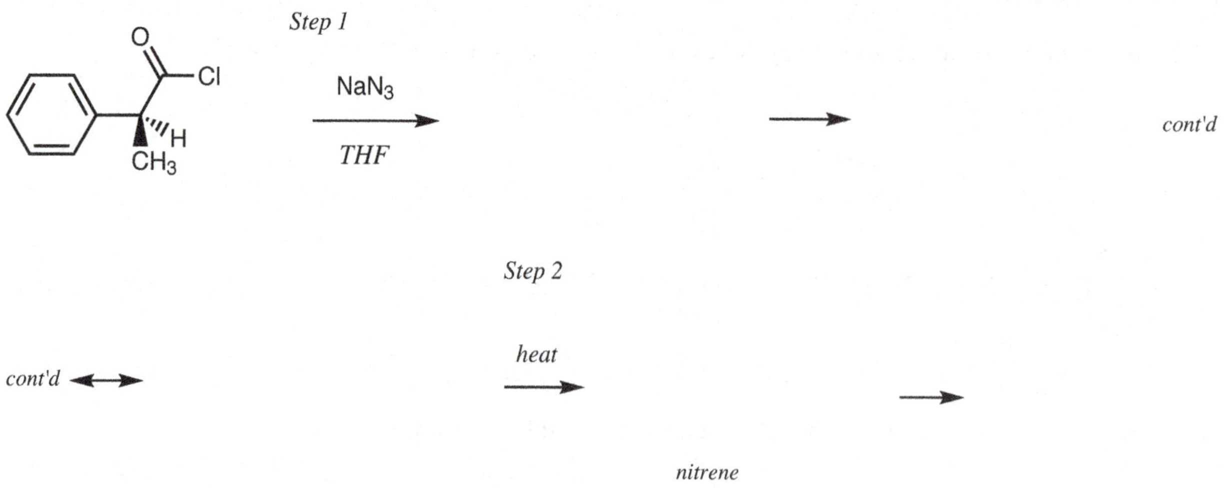

Step 1

NaN$_3$

THF

cont'd

Step 2

cont'd

heat

nitrene

Step 2, hydrolysis of isocyanate to an amine.

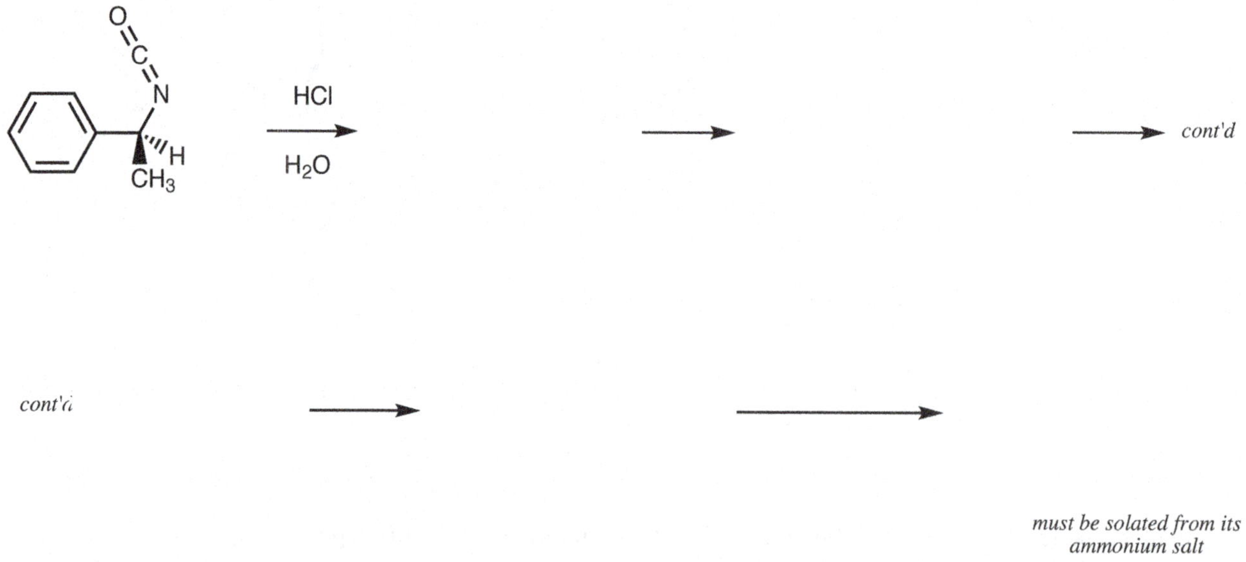

HCl

H$_2$O

cont'd

cont'd

must be isolated from its ammonium salt

Alternate Step 2, hydrolysis of isocyanate to a carbamate. A carbamate is a nitrogen analog of a carbonate ester. The steps are similar to the above reaction with an alcohol replacing the water.

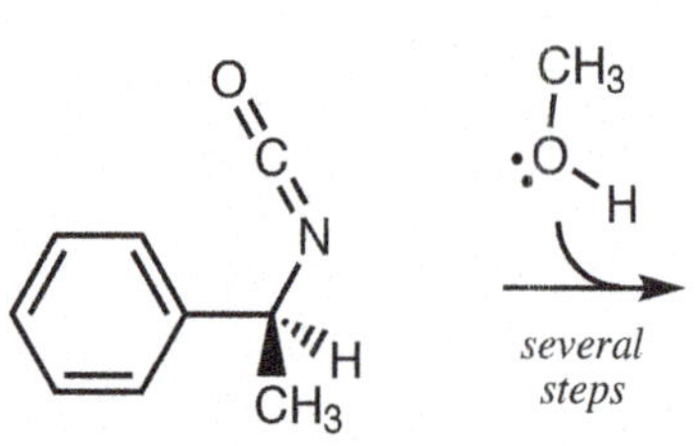

several steps

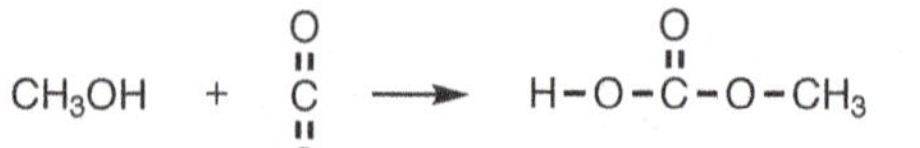

CH$_3$OH + $\overset{O}{\underset{O}{\overset{||}{\underset{||}{C}}}}$ $\longrightarrow$ H–O–$\overset{O}{\overset{||}{C}}$–O–CH$_3$

methyl hydrogen carbonate

Hoffmann Rearrangement

5. Step 1, Hoffmann rearrangement of a primary amide to give the corresponding amine. The reaction of the amide with sodium hydroxide and bromine.

This intermediate may be skipped

Step 2, acidification and decarboxylation to give the amine.

16 — Radical Reactions

For a discussion of radical reactions, see *Notes*.

Bromination (or Chlorination) Reactions

1. Free radical bromination of cyclohexane to give bromocyclohexane.

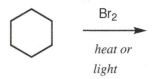

Initiation

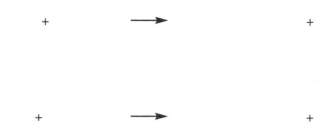

Propagation

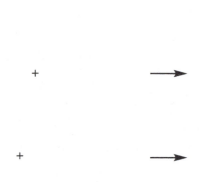

Termination

Allylic Bromination with NBS

2. Free radical bromination of cyclohexene with *N*-bromosuccinimide, an allylic bromination.

Overall reaction

 + → (benzoyl peroxide; *cat. amt.*; *heat or light (h√)*) +

NBS

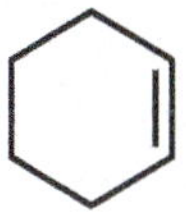

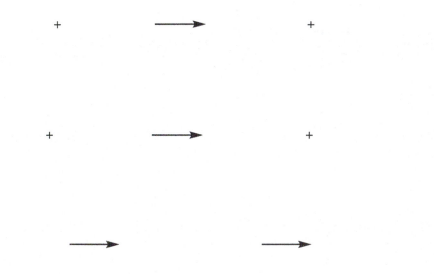

Initiation (See *Notes.*)

→ (heat *or* light) →

benzoyl peroxide

Propagation

+ → +

+ → +

→ →

Termination

+ →

+ *others*

Radical Addition of Hydrogen Bromide

3. Free Radical Addition of HBr/H_2O_2 to propene. (See *Notes.*)

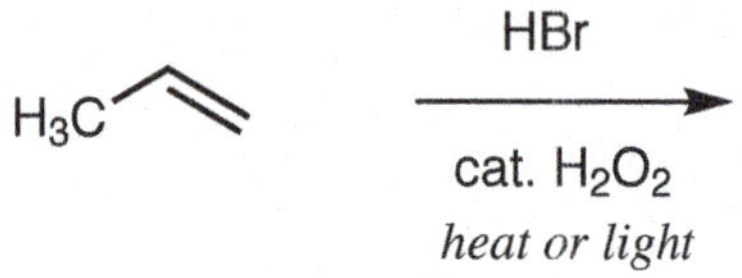
H3C
HBr
cat. H2O2
heat or light

Initiation (See *Notes.*)

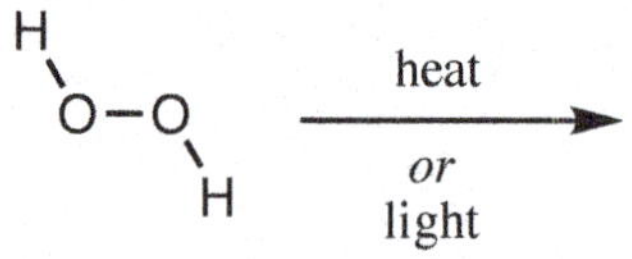
H
O–O
H
heat
or
light

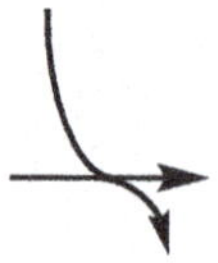

Propagation

+ →

+ → +

Termination

+ →

+ →

others

Benzylic Bromination with NBS

4. Benzylic bromination of ethylbenzene with NBS to give 1-bromo-1-phenylethane.

Overall reaction

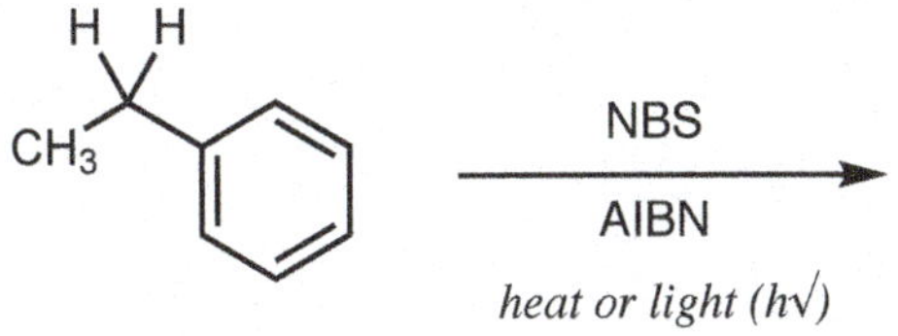

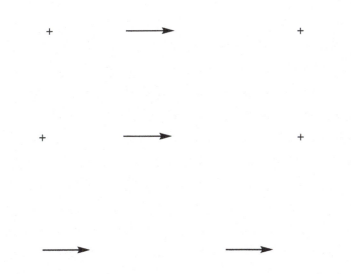

Initiation with azobisisobutyronitrile (AIBN)

Azobisisobutyronitrile

Propagation

+ ⟶ +

+ ⟶ +

⟶ ⟶

Termination

+ ⟶

+ others

1 — Getting Ready for Reactions

About the Atom

Part A

1. The *shorter* the bond length, the stronger the acid.
2. The *longer* the bond length, the stronger the acid.
3. Since we know HF ionizes into H^+ and F^-, we know the hydrogen-electron pair force is weaker than the fluorine-electron pair force.
4. From Number 3, if HI is a stronger acid (weaker proton-electron pair force) than HF, we might conclude the proton-electron pair distance of HI must be greater than that of HF. Coulomb's Law shows that a force decreases inversely with the square of the distance.
5. Similarly, if ethyne is more acidic than ethene or methane, then the proton-electron pair distance of ethyne must be greater than in ethene and in turn methane.
6. If ethyne has the largest proton-electron pair distance and the shortest bond length, then the carbon-electron pair distance must be the shortest. Similarly, methane must have the longest.
7. Given that the atomic radius for carbon is the largest and since CH_4 is the weakest acid of this group, its proton-electron pair force must be the smallest. Therefore, the greater the distance that an electron pair can extend beyond a nucleus, the greater the attraction it may have for a proton or another nucleus.
7. For column one (carbon and fluorine), the bond lengths decrease down the table. Therefore the carbon or fluorine-electron pair distances must decrease. The proton-electron pair distances must increase also to explain the increasing acidity. For the halo-acids, the proton-electron pair distance is masked by the halogens containing differing numbers of electrons and increasing bond lengths. If you look at the *difference* between bond length and halogen radius, it increases from HF to HI. For all compounds, it isn't the HX distance that matters as much as the hydrogen-electron pair distance. The greater that distance, the weaker the bond.

Part B

1. Replacing a hydrogen atom with a methyl group decreases the stability of the carbanion.
2. Replacing hydrogen atoms with methyl groups increases the carbocation stability.
3. If a carbon atom donates electrons, then the carbocation stability should increase as electron donation is increased. It should also decrease the stability of a carbanion. The relative stabilities are consistent with carbon being a better electron donor.
4. Since trifluoromethane is the stronger acid, its hydrogen-electron pair distance should be larger. If the fluorine atoms pulled the electrons away from the central carbon,

it would allow the remaining electrons to be pulled closer to it, thus increasing the proton-electron pair distance.
5. The electrons of a trifluoromethyl anion would be closer, see #4 above. If methyl groups donate electrons, it should have the opposite effect and increase the carbon-electron pair distance. Therefore, the electron pair of a methyl carbanion should be closer than a tert-butyl carbanion.
6. If a pair of electrons on an oxygen atom were pulled away from the nucleus by another atom, then the remaining electrons would feel the attractive force of the oxygen nucleus more strongly. If the electrons are pulled in toward the nucleus, that will increase the proton-electron distance and result in a stronger acid. Therefore, acidity would decrease in the following order, HOCl>HOOH>HOH.
7. If an oxygen atom can pull its electrons in more than a carbon, then the electrons of a CH_3^--carbanion should extend further than a hydroxide anion. If they extend further, they will be more available for reaction.

Part C

1. In the table, they all have the same number of *electrons*.
2. In the table, the columns have the same number of protons, but a proton that was attached to the nucleus of the upper molecule is **in** the nucleus of the lower molecule.
3. For the elements C, N, O, and F, because fluorine has the largest number of protons in its nucleus, it should pull its electrons in the most.
4. For the compounds with C, N, O, and F, because fluorine can pull its electrons into the nucleus the most, it can have the shortest bond length, see 3 above.
5. Since H_2O and HF have the same number of electrons and the properties of electrons are constant, the difference in acidity must be due to the distances of where the electrons are located. If the electrons of fluorine are held closer to the nucleus, we might conclude that as a result they are now further from the hydrogen. If the hydrogen to electron pair distance is larger, that will result in a weaker bond and therefore, greater acidity.
6. If Number 5 is true, then the base strength of an atom would increase if the distance of the electrons to the nucleus were increased. The electrons of nitrogen should be more basic than oxygen or fluorine because there are fewer protons in the nitrogen nucleus to hold them closer. If two nitrogen atoms differed in their basicity, then the less basic nitrogen would have its electrons closer to the nucleus (like oxygen or fluorine).

For further discussion, see *Notes*.

Guide to Drawing Resonance Structures

Resonance Structures

If you are unfamiliar with the use of the curved arrow, refer to the discussion in the *Notes* section. A good place to start pushing electrons is in drawing resonance structures. They have the elements of electron movement, but the problems will be more limited in scope.

Resonance Structures of Anions

The principle for understanding resonance structures is to understand that electrons will operate by a push-pull mode or model. If there's a net negative charge, it will be the electrons of the atom with the negative charge that will push toward the pi bond. We will start our curved arrow with those electrons. Continue to move them toward any neighboring pi bonds (push) to create and break new bonds. You should note that two curved arrows are required to avoid structures with more than eight valence electrons.

For the following examples, add curved arrows, where needed, to show how the electrons move to form the next structure. For 1-6, the first and last structures are the same. In that case, you are converting it back to the starting structure.

1.

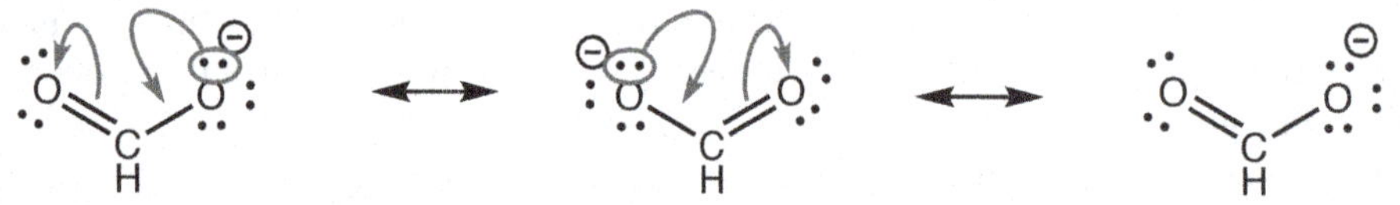

2.

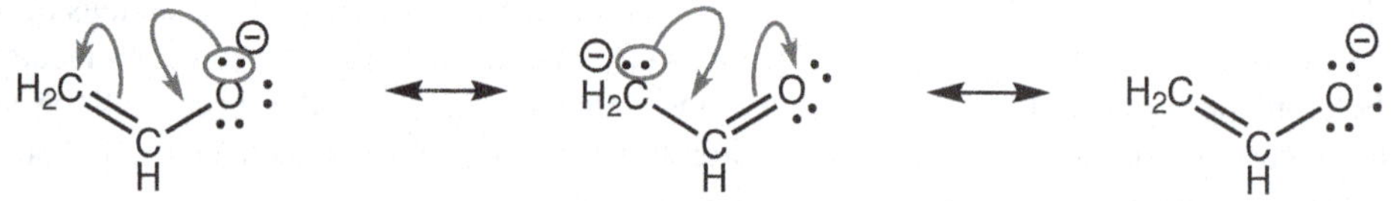

3.

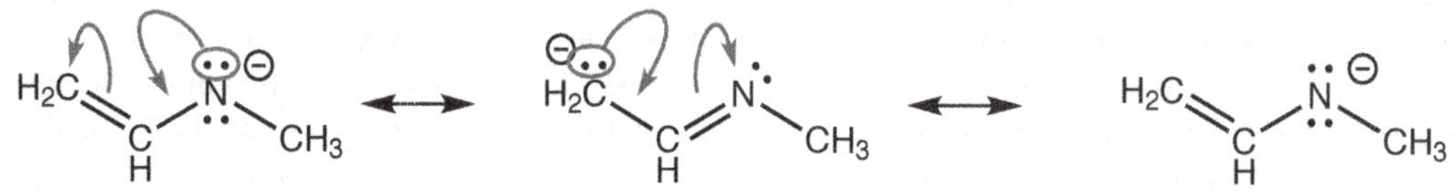

4.

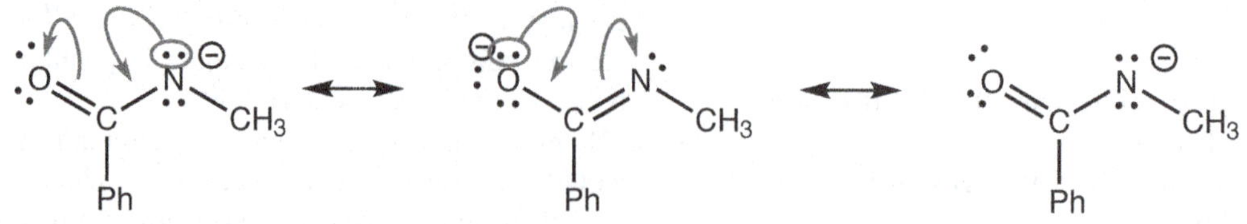

5.

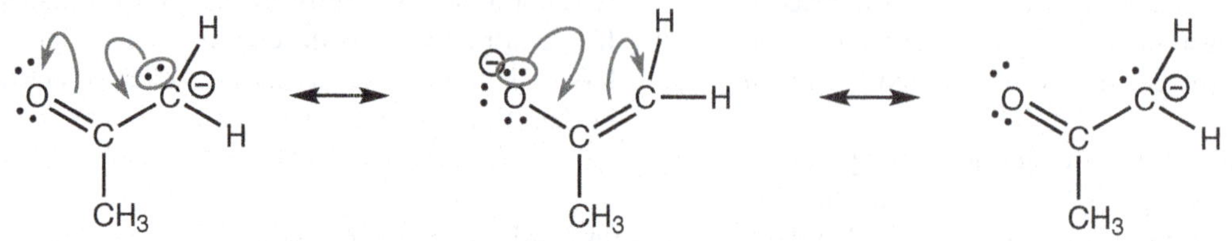

6.

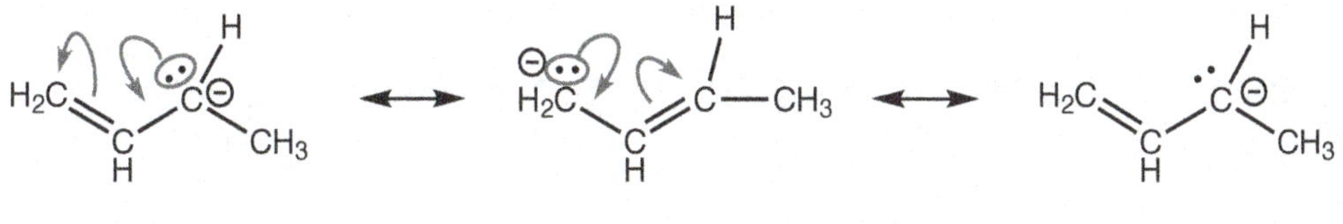

7.

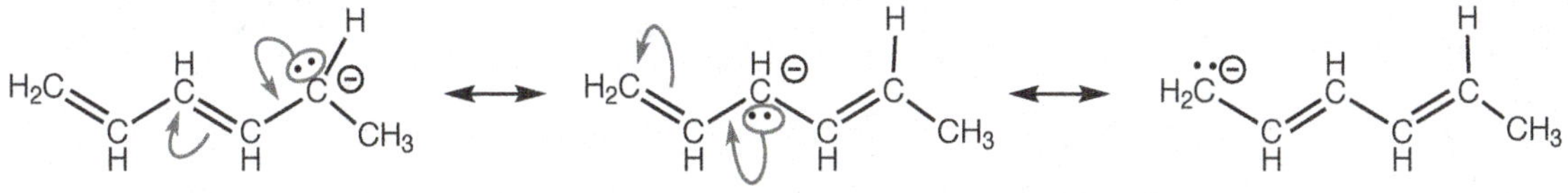

8.

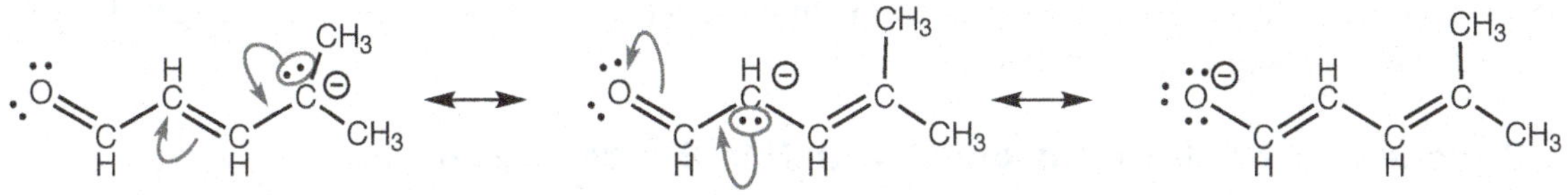

These examples show how the electrons can move, where the resulting charge will form, and how the charge can be distributed. It does not tell you on which atoms the greater charge density might exist nor upon which atom a reaction might next occur. You will note that since the original structure had a negative charge, the only charge that exists on any of the resulting resonance structures is a negative charge.

Resonance Structures of Cations

Conversely, if there is a positive charge, it is the positive charge that will attract (pull) electrons. Start a curved arrow from a neighboring pi or pair of non-bonded electrons and bring it toward the positive charge. You will note that only one curved arrow is necessary to create a new resonance structure. Because the original structure is a cation, completing its octet should not require further electron movement.

Add curved arrows to the following structures. For 9 and 10, the first and last structures are the same. In that case, you are converting it back to the starting structure.

9.

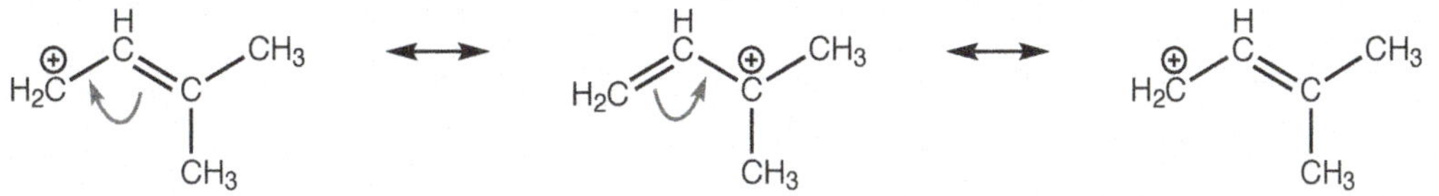

10.

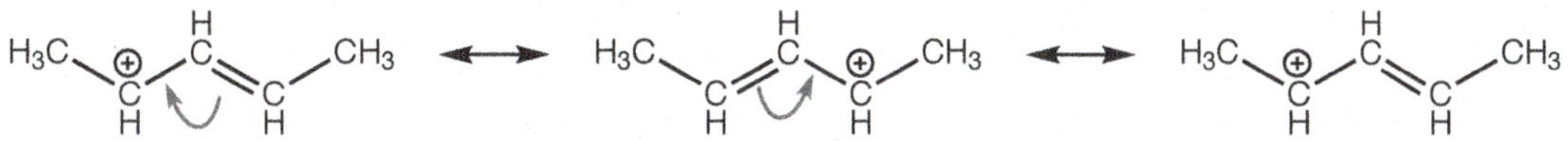

11.

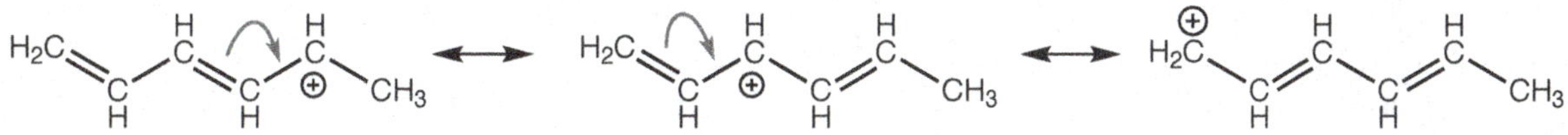

12.

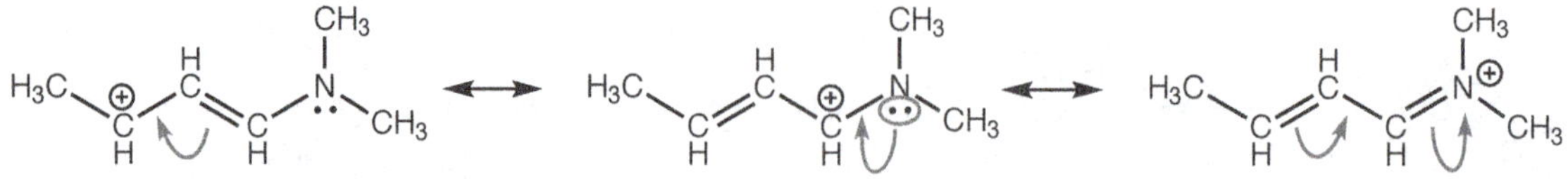

13.

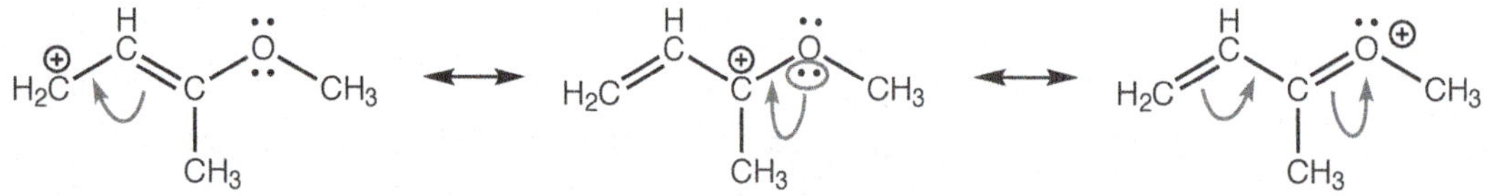

Again, these examples show how the electrons can move, where the resulting charge will form, and how the charge can be distributed. It does not tell you on which atoms the greater charge density might exist nor upon which atom a reaction might next occur. You will note that since the original structure had a positive charge, the only charges that exist on any of the resulting resonance structures are positive charges.

Resonance Structures of Neutral Compounds with Non-Bonded Electrons

If there isn't a charge and there are adjacent non-bonded electrons, then it will be the non-bonded electrons that will move (push) toward a neighboring pi bond. Start a curved arrow with the non-bonded electrons and direct them to the neighboring double bond.

Add curved arrows to the following structures. For 14-17, the first and last structures are the same. In that case, you are converting it back to the starting structure.

14.

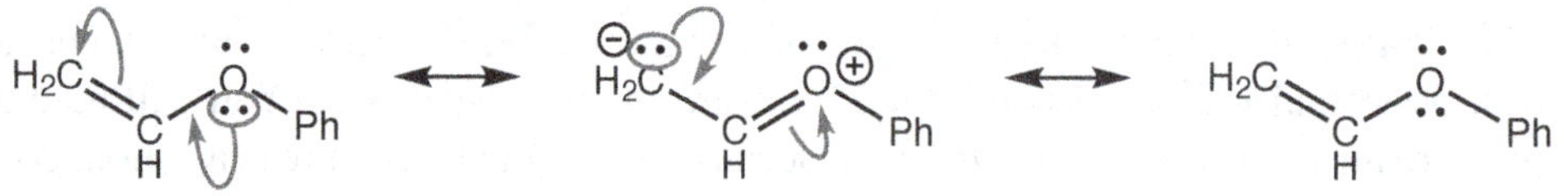

15.

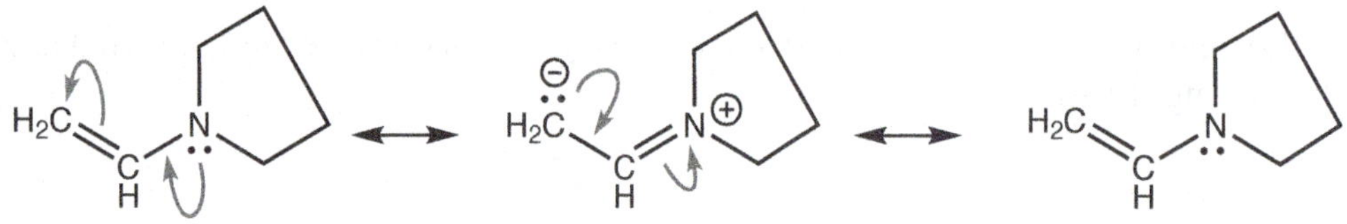

16.

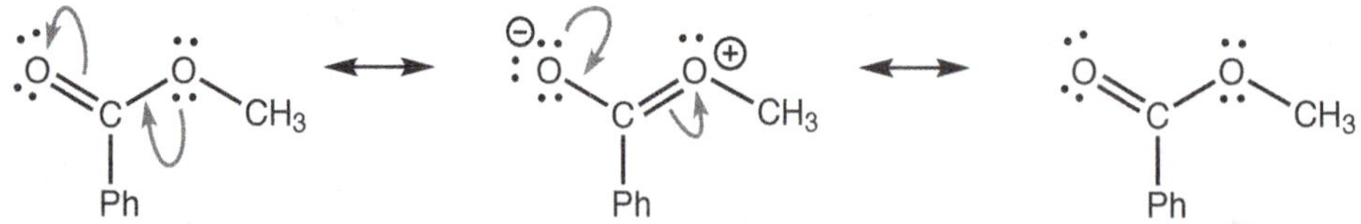

17.

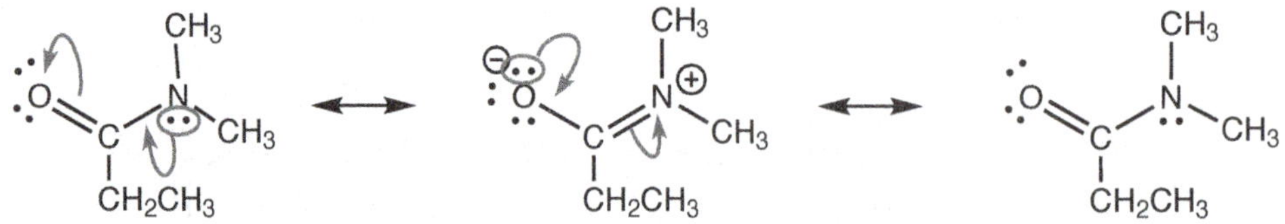

18.

19.

These examples show how the electrons can move, where the resulting charges will form, and how the charges can be distributed. It does not tell you on which atoms the greater charge densities might exist nor upon which atoms a reaction might next occur. You will note that since the original structure was neutral, the net charges that exist on any of the resulting resonance structures are also neutral and only two atoms have a charge.

Resonance Structures of Neutral Compounds without Non-Bonded Electrons

If there is no charge or neighboring non-bonded electrons, then we will push the pi electrons of a double bond toward the least substituted carbon or the most electron withdrawing atom, see Example 26. Start a curved arrow with the pi electrons of a double bond and direct them from the most to the least substituted carbon or for a C=O bond, toward the oxygen atom.

20.

21.

This is the same example as above, however the arrows are pointing in the opposite direction. If you are uncertain in which direction the electrons might move, a good strategy is to draw an arrow in the opposite direction and then to compare the results of the two possibilities. Compare the result below with the one above. Which is the more stable? If you do not recognize the lower arrangement as a lesser contributor, you may need to refer to your textbook for the rules of carbanion, carbocation, and resonance stability. Examples 23 and 24 are similar, which is more stable?

22.

23.

24. Examples 23 is preferred over Example 24.

25.

26.

In these examples, we have incorporated chemical principles. The carbocations that are the most substituted are the most stable. Coinciding with this principle is that carbanions with the least substitution (or a heteroatom, last example) are the most stable.

Resonance Structures of Radicals

Radicals, compounds with unpaired electrons, are less stable than those with paired electrons. The fate of radical reactions is to form a paired-electron bond. However, sharing unpaired electrons with neighboring non-bonded electrons or pi-bonds can attain added stability. Also note the curved arrow has a single barb indicating the movement of a single electron. Two arrows are required for a pair of electrons.

27.

28.

2 — Acid-Base Chemistry

Bronsted-Lowrey Acids and Bases

Acid-base reactions are often the first intermolecular reaction you will encounter. A proton will be exchanged from the strongest acid to the strongest base.

In these examples, you must note the conjugate acid and conjugate base that result in each reaction. You should note the use and meaning of the curved arrows. If the example does not contain a curved arrow, you must supply one. The meaning of the curved arrow is important for you to understand.

- The rule for predicting the product of an acid-base reaction is simple. A reaction will generally give the product that is the weakest base (or conjugate base). The base strength of a compound is related to the acidity of the acid, the stronger the acid, the weaker the base, or the corollary, the weaker the acid, the stronger the base. In order to compare the base strengths, the acidities of the acid and conjugate acid must be determined first.

- Look at each example and write the pK_a under each acid (on the left) and conjugate acid (on the right). Be careful that you correctly identify the acid and the corresponding pK_a. Strong acids have a small pK_a and weak acids have a large pK_a. You may need to use a table to find some values.

- For each acid (or conjugate acid), write the corresponding pK_a beneath it. Label each base and use the pK_a of the acid or conjugate acid to find the weakest base. The strongest acid corresponds with the weakest base. In example 1, the pK_a of HF is 3.2. It is placed under HF. The pK_a of acetic acid is 4.75 and 4.75 is placed under it. Because HF is the strongest acid (lowest pK_a), the label "weakest base" is placed under its conjugate base. The equilibrium will shift to the right.

- Label the equilibrium of each reaction, L(eft) or R(ight). Example 1, R.

1. For this example, the acids and bases are labeled. HF is the strongest acid as it has the lower pKa. Therefore, F⁻, its conjugate base is the weakest base.

Base	Acid	Conjugate Acid	Conjugate Base
L/**R**	pK_a 3.2	pK_a 4.75	weakest base

Notice the curved arrows. They describe the reaction that is taking place. We could write the following sentences to describe the curved arrows.

A bond is being made between the oxygen and hydrogen atom with the electrons from the oxygen atom.
A bond is being broken between the hydrogen and the fluorine atom with the electrons remaining attached to the fluorine atom.

2. Label acids, bases, and conjugate acids and bases.

L/**R**	base	pK_a 3.2	pK_a 15.7	weakest base

Notice the curved arrows. We could write the following sentences to describe the curved arrows.

A bond is being made between the oxygen and hydrogen atom with the electrons from the oxygen atom.
A bond is being broken between the hydrogen and the fluorine atom with the electrons remaining attached to the fluorine atom.

Continue by completing the equation, adding curved arrows, pK$_a$ values, indicate the weakest base, the direction of the equilibrium, and write a sentence(s) describing any bonds being made or broken.

3.

L/R   weakest base   3.2   -1.7   base

A bond is formed between the oxygen and hydrogen with the electrons from the oxygen. The bond between the hydrogen and fluorine is broken with the electrons remaining attached to the fluorine.

4.

L/**R**   7.0   base   9.2   weakest base

A bond is formed between the nitrogen and hydrogen with the electrons from the nitrogen. The bond between the hydrogen and sulfur is broken with the electrons remaining attached to the sulfur.

5.

L/R   3.2   weakest base   base   -8

A bond is formed between the chloride and hydrogen with the electrons from the chloride. The bond between the hydrogen and fluorine is broken with the electrons remaining attached to the fluorine.

6.

L/R   10.0   base   weakest base   15.7

A bond is formed between the oxygen and hydrogen with the electrons from the oxygen. The bond between the hydrogen and oxygen is broken with the electrons remaining attached to the oxygen.

7.

L/R   4.75   base   weakest base   10.5

A bond is formed between the nitrogen and hydrogen with the electrons from the nitrogen. The bond between the hydrogen and oxygen is broken with the electrons remaining attached to the oxygen.

8.

L/**R**   16.0   weakest base   base   10.5

A bond is formed between the nitrogen and hydrogen with the electrons from the nitrogen. The bond between the hydrogen and oxygen is broken with the electrons remaining attached to the oxygen.

Continue by completing the equation, adding curved arrows, pK$_a$ values, indicate the weakest base, the direction of the equilibrium, and write a sentence(s) describing any bonds being made or broken.

9.

L/R   22   weakest base   base   19

A new bond is formed between the oxygen and hydrogen with the electrons from the oxygen. The bond between the hydrogen and carbon is broken with the electrons remaining attached to the carbon.

10.

L/R   base   38   50   weakest base

A bond is formed between the carbon and hydrogen with the electrons from the carbon. The bond between the hydrogen and nitrogen is broken with the electrons remaining attached to the nitrogen.

11.

L/R   weakest base   9.2   -1.7   base

A bond is formed between the oxygen and hydrogen with the electrons from the oxygen. The bond between the hydrogen and nitrogen is broken with the electrons remaining attached to the nitrogen.

12.

L/R   base   4.6   9.2   weakest base

A new bond is formed between the nitrogen and hydrogen with the electrons from the nitrogen. The bond between the hydrogen and nitrogen is broken with the electrons remaining attached to the nitrogen.

13.

L/R   4.75   base   weakest base   15.7

A bond is formed between the oxygen and hydrogen with the electrons from the oxygen. The bond between the hydrogen and oxygen is broken with the electrons remaining attached to the oxygen.

14.

L/R   24   base   weakest base   50

A bond is formed between the carbon and hydrogen with the electrons from the carbon. The bond between the hydrogen and carbon is broken with the electrons remaining attached to the carbon.

Continue by completing the equation, adding curved arrows, pK$_a$ values, bases, and label the direction of equilibrium.

15.

L/R   24   weakest base   base   16

16.

L/R   11.7   base   weakest base   16

17. Sometimes we may be unsure how a reaction might proceed. Will H$_2$S and (CH$_3$)$_2$NH react together? What will the products be if they do? In that case, draw ALL of the possible products and analyze the results. Complete this problem as before.

From the individual equilibria, can you predict the overall result?

weakest base   38   <-5   base

7.0   base   weakest base   10.7

18. What is the equilibrium between HOCH$_3$ and CH$_3$NHCH$_3$? First, determine the individual equilibria and then predict the overall result.

weakest base   38   -2.2   base

15.5   weakest base   base   10.7

Lewis Acids and Bases

The prior exercise showed a hydrogen atom accepting electrons. With Lewis acids, other atoms can also accept electrons. See your text for further discussion.

Add structures, non-bonded electrons, curved arrows, and formal charges as needed to complete the following.

19.

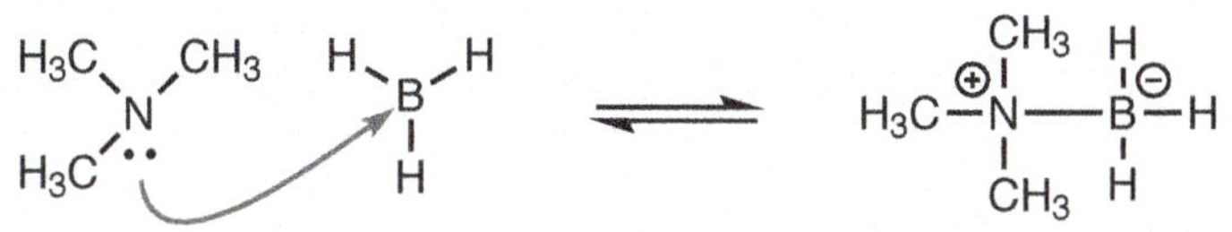

20.

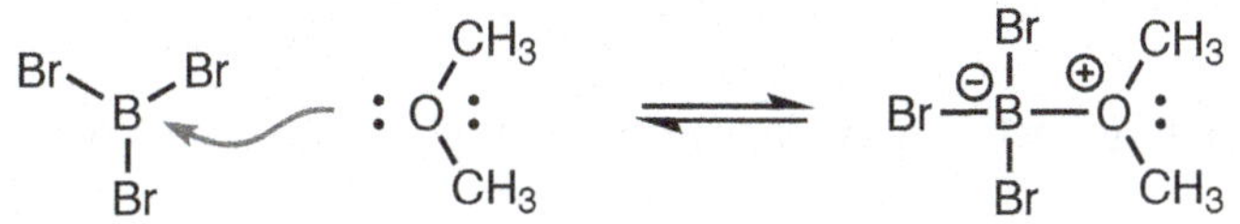

21.

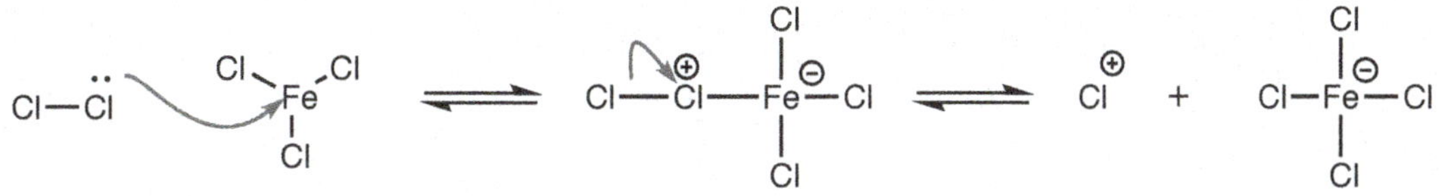

22.

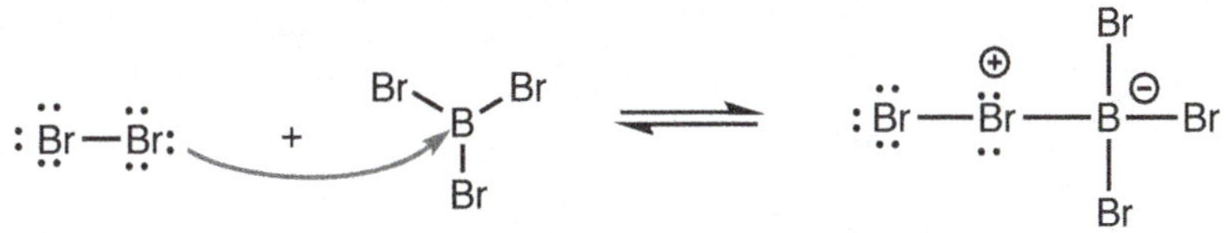

23.

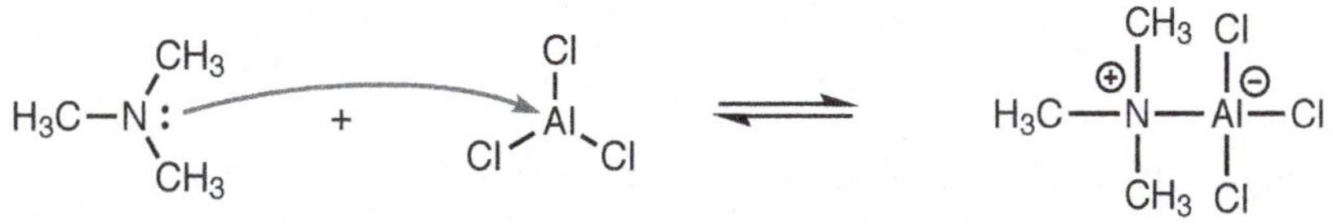

24.

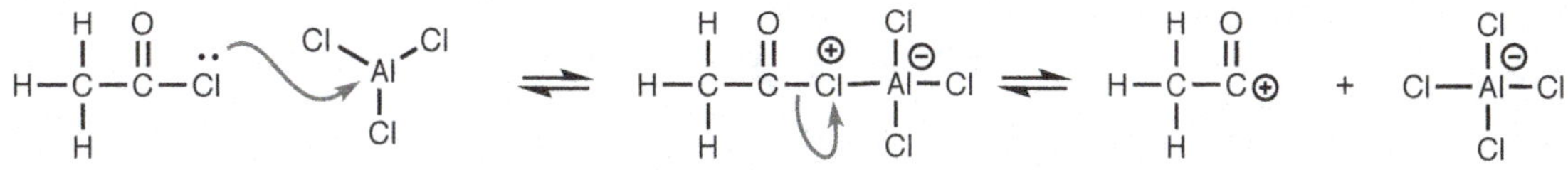

25.

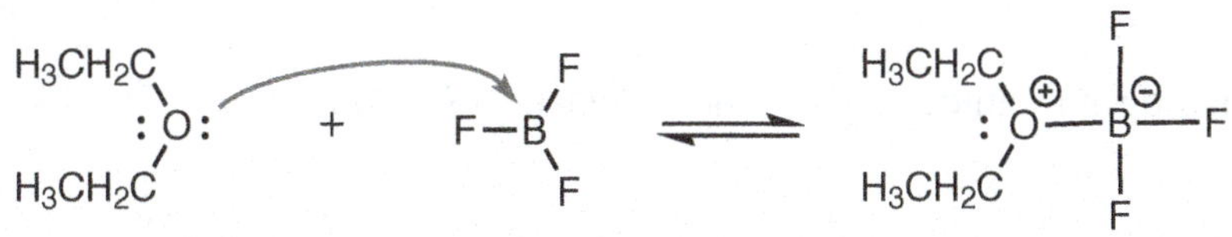

26.

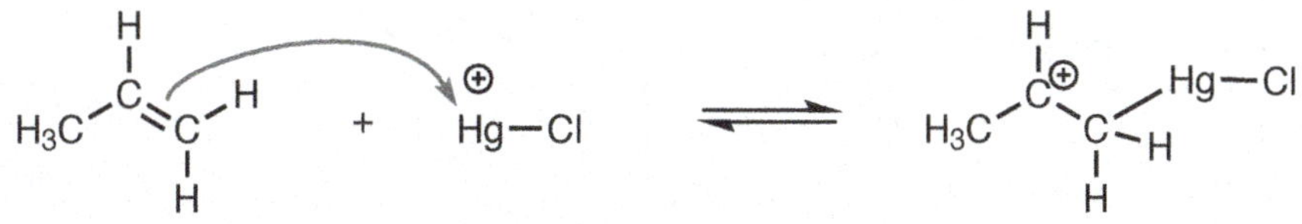

3 — Substitution Reactions

S$_N$2 Substitution Reactions

Add curved arrows to the following reactions.

1. An S$_N$2 reaction of 1-chlorobutane with sodium iodide to give 1-iodobutane. (See *Notes*.)

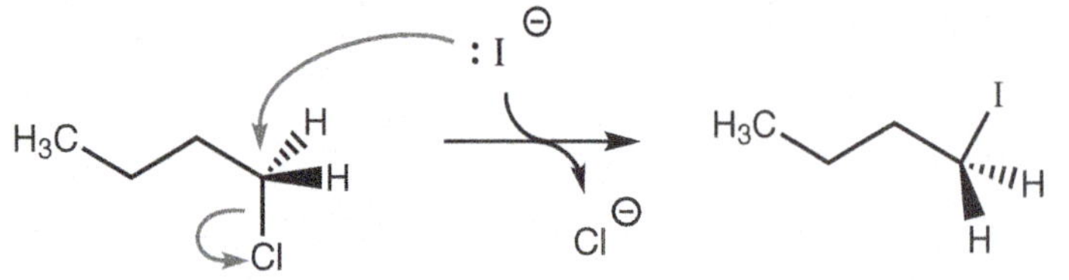

2. An S$_N$2 reaction of 1-bromobutane with ethoxide to give 1-ethoxybutane (butyl ethyl ether). (See *Notes*.)

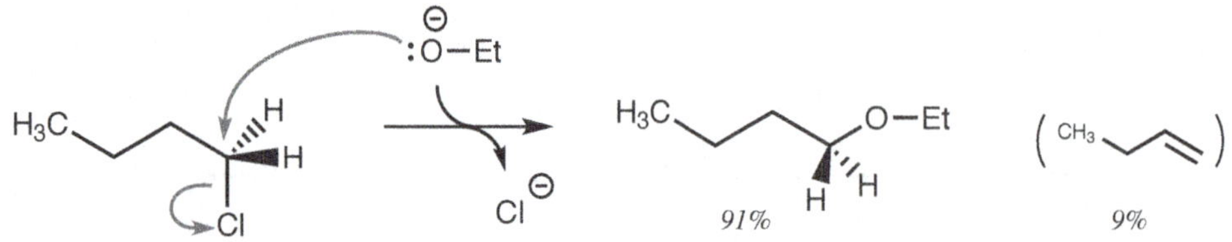

3. An S$_N$2 reaction of *(R)*-2-bromobutane with thiocyanate to give *(S)*-2-thiocyanatobutane. (See *Notes*.)

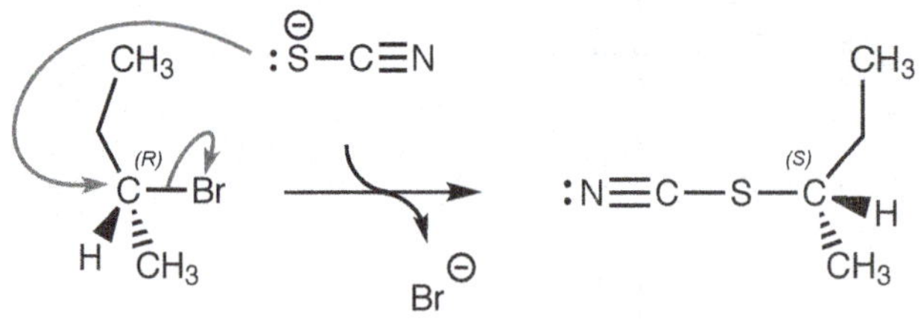

4. An S$_N$2 reaction of *(S)*-2-bromobutane with acetate to give *(R)*-sec-butyl acetate. (See *Notes*.)

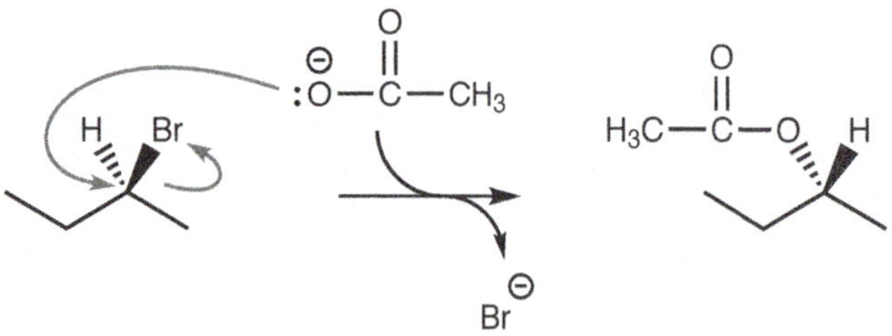

5. An S$_N$2 reaction of 1-butanol with hydrogen bromide to give 1-bromobutane. (See *Notes*.)

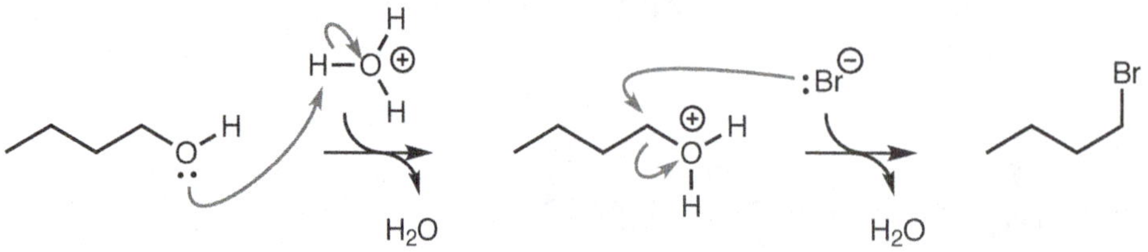

6. An S$_N$2 reaction of a triflate with sodium methanethiolate to give a thioether with inversion. (See *Notes.*)

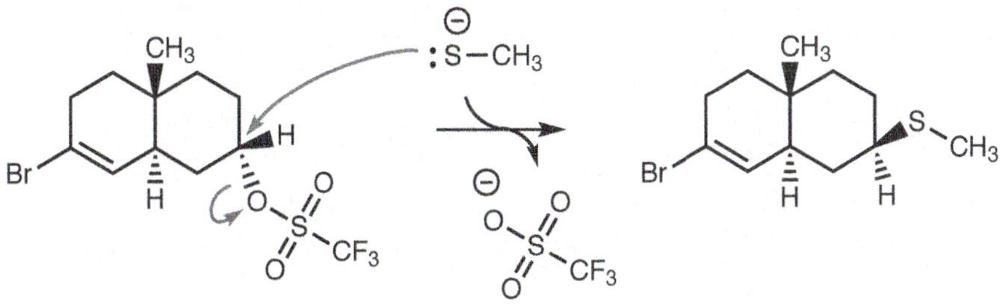

7. An S$_N$2 reaction of *cis*-1-chloro-4-methylcyclohexane with azide to give *trans*-1-azido-4-methylcyclohexane.

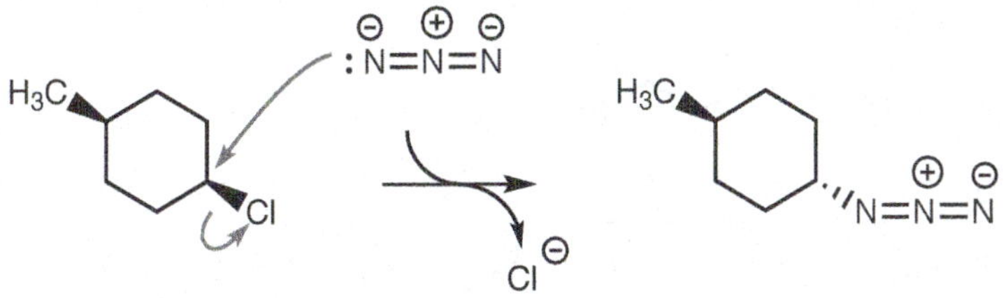

8. An S$_N$2 reaction of a ditosylate with one equivalent of cyanide to give a mono-nitrile. (See *Notes.*)

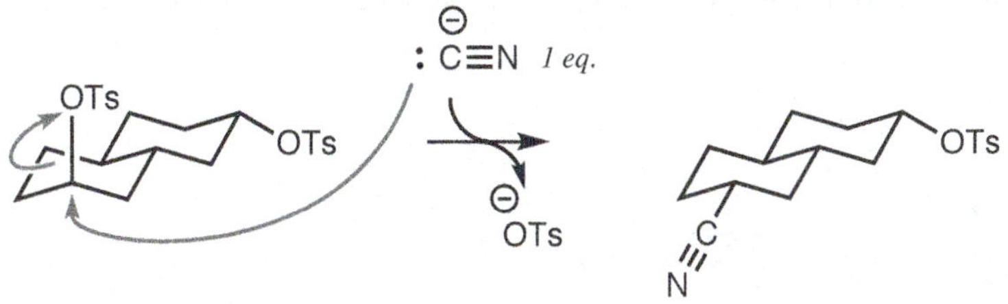

9. An S$_N$2 reaction of 1-bromo-3-chloropropane with cyanide to give 4-chlorobutanenitrile. (See *Notes.*)

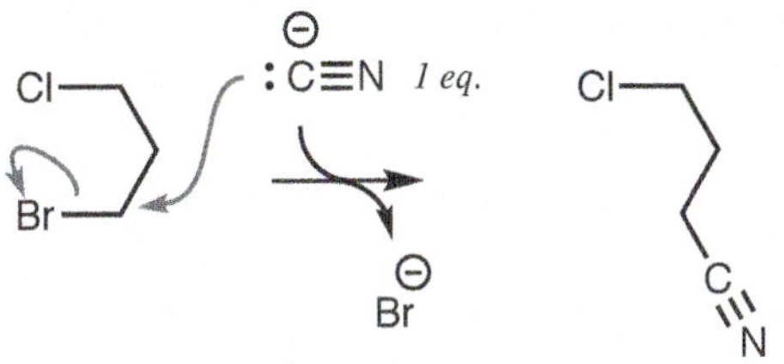

10. An S$_N$2 reaction of 1-bromobutane with acetylide anion to give 1-hexyne. (See *Notes.*)

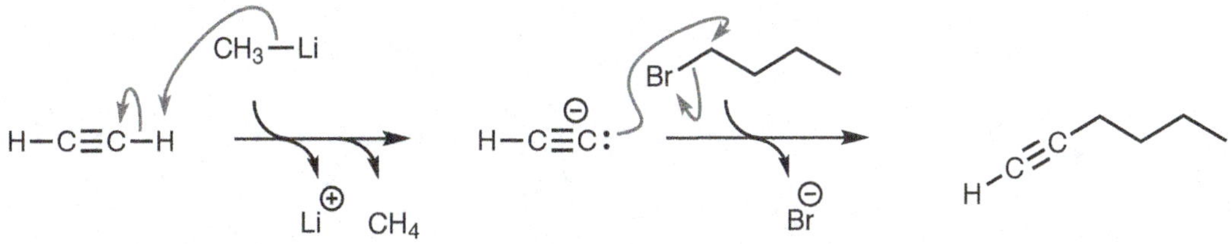

11. An S_N2 reaction of methyl phenylacetate enolate with iodopropane to give methyl 2-phenylpentanoate. (see *Notes*)

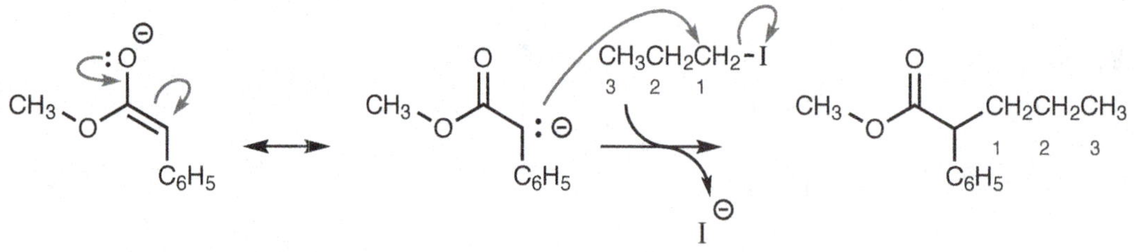

12. A Gabriel amine synthesis with formation of phthalimide anion (potassium carbonate) and alkylation with 1-bromo-2-butene. (See *Notes*.)

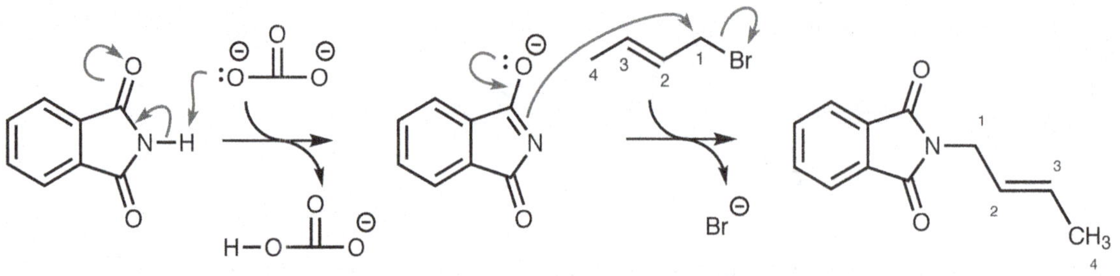

13.   An S_N2 reaction of bromomethane with phenoxide to give methoxybenzene (anisole). (See *Notes*.)

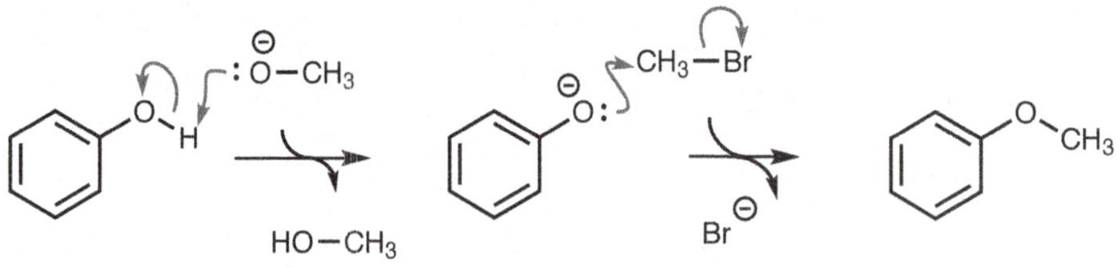

14. An S_N2 reaction of benzyl bromide with sodium ethoxide to give ethyl benzyl ether.

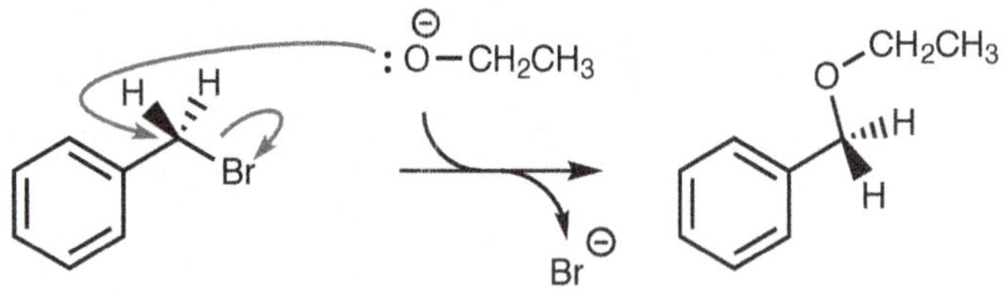

15. An S_N2 cleavage reaction of *p*-nitroanisole to give iodomethane and *p*-nitrophenol. (See *Notes*.)

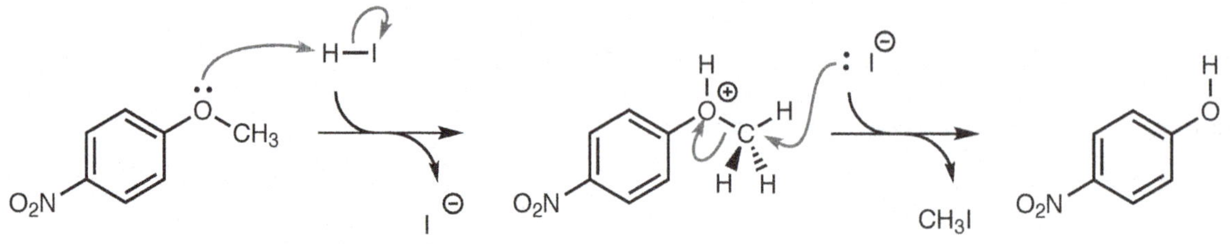

16. An S_N2 reaction of 1-bromo-2-butanol with sodium hydroxide to give 2-ethyloxirane (butylene oxide).

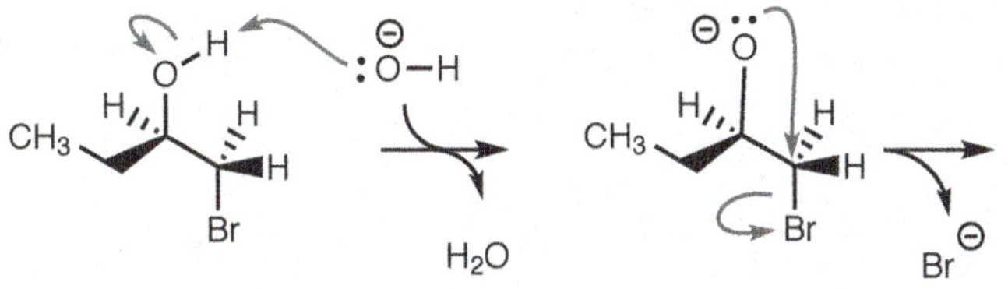

17. An S_N2 reaction of the anion of 1-butyne with *(R)*-2-ethyloxirane to give *(R)*-hept-5-yn-3-ol. (See *Notes*.)

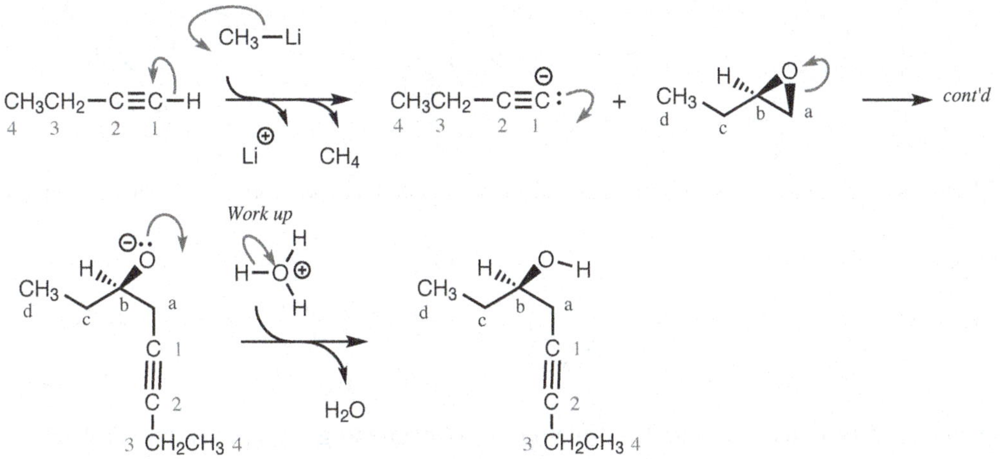

18. An acid catalyzed opening of 2-ethyloxirane with methanol to give 2-methoxybutan-1-ol. (See *Notes*.)

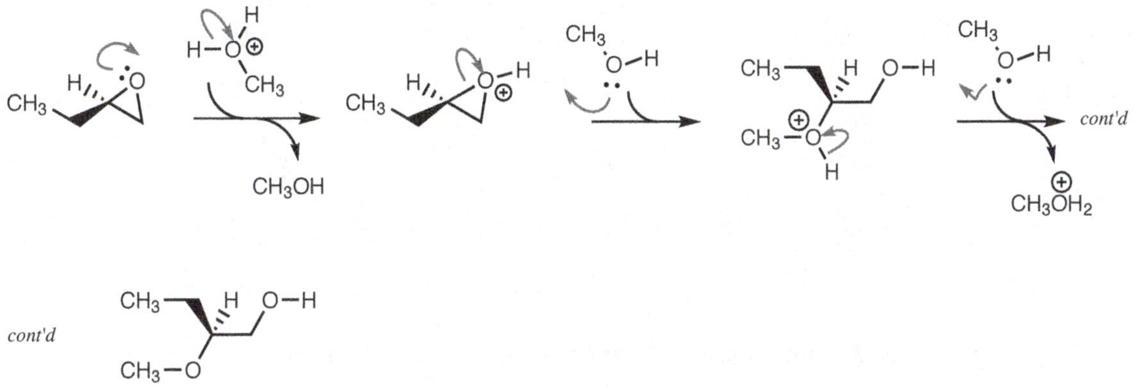

19. An S_N2 reaction of the enolate of acetophenone with *(R)*-2-methyloxirane (propylene oxide) to give *(R)*-4-hydroxy-1-phenylpentan-1-one. (See *Notes*.)

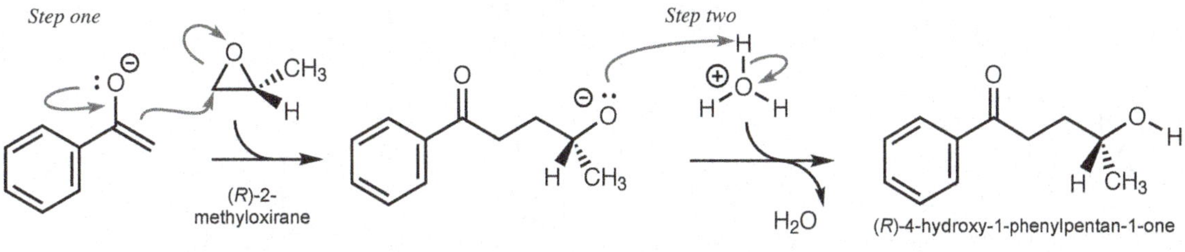

S_N1 Substitution Reactions

20. An S_N1 solvolysis reaction of *t*-butyl iodide to give *t*-butyl alcohol. (See *Notes.*)

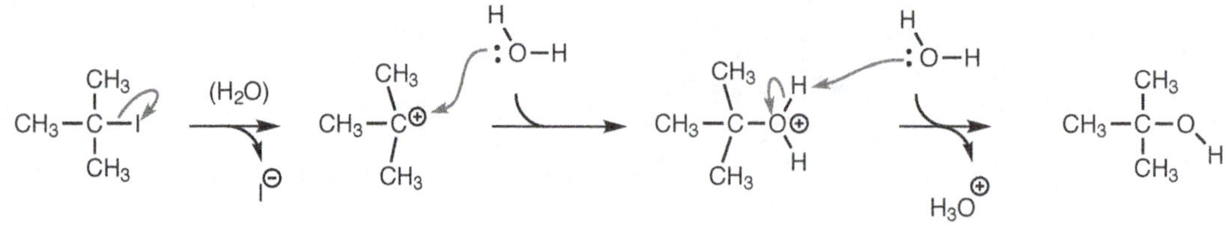

21. An S_N1 solvolysis reaction of *(R)*-(1-chloroethyl)benzene to give *rac*-1-phenylethanol.

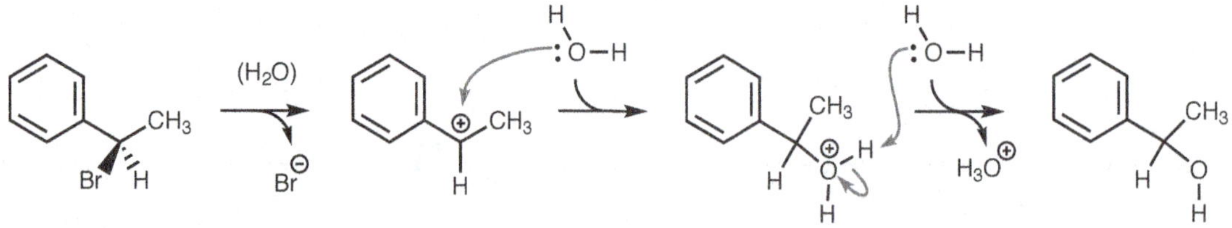

22. An S_N1 reaction of 1-methylcyclohexanol with hydrogen chloride to give 1-chloro-1-methylcyclohexane. (see *Notes*)

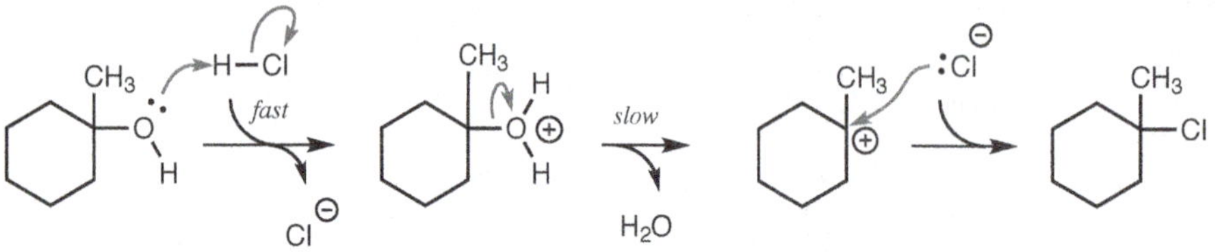

23. An S_N1 solvolysis reaction of *(1S,3R)*-1-bromo-1,3-dimethylcyclohexane to give *(1S,3R)*- and *(1R,3R)*-1,3-dimethylcyclohexanol. (See *Notes.*)

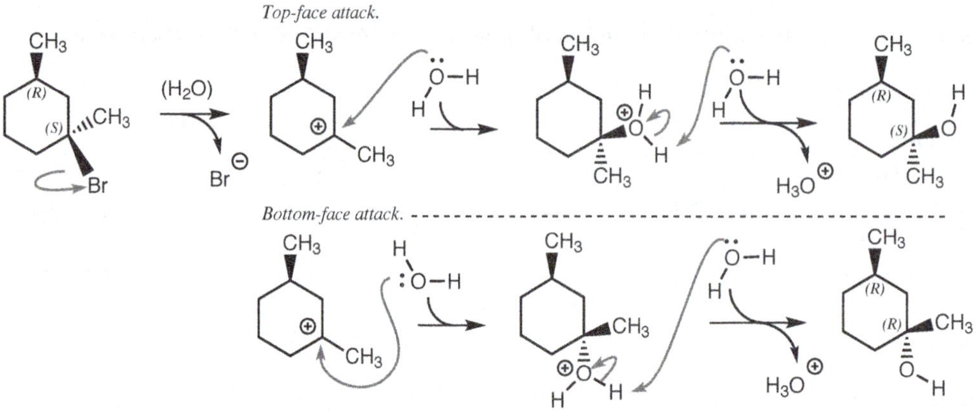

24. An S_N1 solvolysis reaction of 2-bromo-3-methylbutane to give 2-methyl-2-butanol. .(See *Notes.*)

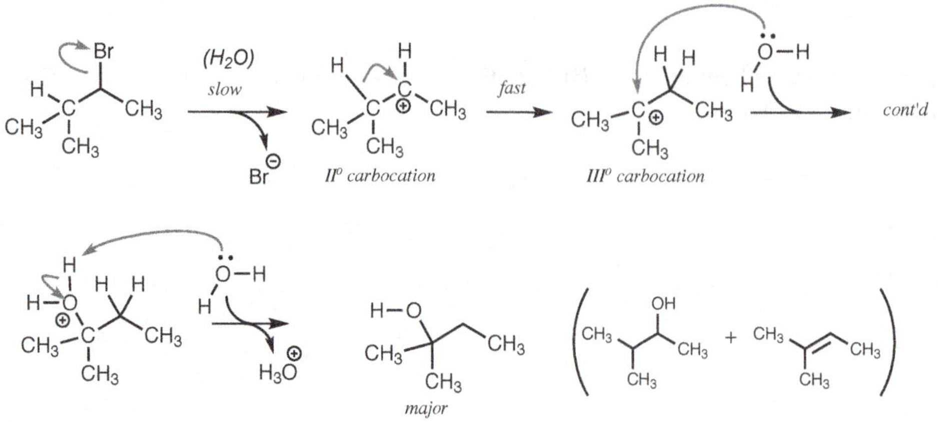

25. An S_N1 reaction of 2-methylbut-3-en-2-ol with hydrogen bromide to give 1-bromo-3-methyl-2-butene. (see *Notes*)

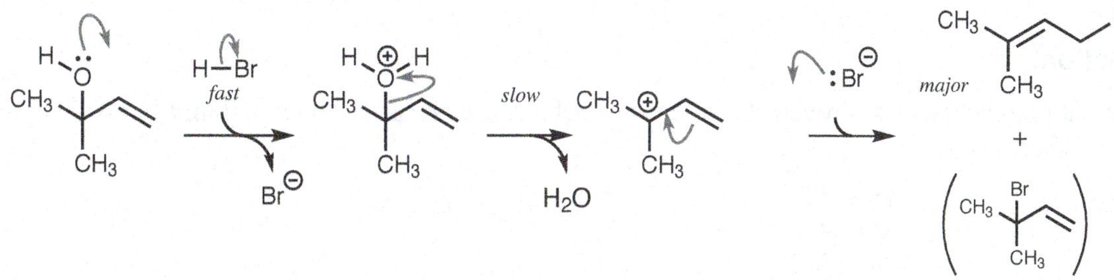

26. An S_N1 solvolysis reaction of 1-bromo-3-methyl-2-butene to give 2-methyl-3-buten-2-ol. (See *Notes*.)

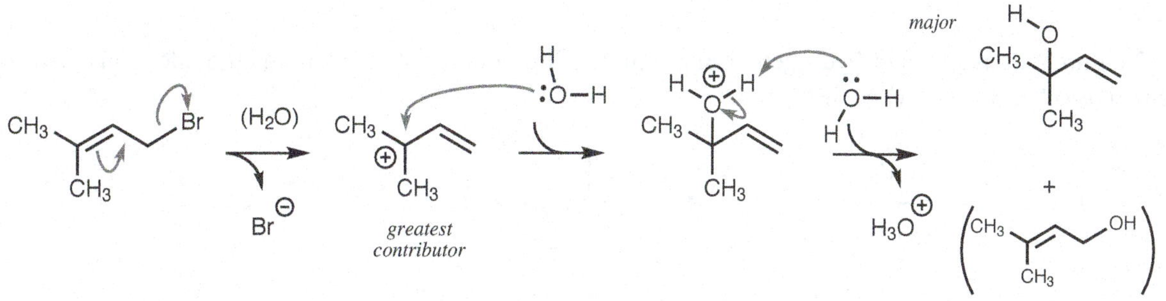

27. An S_N1 reaction of 2-methylcyclopentanol with hydrogen bromide to give 1-bromo-1-methylcyclopentane.

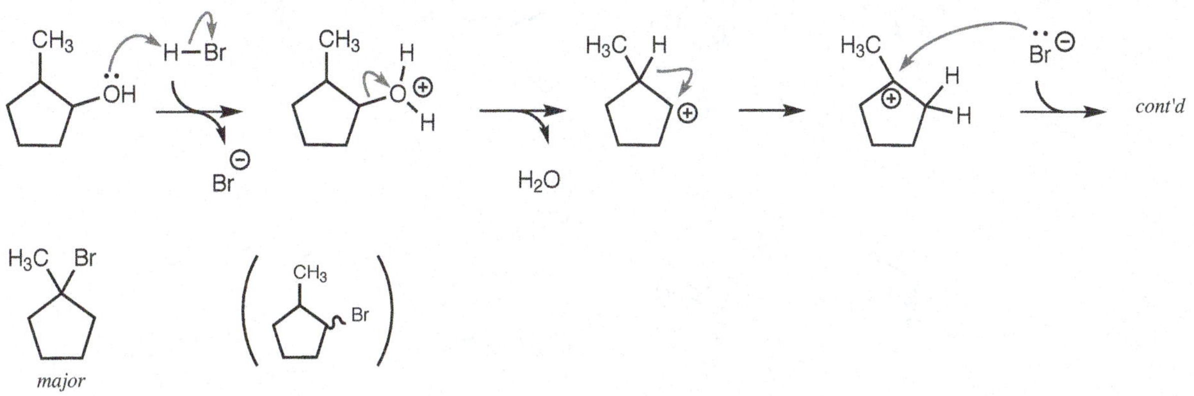

28. An S_N1 solvolysis reaction of 3-chlorocyclopentene to give 3-methoxycyclopentene. (See *Notes*.)

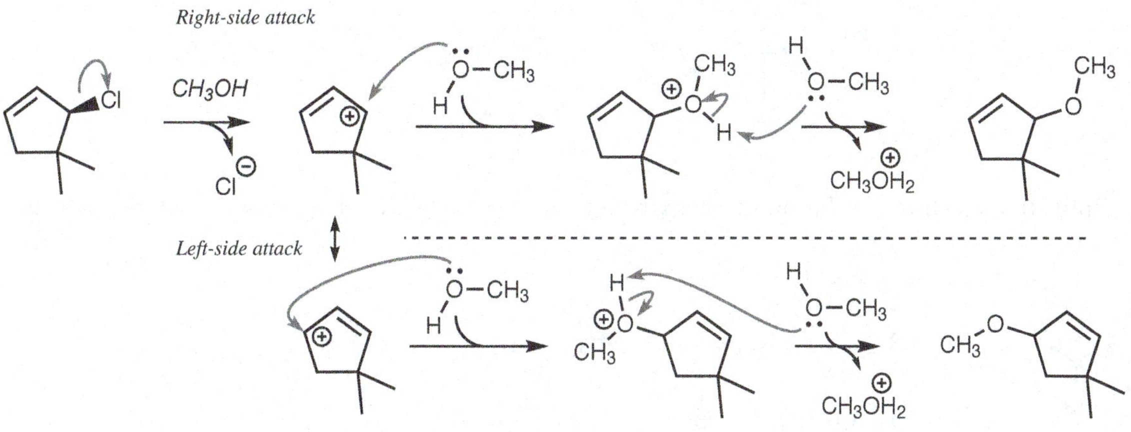

4 — Elimination Reactions

Alkene Formation

1. An E2 elimination reaction of hydrogen chloride from 1-chlorooctadecane with potassium *t*-butoxide to give 1-octadecene. (See *Notes*.)

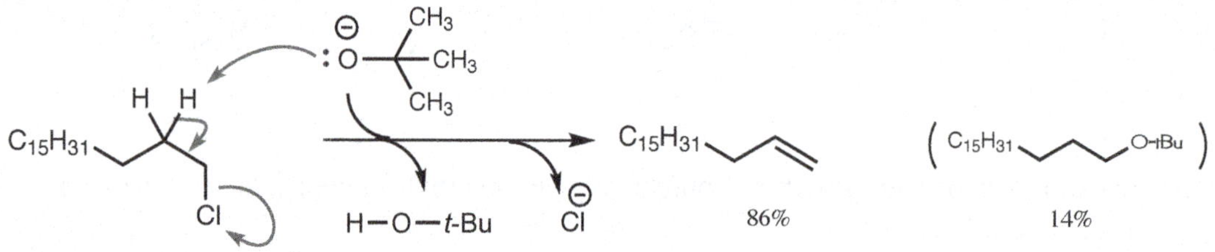

2. An E2 elimination reaction of hydrogen bromide from 2-bromobutane with sodium ethoxide to give *trans*-2-butene plus other butenes. (See *Notes*.)

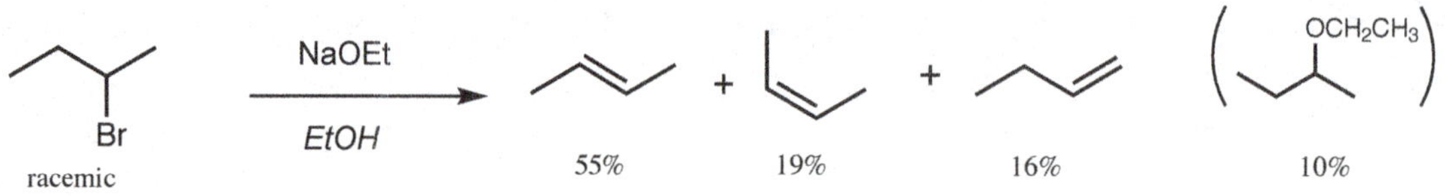

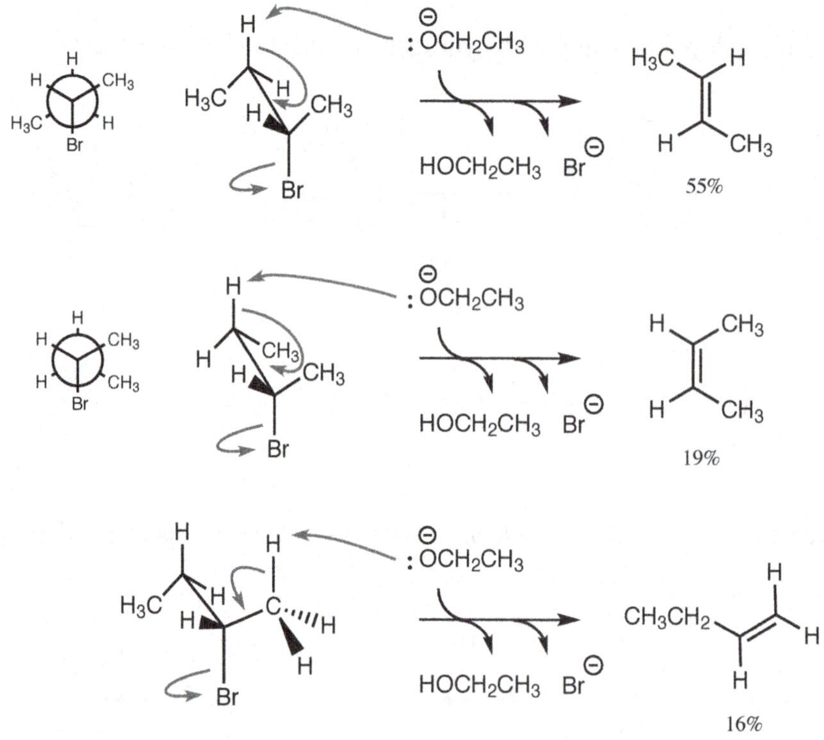

3. An E2 elimination reaction of 2-bromo-2-methylbutane to give 2-methyl-2-butene, a Zaitsev product. (See *Notes*.)

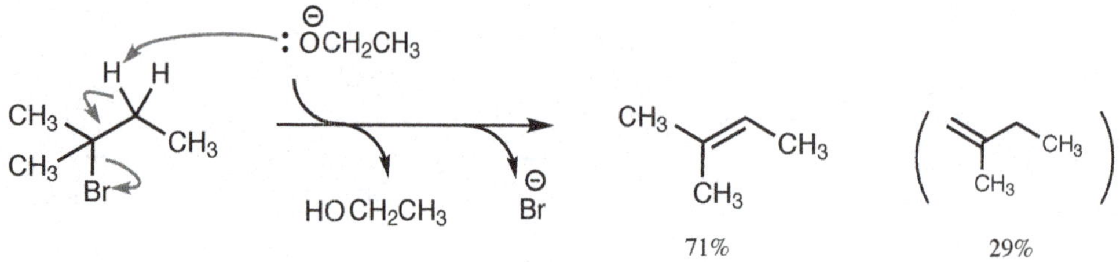

4. An E2 elimination reaction of (1R,2R)- or (1S,2S)-1-bromo-1,2-diphenylpropane to give (Z)-1,2-diphenylpropene. (See *Notes*.)

5. An E2 elimination reaction of *trans*-1-chloro-2-isopropylcyclohexane to give 3-isopropylcyclohexene. (see *Notes*)

stereoview

6. An E2 elimination reaction of *cis*-1-chloro-2-isopropylcyclohexane to give 3-isopropylcyclohexene. (See *Notes*.)

stereoview   *major*

7. A competitive E2 elimination reaction of *cis*- and *trans*-1-bromo-4-*t*-butylcyclohexane and one equivalent of *t*-butoxide to give 4-*t*-butylcyclohexene and unreacted bromocyclohexane. (See *Notes*.)

1 mole   *1 mole*   *1 mole*   *1 mole*

8. An E2 elimination reaction of 1-chloro-1-methylcyclohexane with ethoxide to give cyclohexene. (See *Notes*.)

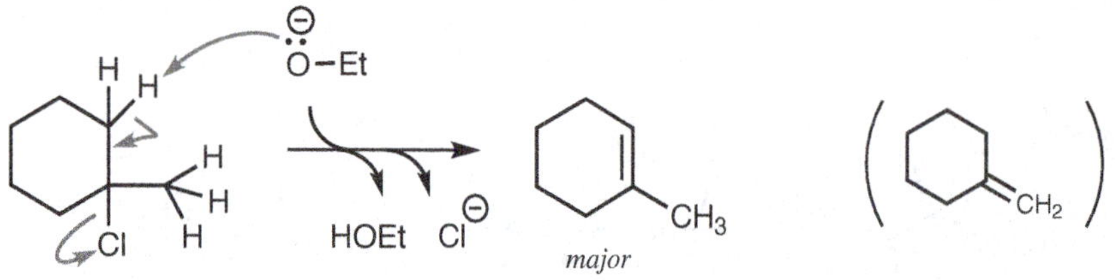

9. An E2 elimination reaction of 1-chloro-1-methylcyclohexane with *t*-butoxide to give methylenecyclohexane. (See *Notes*.)

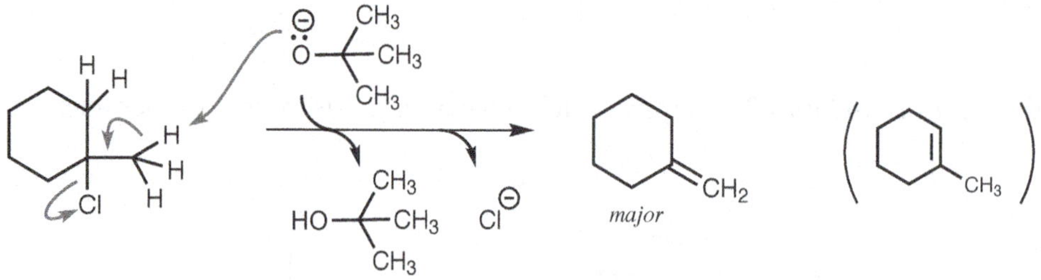

10. An E1cb elimination reaction of 3-chloro-3-methylcyclohexanone with *t*-butoxide to give 3-methyl-2-cyclohexenone. (See *Notes*.)

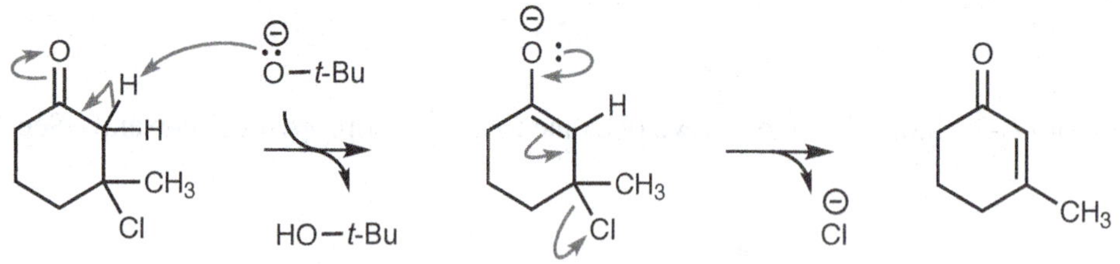

11. An E2 elimination reaction of 4-bromo-cyclohexene with *t*-butoxide to give 1,3-cyclohexadiene. (See *Notes*.)

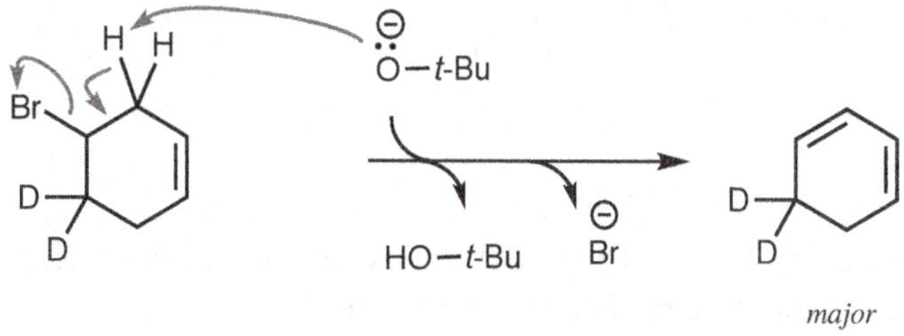

12. An E2 elimination reaction of 3-bromo-1-cyclohexene with *t*-butoxide to give 1,3-cyclohexadiene. (See *Notes*.)

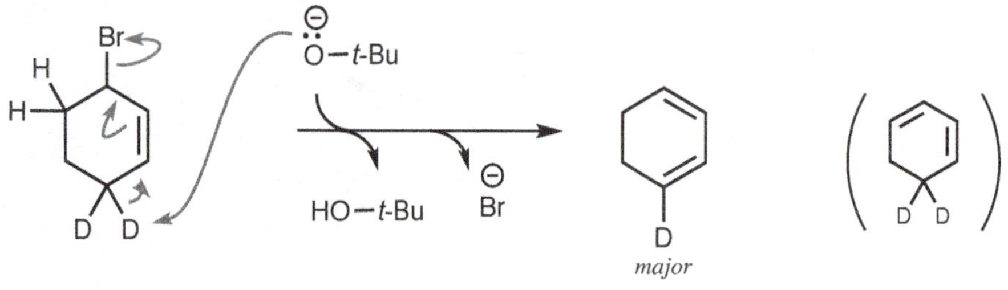

13. An E1 elimination reaction of 2-methylcyclopentanol by treatment with sulfuric acid to give methylcyclopentene.

slow
HSO4
H2O
H3O

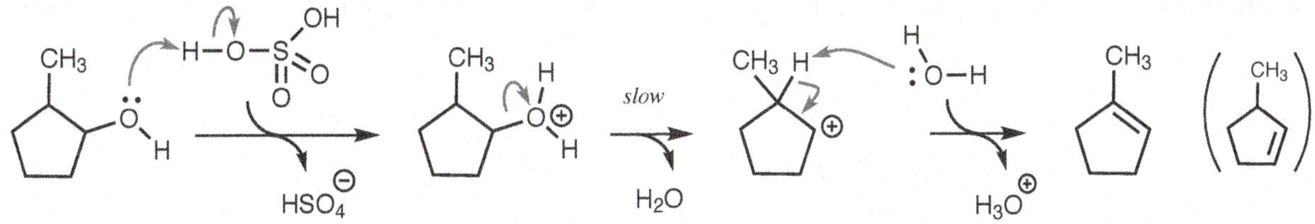

14. An E1 elimination reaction of 3-methyl-3-pentanol by treatment with acid to give pentenes. (See *Notes.*)

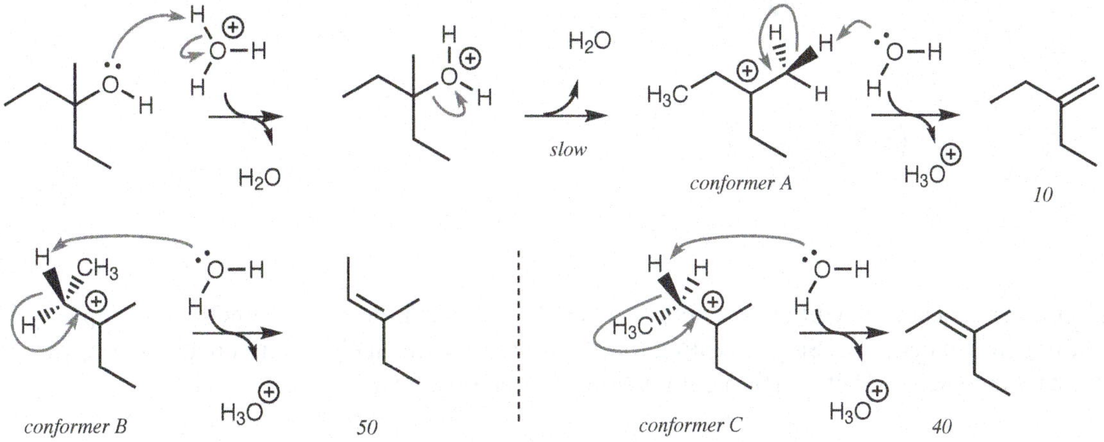

H2O
slow
conformer A
10
conformer B
50
conformer C
40

15. An E1 solvolysis-elimination reaction of a benzyl bromide in aqueous base to give methyl stilbenes. (See *Notes.*)

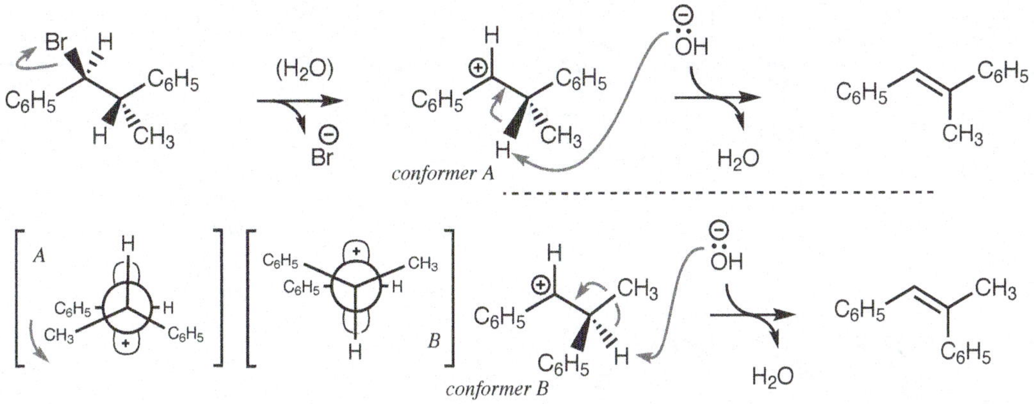

(H2O)
conformer A
conformer B
A
B
H2O

16. An E1 elimination reaction of 2-cyclobutyl-2-propanol and sulfuric acid to give 1,2-dimethylcyclopentene. (See *Notes.*)

OSO3H
H2O
cont'd
HOSO3H

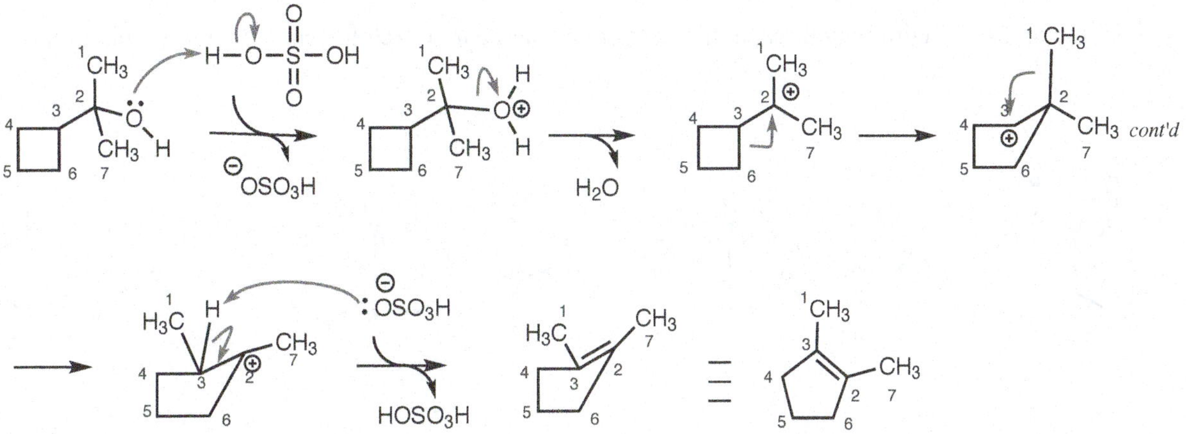

17. A Hofmann elimination reaction of a trimethylamine to give a 1-alkene. Reaction of *N,N*-dimethyl-2-pentanamine with iodomethane, silver oxide and elimination to give 1-pentene, the Hofmann elimination product. (See *Notes*.)

major

18. A Cope elimination reaction of a dimethylamine-*N*-oxide. Step 1, reaction of *N,N*-dimethyl-2,3-diphenylbutan-2-amine with hydrogen peroxide. Step 2, heating the *N*-oxide results in an elimination reaction to give *cis*-α,β-dimethylstilbene, the Zaitsev (syn) elimination product. (See *Notes*.)

major

19. A selenoxide oxidation-elimination reaction of a cyclohexanone to a cyclohexenone, a *syn*-elimination. (See *Notes*.)

Acetylene Formation

20. A synthesis of 3-hexyne from *trans*-3-hexene by bromination and two elimination reactions. (See *Notes*.)

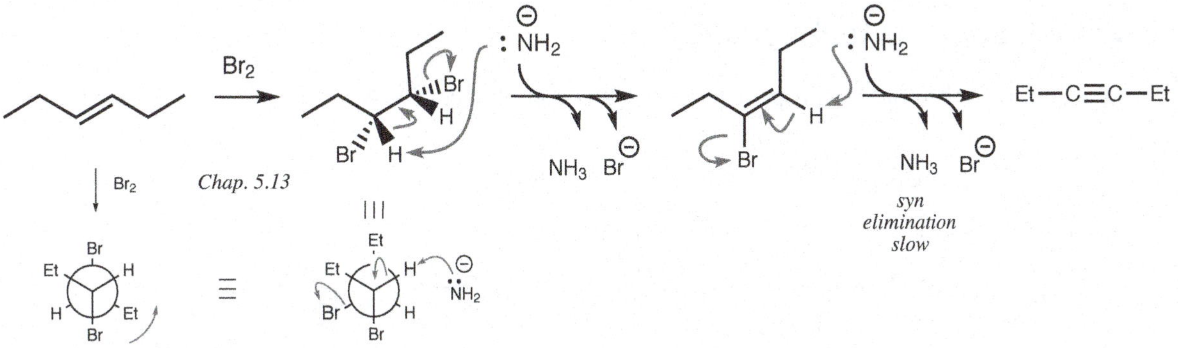

21. A synthesis of 3-hexyne from *cis*-3-hexene by bromination and two elimination reactions. (See *Notes*.)

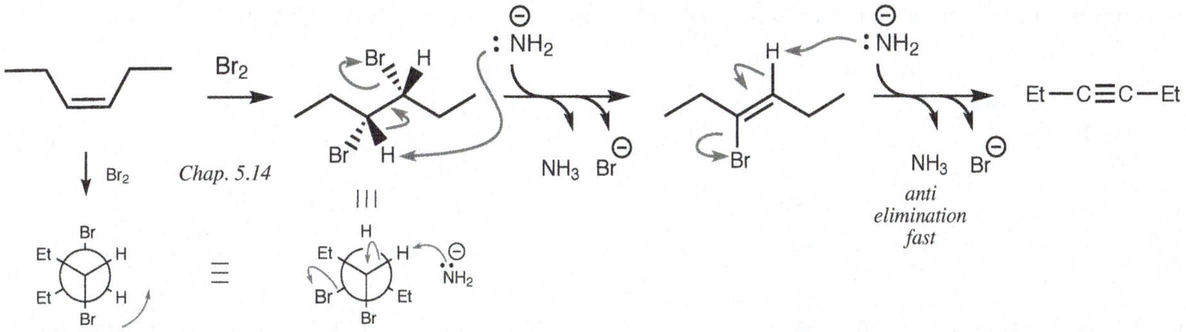

22. An E2 elimination reaction of 1,1-dibromopentane with *t*-butoxide to give 1-pentyne. (See *Notes*.)

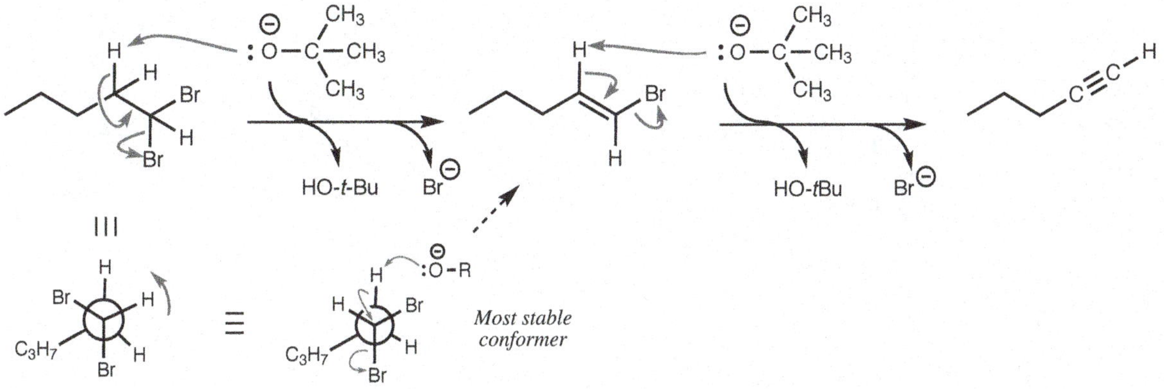

23. An E2 elimination reaction of 2,2-dibromopentane with LDA to give 1-pentyne. (See *Notes*.)

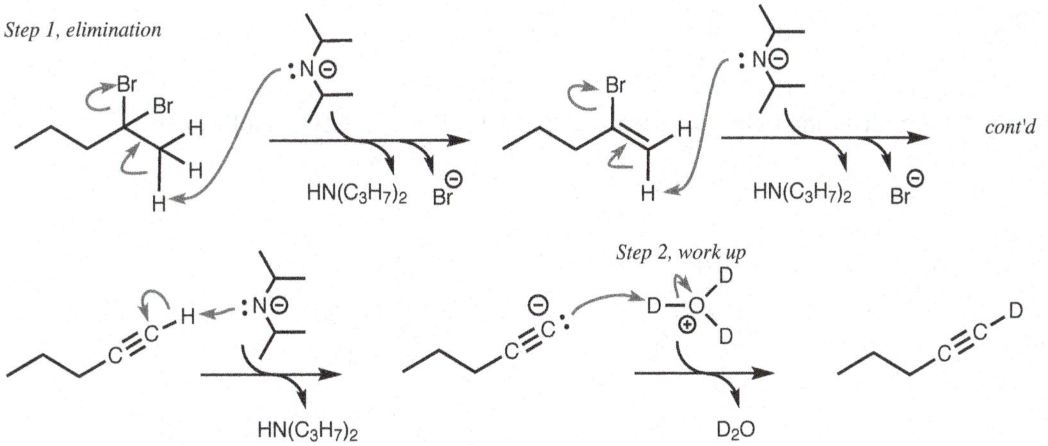

5 — Electrophilic Addition to Alkenes and Alkynes

Addition of HX and H₂O to Alkenes

1. Addition of hydrogen bromide to propene to give 2-bromopropane. (See *Notes*.)

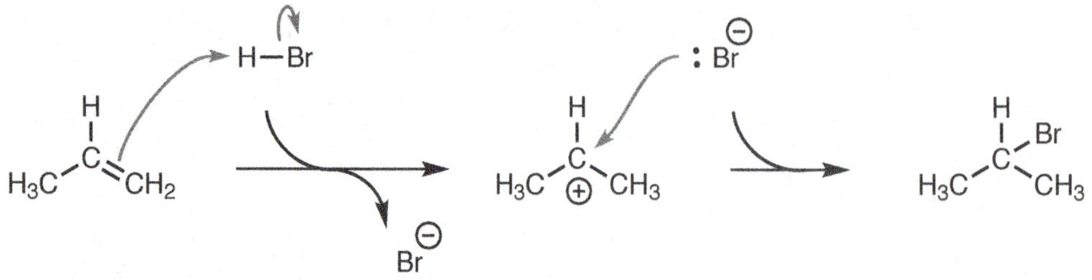

2. Acid catalyzed addition of water to methylcyclopentene to give 1-methylcyclopentanol. (See *Notes*.)

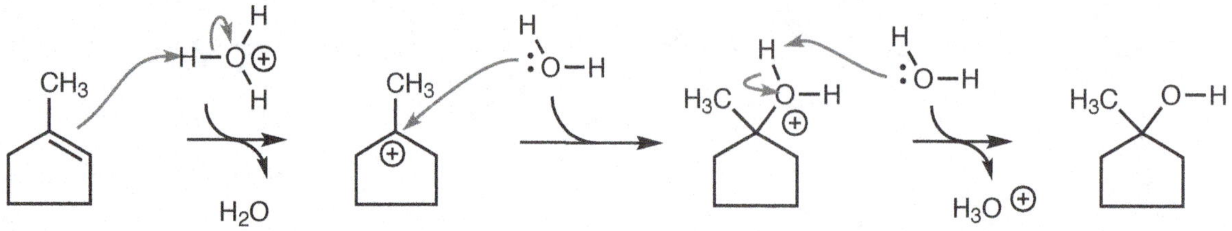

3. Addition of acetic acid to propene catalyzed by sulfuric acid to give 2-propyl ethanoate (isopropyl acetate). (See *Notes*.)

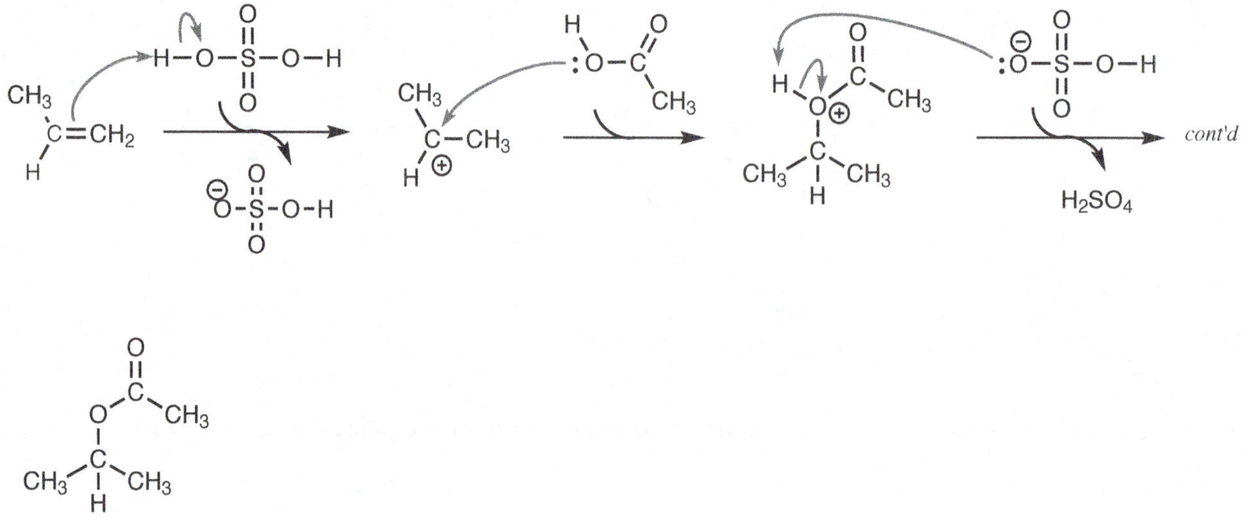

4. Addition of hydrogen bromide to methylcyclohexene to give 1-bromo-1-methylcyclohexane.

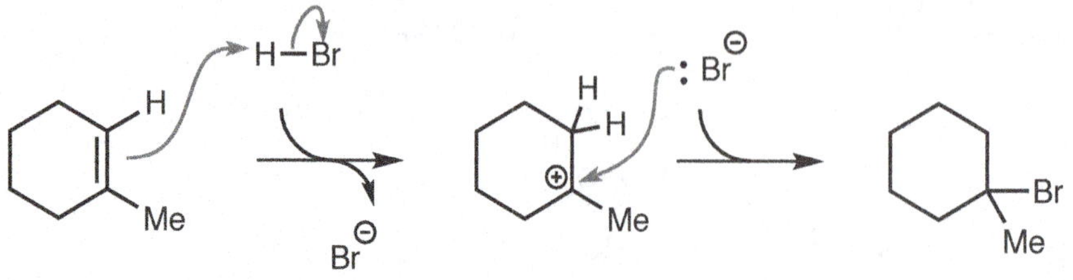

5. Addition of hydrogen chloride to *(E)*-3-hexene to give *(R)*- and *(S)*-3-chlorohexane.

(*R*)-3-chlorohexane   (*S*)-3-chlorohexane

6. Addition of hydrogen chloride to *(Z)*-3-hexene to give *(R)*- and *(S)*-3-chlorohexane.

(*S*)-3-chloro-hexane

(*R*)-3-chloro-hexane

7. Addition of hydrogen bromide to 3-methyl-1-butene to give after rearrangement, 2-bromo-2-methylbutane. (See *Notes*.)

II° carbocation   *III° carbocation*   55%

45%

8. Addition of HBr to 2-cyclobutylpropene to give, after rearrangement, 1-bromo-1,2-dimethylcyclopentane. (See *Notes*.)

Number the atoms to avoid confusion.

cont'd

mixture of diastereomers

9. Addition of hydrogen chloride to allylbenzene to give, after rearrangement, (1-chloropropyl)benzene. (See *Notes*.)

cont'd

10. Addition of hydrogen chloride to 2-methyl-1,3-butadiene (isoprene) to give, 3-chloro-3-methyl-1-butene, the kinetic product, or 1-chloro-3-methyl-2-butene, the thermodynamic product. (See *Notes*.)

III° greater contributor

Kinetic product

Thermodynamic product

11. Addition of bromine to 2-methyl-1,3-butadiene to give 3,4-dibromo-3-methylbut-1-ene, the kinetic product and 1,4-dibromo-2-methylbut-2-ene, the thermodynamic product. (See *Notes*.)

III° greater contributor

Kinetic product

Thermodynamic product

Bromination

12. Bromination of cyclohexene to give *trans*-1,2-dibromocyclohexane. (See *Notes*.)

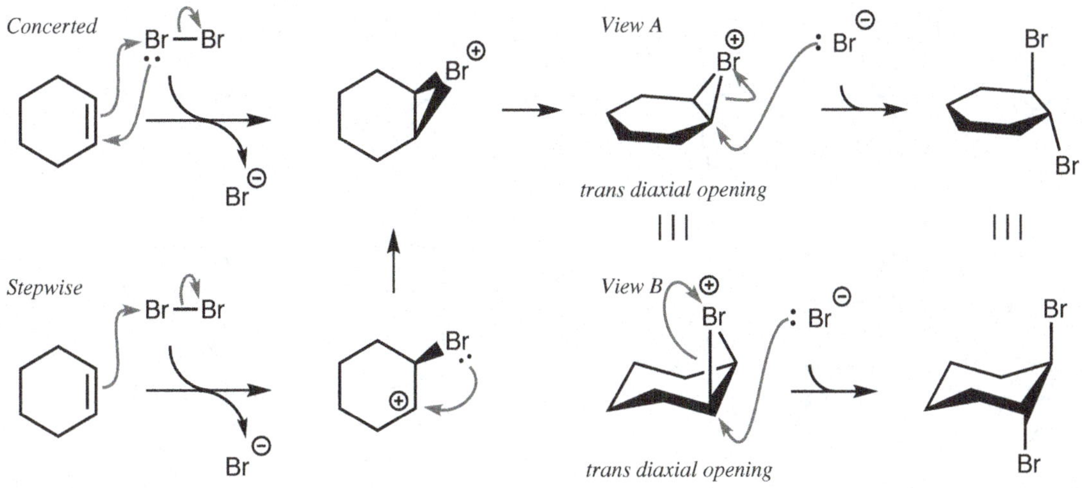

13. Bromination of *trans*-2-butene to give (erythro) *(2R,3S)*- and *(2S,3R)*-2,3-dibromobutane.

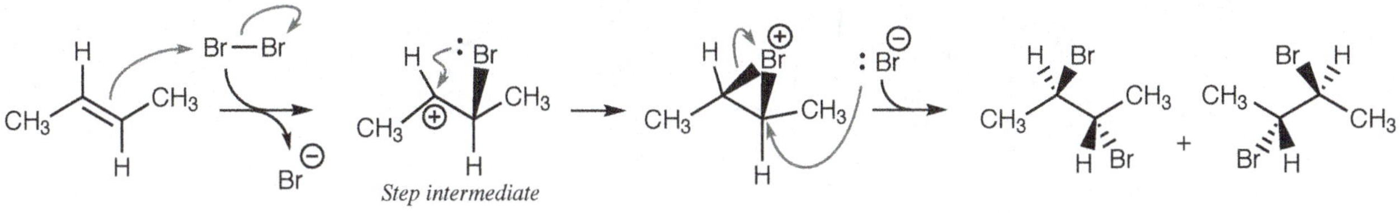

14. Bromination of *cis*-2-butene to give (threo) *(2R,3R)*- and *(2S,3S)*-2,3-dibromobutane.

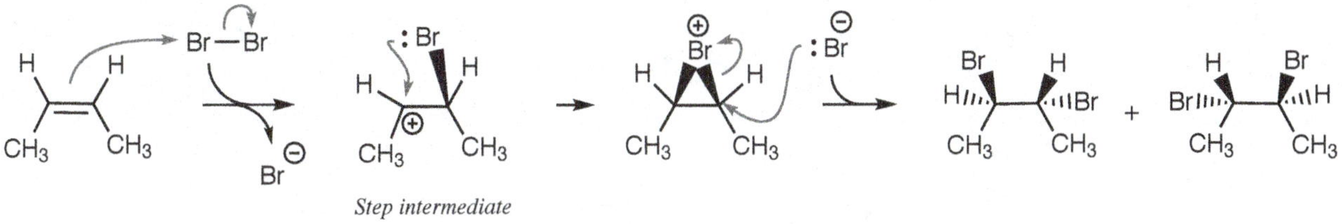

15. Bromination of methylcyclohexene to give (1R,2R)- and (1S,2S)-2-bromo-1-methylcyclohexanol. (See *Notes*.)

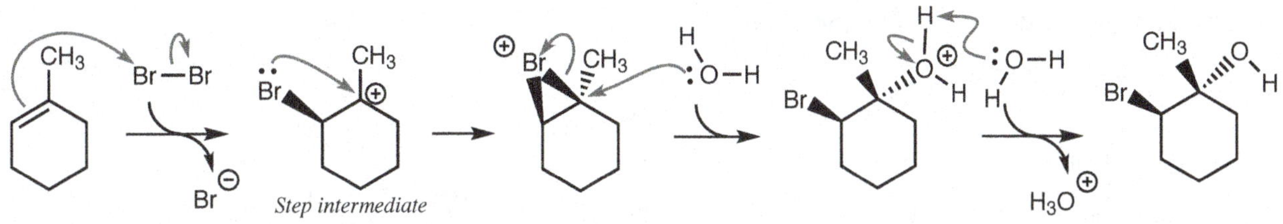

Oxymercuration

16. Step 1, oxymercuration of 3-methyl-1-butene to give 3-methyl-2-butanol. (See *Notes*.)

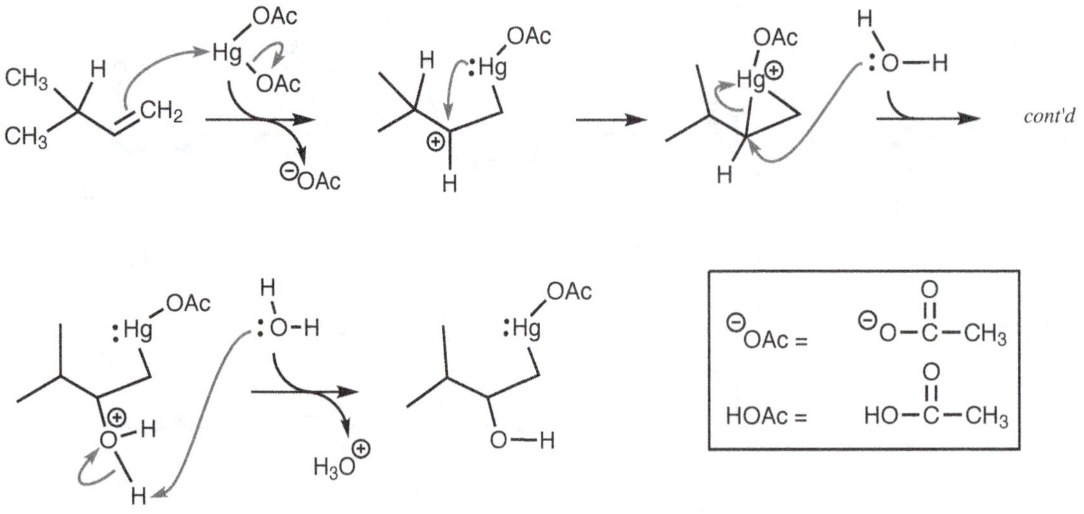

Step 2, reductive demercuration. (See *Notes*.)

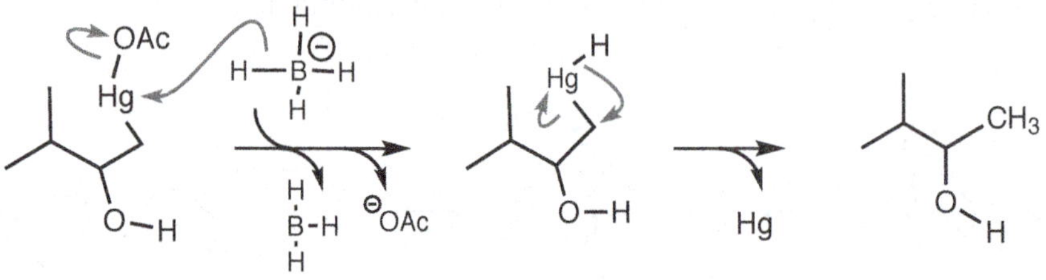

17. Step 1, oxymercuration of 1-methylcyclohexene to give 1-methylcyclohexanol. (See *Notes*.)

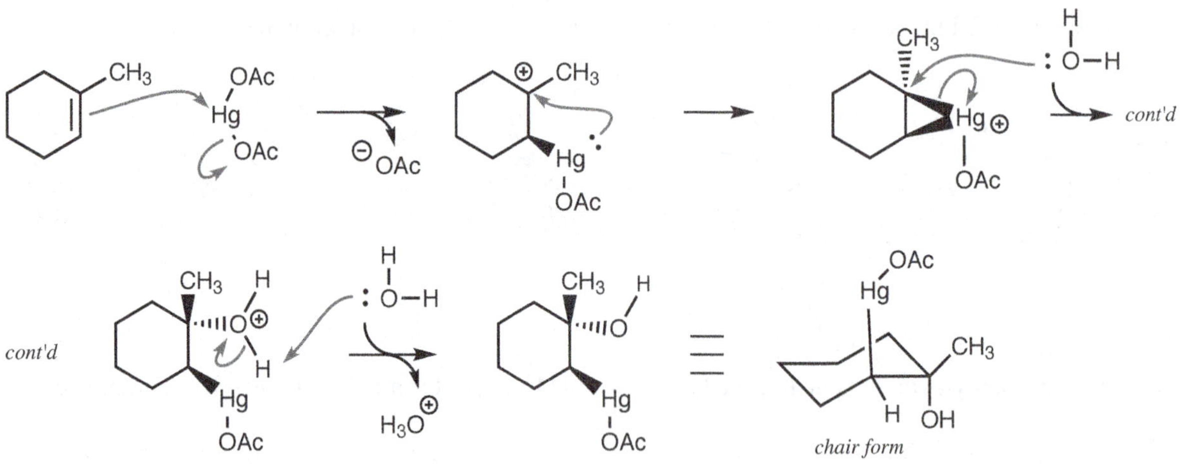

Step 2, reductive demercuration. (See *Notes*.)

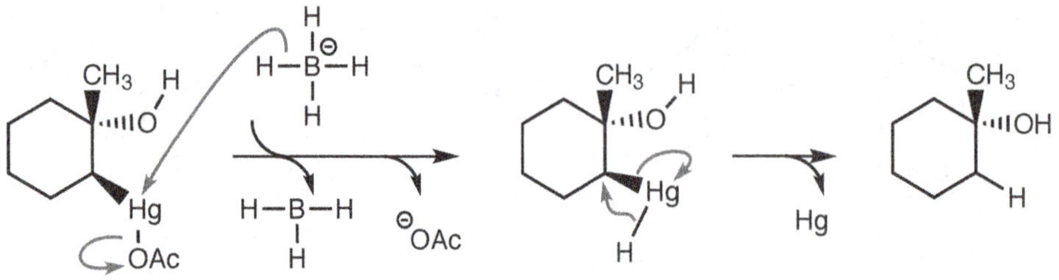

Hydroboration-Oxidation of Alkene

18. Hydroboration-oxidation of propene to give 1-propanol.

Step 1, hydroboration. Each bracket represents one of three hydroboration steps. (See *Notes.*)

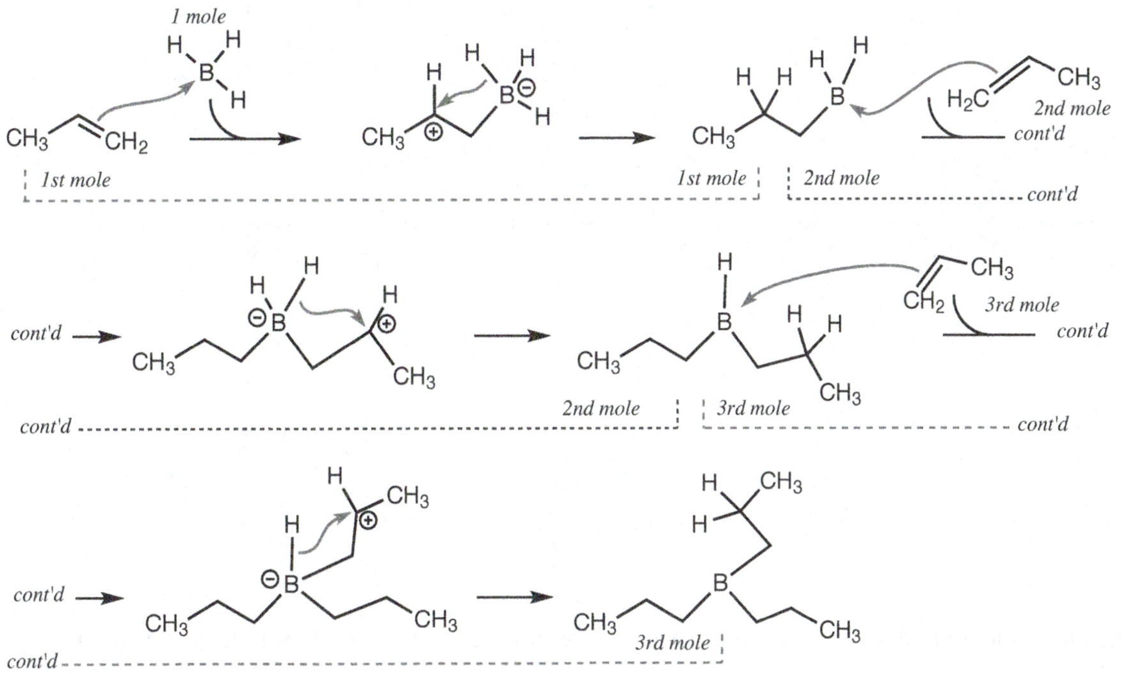

Step 2, oxidation. Each bracket represents one of three oxidation steps. (See *Notes.*)

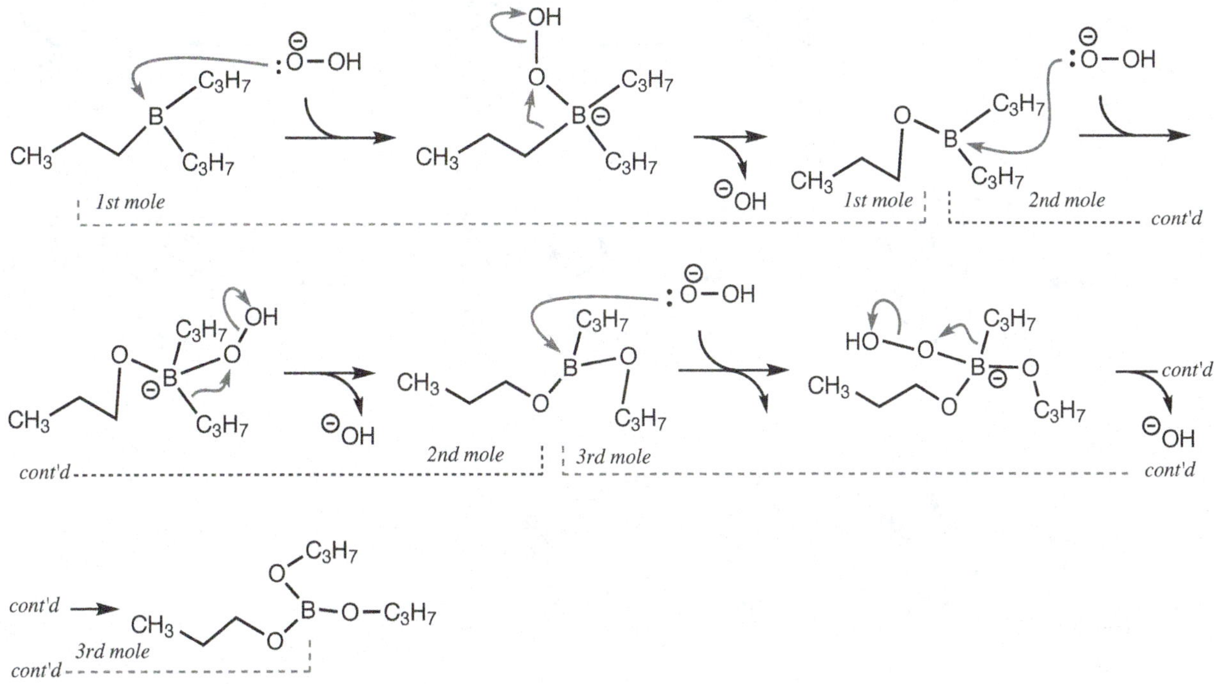

Step 3, borate ester hydrolysis to 1-propanol. Each bracket represents one of three hydrolysis steps. (See *Notes*.)

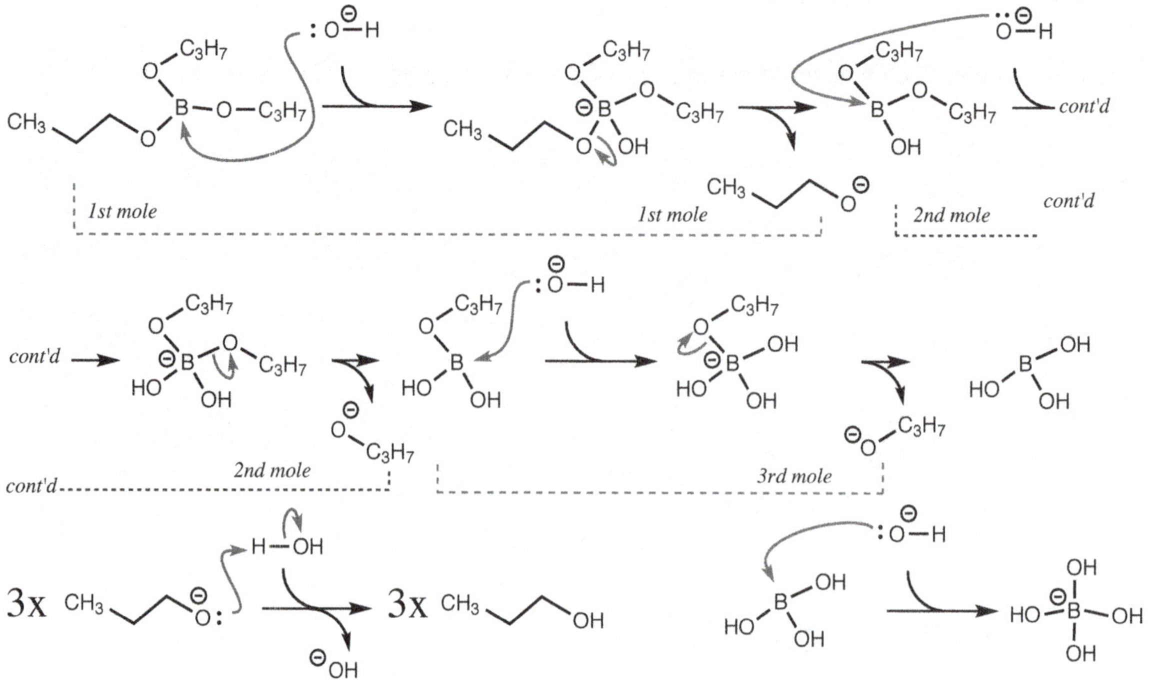

19. Hydroboration/oxidation of 1-methylcyclohexene to give *trans*-2-methylcyclohexanol. (See *Notes*.)

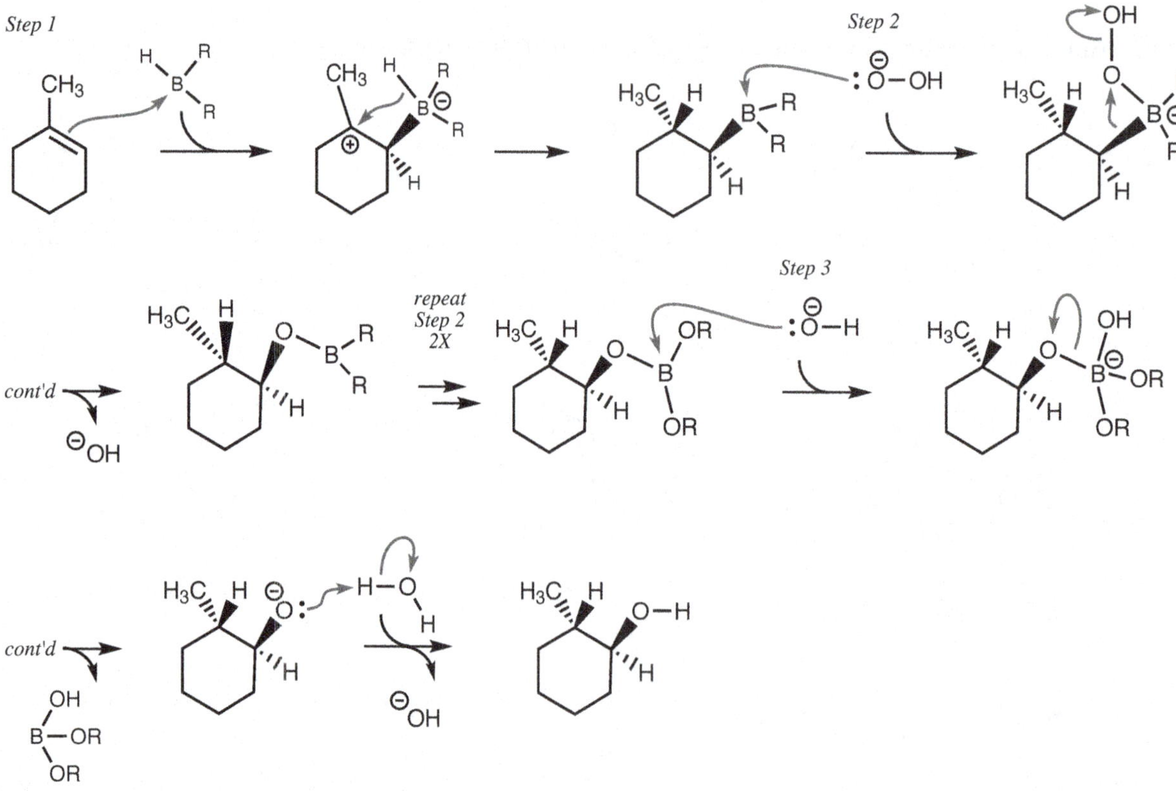

Carbon-Carbon Triple Bond Electrophilic Reactions

Addition to an Internal Acetylene

20. Addition of HCl to 2-butyne (dimethyl acetylene) to give (*E*)-2-chloro-2-butene and 2,2-dichlorobutane. (See *Notes*.)

2nd Equivalent of HCl

21. Addition of bromine to ethynylcyclopentane to give *(E)*-(1,2-dibromovinyl)cyclopentane and (1,1,2,2-tetrabromoethyl)cyclopentane. (See *Notes*.)

2nd Equivalent of bromine.

22. Sulfuric acid catalyzed hydration of 1-propynylbenzene to give 1-phenyl-1-propanone. (See *Notes*.)

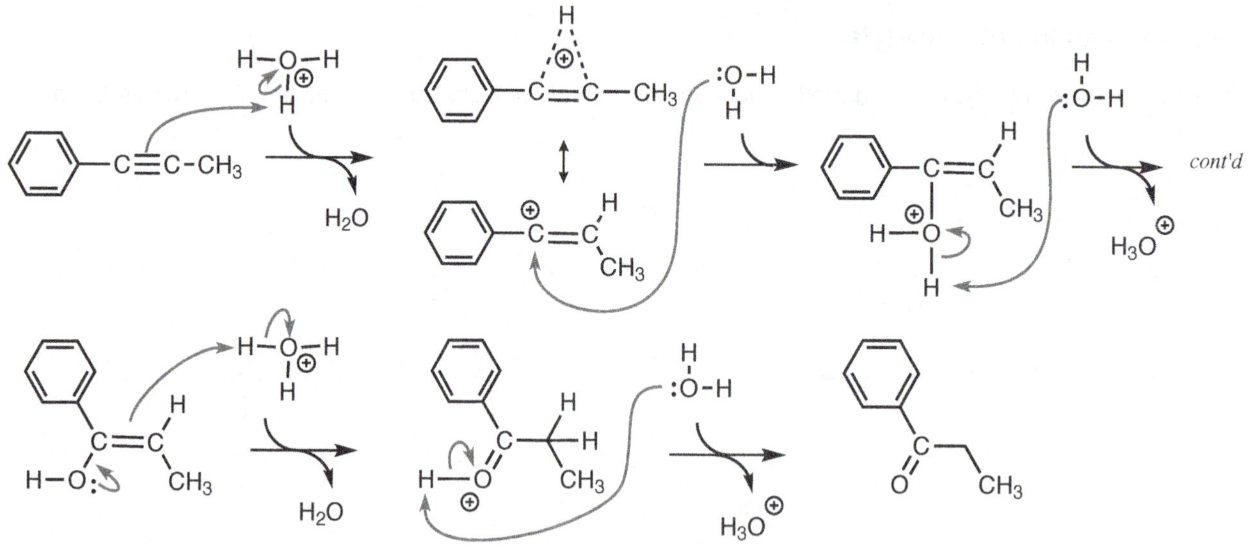

Addition to a Terminal Acetylene

23. Mercury catalyzed hydration of propyne (methyl acetylene) to give 2-propanone (acetone). (See *Notes*.)

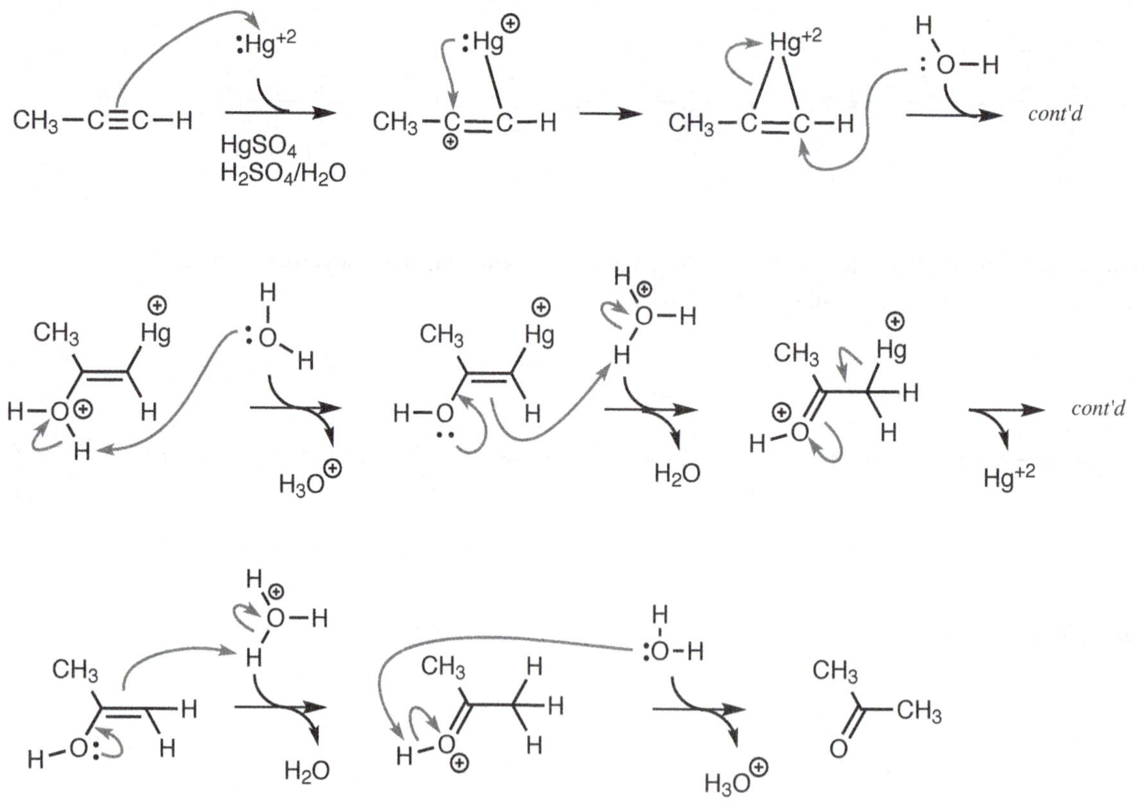

Disiamylborane Hydroboration–Oxidation of an Acetylene

24. Hydroboration-oxidation of phenylacetylene with disiamylborane to give phenylacetaldehyde. (See *Notes*.)

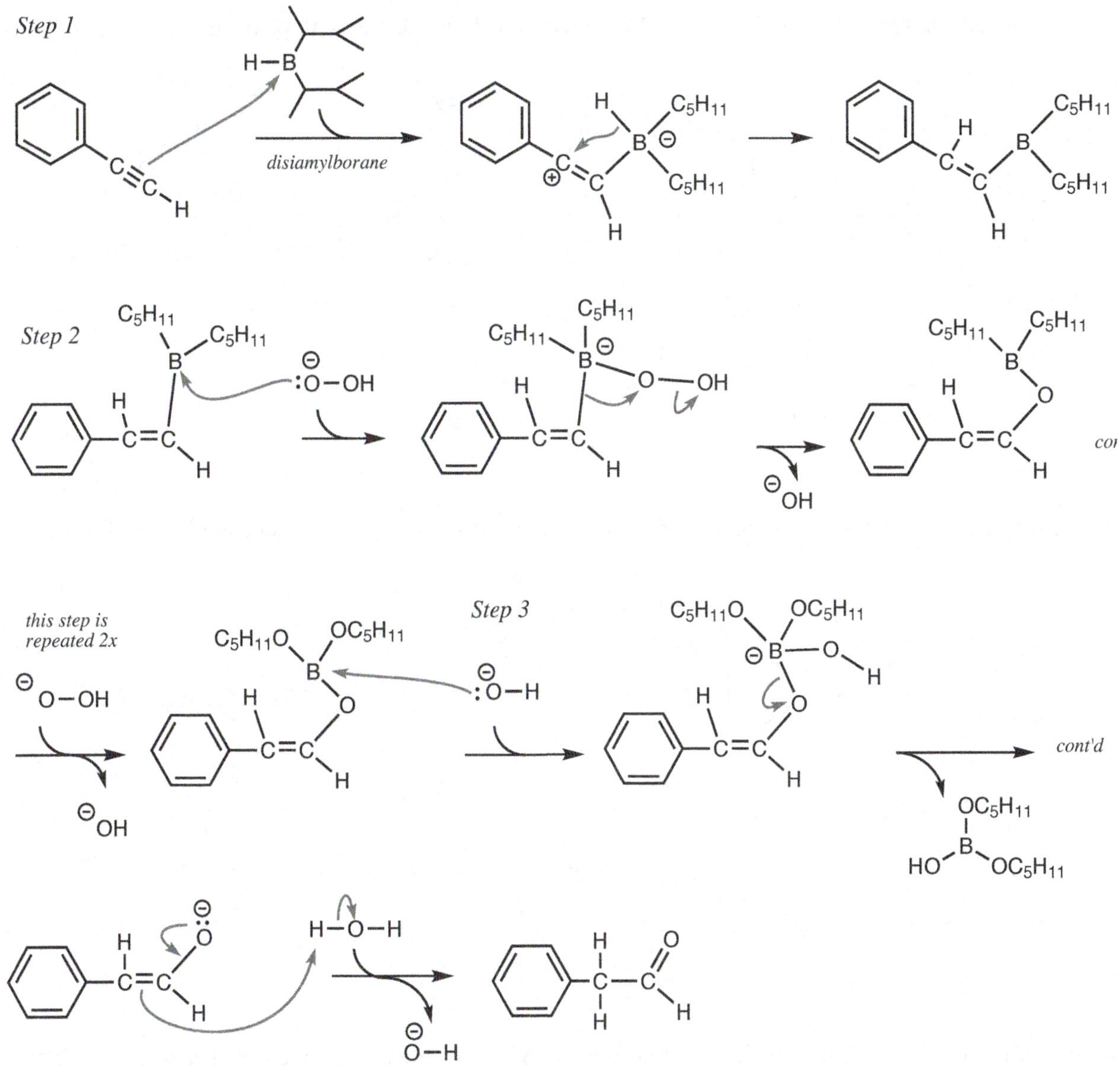
Step 1
disiamylborane
Step 2
cont'd
this step is repeated 2x
Step 3
cont'd

6 — Rearrangement Reactions

Baeyer-Villiger Oxidation

1. Acid catalyzed Baeyer-Villiger oxidation of 2,2-dimethylcyclopentanone with peracetic acid. (See *Notes*.)

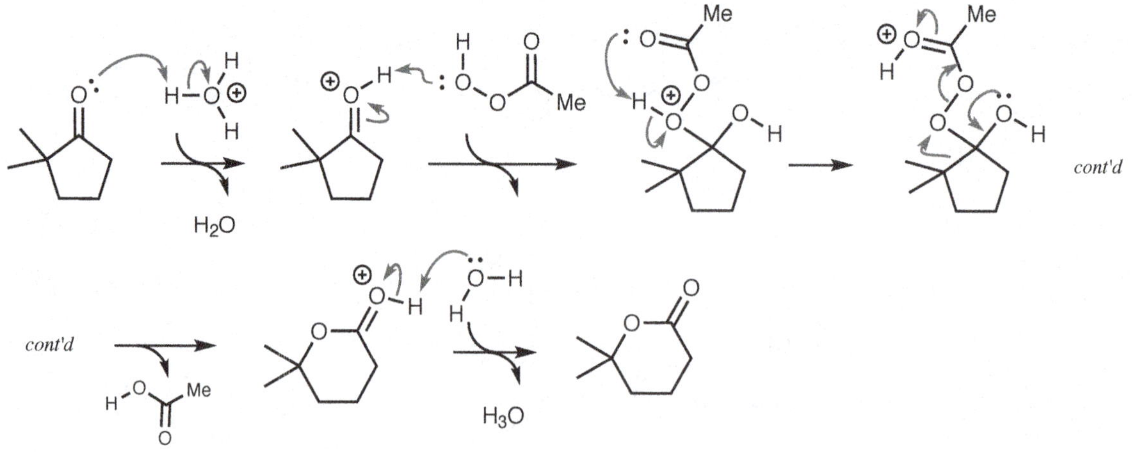

2. Acid catalyzed Baeyer-Villiger oxidation of *o*-methoxyacetophenone with peracetic acid. (See *Notes*.)

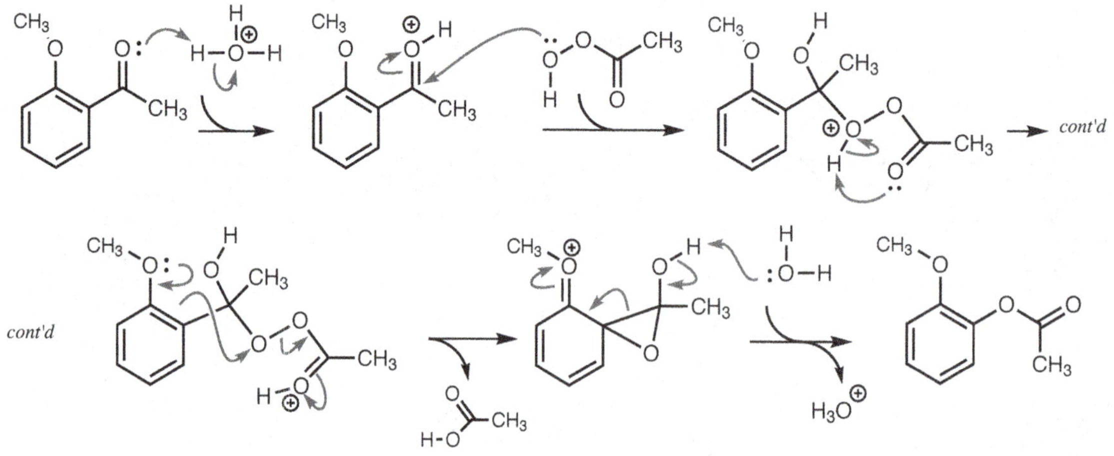

3. Acid catalyzed Baeyer-Villiger oxidation of benzaldehyde with peracetic acid to give benzoic acid. (See *Notes*.)

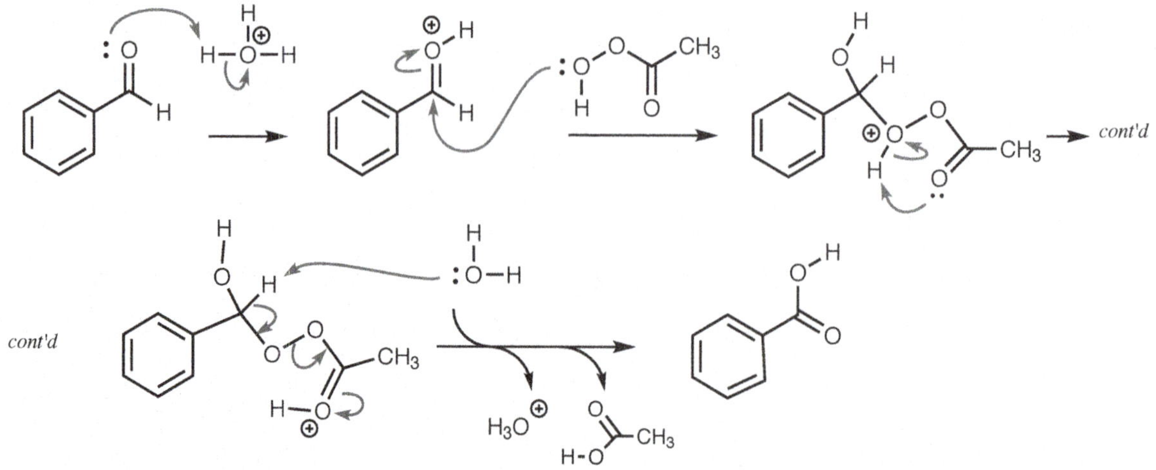

4. Baeyer-Villiger oxidation of a benzophenone with trifluoroperacetic acid. (See *Notes.*)

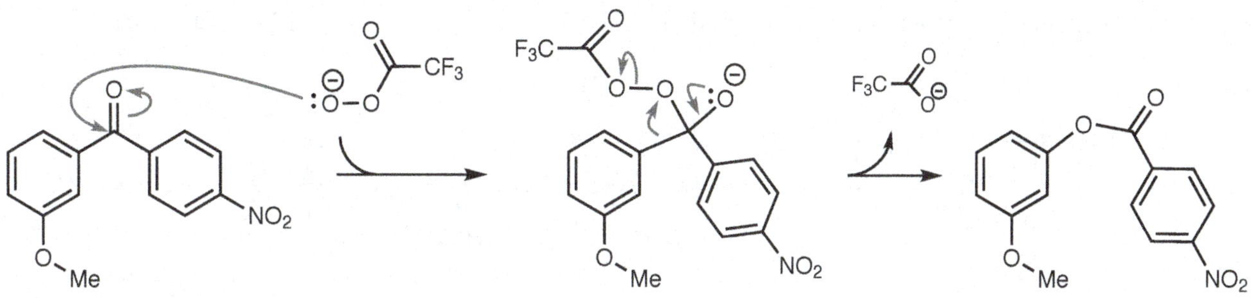

5. Baeyer-Villiger oxidation of *o*-methoxyacetophenone with *m*-chloroperoxybenzoic acid. (See *Notes.*)

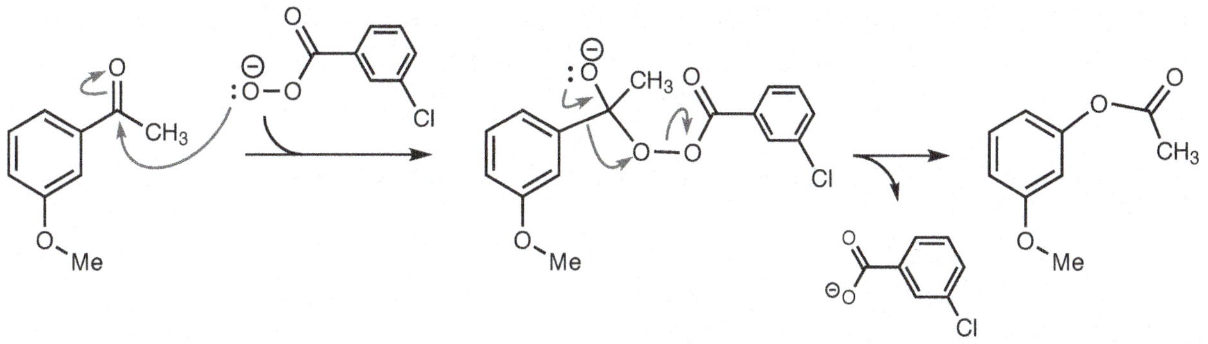

6. Baeyer-Villiger oxidation of bicyclic phenyl ketone with trifluoroperacetic acid. (See *Notes.)*

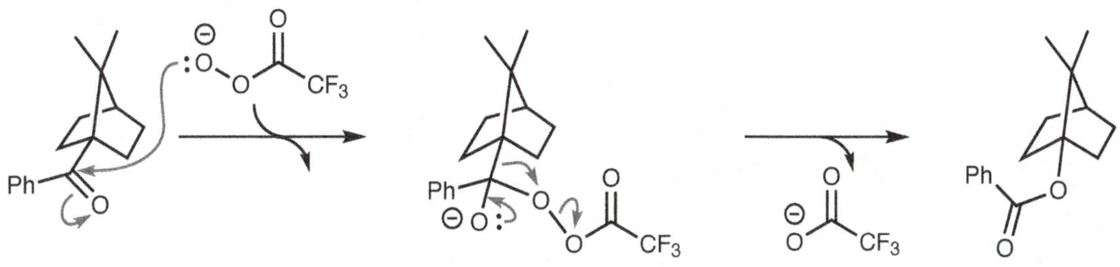

7. Baeyer-Villiger oxidation of *p*-chlorobenzaldehyde with peracetic acid to give *p*-chlorobenzoic acid. (See *Notes.)*

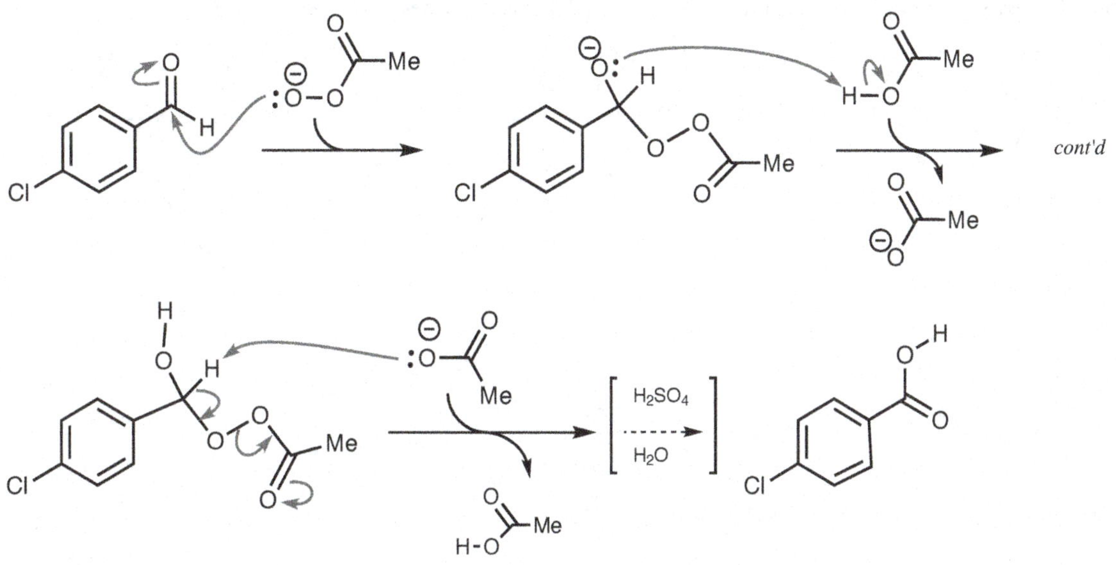

cont'd

Pinacol Rearrangement

8. Rearrangement of pinacol to pinacolone, methyl *t*-butyl ketone or 3,3-dimethyl-2-butanone. (See *Notes*.)

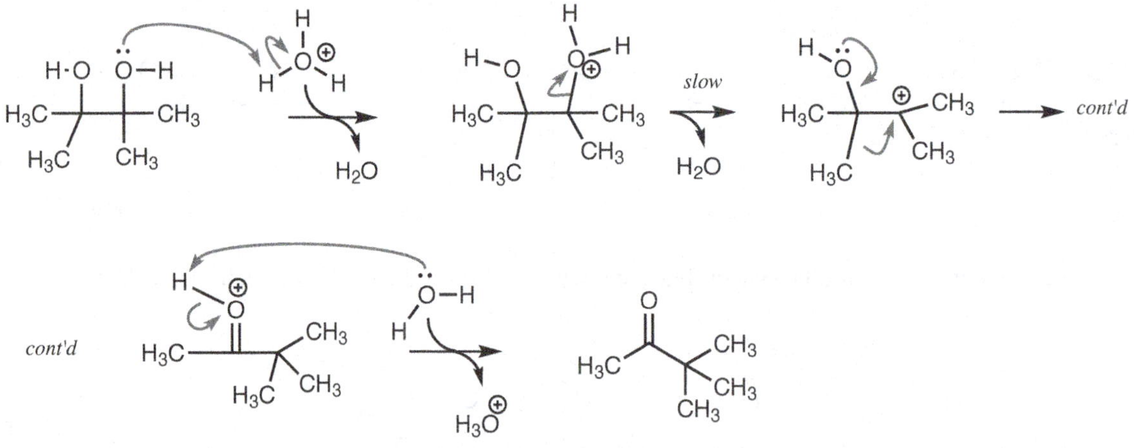

9. Rearrangement of 1,2-dimethyl-1,2-cyclohexanediol with acid to a methyl ketone. (See *Notes*.)

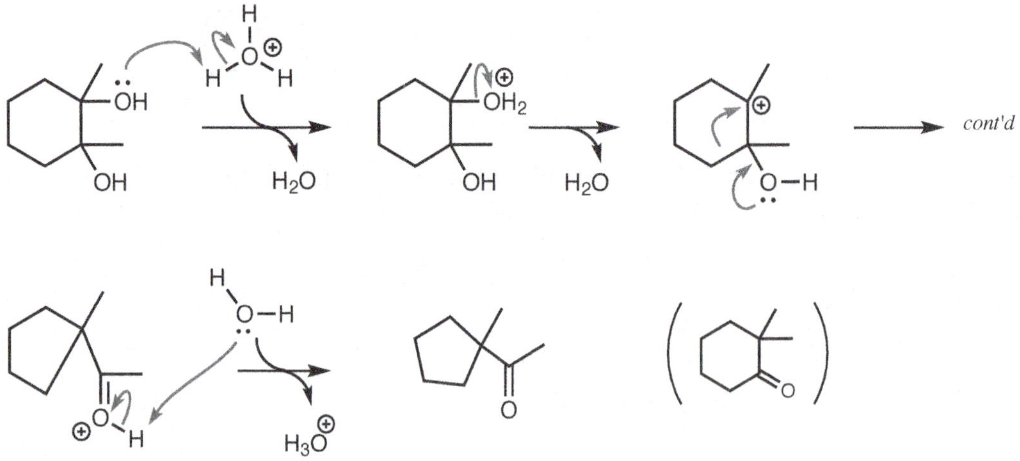

Benzilic Acid Rearrangement

10. Reaction of benzil with hydroxide to give benzilic acid after rearrangement. (See *Notes*.)

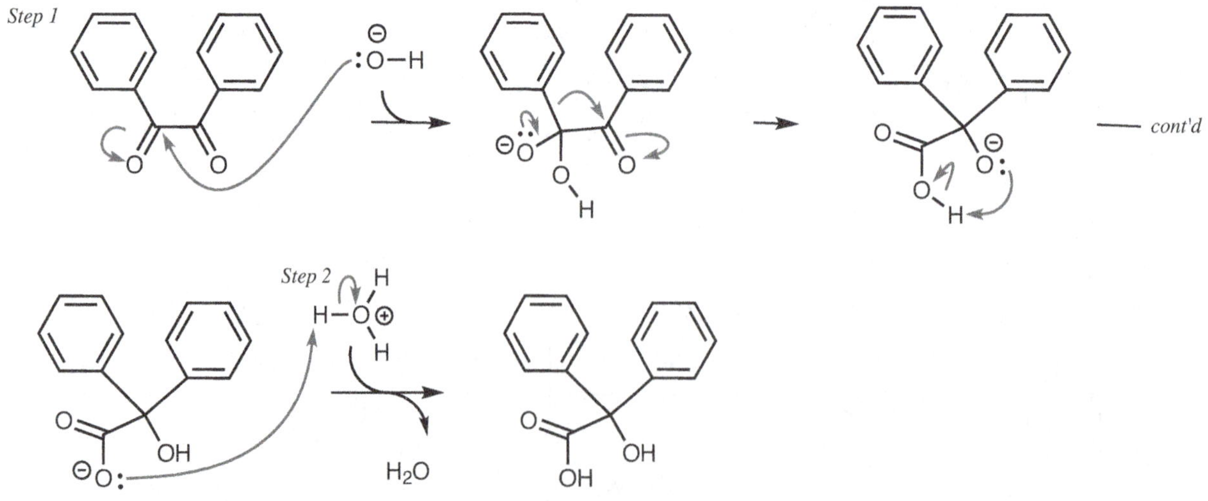

Dakin Reaction

11. Reaction of an *o*- or *p*-hydroxybenzaldehyde with basic hydrogen peroxide to give a phenol. (See *Notes*.)

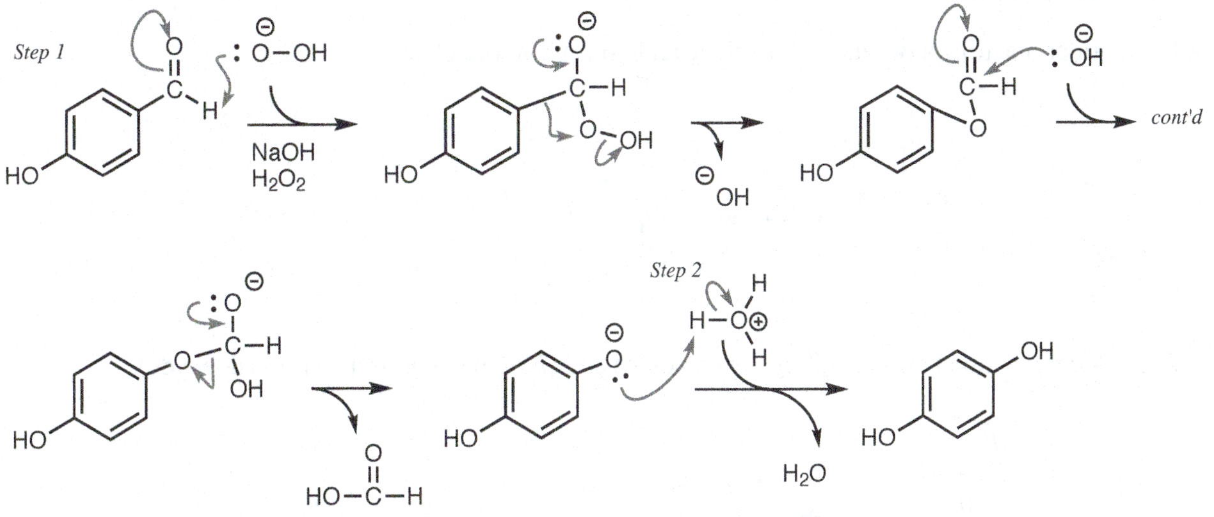
Step 1
NaOH
H_2O_2
cont'd
Step 2

Acetone from Cumene

12. Reaction of isopropylbenzene (cumene) to acetone and phenol.

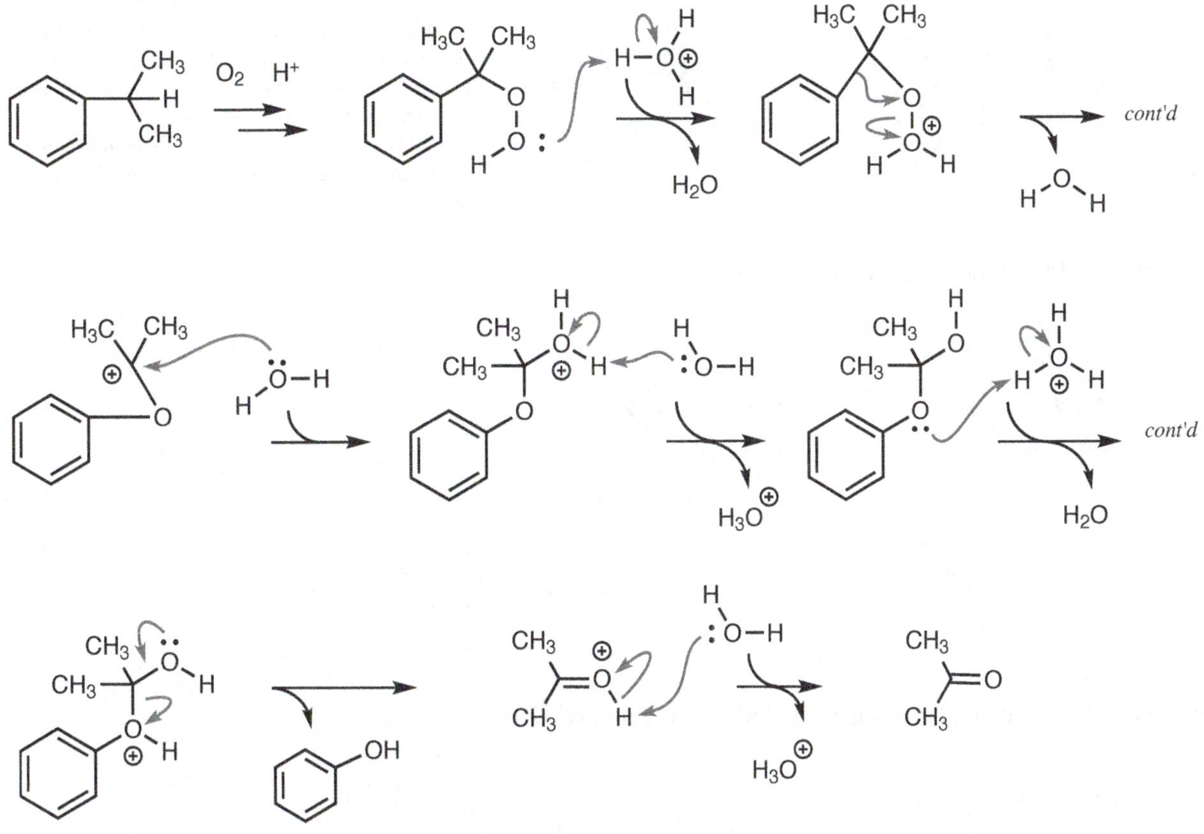
O_2 H^+
cont'd
cont'd

7 — Electrocyclic Reactions

Diels Alder Reactions

1. A Diels-Alder reaction between 1,3-butadiene and 2-propenal (acrolein).

2. A Diels-Alder reaction between 1,3-cyclopentadiene and (*E*)-2-butenal (trans-crotonaldehyde).

redraw

The preferred orientation of the dienophile is below. You should check with your text for further details.

This is the same example as above. It is more difficult to negotiate the atom movements, congestion, and drawing the final product all at once. I wrote the upper example in two steps as it is easier to visualize and convert it to the bicyclic product.
The preferred orientation has the electron-withdrawing group of the dienophile overlapping with the diene, if possible.

3. The Diels-Alder dimer of cyclopentadiene.

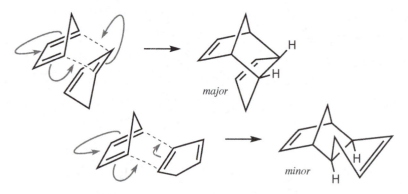

major

minor

4. A Diels-Alder reaction between 1,3-butadiene and methyl (*Z*)-2-butenoate.

5. A Diels-Alder reaction between 2-methyl-1,3-butadiene and *N*-methylmaleimide.

H3C–N
CH3

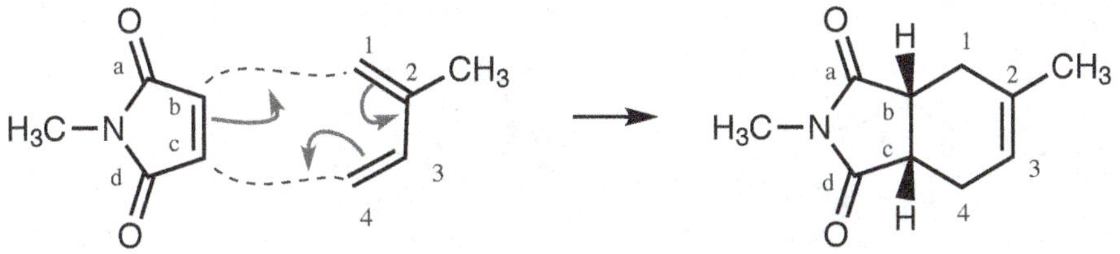

6. A Diels-Alder reaction between furan and but-3-en-2-one (methyl vinyl ketone, MVK).

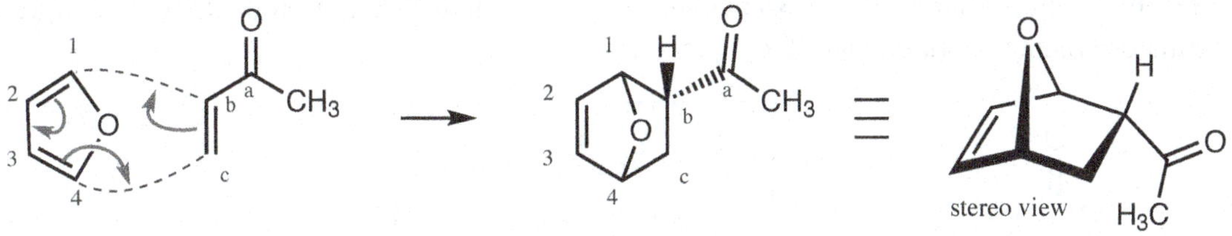

CH3
stereo view
H3C

7. A Diels-Alder reaction between cyclopentadiene and dimethyl acetylenedicarboxylate.

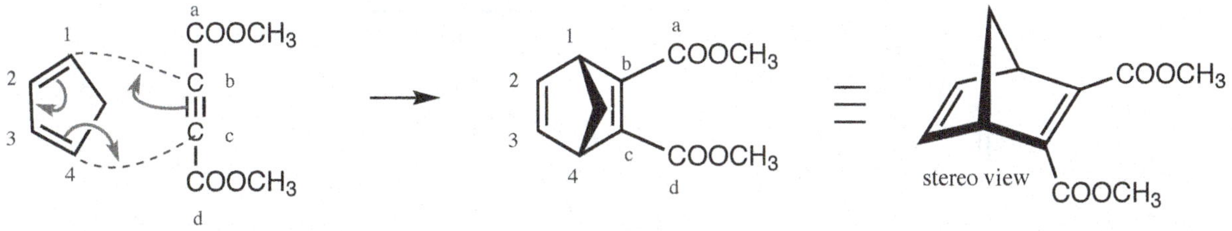

COOCH3
stereo view

8. A reverse-forward Diels-Alder reaction between cyclopentadiene and maleic anhydride.

heat

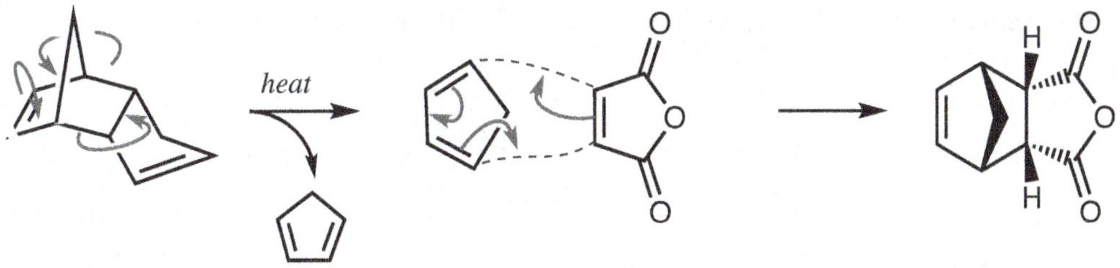

9. A reverse-forward Diels-Alder reaction between butadiene (sulfone) and maleic anhydride.

heat
SO2

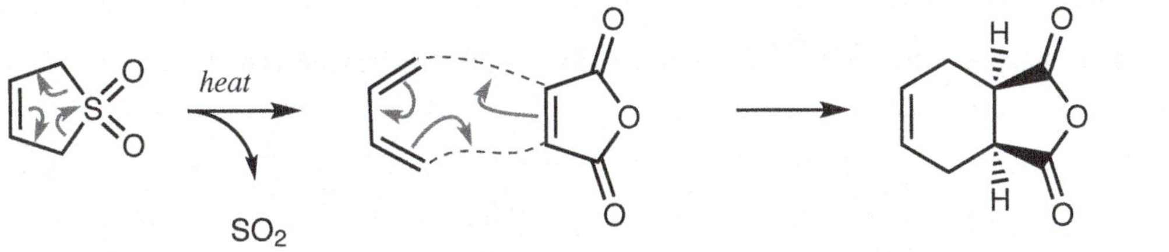

10. A Diels-Alder reaction between (*2E,4E*)-hexa-2,4-diene and maleic anhydride.

CH3 CH3

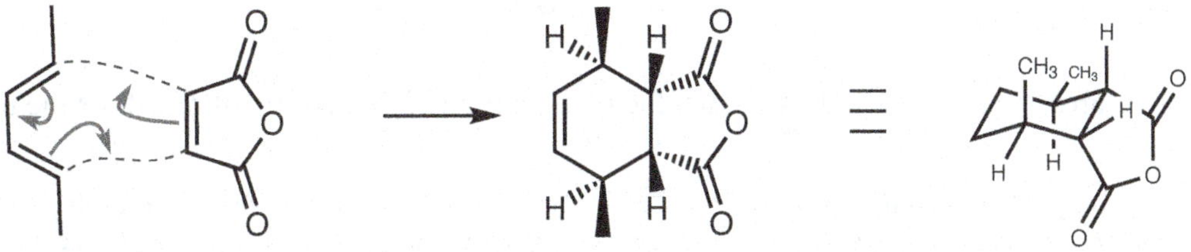

11. Draw the Diels-Alder product for the reaction of 1,3-cyclohexadiene and 2-butenal (crotonaldehyde).

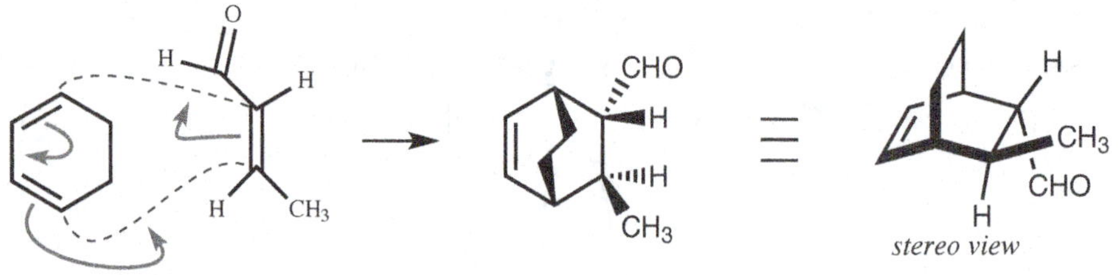

stereo view

12. A Diels-Alder reaction between 1-methoxy-1,3-butadiene and MVK gives a major product. To determine the structure of the major product, work down to Example 15.

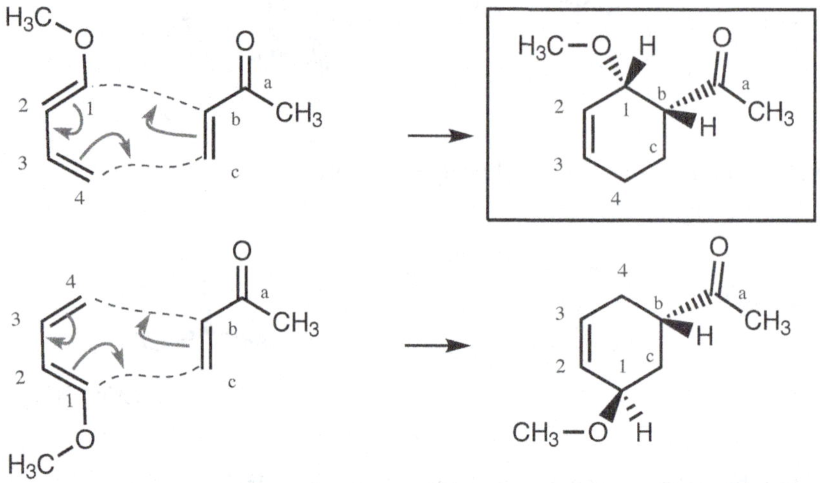

13. Examine the electrostatic charges of the starting materials by drawing the resonance structures.

Add curved arrows

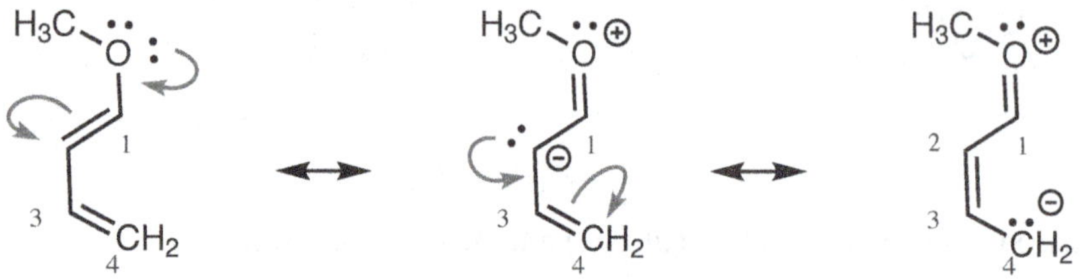

14. Examine the electrostatic charges of the starting materials by drawing the resonance structures.

Add curved arrows

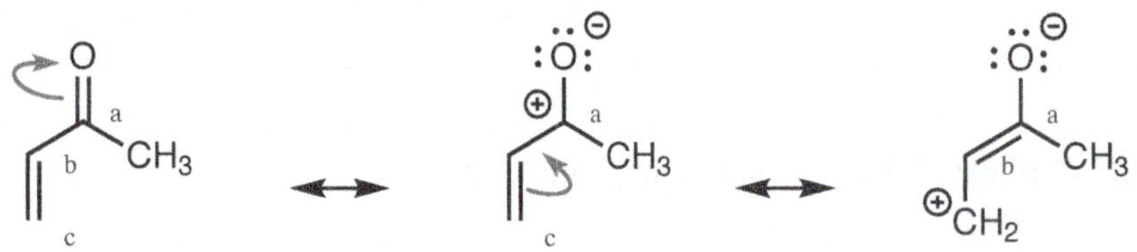

15. How do resonance structures shown in 13 and 14 predict the product for 12? Which structure will be the major product?

> *The boxed structure is the preferred product. It matches a Coulombic attraction of the charges in the resonance structures (which might contribute to the transition state). In a completely concerted reaction, there should be little build up of charge and both products could result.*

Other Electrocyclic Reactions

16. A 3+2 cycloaddition between cyclopentene and benzonitrile oxide. How many pairs of electrons (curved arrows) move to complete the formation of product. Compare the number of electrons that move in this reaction with the Diels-Alder reaction. Is it the same?

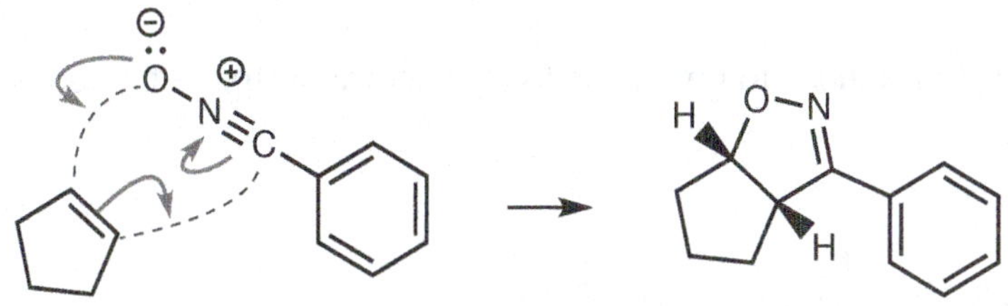

17. A Claisen rearrangement (electrocyclic) reaction to transfer a group from oxygen to carbon.

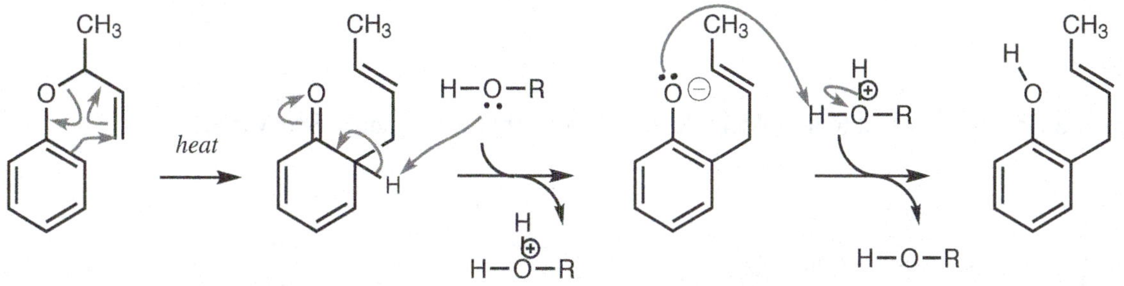

18. A double Claisen rearrangement (electrocyclic) reaction to transfer a group from oxygen to carbon to another carbon. Count the number of electrons that move in each step.

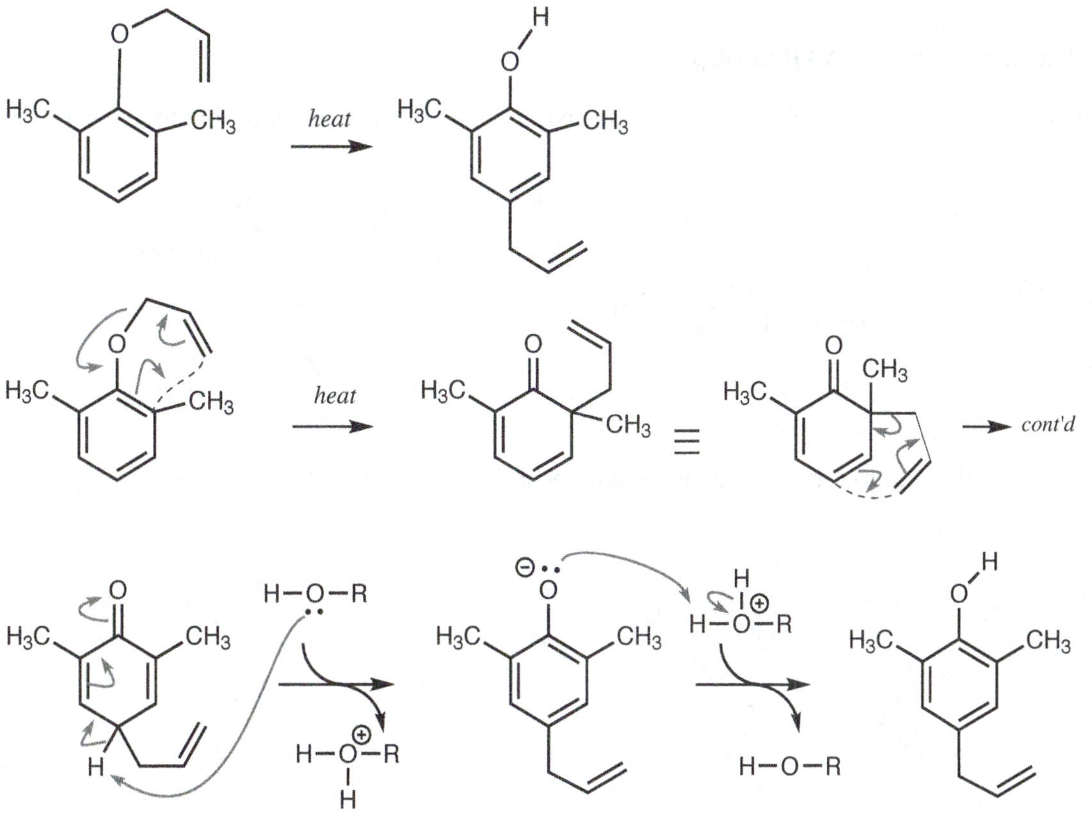

8 — Carbonyl Addition and Addition-Elimination Reactions

Additions by C, N, and O. Addition by hydrogen nucleophiles are discussed in Chapter 11, Reduction.

Grignard Addition to a Carbonyl Group

1. Addition of methyl magnesium bromide to cyclohexanone to give 1-methylcyclohexanol (for formation of Grignard reagents, see *Notes*).

2. Addition of a Grignard reagent to acetaldehyde to give 6-methyl-2-heptanol. (See *Notes*.)

Alkyllithium Addition to a Carbonyl Group

3. Addition of ethyllithium to benzaldehyde to give 1-phenylpropanol. (for formation of lithium reagent, see *Notes*).

4. Addition of propynyllithium to acetone to give 2-methylpent-3-yn-2-ol. (see Chapter 2.14)

Wittig Reaction

5. Wittig reaction, Step 1, formation of Wittig reagent. (See *Notes*.)

Step 2, reaction with benzaldehyde

The Wittig reaction stereochemistry was controlled by the configuration of the faster addition to the carbonyl group.

6. Wittig reaction, Horner-Wadsworth-Emmons modification, Step 1, Arbusov reaction. (See *Notes*.)

Step 2, reaction with benzaldehyde

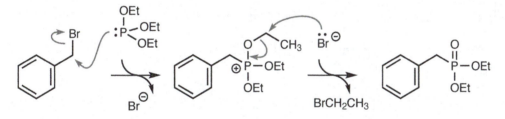

This is the most stable equilibrium addition product. This geometry determines whether the final product will be E or Z.

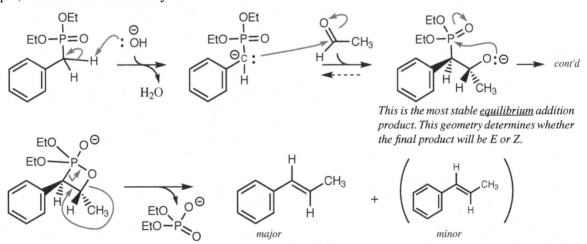

Addition-Elimination Reactions (Reversible Additions)

Ketal Formation and Hydrolysis

7.   Acid catalyzed ketalization of cyclohexanone. (See *Notes*.)

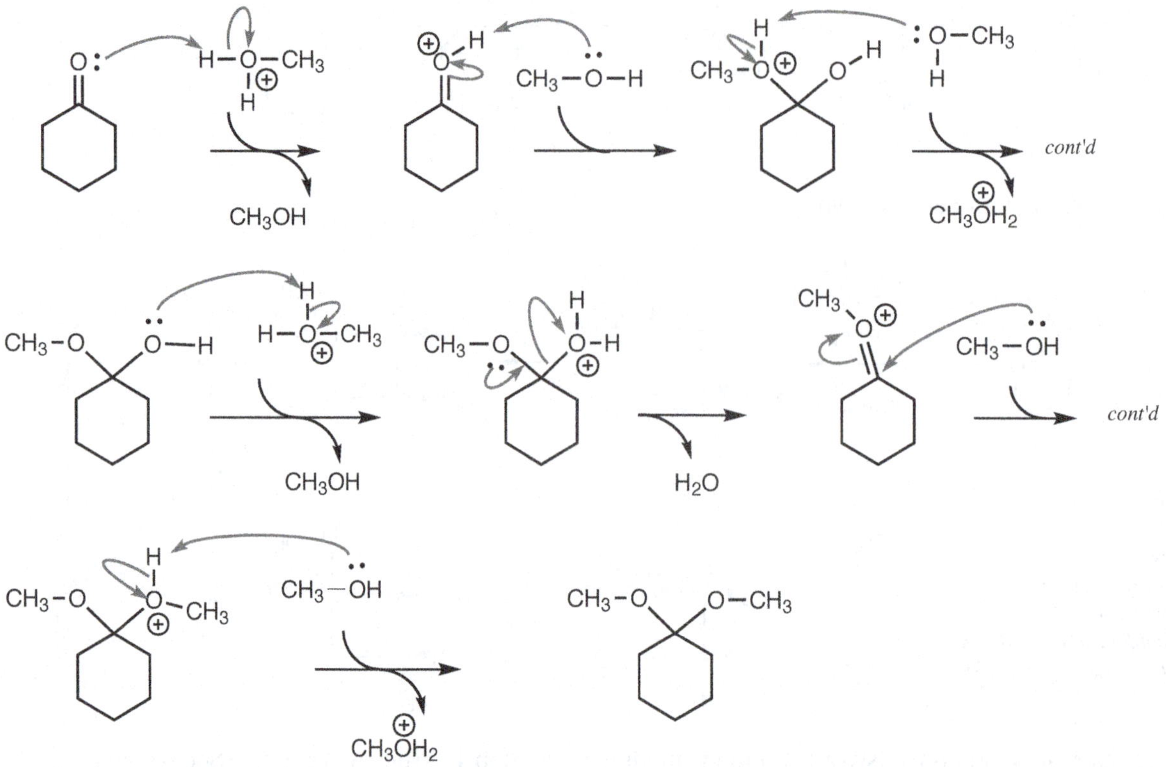

8.   Acid catalyzed hydrolysis of the dioxolane acetal of benzaldehyde. (See *Notes*.)

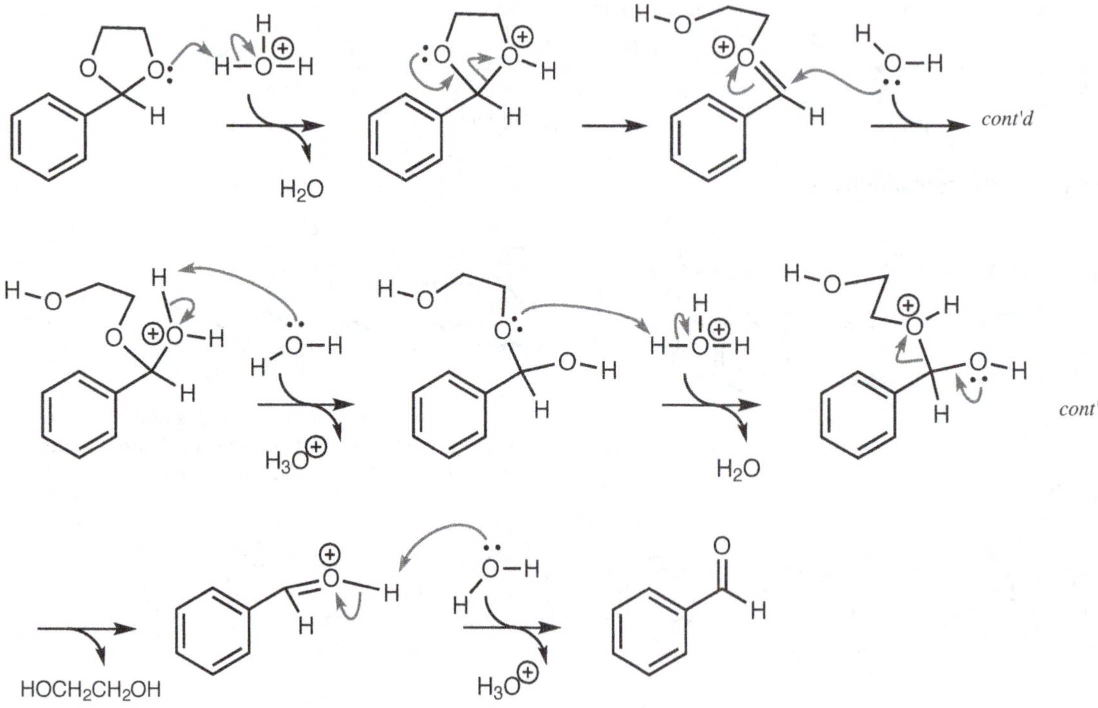

Oxime Formation

9. Formation of the oxime of cyclohexanone. (See *Notes*.)

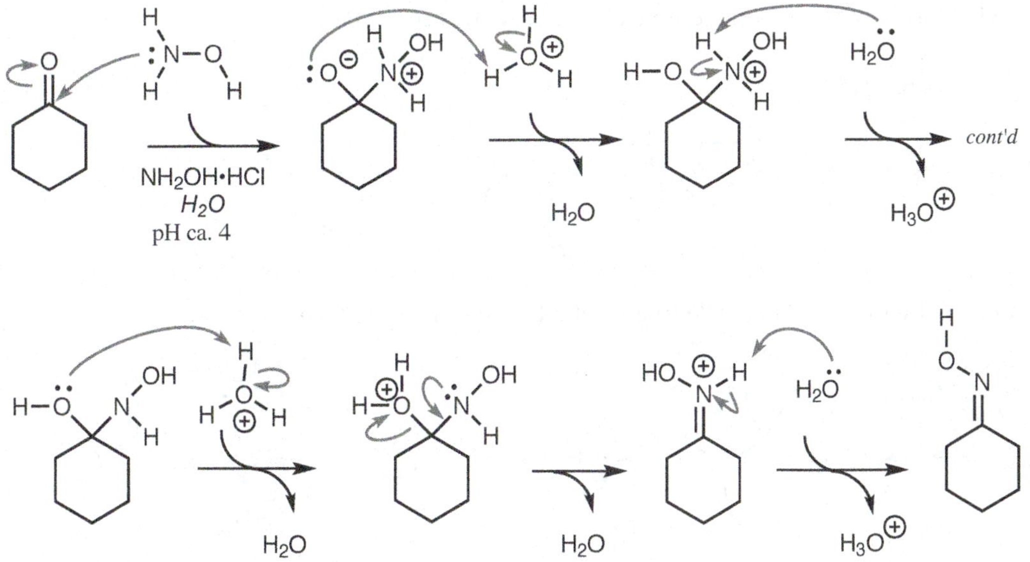

Other Additions to a Carbonyl Group

10. Formation of the cyanohydrin (2-hydroxy-2-methylpropanenitrile) from acetone. (See *Notes*.)

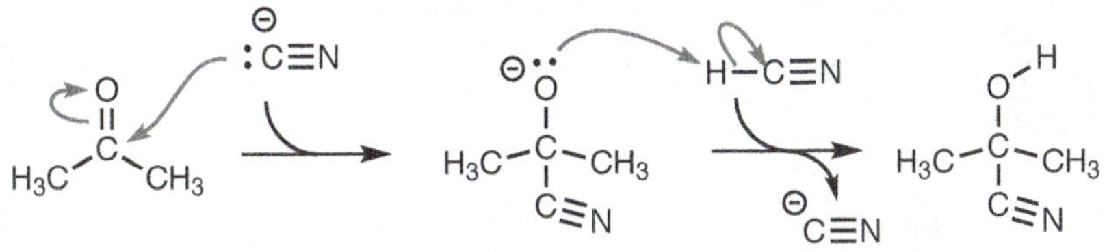

11. Reversion of the cyanohydrin (2-hydroxy-2-methylpropanenitrile) to form acetone. (See *Notes*.)

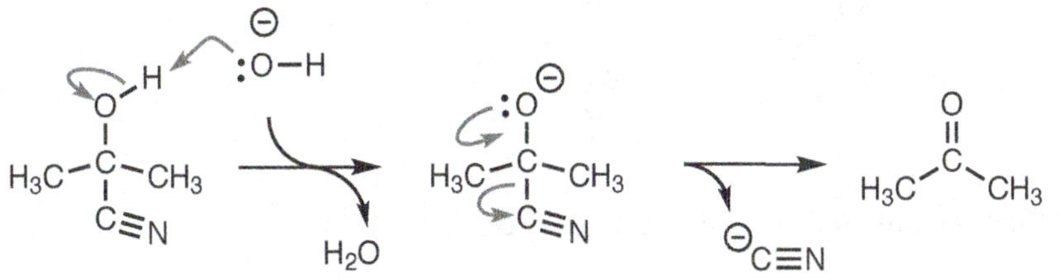

12. Addition of ethyl acetate enolate to 4-bromobenzaldehyde to give a benzyl alcohol.

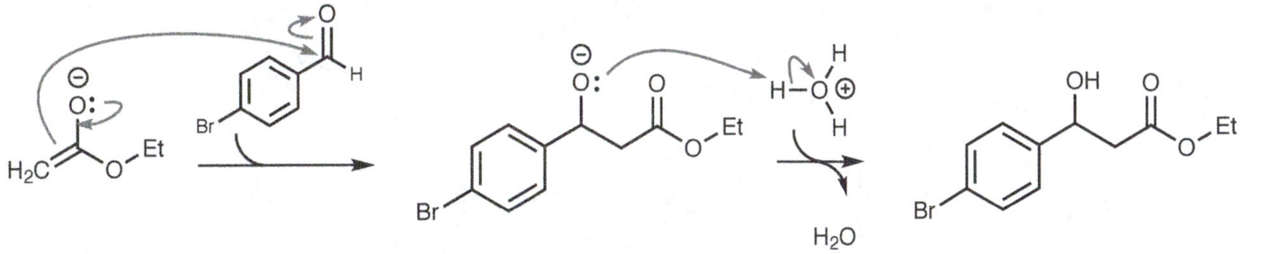

Reactions of Acyl Chlorides, Anhydrides, Esters, and Amides

Esters from Acid Chlorides or Anhydrides

13. Reaction of benzoyl chloride with ethoxide to give ethyl benzoate. (See *Notes*.)

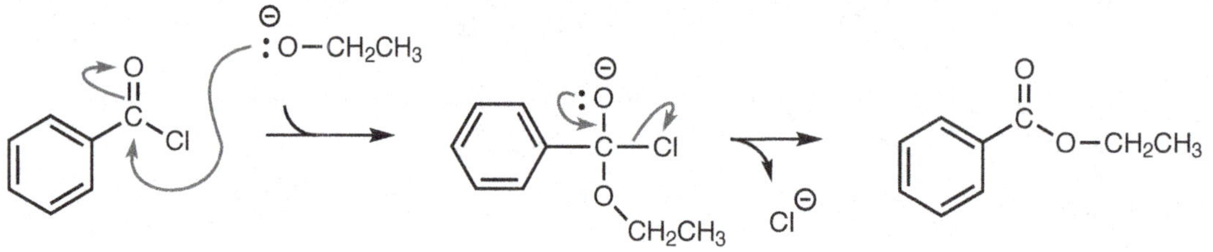

14. Direct reaction of ethanol with acetyl chloride to give ethyl acetate. (See *Notes*.)

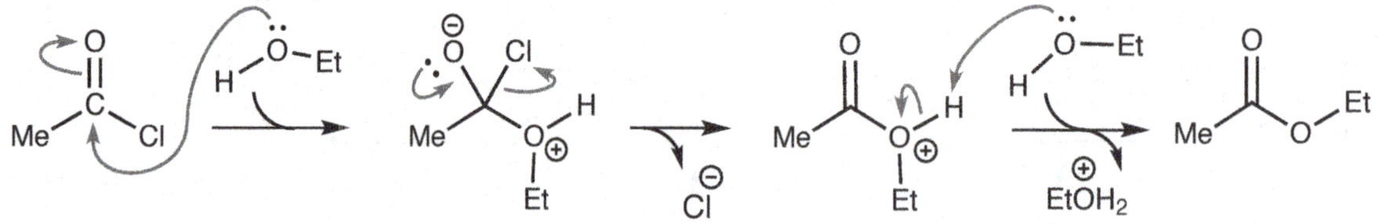

15. Pyridine catalyzed acylation with benzoyl chloride to give ethyl benzoate. (See *Notes*.)

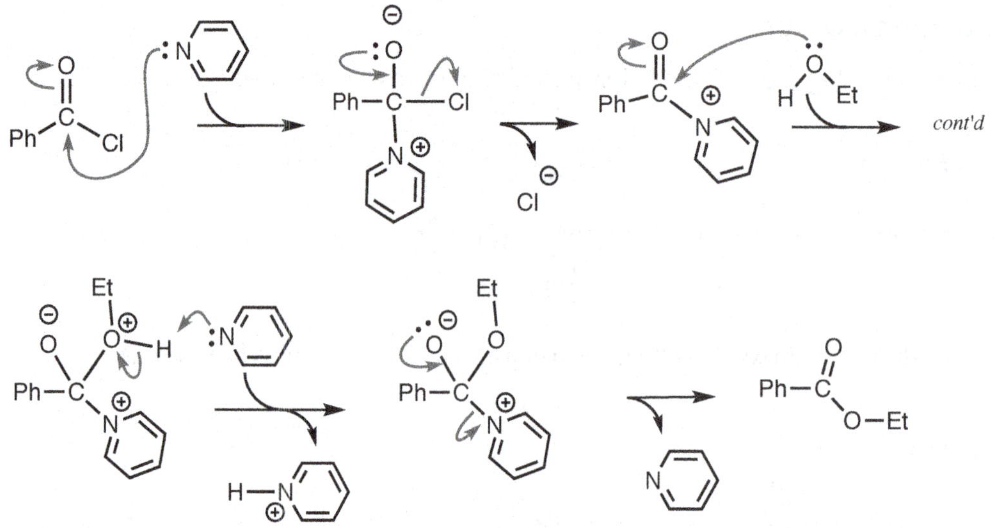

16. Reaction of acetic anhydride with ethanol catalyzed by sulfuric acid. (See *Notes*.)

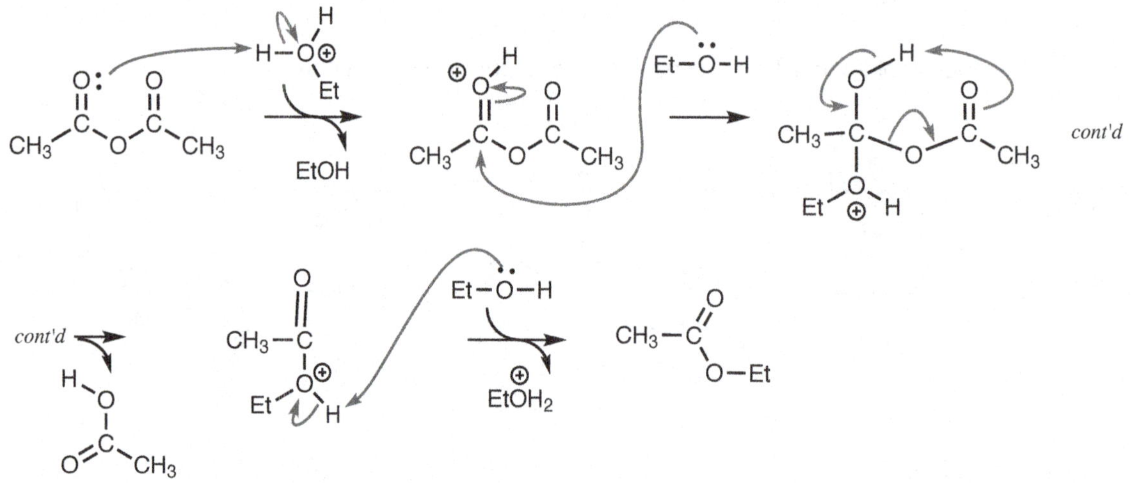

Amides from Acid Chlorides or Anhydrides

17. Reaction of benzoyl chloride with ethylamine to give *N*-ethylbenzamide. (See *Notes*.)

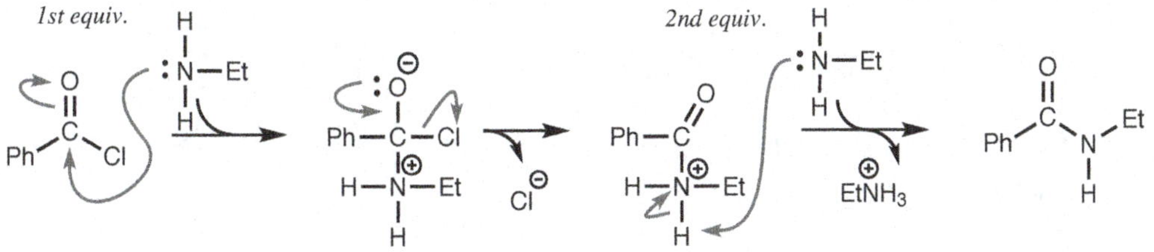

18. Reaction of acetic anhydride with aniline (and pyridine) to give acetanilide. (See *Notes*.)

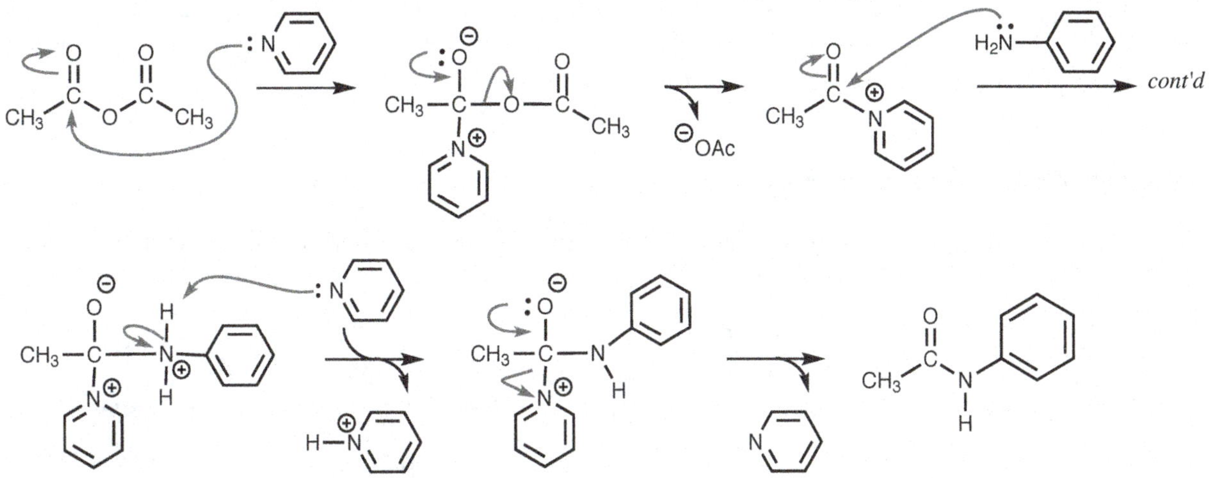

19. Reaction of acetic anhydride with aniline to give acetanilide. (See *Notes*.)

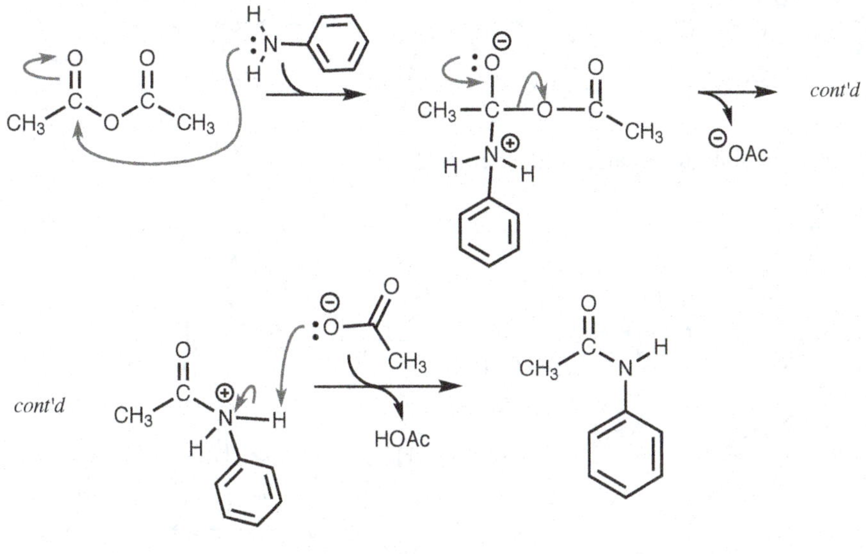

Ester from Acid with Mineral Acid Catalysis (Fischer Esterification)

20. Acid catalyzed esterfication. Formation of methyl benzoate from benzoic acid and methanol with sulfuric acid or hydrogen chloride. (See *Notes*.)

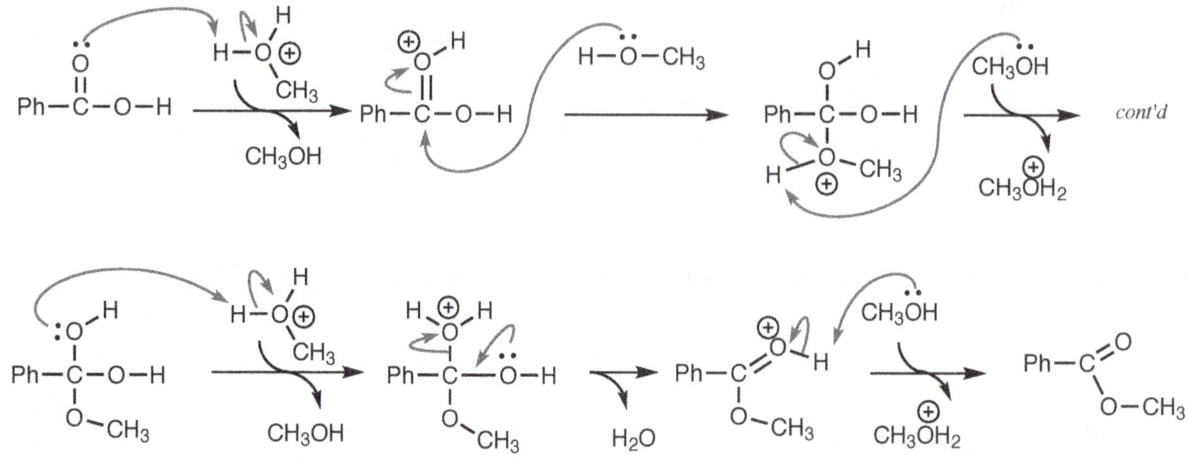

Acid Catalyzed Hydrolysis of an Ester

21. Acid catalyzed hydrolysis of methyl pentanoate (valerate) to pentanoic (valeric) acid plus methanol. (See *Notes*.)

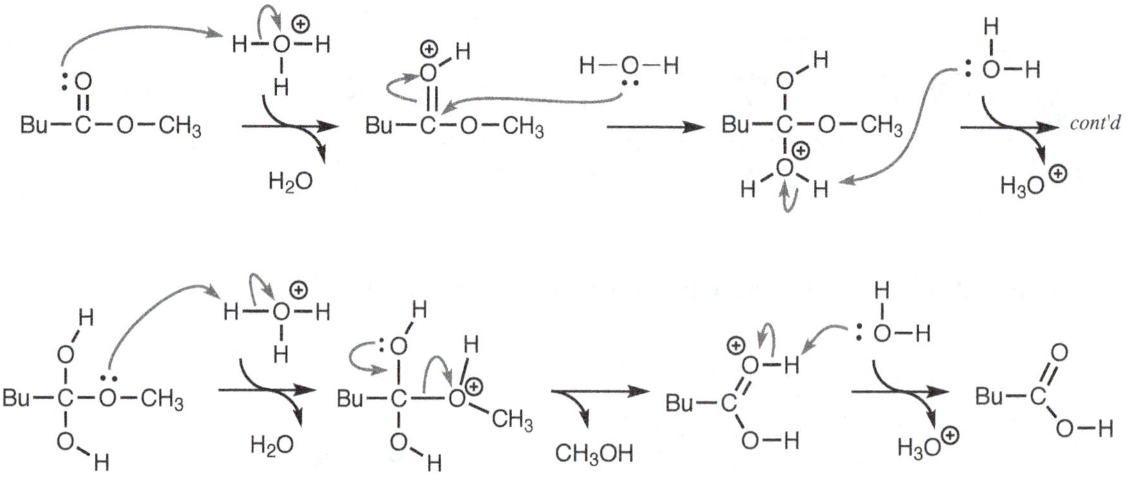

22. Acid catalyzed conversion of *t*-butyl pivalate to pivalic acid. (See *Notes*.)

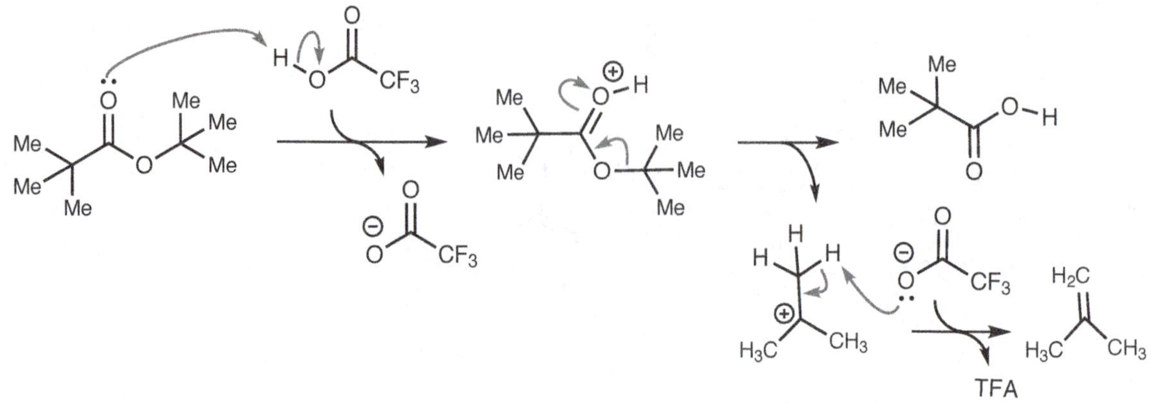

Base Hydrolysis of an Ester (Saponification)

23. Base hydrolysis of octyl isobutyrate to give octanol and isobutyric acid. Step 1, treatment with base. (See *Notes*.)

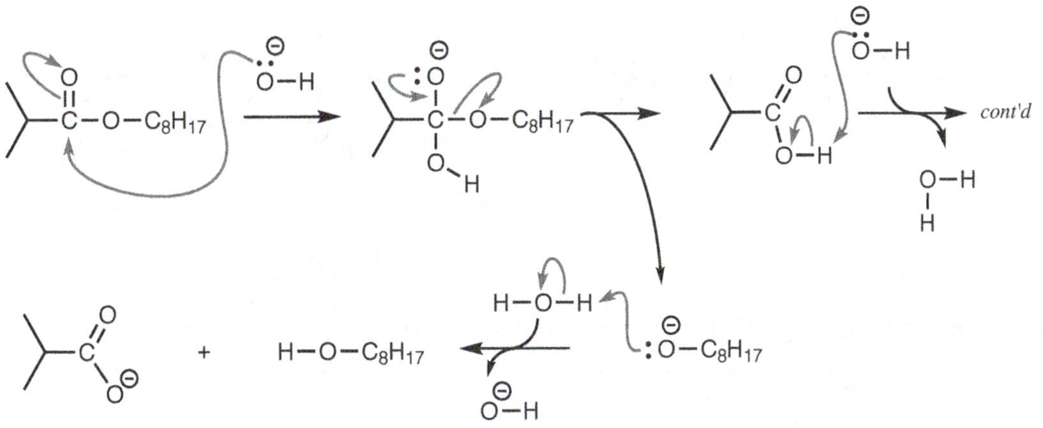

Step 2, acidification of isobutyrate and isolation of isobutyric acid.

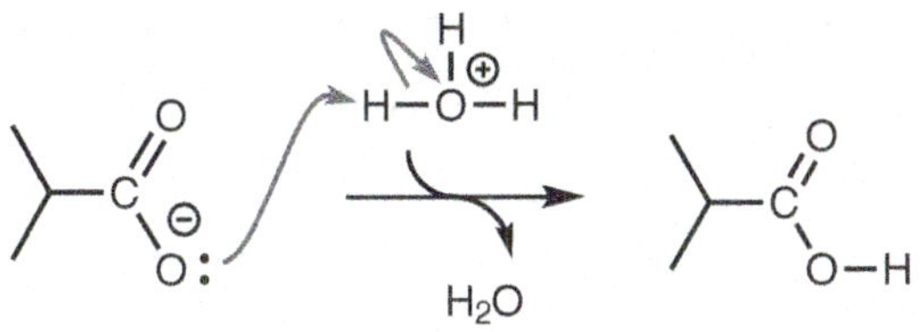

Hydrolysis of an Amide

24. Base hydrolysis of *N*-butylacetamide to give *n*-butylamine and acetic acid. (See *Notes*.)

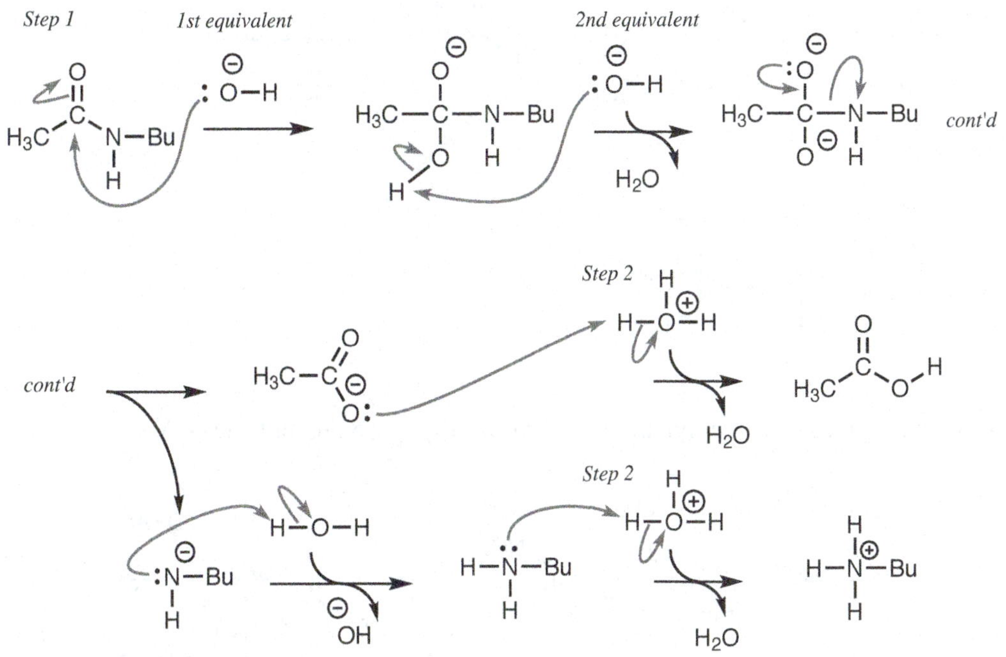

25. Acid catalyzed hydrolysis of *N,N*-dimethylacetamide to give acetic acid and dimethylammonium chloride. (See *Notes.*)

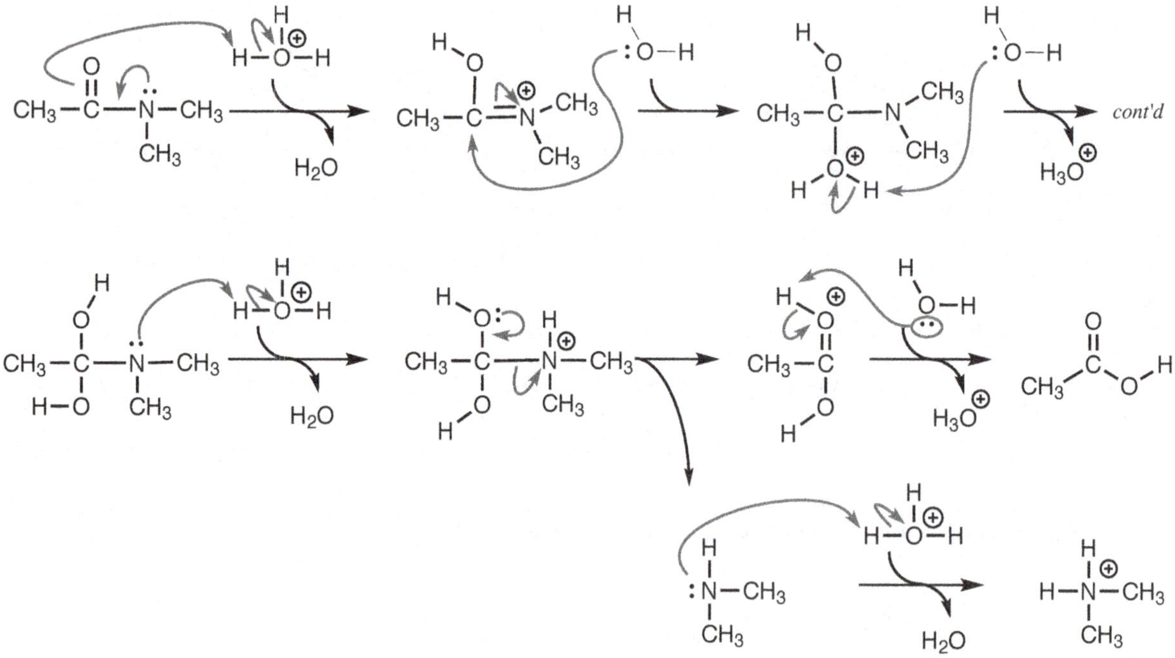

Reactions of Esters

26. Addition of methyl magnesium bromide to ethyl benzoate to give 2-phenyl-2-propanol. (See *Notes.*)

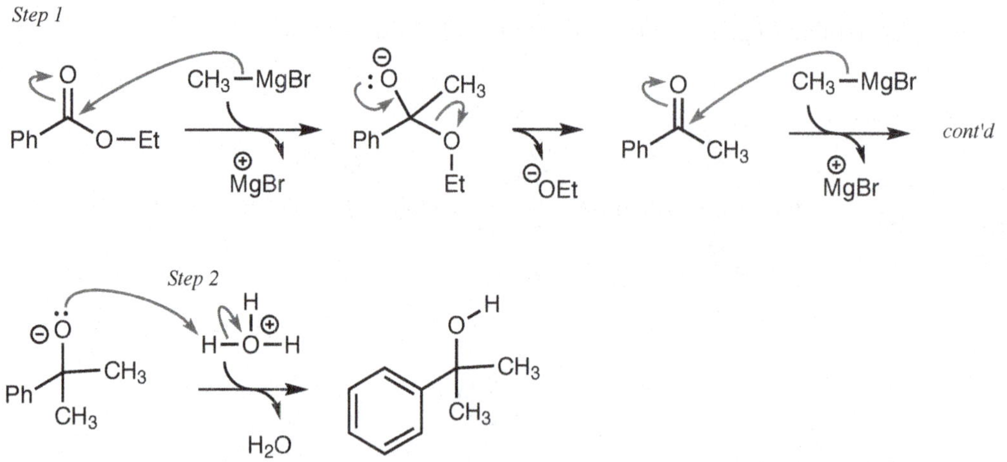

Step 1

Step 2

27. Reaction of 4-nitrophenyl propionate with ethyl amine to give *N*-ethyl propionamide. (See *Notes.*)

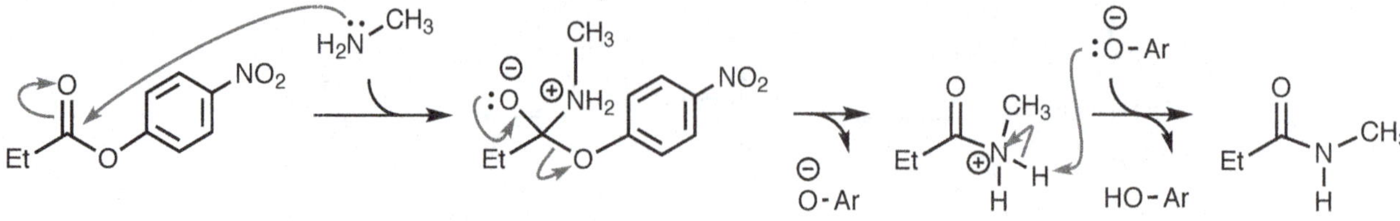

Reactions of Nitriles

28. Acid catalyzed hydrolysis of benzonitrile to give benzoic acid and ammonium chloride. (See *Notes.*)

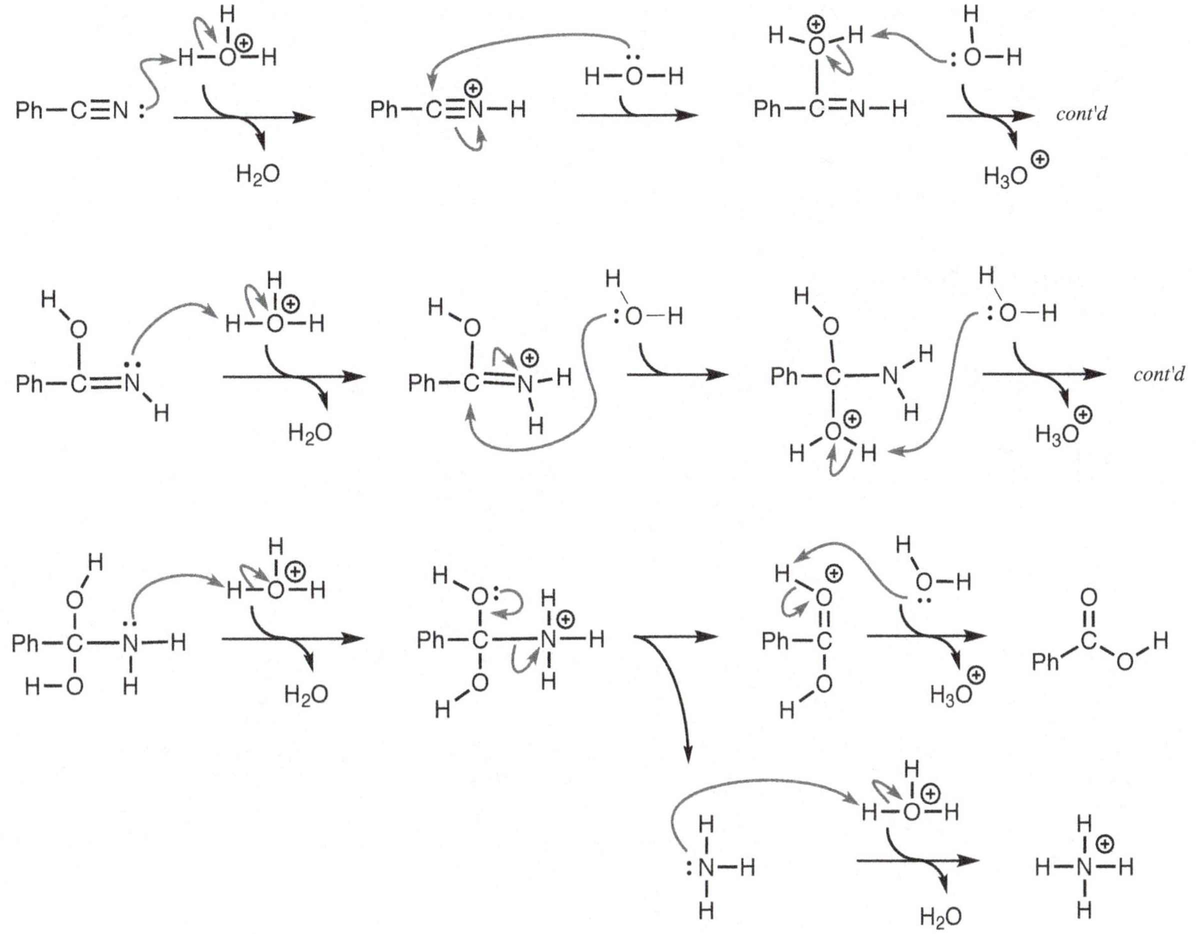

29. Base catalyzed hydrolysis of a nitrile, conversion of cyclopentanecarbonitrile to cyclopentanecarboxamide. (See *Notes.*)

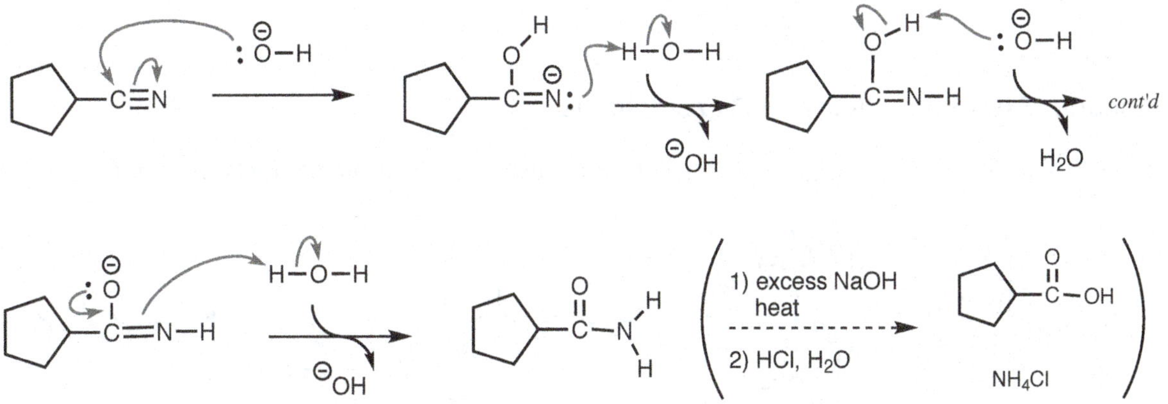

30. Addition of phenyllithium to a nitrile to give, after hydrolysis, cyclohexyl phenyl ketone. (See *Notes*.)

Step 1, addition to the nitrile.

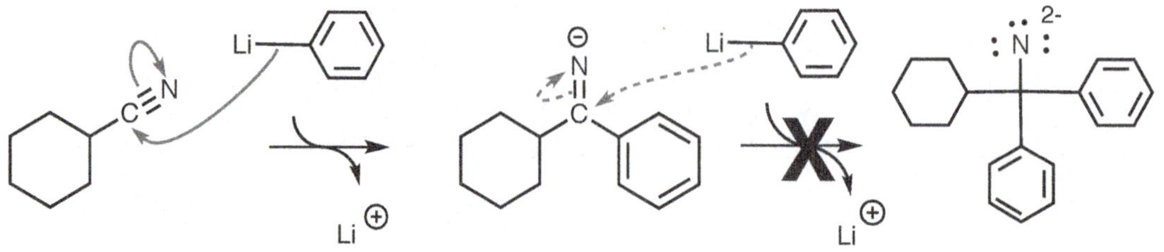

Step 2, hydrolysis of the imine.

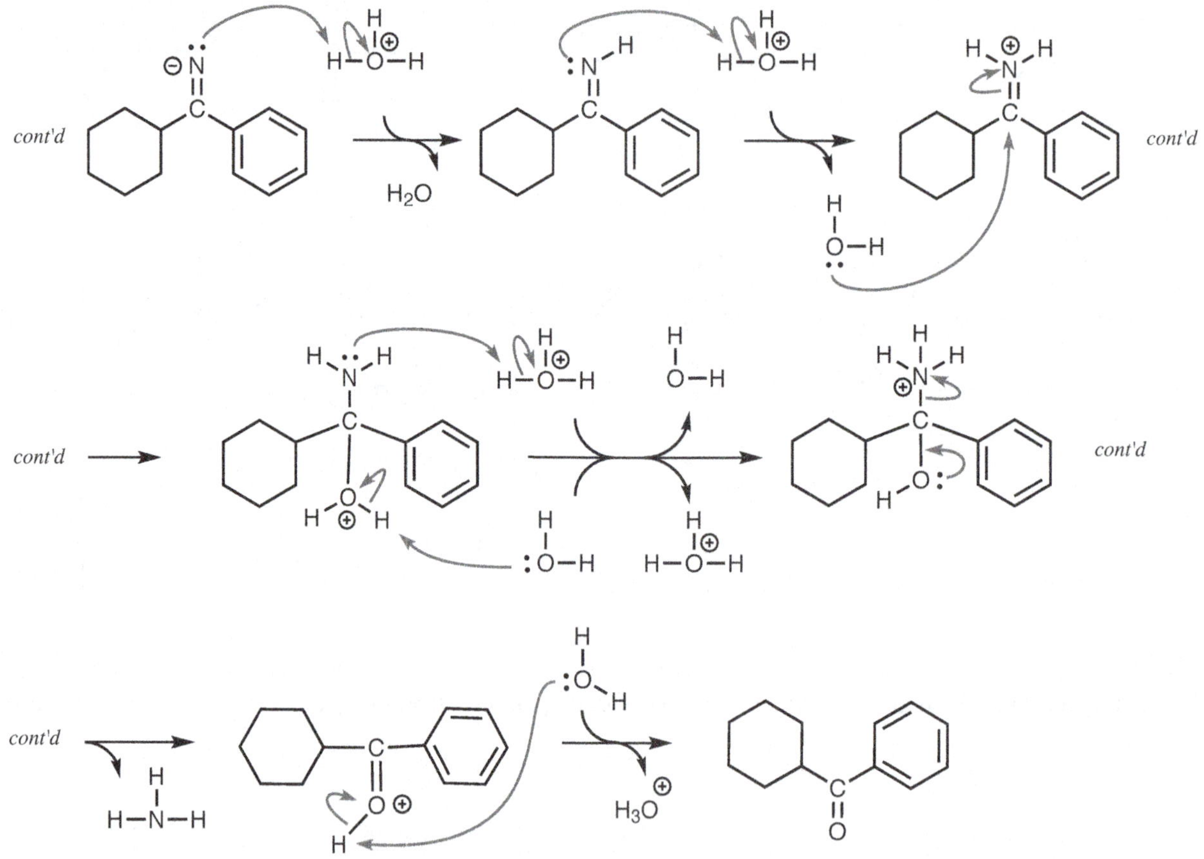

Miscellaneous, Ester Formation with Diazomethane

31. Reaction of diazomethane with a carboxylic acid to form methyl 2-butenoate (methyl crotonate). (See *Notes*.)

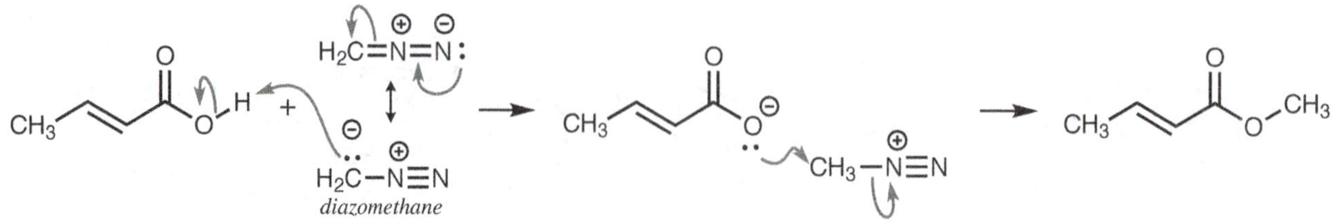

9 — Reactions of Enols and Enolates

Aldol Reaction

1. Base catalyzed aldol condensation of butanal (butyraldehyde). (See *Notes*.)

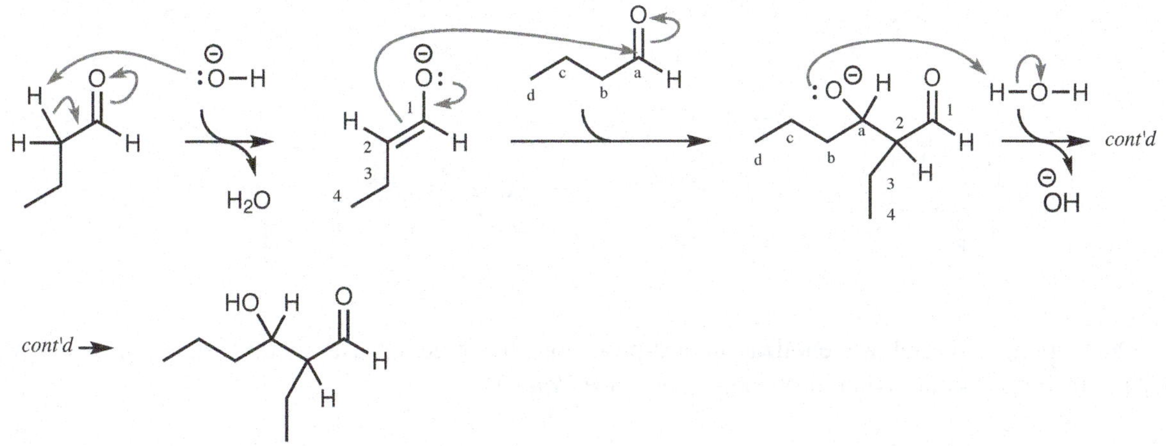

Base catalyzed dehydration step. Under concentrated base or heating, the dehydration reaction can occur spontaneously.

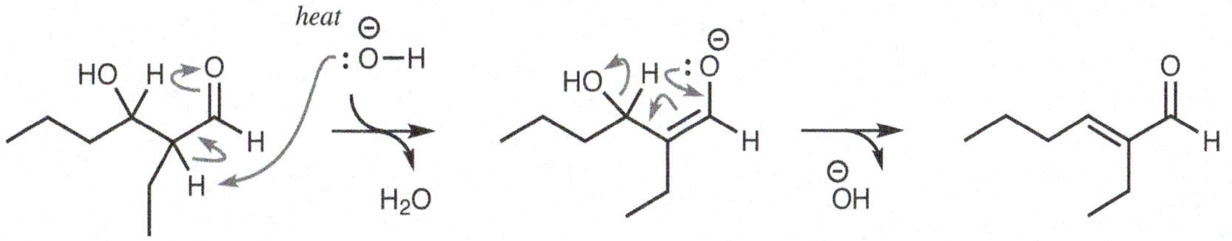

2. Directed kinetic aldol condensation of 2-methylcyclohexanone with propanal (propionaldehyde). Step 1, enolate formation; step 2, reaction with propanal; step 3, neutralization. (See *Notes*.)

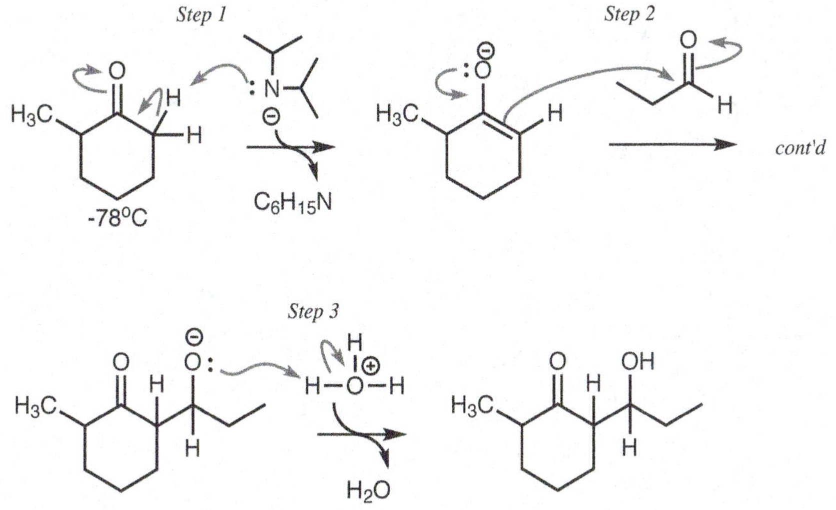

3. Base catalyzed mixed or crossed aldol condensation of acetone and benzaldehyde.

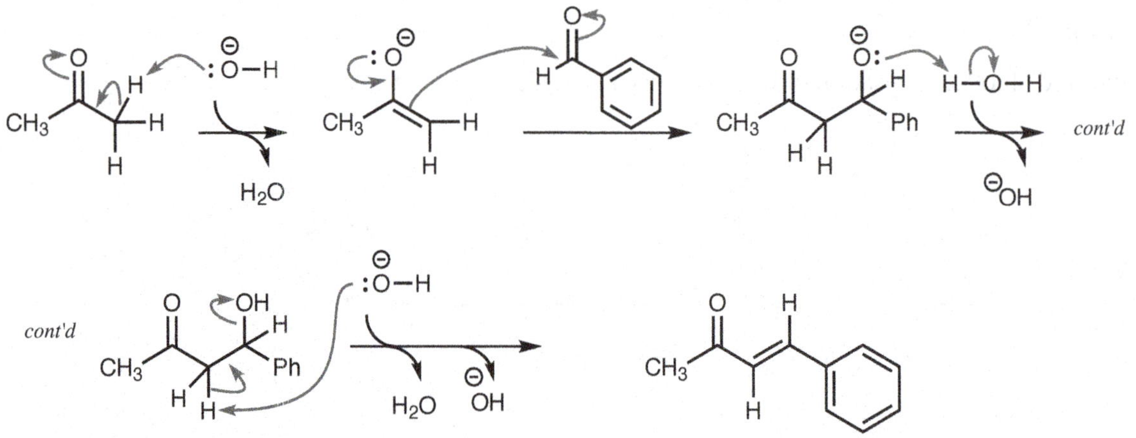

4. Mannich reaction, acid catalyzed enolization of 2-propanone in a reaction with diethylamine, formaldehyde, and 2-propanone to give 4-(diethylamino)butan-2-one. (See *Notes*.)

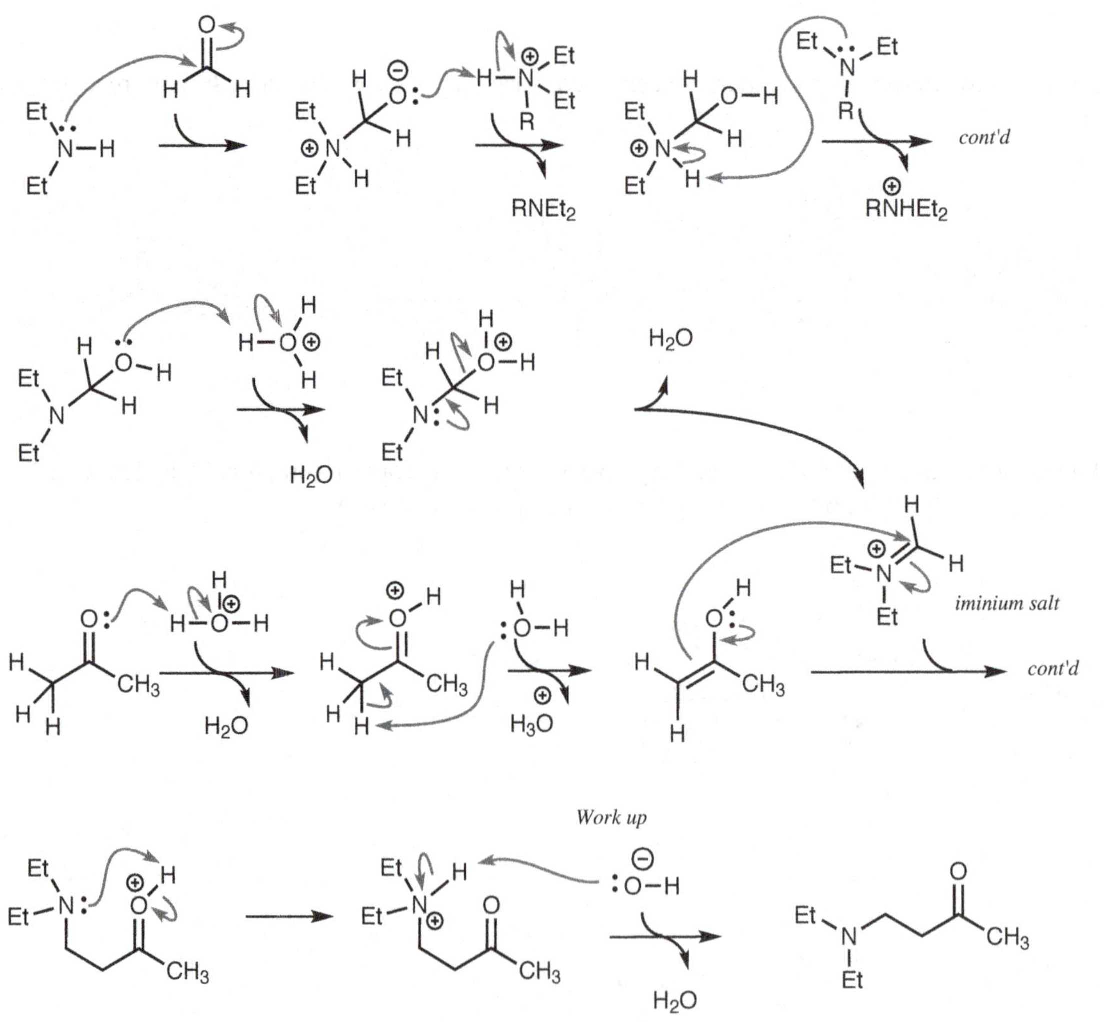

Claisen Condensation

5. Ethoxide catalyzed Claisen condensation of ethyl acetate to ethyl acetoacetate (ethyl 3-oxobutanoate). (See *Notes.*)

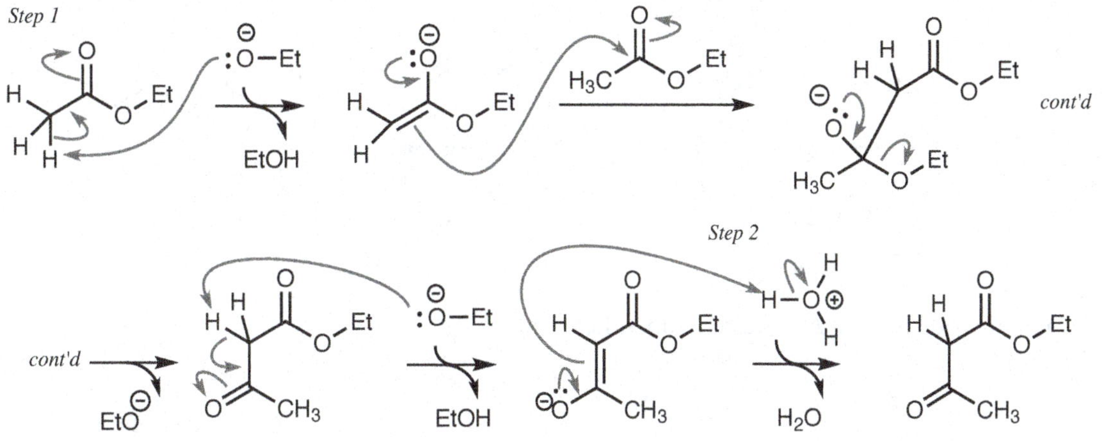
Step 1
EtOH
cont'd
cont'd
Step 2
EtOH
H_2O

6. Ethoxide catalyzed crossed-Claisen condensation of cyclohexanone and ethyl formate. (See *Notes.*)

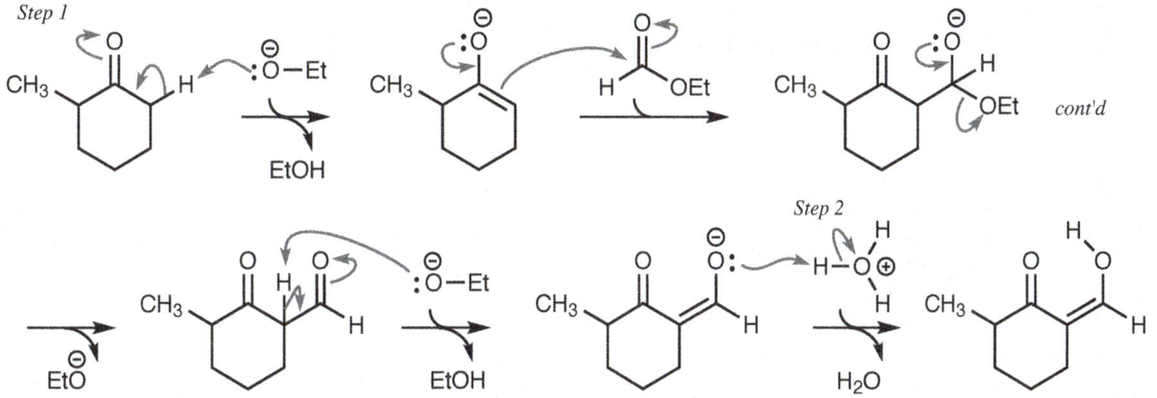
Step 1
EtOH
cont'd
Step 2
EtOH
H_2O

7. Ethoxide catalyzed crossed-Claisen condensation of ethyl acetate and ethyl benzoate. (See *Notes.*)

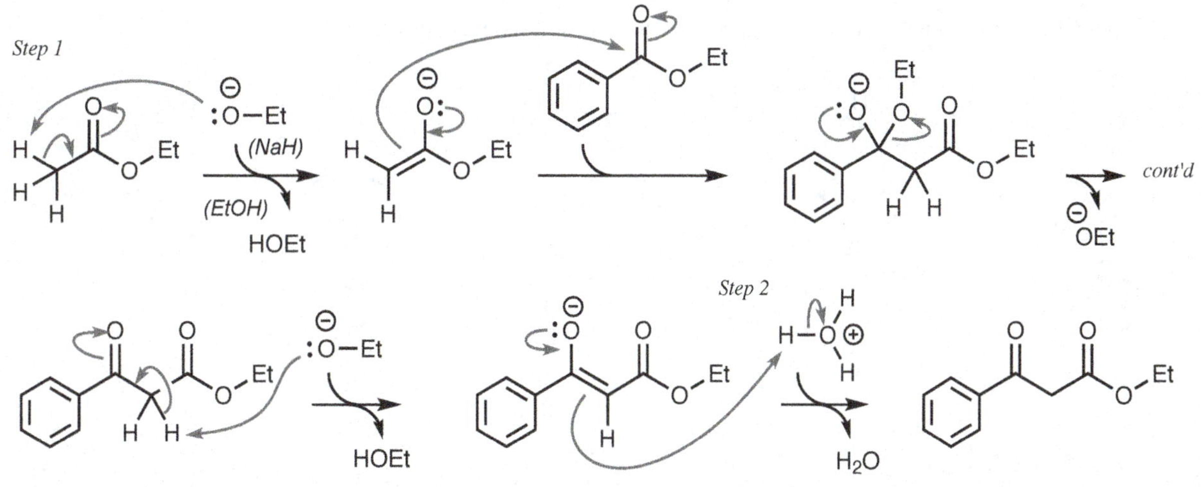
Step 1
(NaH)
(EtOH)
HOEt
cont'd
Step 2
HOEt
H_2O

Ethoxide propagation

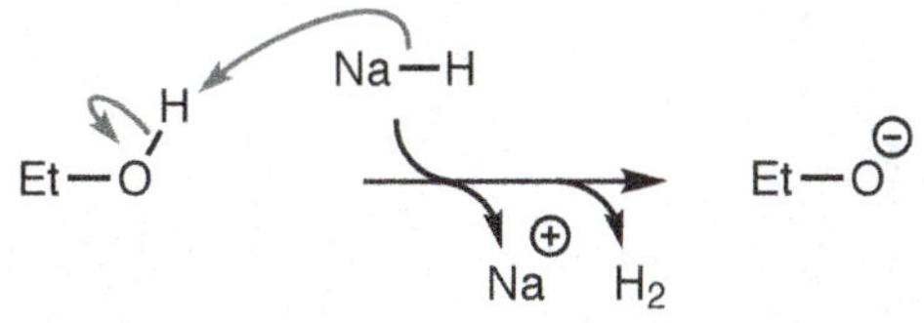
Na—H
Na
H_2

Acetoacetate Synthesis

8. Step 1, S$_N$2 alkylation of acetoacetate.

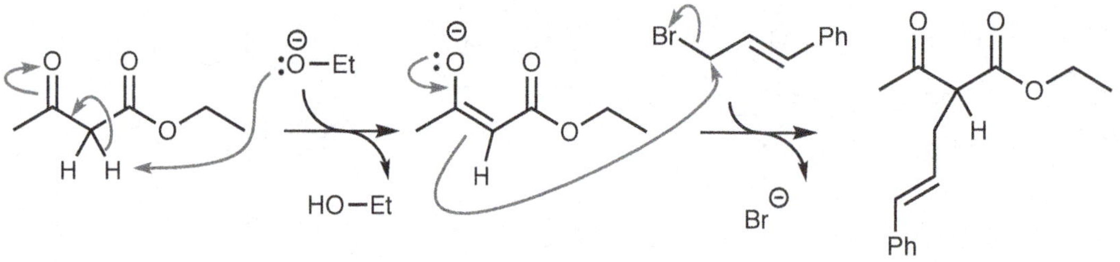

Step 2, sodium hydroxide hydrolysis of ester (saponification).

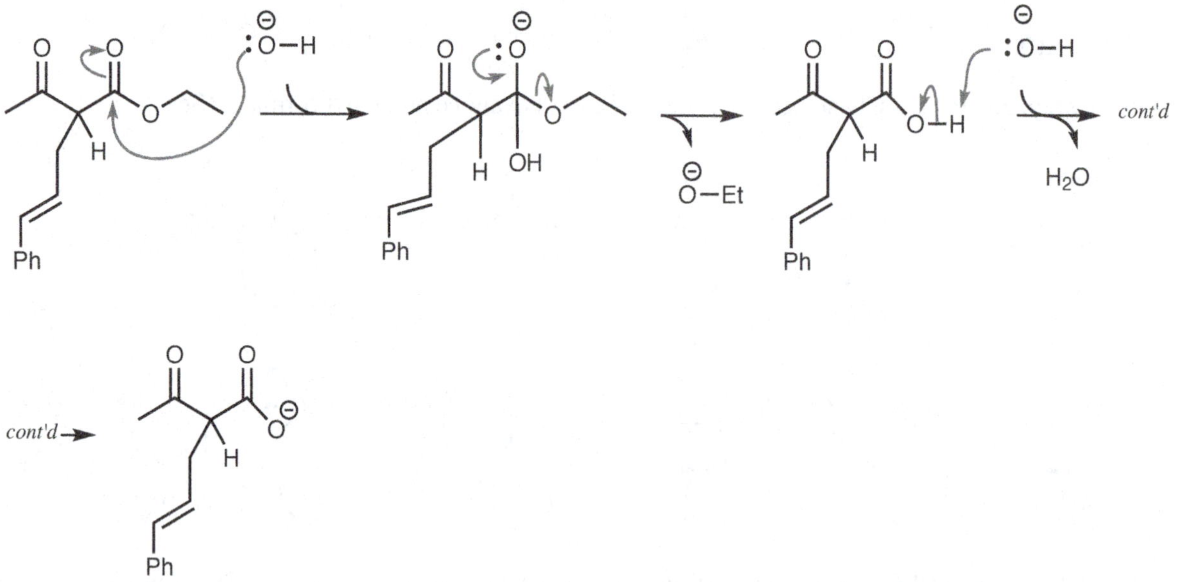

Step 3, acidification of carboxylate, decarboxylation, and tautomerization.

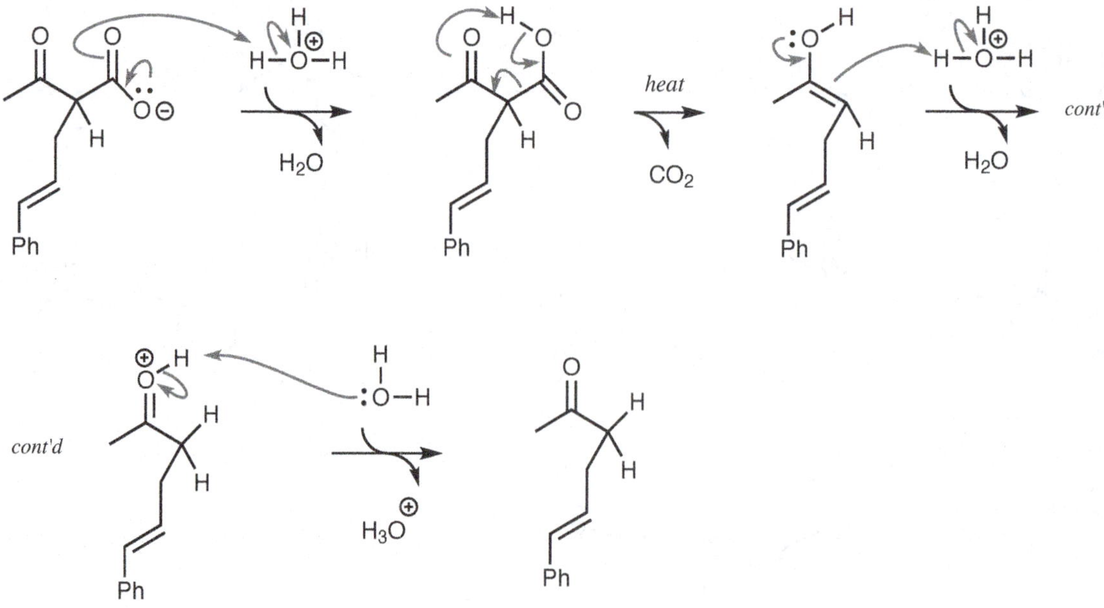

Enolate Alkylation Reactions

9. Enolization and alkylation of ethyl propionate with benzyl bromide. (See *Notes*.)

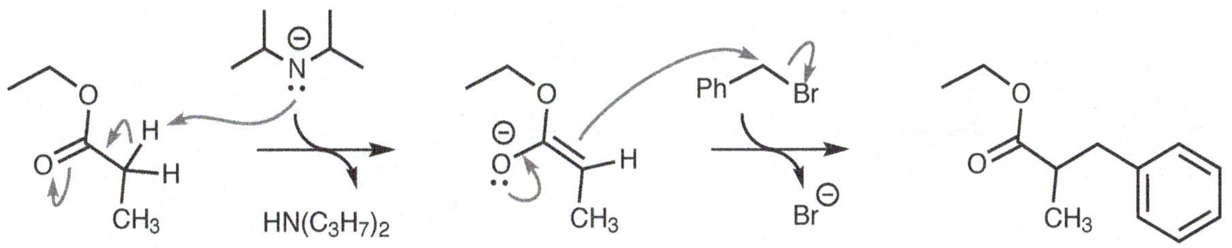

10. Sequential alkylation of a dianion of a *beta*-ketoester. (See *Notes*.)

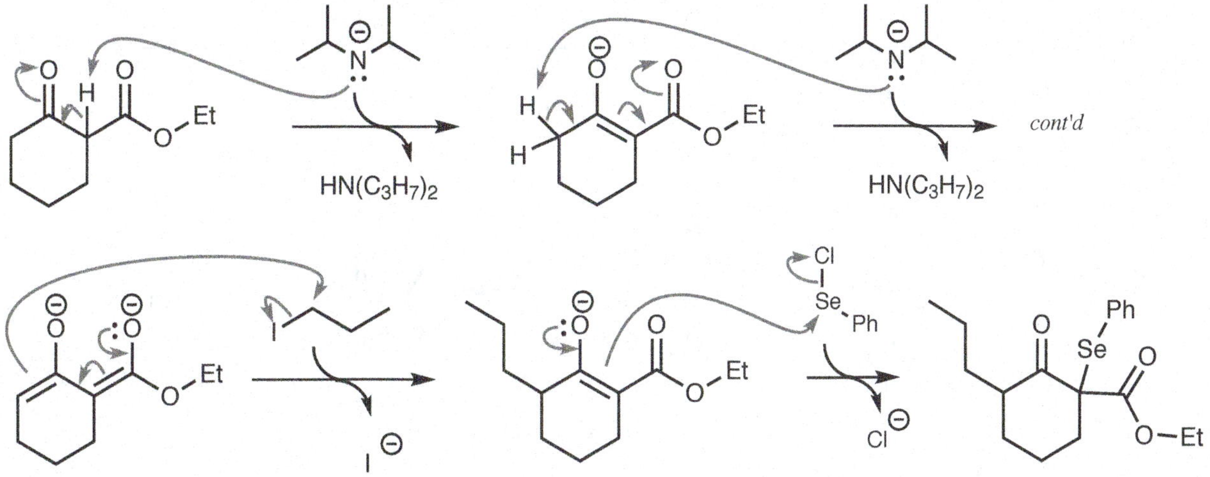

11. A retro-Claisen reaction from a *beta*-ketoester. (See *Notes*.)

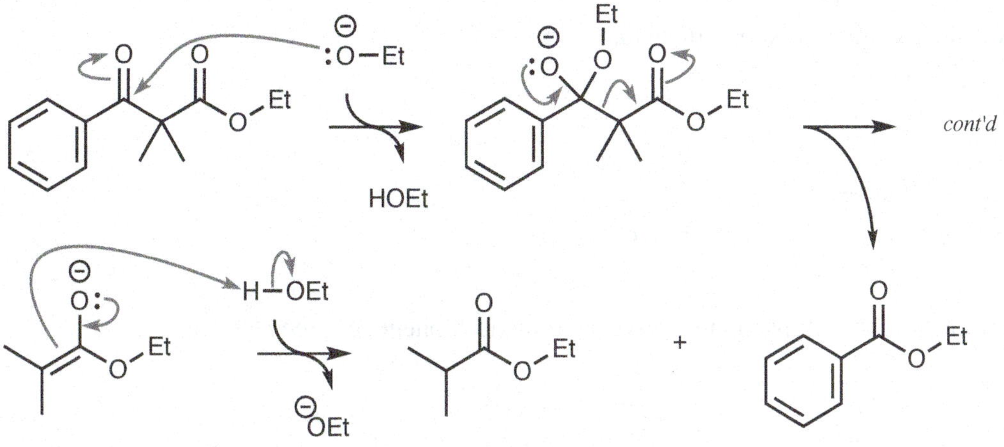

12. Enolization and alkylation of phenylacetonitrile with methyl iodide. (See *Notes*.)

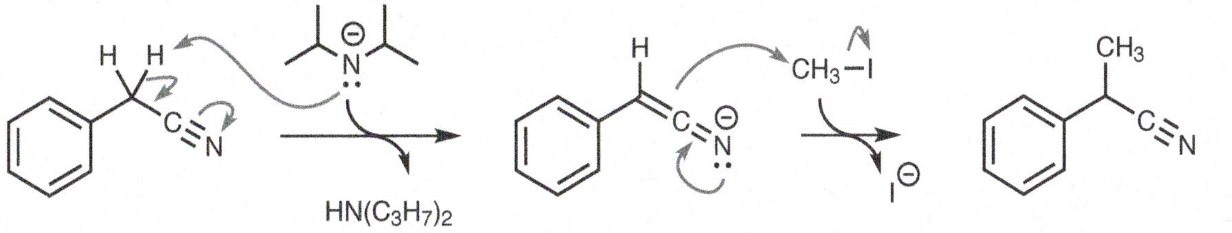

Halogenation of Carbonyl Compounds

13. Basic bromination of 3-methyl-2-butanone with sodium hydroxide and bromine, bromoform reaction. (See *Notes*.)

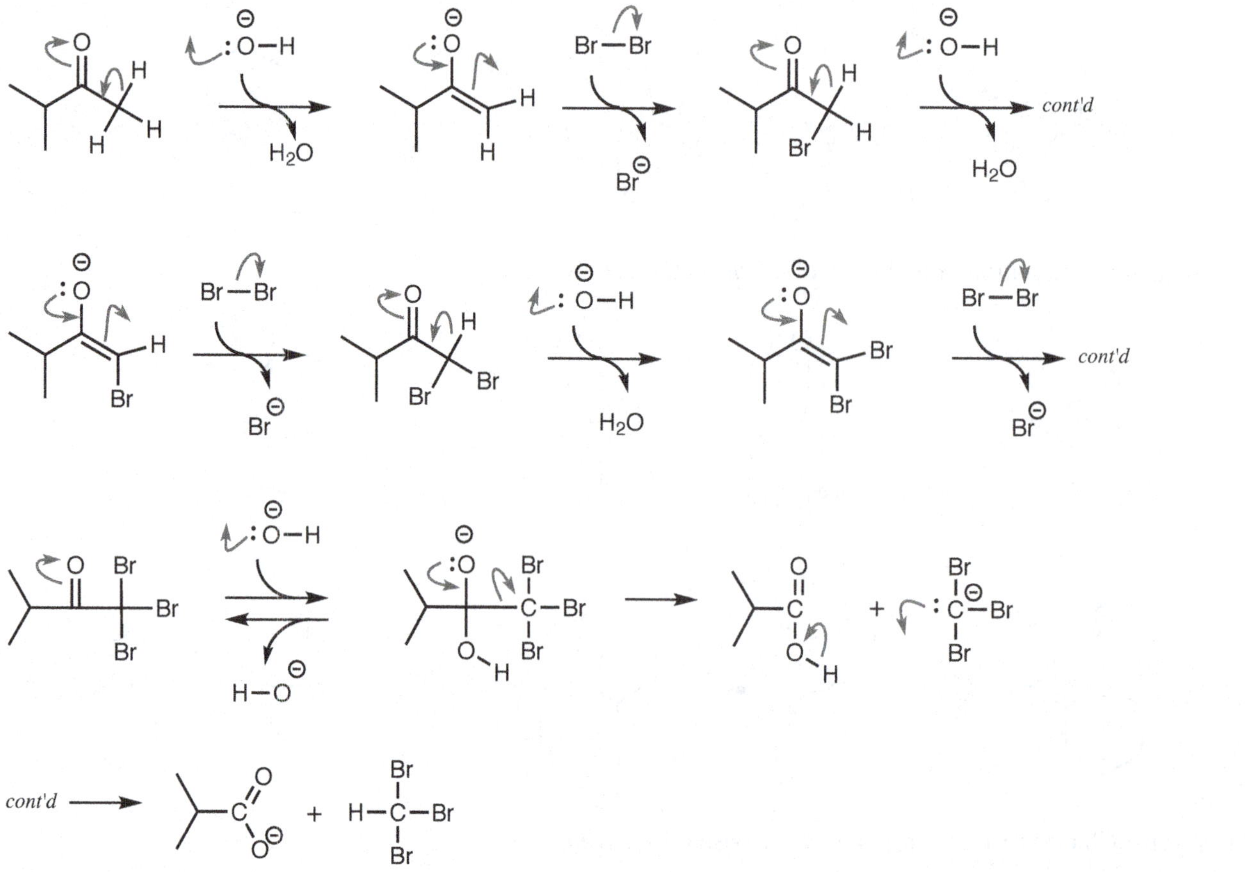

Step 2, acidification and isolation of isobutyric acid.

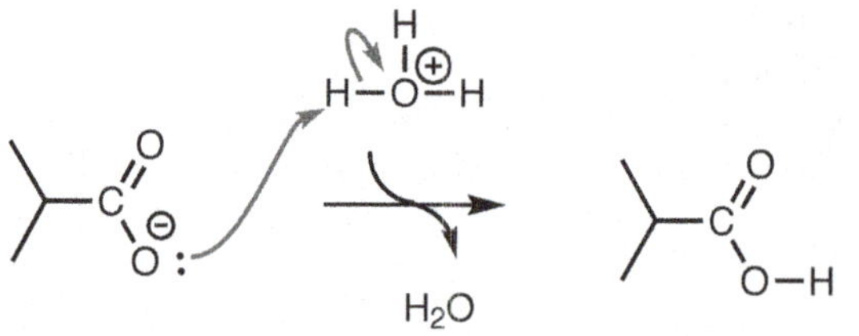

14. Acid catalyzed bromination of acetophenone to give α-bromoacetophenone. (See *Notes*.)

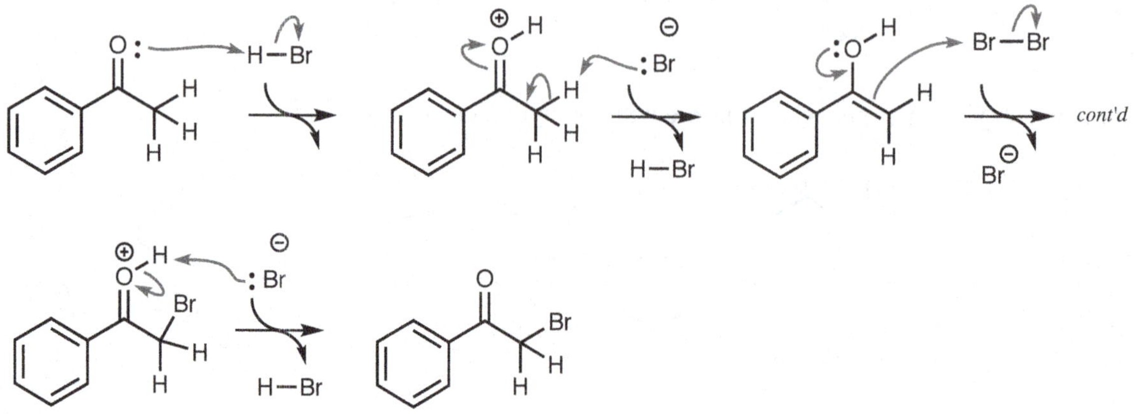

Michael or 1,4-Conjugate Addition Reaction

15. Michael addition reaction of dimethylamine to methyl vinyl ketone. (See *Notes*.)

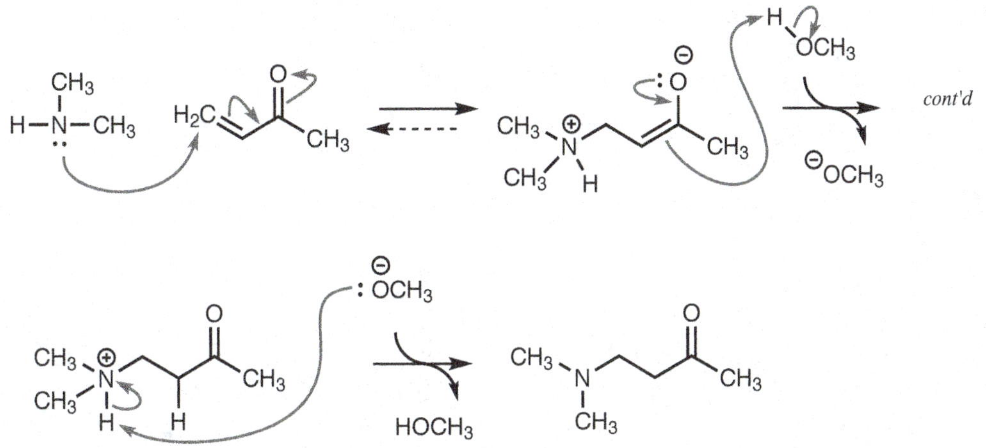

16. Michael addition reaction of ethyl acetoacetate to cyclohexenone. (See *Notes*.)

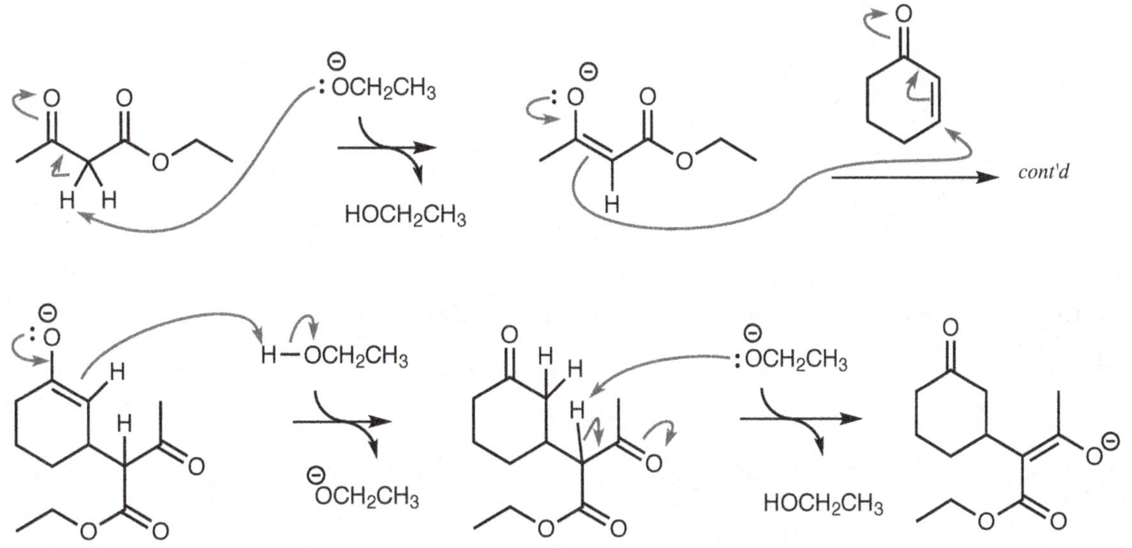

Step 2, neutralization of reaction mixture.

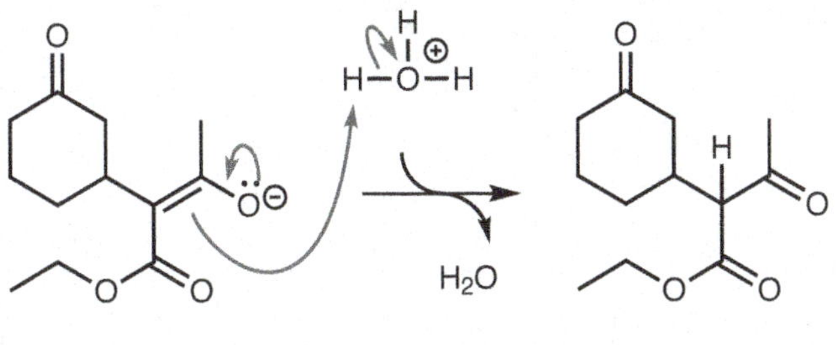

Enamine Alkylation

17. Step 1, formation of the morpholine enamine of cyclohexanone. (See *Notes*.)

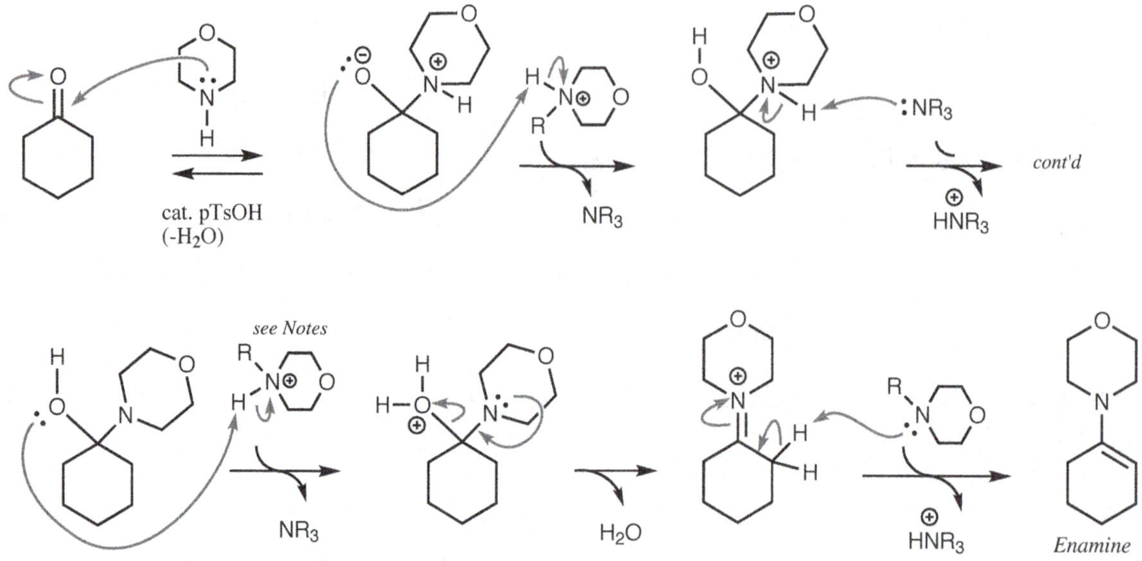

cat. pTsOH
(-H₂O)

NR₃

HNR₃

see Notes

NR₃

H₂O

HNR₃

Enamine

Step 2, enamine alkylation with methyl iodide.

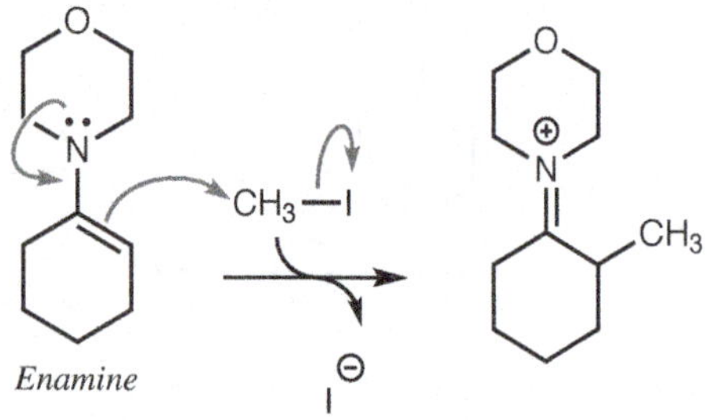

Enamine

Step 3, hydrolysis of iminium salt and isolation of substituted cyclohexanone.

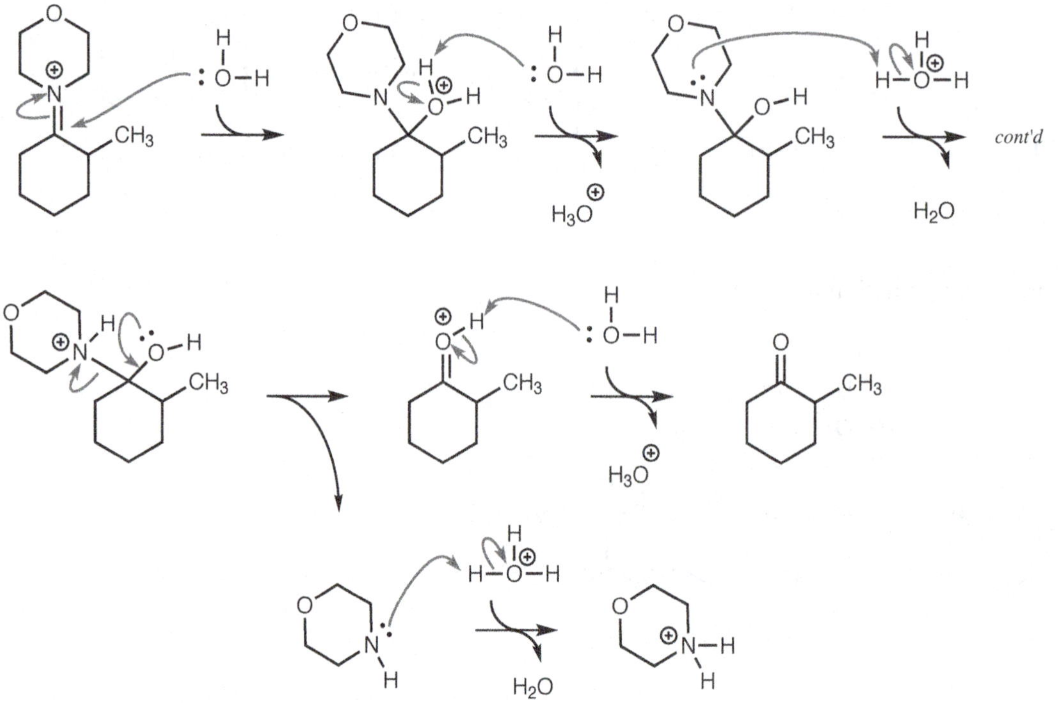

cont'd

H₂O

10 — Dehydration/Halogenation Agents

1. Reaction of cyclohexanecarboxamide with acetic anhydride to give cyclohexanecarbonitrile.

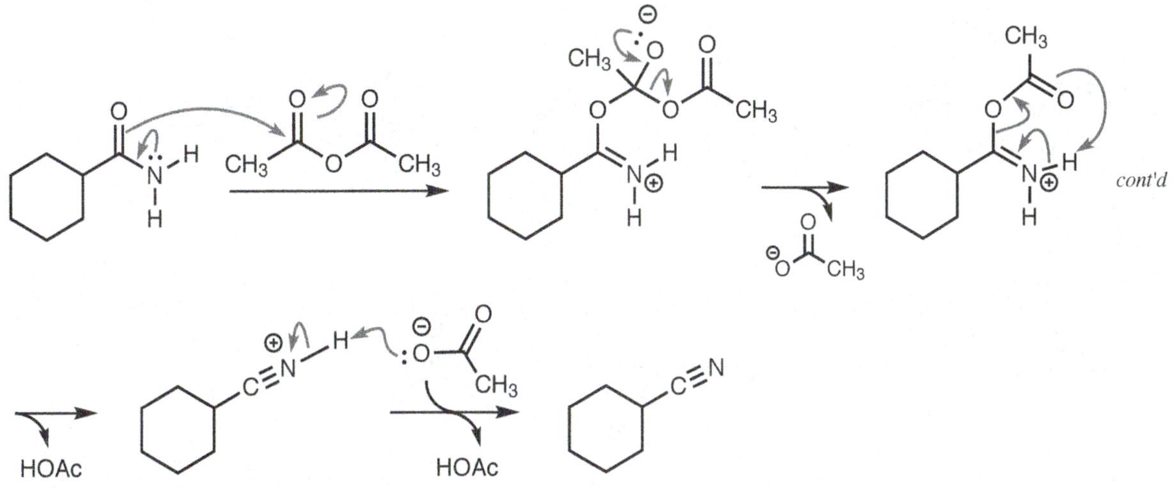

2. S_N2 reaction of thionyl chloride with 1-butanol to give chlorobutane.

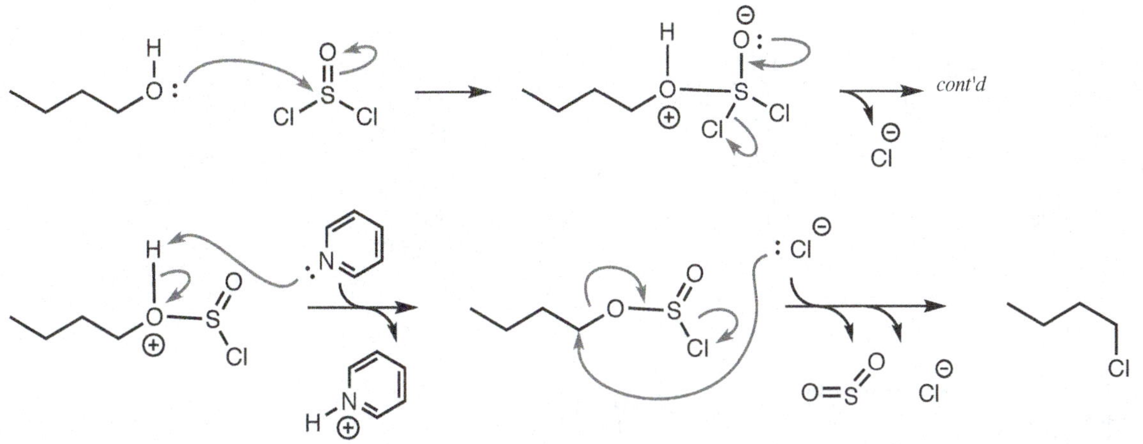

3. Reaction of 1-butanol with tosyl chloride and pyridine to give butyl tosylate. (See *Notes*.)

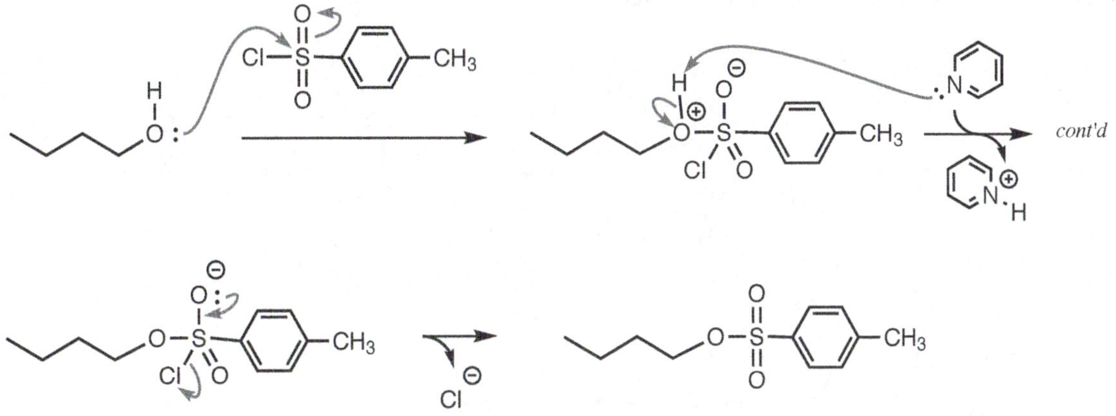

Carboxylic Acid with Thionyl Chloride

4. Conversion of acetic acid to acetyl chloride with thionyl chloride.

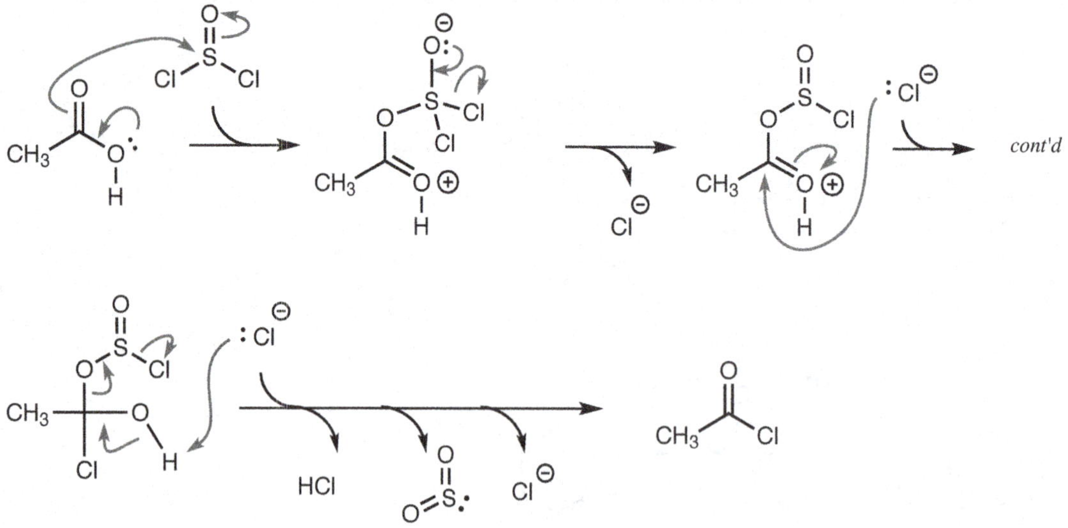

5. Conversion of acetic acid to acetyl chloride with thionyl chloride. Part 1, reaction of thionyl chloride with DMF to form iminium salt. (See Notes.)

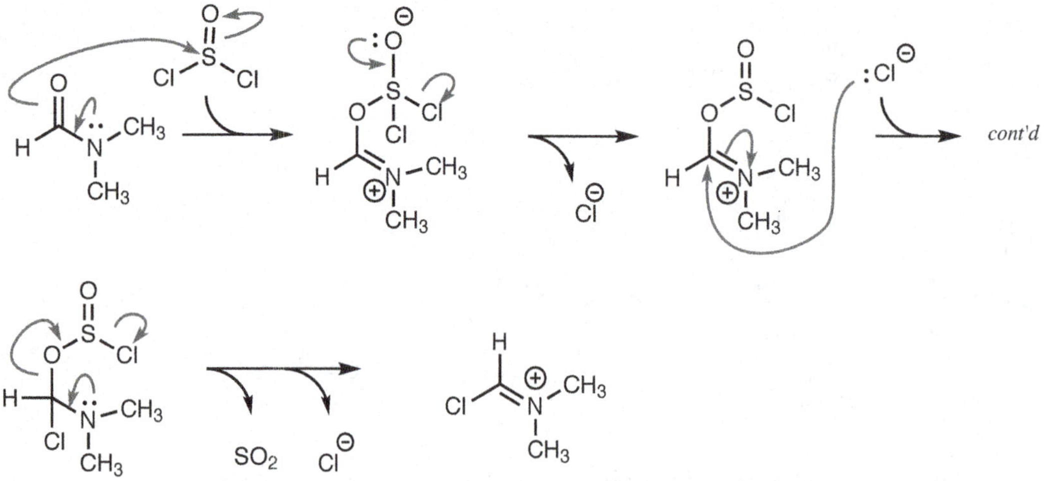

Part 2, reaction of iminium salt with carboxylic acid

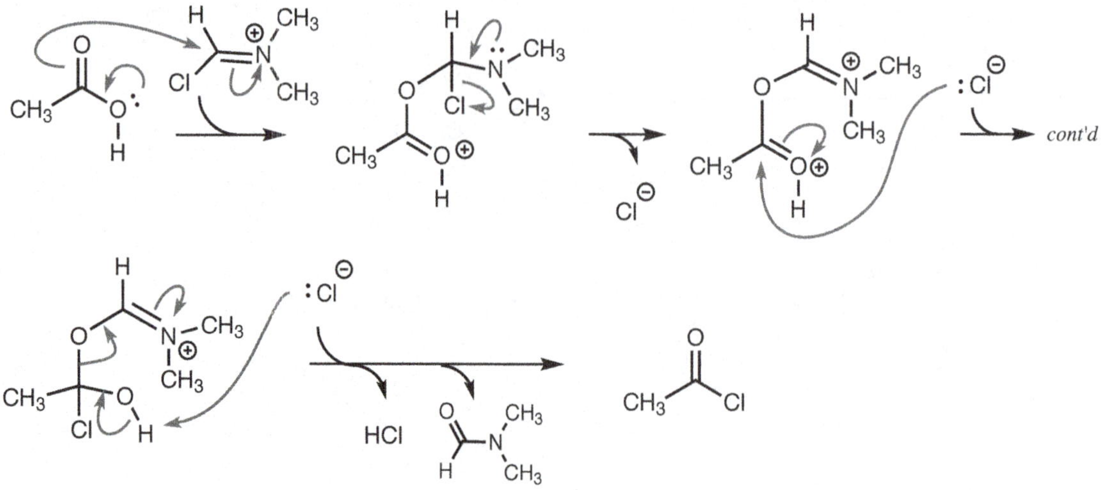

Halide from an Alcohol with a Phosphorus Reagent

6. Reaction of isobutyl alcohol with phosphorus tribromide to give isobutyl bromide.

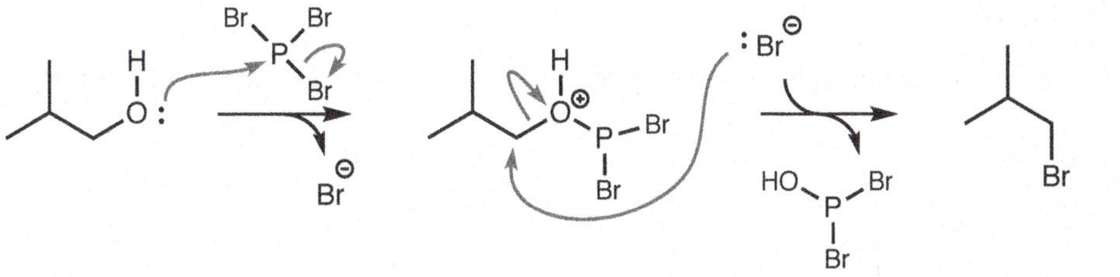

7. Reaction of triphenylphosphine, carbon tetrachloride and cyclopentanol to give chlorocyclopentane.

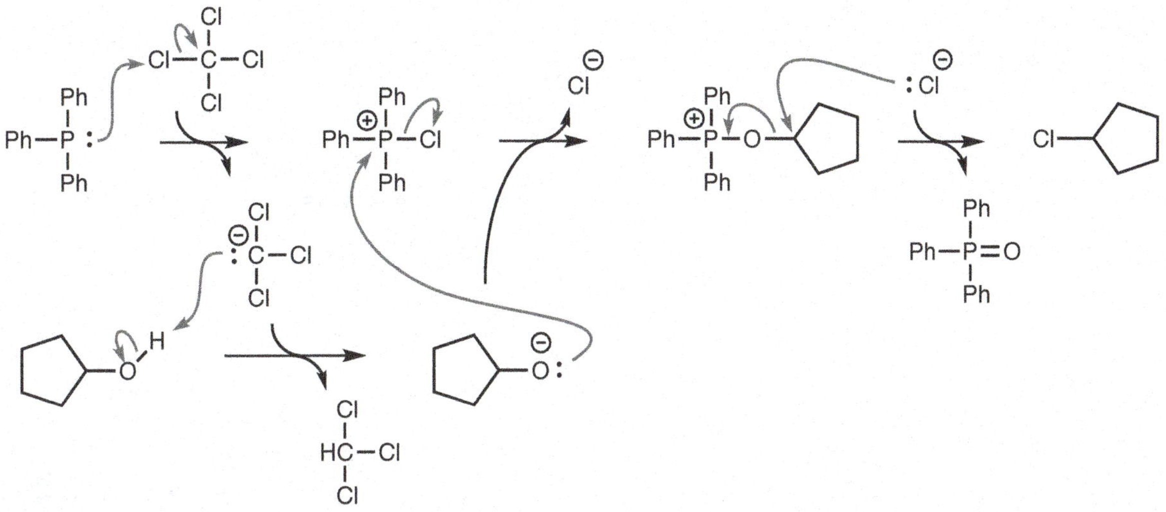

Ester from an Alcohol and Carboxylic Acid with a Phosphorus Reagent

8. Mitsunobu reaction with triphenyl phosphine and diethyl azodicarboxylate to prepare an ester from an acid and an alcohol.

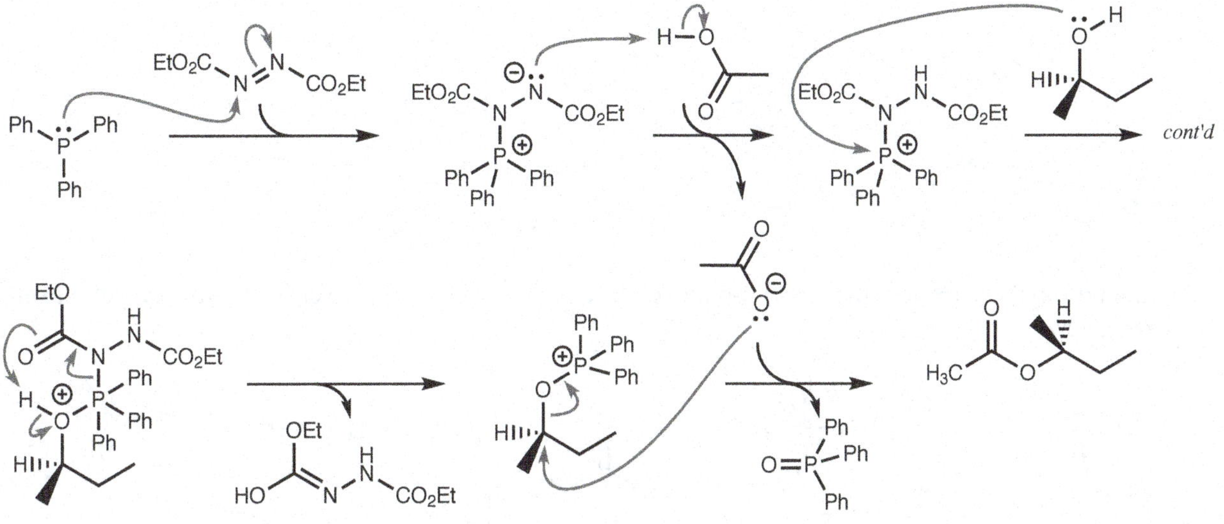

11 — Reduction Reactions

Sodium Borohydride Reductions

1. Sodium borohydride reduction of 2-methylpropanal (isobutyraldehyde) to 2-methyl-1-propanol (isobutyl alcohol).

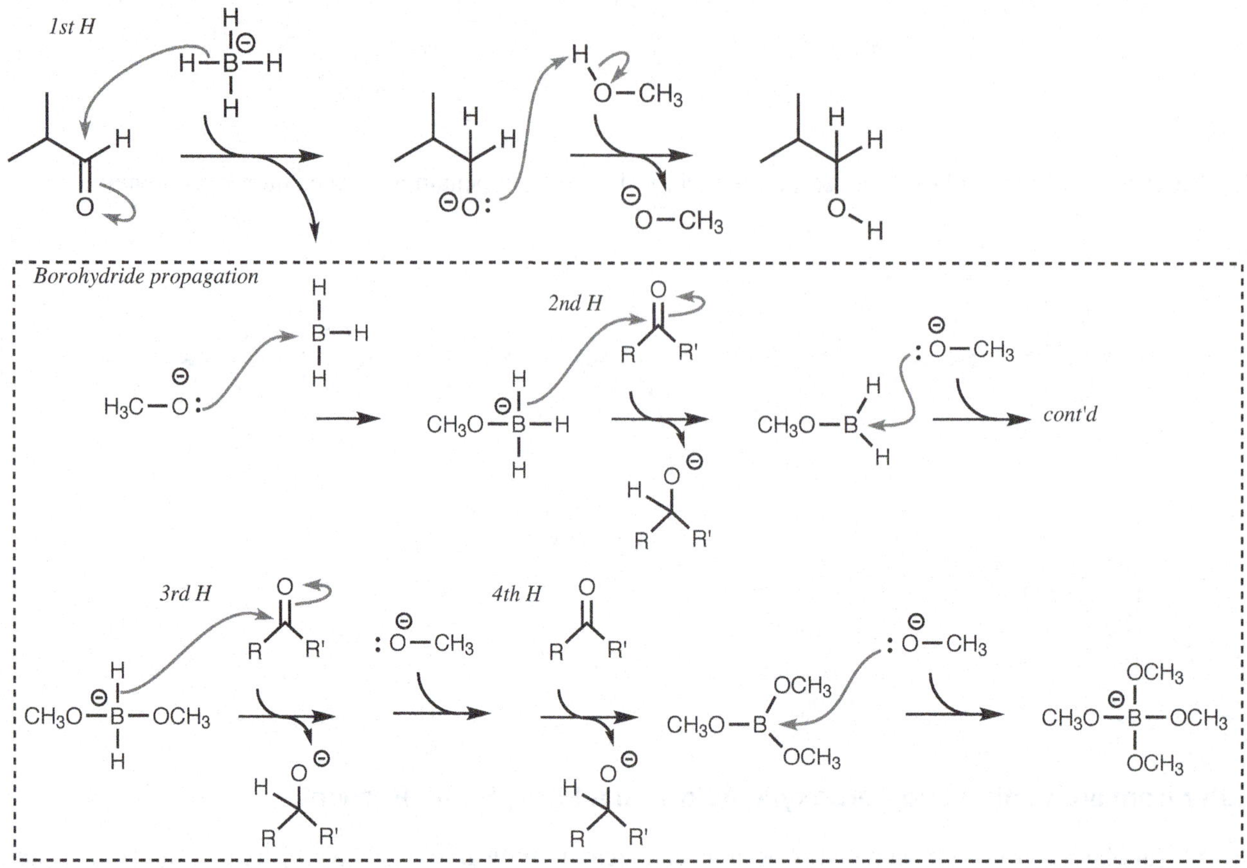

2. Sodium borohydride reduction of cyclopentanone to cyclopentanol.

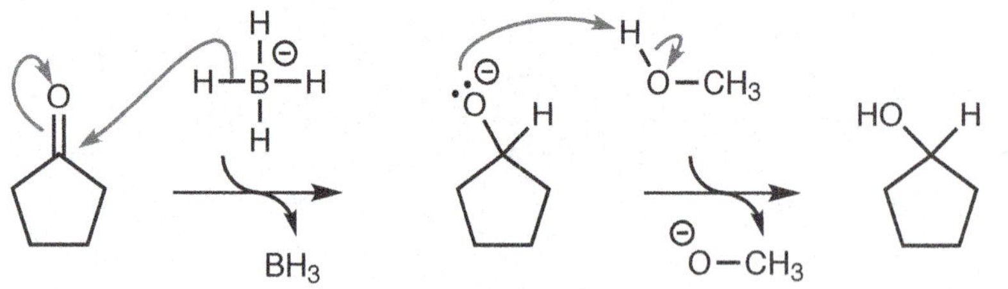

3. Sodium borohydride reduction of ethyl 4-oxocyclohexanecarboxylate to give ethyl 4-hydroxycyclohexanecarboxylate. (See *Notes*.)

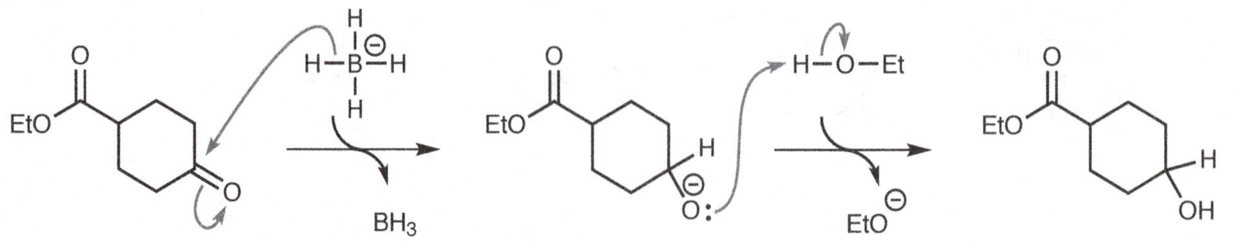

Lithium Aluminum Hydride Reductions

4.   Lithium aluminum hydride reduction of acetophenone to 1-phenylethanol.

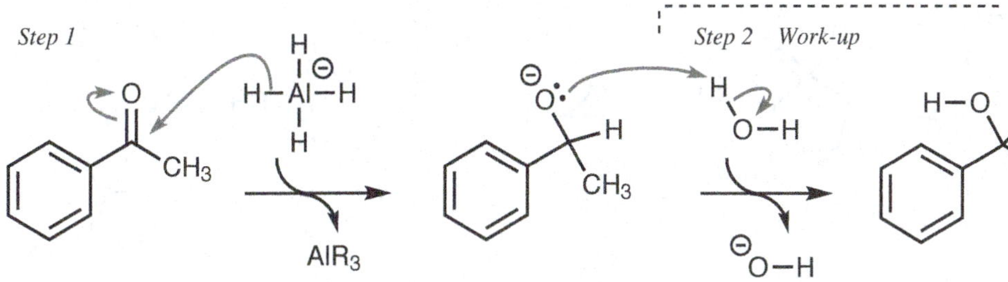

5.   Lithium aluminum hydride reduction of ethyl butanoate (butyrate) to give ethanol and butanol. (See *Notes*.)

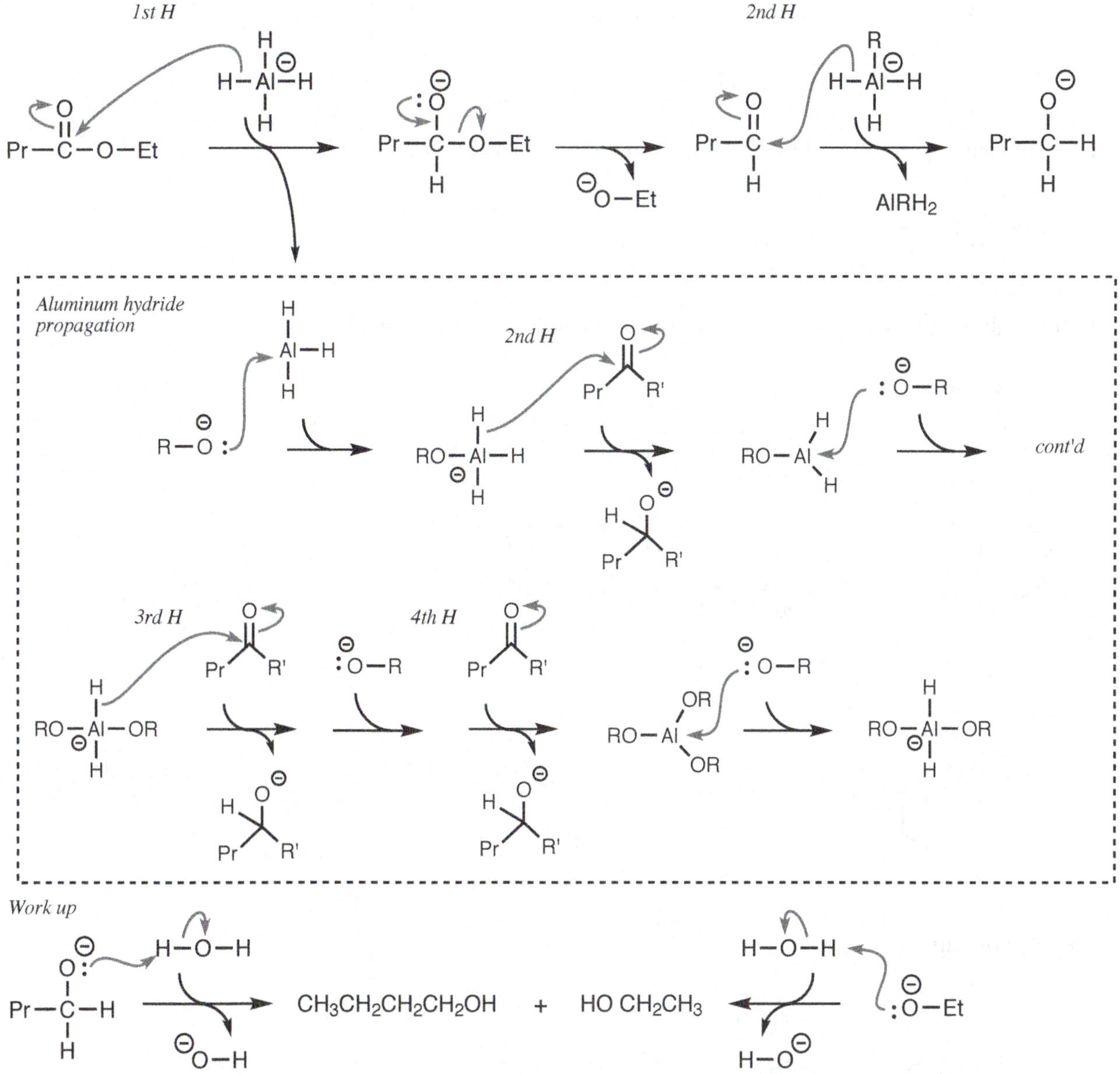

6. Lithium aluminum hydride reduction of a cyclic amide (1-methylpyrrolidin-2-one) to give a cyclic amine (1-methylpyrrolidine). (See *Notes*.) See Example 5 for regeneration of aluminum (IV) reductant.

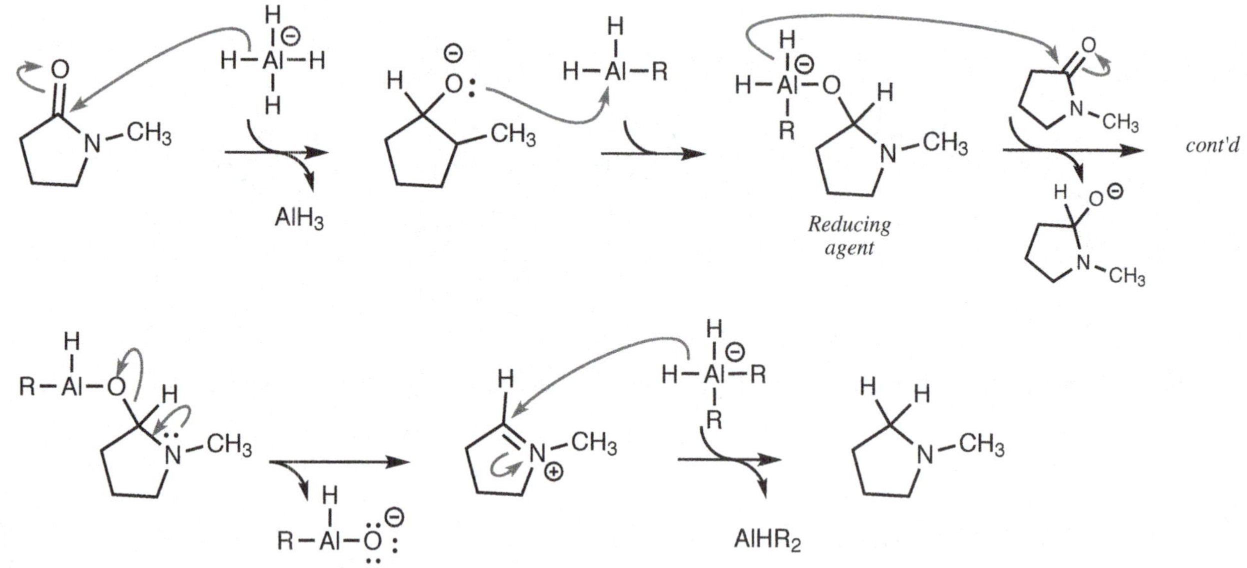

AlH₃

Reducing
agent

cont'd

AlHR₂

Preferred leaving group for lithium aluminum hydride reductions.

7. Lithium aluminum hydride reduction of propionyl anilide to *N*-propylaniline.

Step 1, reduction.

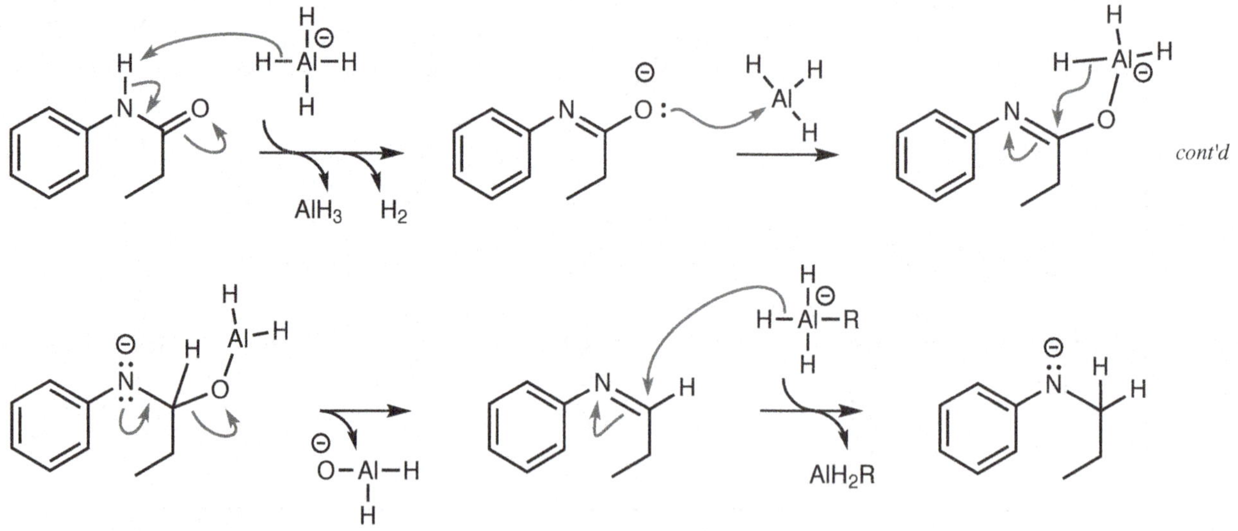

AlH₃   H₂

cont'd

AlH₂R

Step 2, work up.

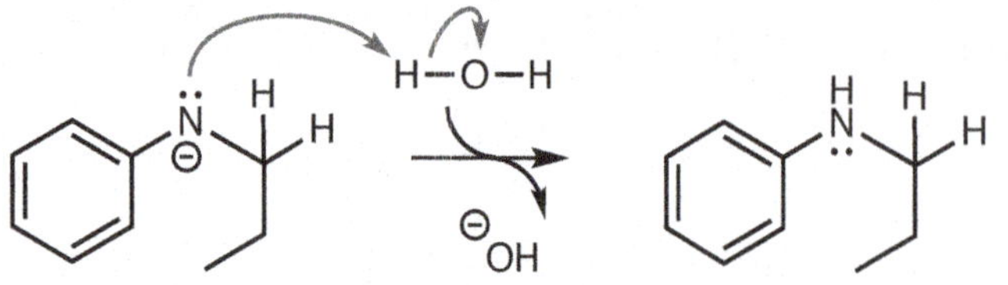

8. Lithium aluminum hydride reduction of phenylacetic acid to 2-phenylethanol. (See *Notes*.)

Step 1, reduction.

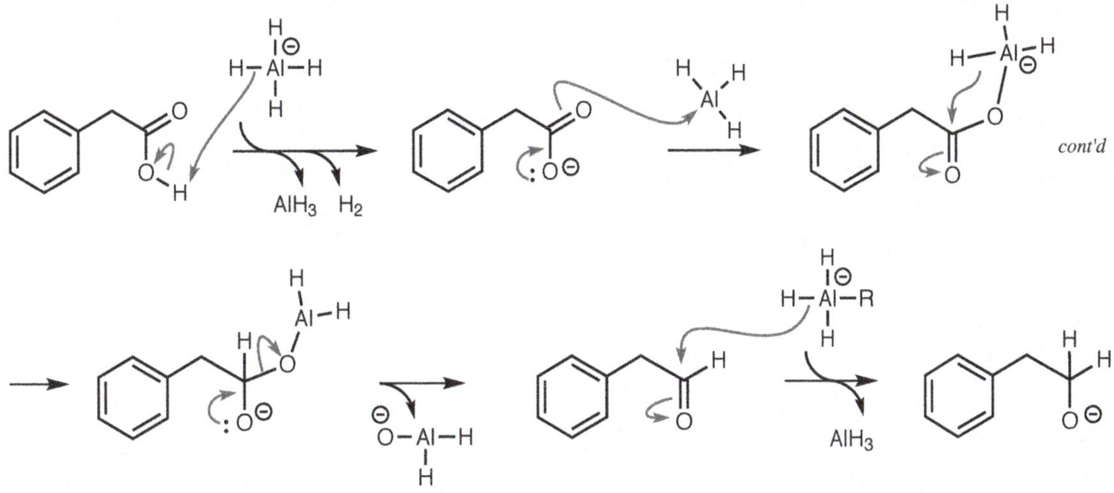

Step 2, work up.

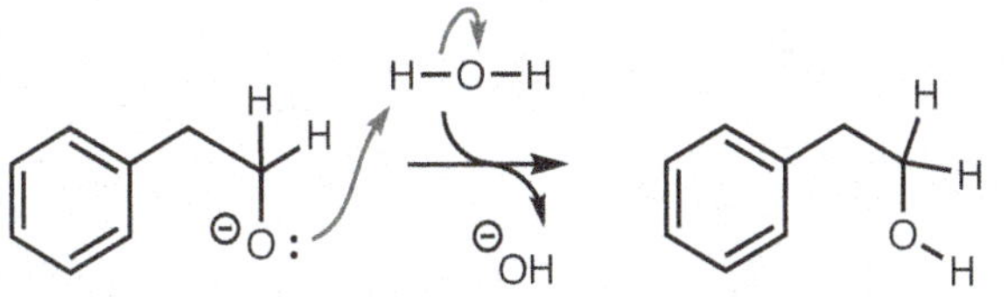

Reductive Amination

9. Step 1, formation of imine from benzaldehyde and ethylamine. (See *Notes*.)

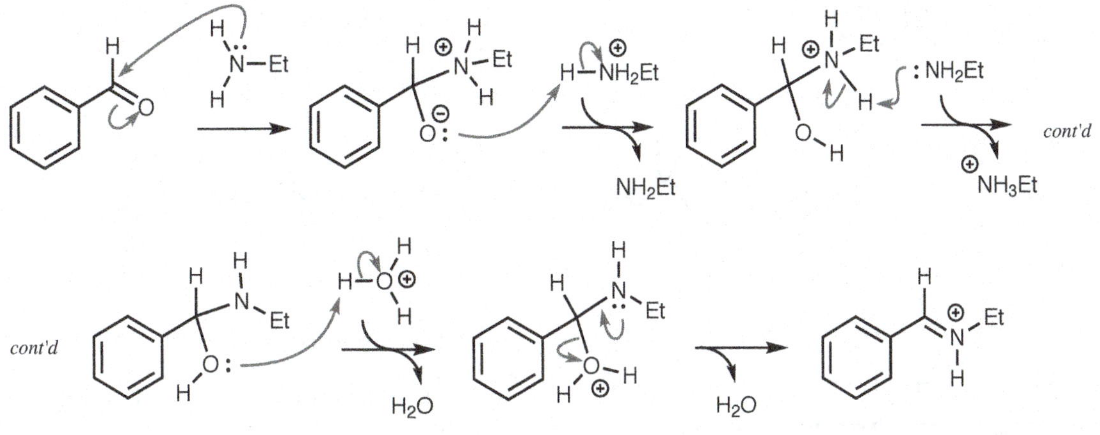

Step 2, reduction

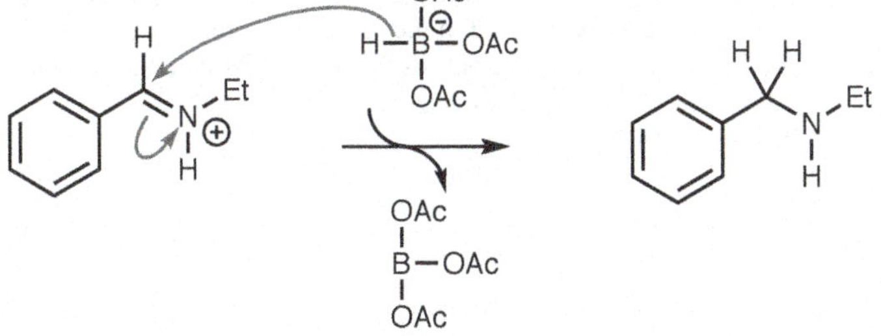

Even though this is written as a two-step process, which it is, both steps can be carried out at the same time. The imine, as it forms, can be reduced.

Triacetoxyborohydride is one of several reagents that one can use. Others are sodium cyano-borohydride (NaBH$_3$CN) or hydrogen with a palladium or nickel catalyst.

Diisobutylaluminum Hydride Reduction of an Ester

10. Step 1, addition of DIBAH or DIBAL(H) (diisobutylaluminum hydride) to ethyl isobutyrate. Reduction to give isobutyraldehyde (2-methylpropanal).

Step 2, work-up

Reduction of Alkyne with Sodium and Ammonia

11. Sodium and ammonia *trans*-reduction of 2-butyne to *trans*-2-butene.

Note, use a single barbed arrow for one-electron transfers reaction.

Which is the stronger base?

The pK$_a$ of an sp^2 carbanion pK$_a$ is 44 and NH$_3$ is 36.

The alkenyl anion is the stronger base, therefore it will abstract a proton from ammonia, the stronger acid in the reaction mixture. The equilibrium favors formation of sodium amide as the conjugate base.

Wolff Kischner Reduction

12. Reaction of the ketone with hydrazine under basic conditions to form the hydrazide.

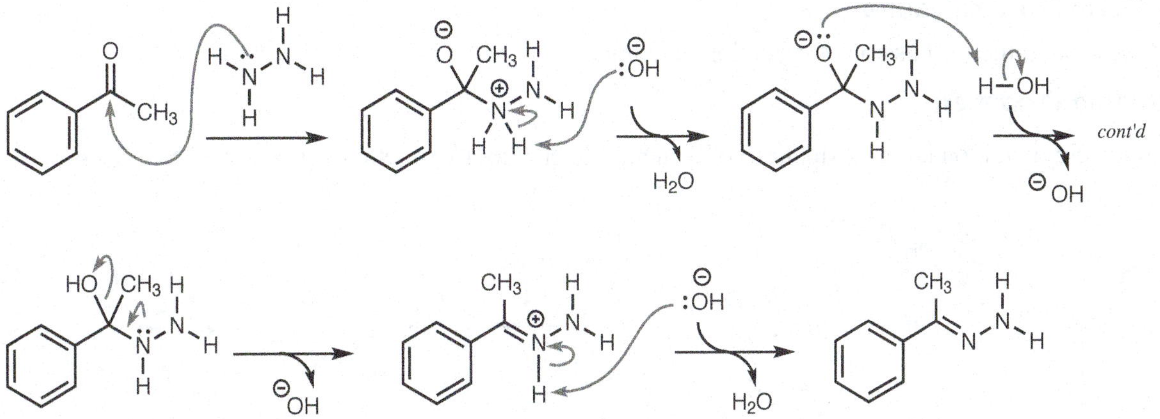

Reaction of the *in situ* formed hydrazide with KOH to form the methylene.

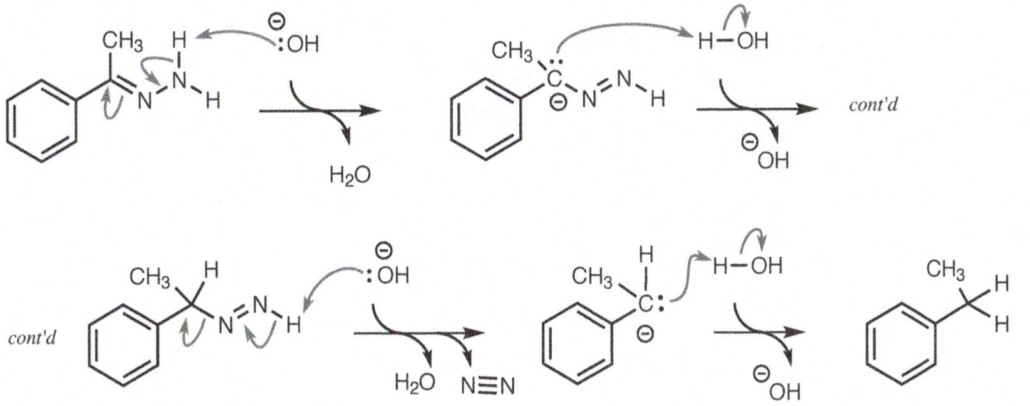

Catalytic Reduction of Nitrobenzene

Catalytic reduction of nitrobenzene to aniline. This isn't a mechanism. It shows that hydrogen can add across N-O single and double bonds. It is a very facile reduction. One can see that the oxygen atoms form water and the reduction takes three moles of hydrogen.

13. Catalytic reduction of nitrobenzene to aniline.

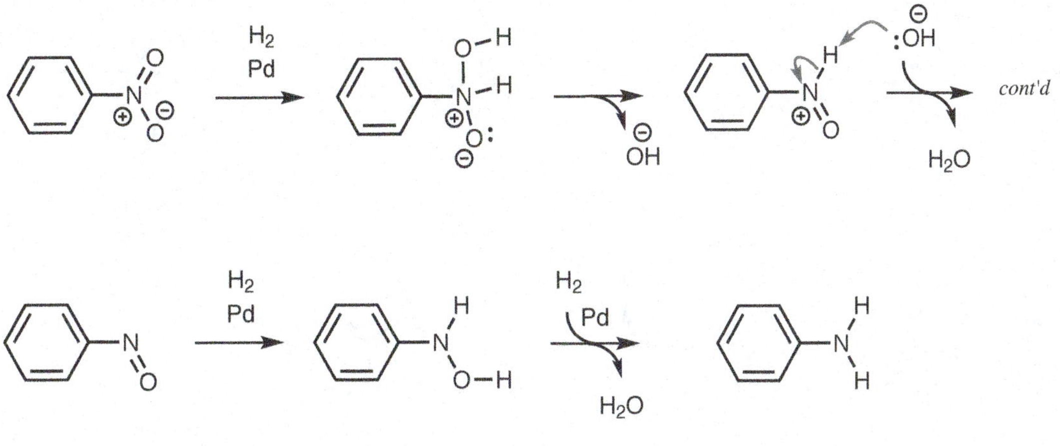

12 — Oxidation Reactions

General Form For Oxidations

See *Notes* for further discussion of oxidation reactions.

Chromic Acid Oxidation

1. Chromic acid oxidation (Jones oxidation) of 3-methyl-2-butanol to 3-methyl-2-butanone. (See *Notes*.)

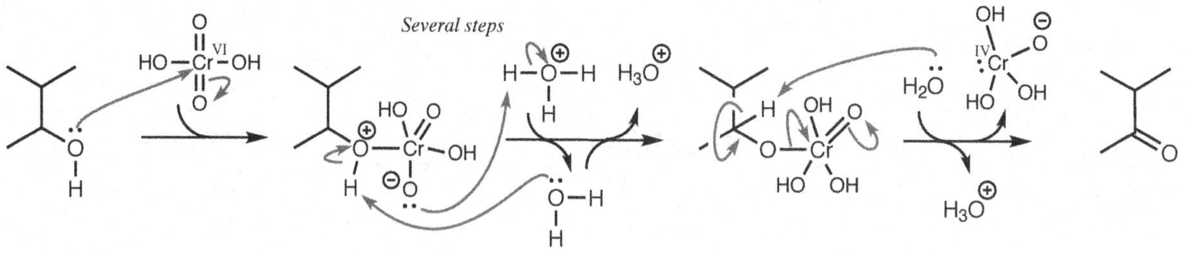

2. Chromic acid oxidation of isobutyl alcohol (2-methyl-1-propanol) to isobutyric acid (2-methylpropanoic acid). (See *Notes*.)

    Step 1, chromate ester formation and oxidation to aldehyde.

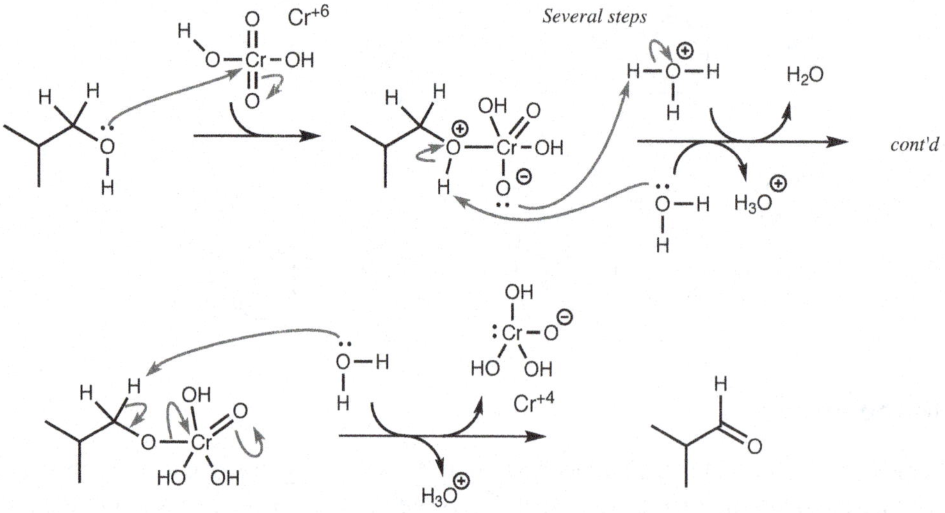

Step 2, hydration of aldehyde and oxidation of hydrate to isobutyric acid (2-methylpropanoic acid).

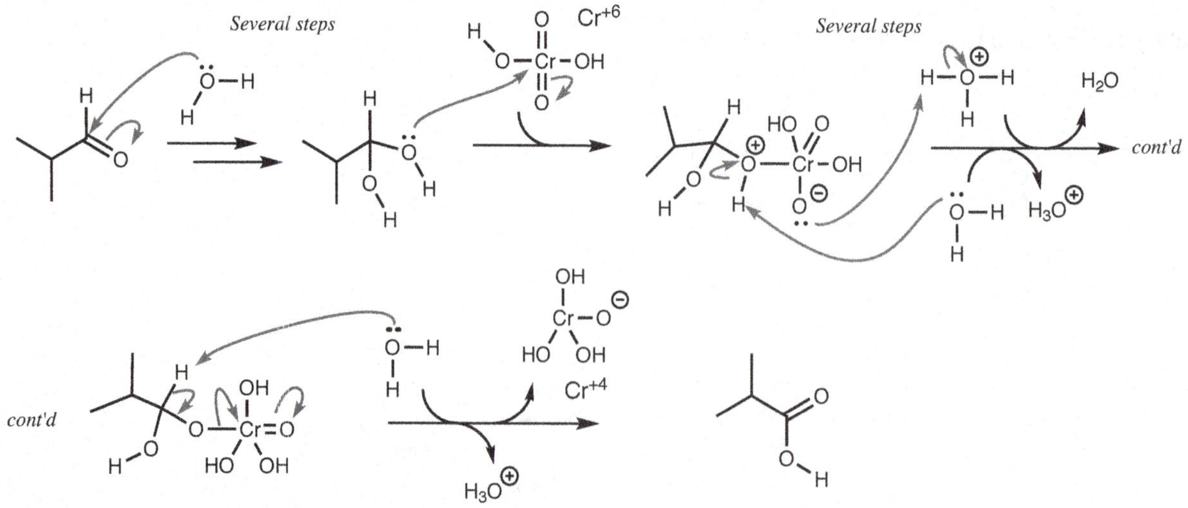

PCC, Tollens, Hypochlorite, *m*CPBA, and Sulfonium Based Oxidations

3. Pyridinium chlorochromate (PCC) oxidation of benzyl alcohol to benzaldehyde.

4. Tollens oxidation of 2-methylpropanal (isobutyraldehyde) with silver oxide (or hydroxide).

5. Oxidation of cyclohexanol to cyclohexanone with sodium hypochlorite (NaOCl, bleach).

6. Peracid epoxidation of *trans*-2-butene with *m*-chloroperoxybenzoic acid (MCPBA) to give an epoxide, (*2R,3R*)-2,3-dimethyloxirane. (See *Notes*.)

7. Step 1, Swern oxidation, preparation of chlorosulfonium salt.

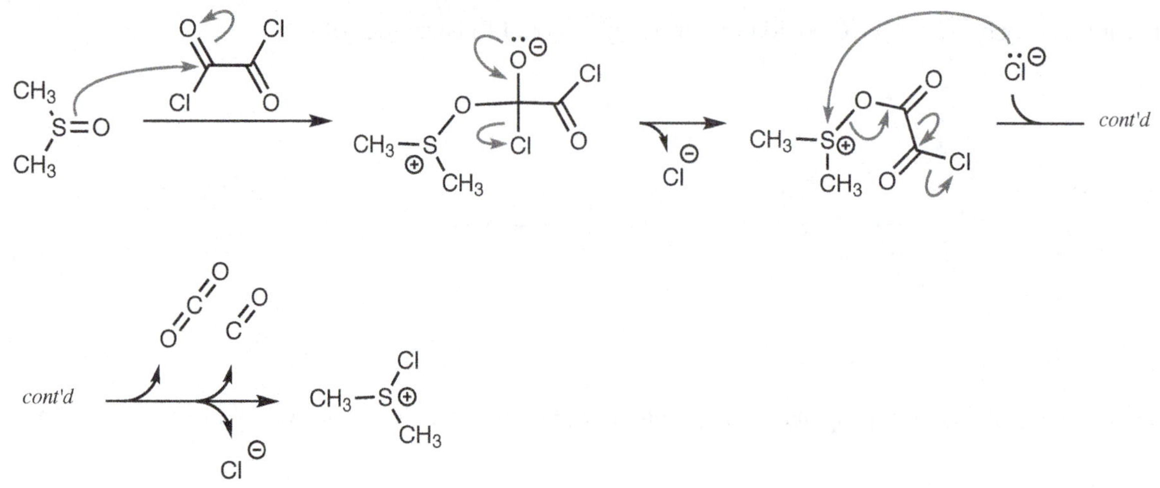

Step 2, Swern oxidation, oxidation of alcohol.

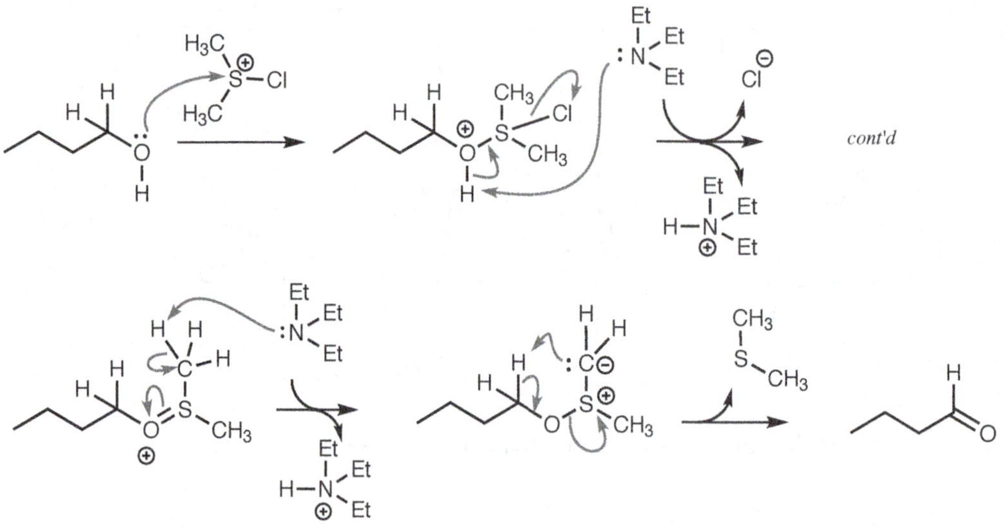

8. Corey-Kim (chlorosulfonium salt) oxidation of cyclopentanol to cyclopentanone. (See *Notes*.)

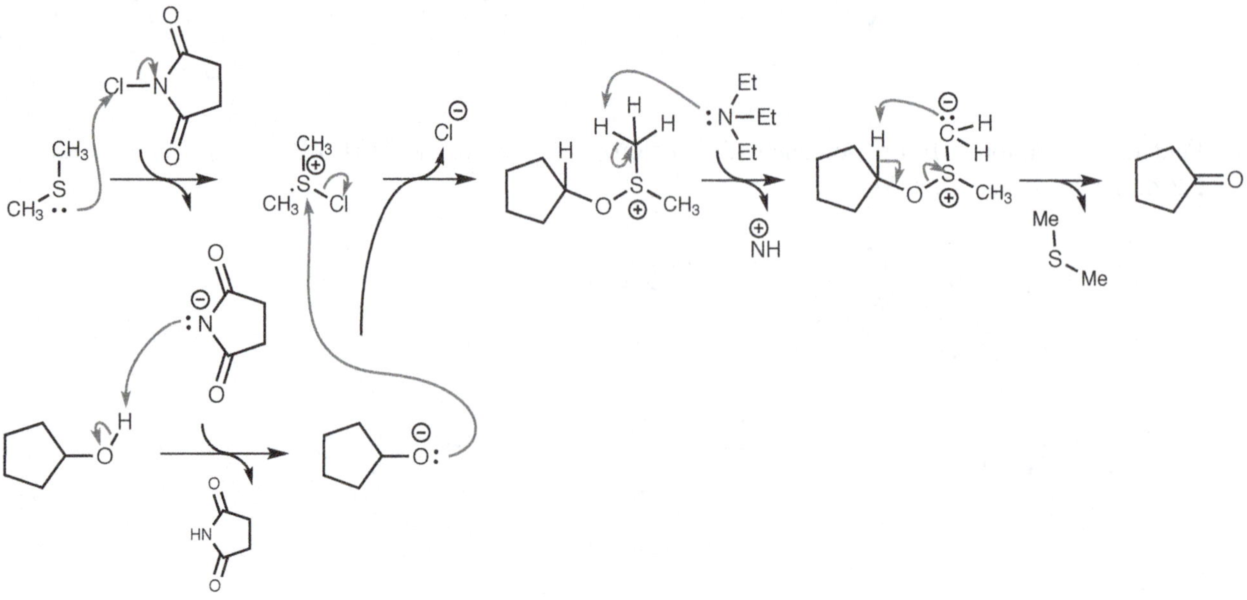

Ozone Oxidation

9. Step 1, ozone reaction with an alkene to give an ozonide. (See *Notes.*)

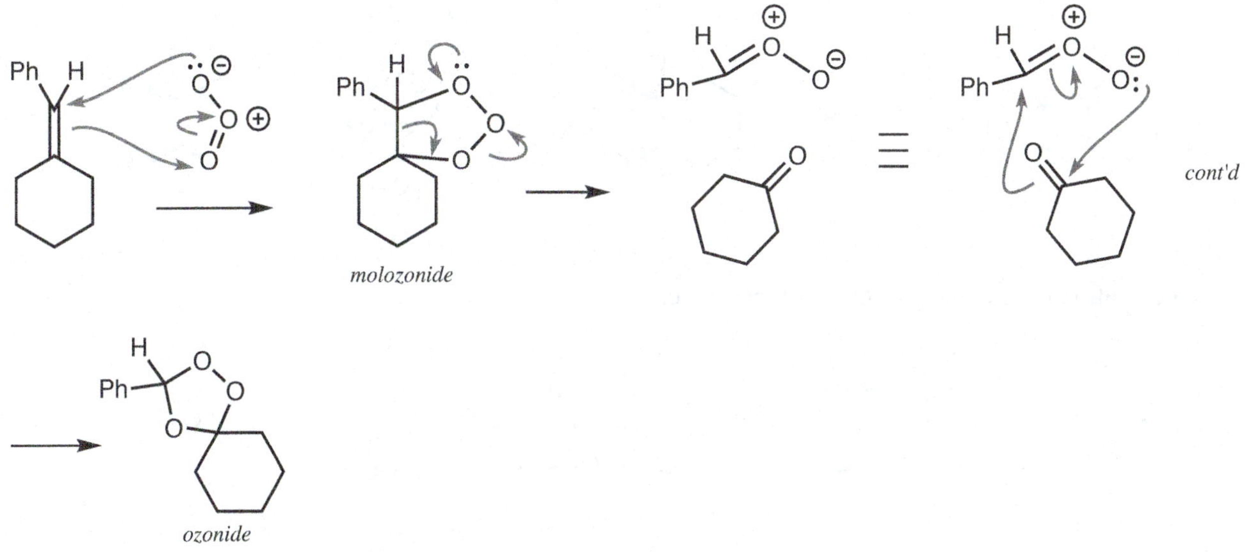

molozonide

cont'd

ozonide

Step 2, reduction of ozonide to two carbonyl compounds.

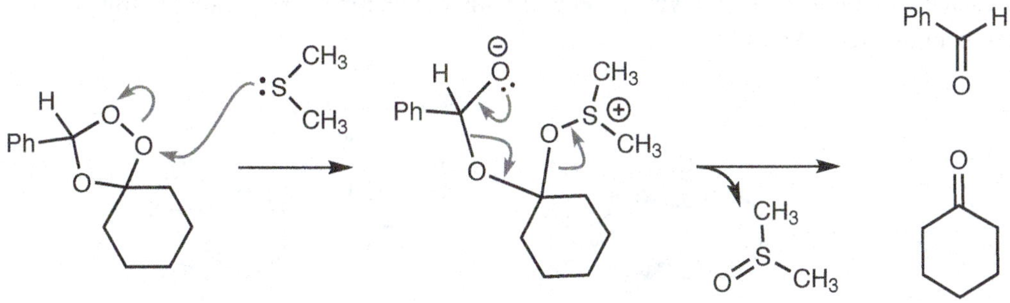

Alternate Step 2, oxidative destruction of ozonide.

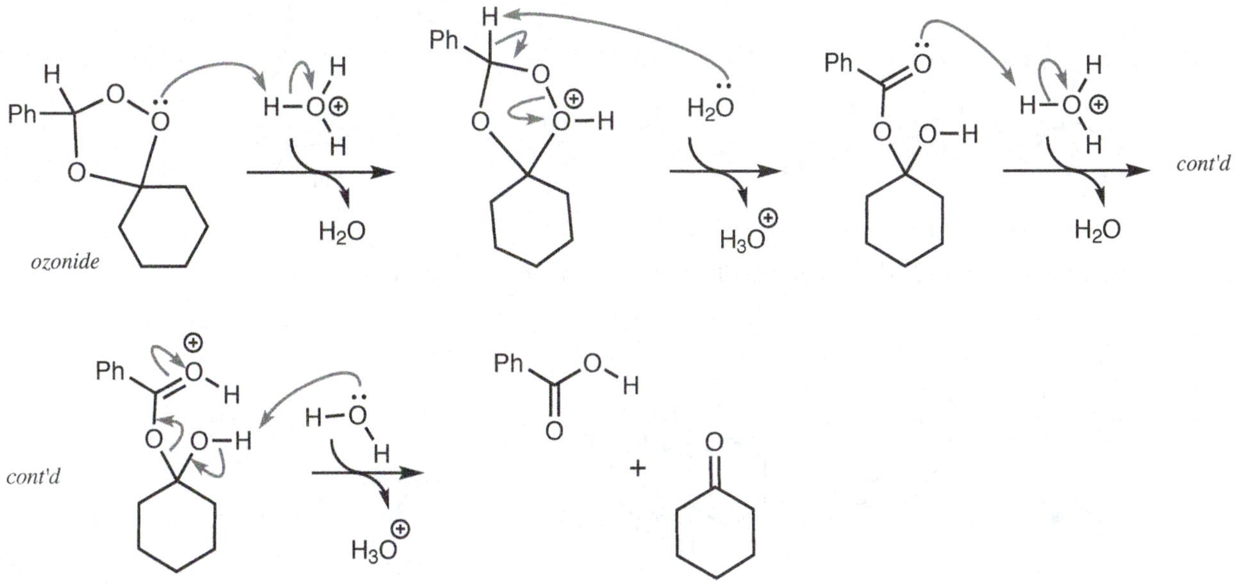

ozonide

cont'd

cont'd

Osmium Tetroxide, Potassium Permanganate, and Periodate Oxidations

10. Osmium tetroxide oxidation of cyclohexene to give *cis*-1,2-cyclohexanediol. (See *Notes*.)

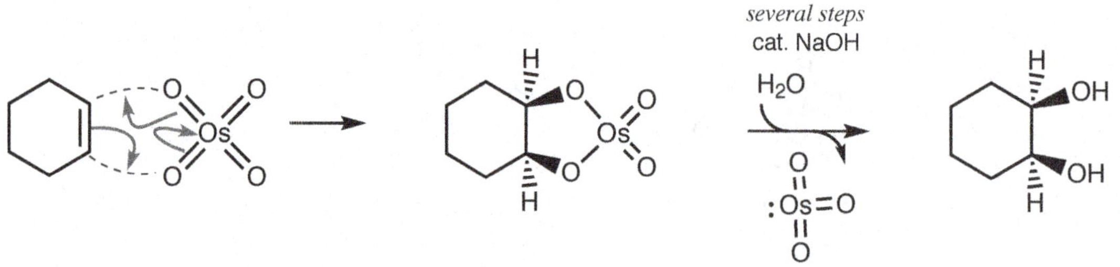

Regeneration of oxidant with *tert*-butylperoxide.

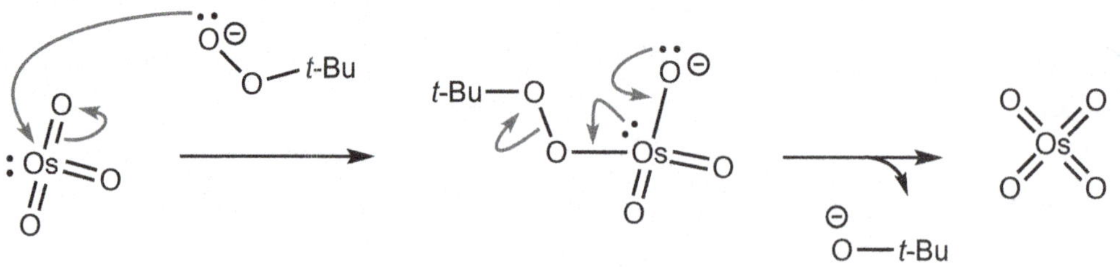

11. Potassium permanganate oxidation of methylcyclohexene to 1-methylcyclohexane-1,2-diol at low temperature or cleavage at high temperature. (See *Notes*.)

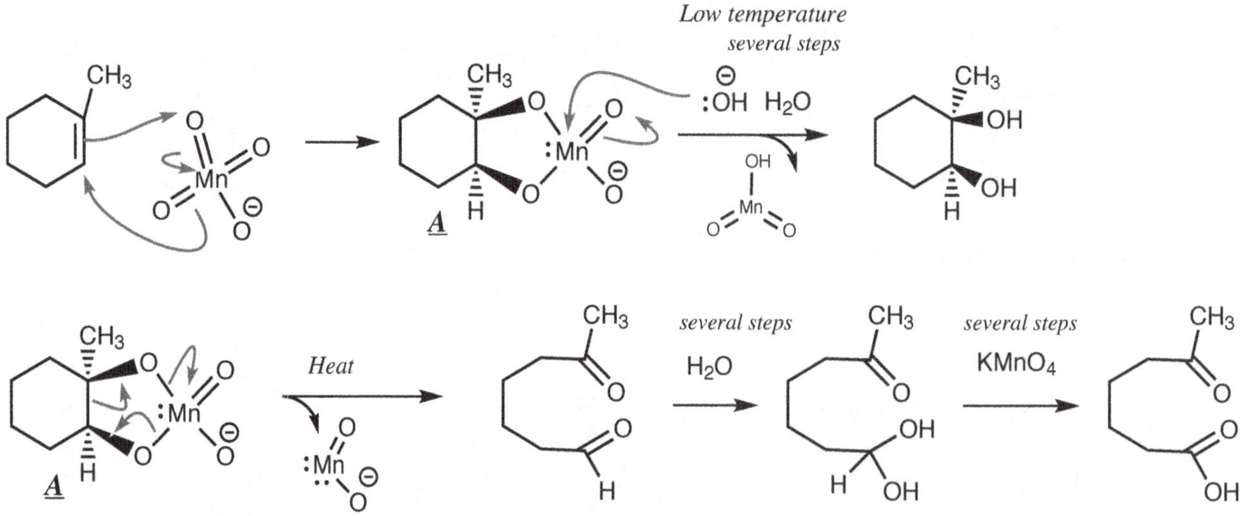

12. Periodate cleavage of a 1,2-diol to give a dicarbonyl compound. (See *Notes*.)

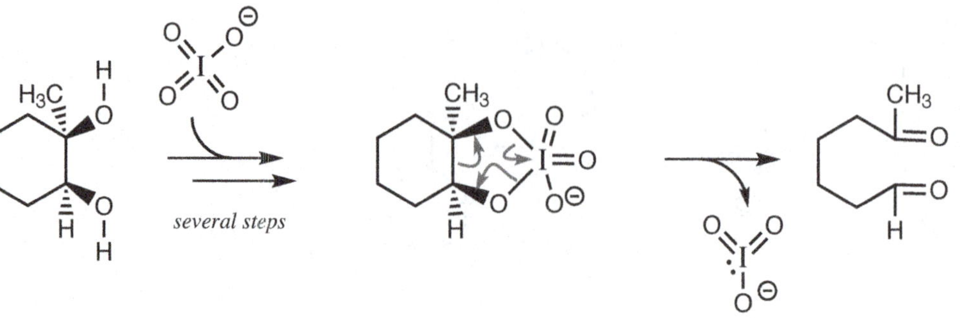

13 — Organometallic Reactions

The (transition) organometallic reactions in this chapter are of increasing importance. The reactions may recount the steps more than explain a reaction. Significantly absent is the concept of 18 valence electrons about the catalytic palladium atom.

Acyclic Heck Reaction

1. Step 1, reduction of palladium (II) to zero valent palladium with propene.

Step 2, the catalytic cycle (oxidative-addition, *syn*-addition, *syn*-elimination, and reductive-elimination) with 2-bromopropene and propene.

The alkene pi-electrons probably complex and are donated to the palladium before the syn addition takes place.

Cyclic Heck Reaction

2. Step 1, reduction of palladium (II) to zero valent palladium with cyclopentene.

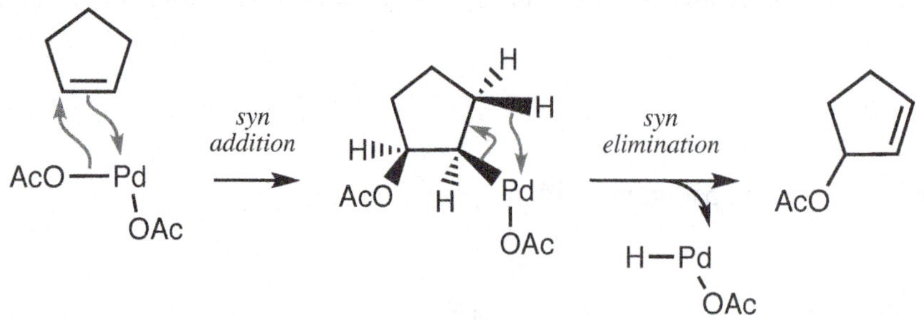

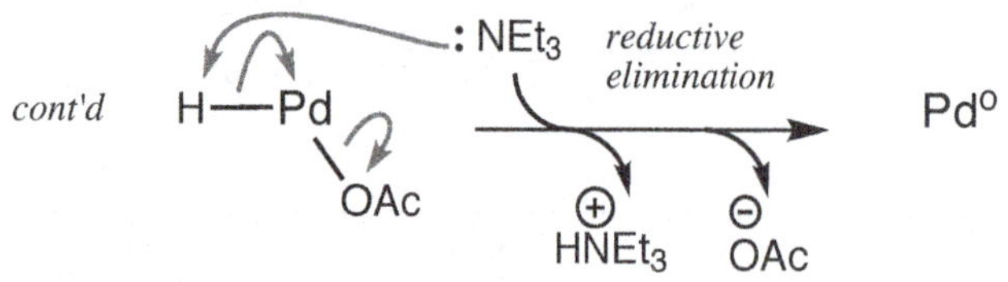

Step 2, the catalytic cycle (oxidative-addition, *syn*-addition, *syn*-elimination, and reductive-elimination) with iodobenzene and cyclopentene.

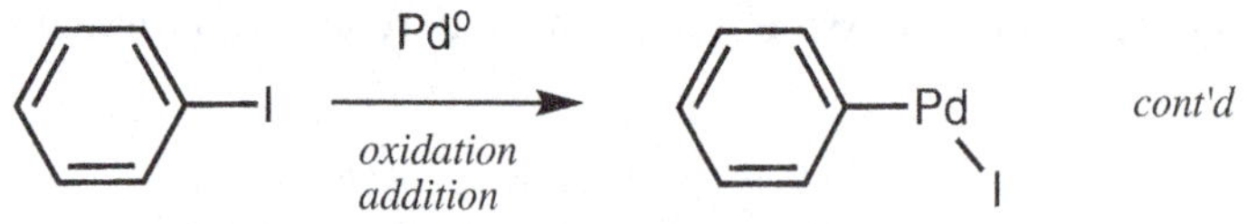

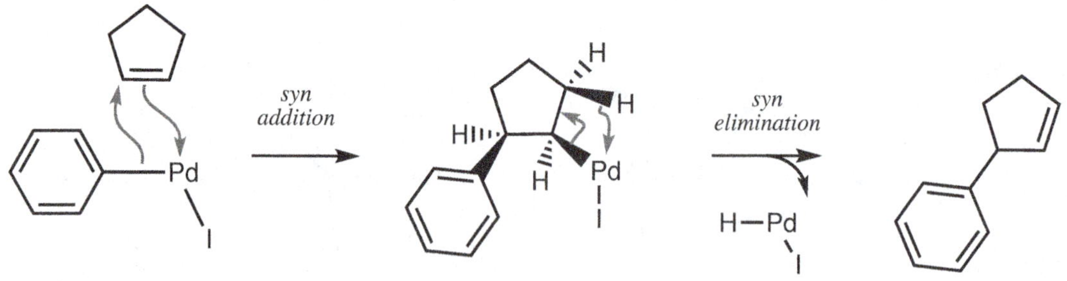

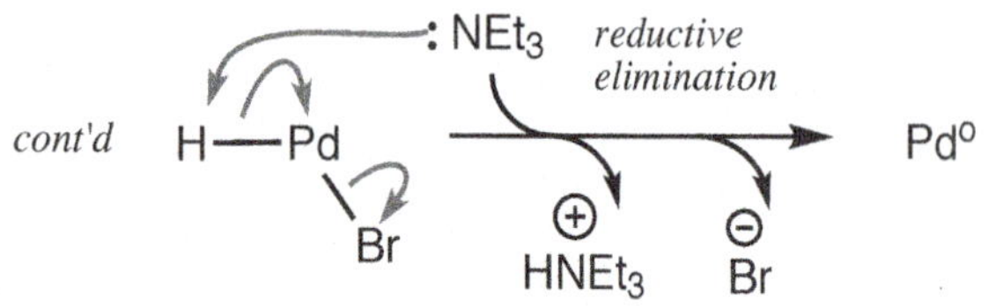

Catalytic Reduction of an Alkene

3.   Catalytic hydrogenation of *cis*-3-hexene to hexane.

If there isn't enough hydrogen to complex with the palladium, then the *cis*-addition can reverse. If you started with a *cis*-alkene, you would end up with the *trans*. This is how *trans*-fatty acids are formed in vegetable oils.

Gilman Reagent

4.   Formation of Gilman reagent, lithium dimethylcuprate.

Coupling of Gilman reagent, lithium dimethylcuprate with (*Z*)-1-bromopentene to give *cis*-2-hexene. (See *Notes*.)

5.   A 1,4-conjugate addition of lithium dimethylcuprate to 2-cyclohexenone to give 3-methylcyclohexanone.

6.   Reaction of lithium dimethylcuprate to benzoyl chloride to give acetophenone.

14 — Aromatic Substitution Reactions

Electrophilic Aromatic Substitution of Benzene

1. Friedel-Crafts acylation of benzene.

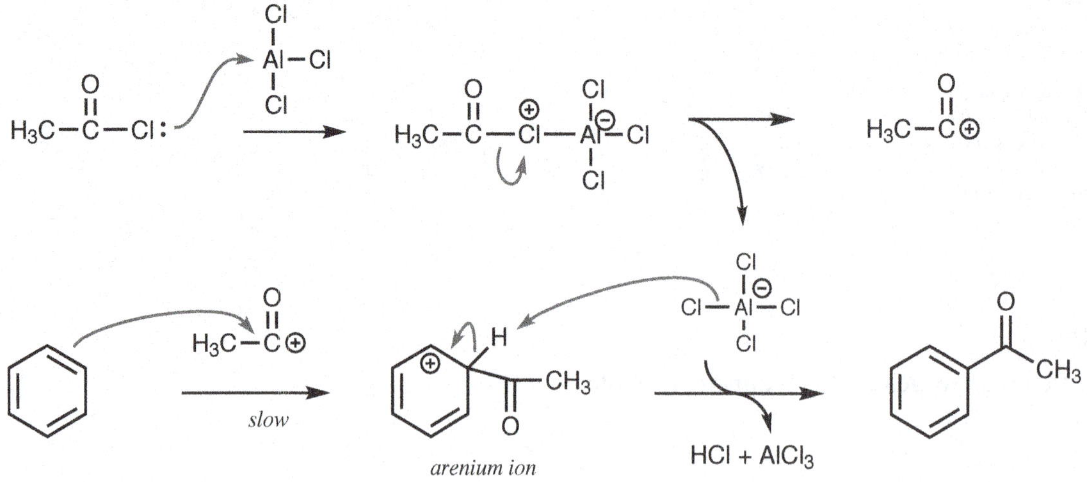

arenium ion

2. Friedel Crafts alkylation of benzene. (See *Notes*.)

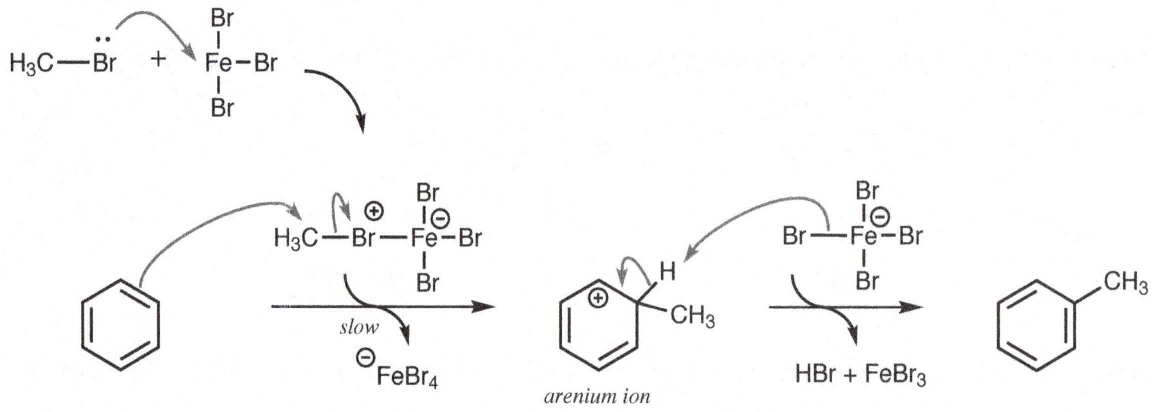

arenium ion

3. Ferric bromide bromination of benzene.

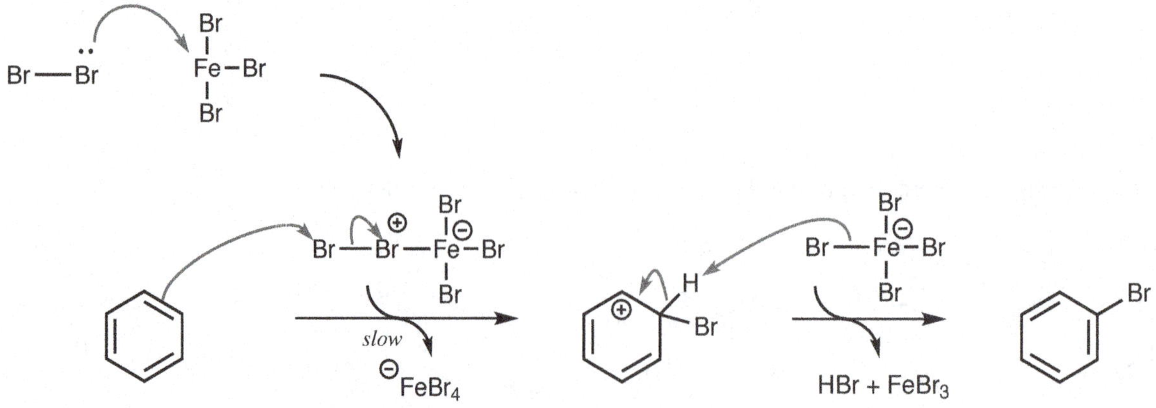

4. Nitration of benzene with nitric and sulfuric acids.

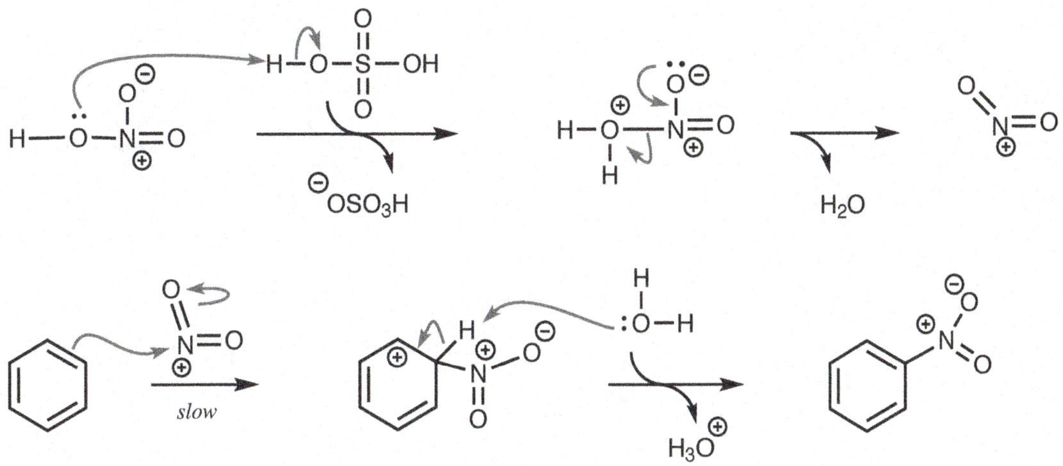

Electrophilic Substitution of Substituted Aromatic Compounds

5. Bromination of acetophenone to give *m*-bromoacetophenone.

    (For 5, 6, and 7, also answer the question, "Will it react faster than benzene?")

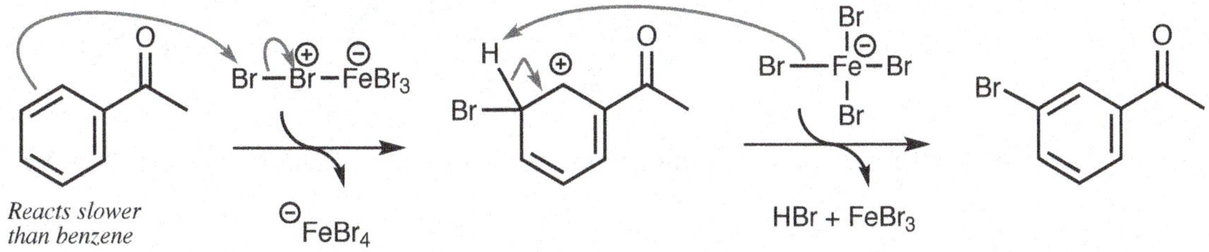

*Reacts slower
than benzene*

6. Nitration of methyl *p*-chlorobenzoate to give methyl 4-chloro-3-nitrobenzoate.

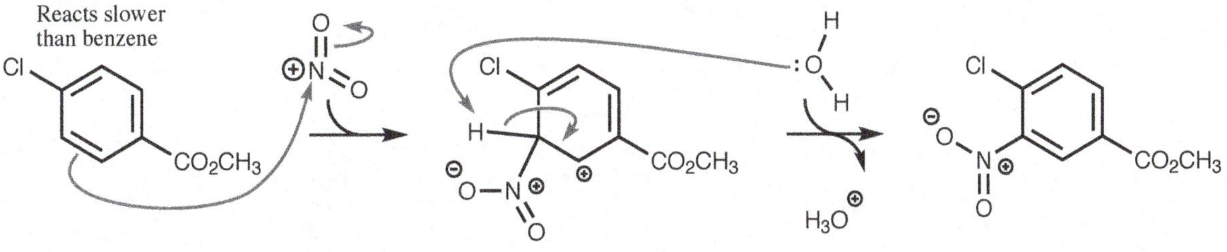

*Reacts slower
than benzene*

7. Friedel-Crafts acylation of phenyl acetate with acetyl chloride and aluminum chloride.

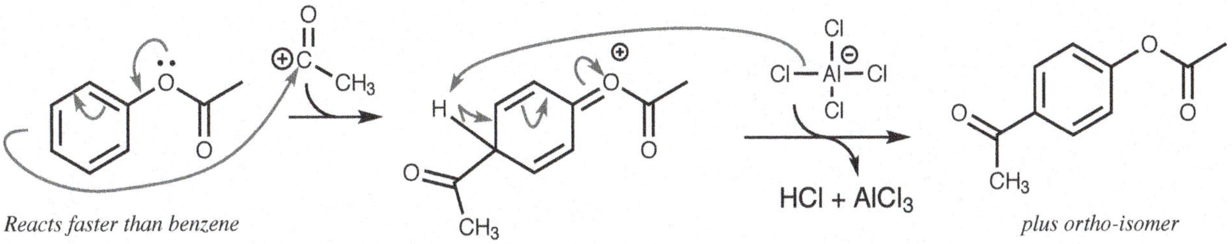

Reacts faster than benzene   *plus ortho-isomer*

8. Triflic acid catalyzed acetylation of toluene to give *o*- and *p*-methylacetophenone.

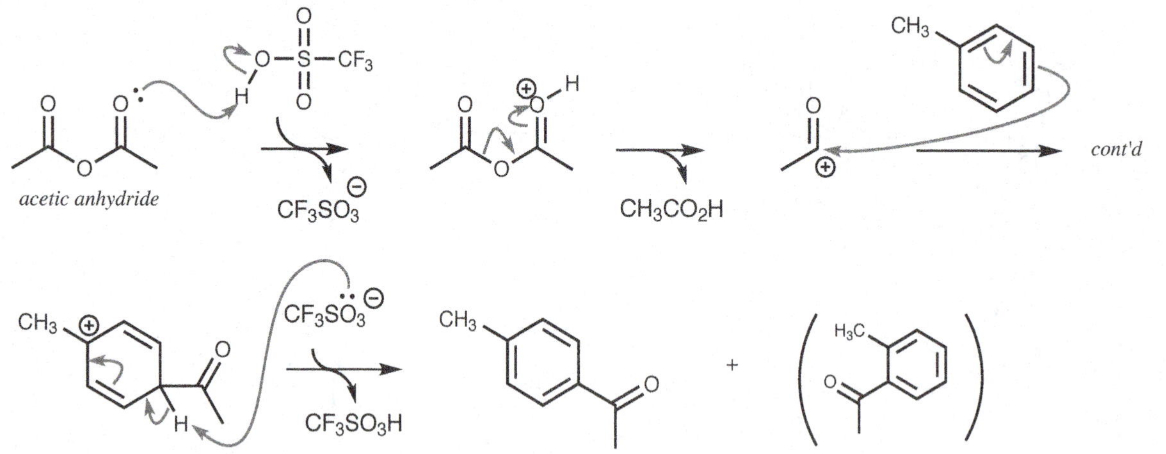

acetic anhydride

9. Aluminum chloride catalyzed chlorination of methyl 3-methoxybenzoate (methyl *m*-anisate) to give methyl 2-chloro-5-methoxybenzoate.

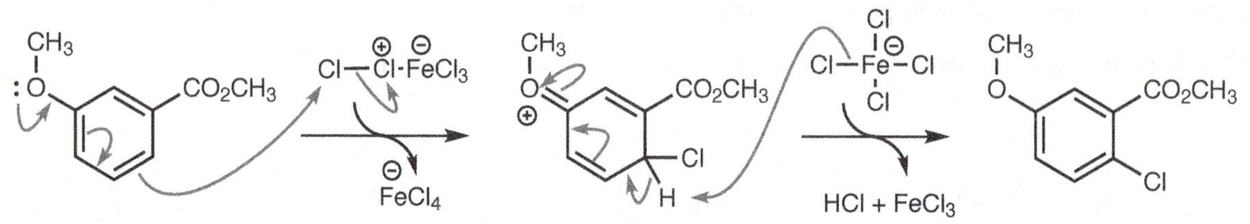

10. Friedel-Crafts alkylation of 4-nitro-*N*-*p*-tolylbenzamide with two equivalents of chloromethane. (See *Notes*.)

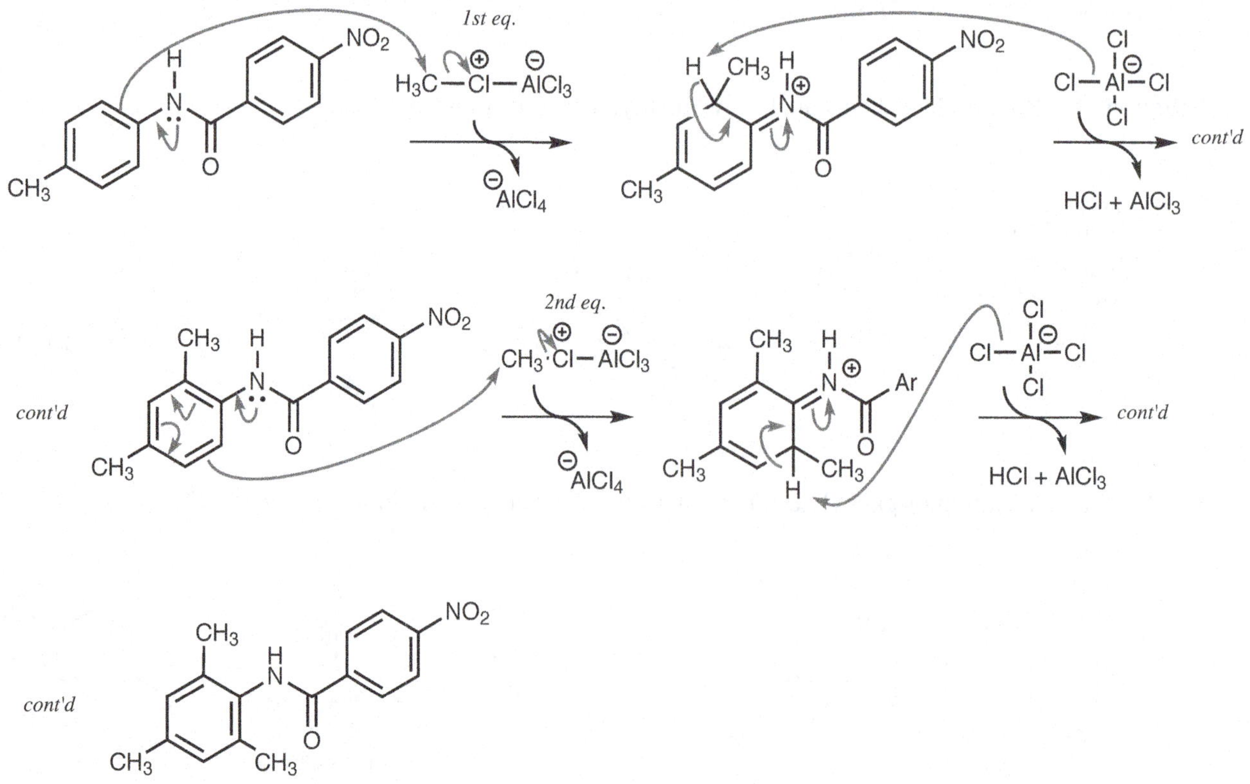

Nucleophilic Aromatic Substitution

11. Nucleophilic aromatic substitution of 1-fluoro-4-nitrobenzene with ammonia to give 4-nitroaniline.

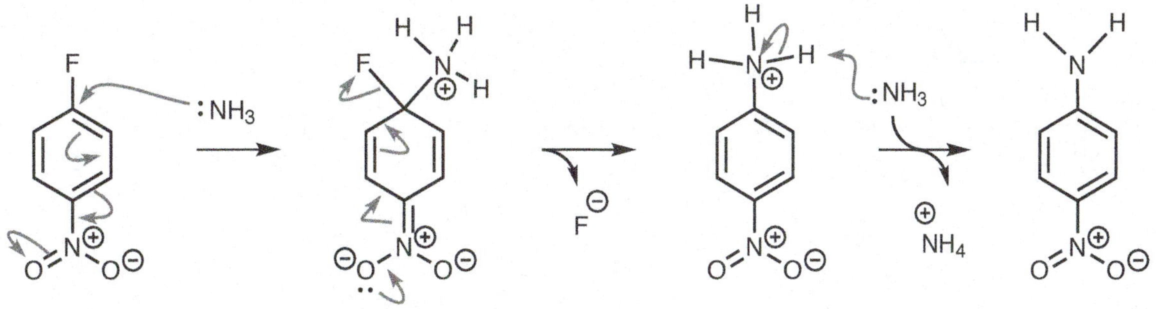

12. Nucleophilic aromatic substitution of 1,2-difluoro-4-nitrobenzene with sodium methoxide to give 2-fluoro-1-methoxy-4-nitrobenzene.

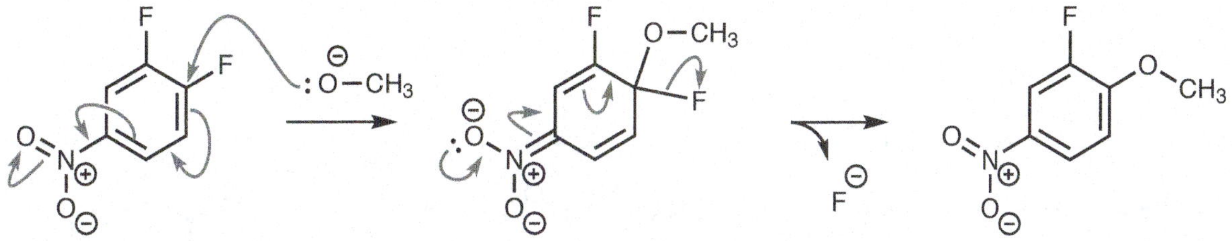

Benzyne Reaction

13. Reaction of 4-bromoanisole with sodium amide (sodamide, NaNH$_2$) to give 3- and 4-methoxyaniline via a benzyne intermediate.

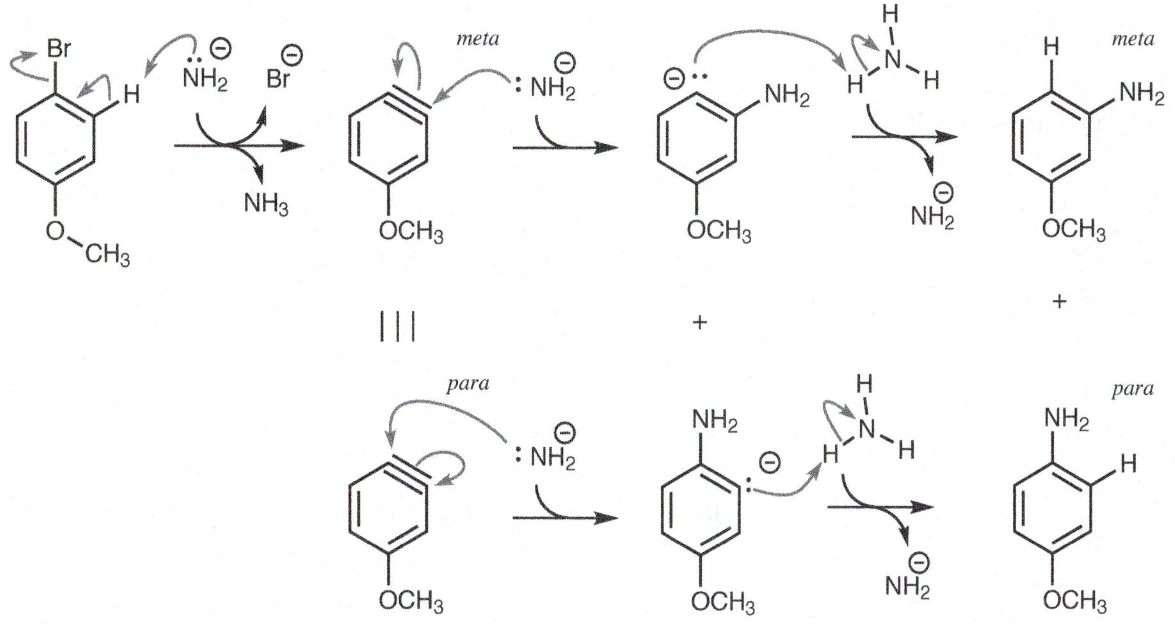

Diazonium Chemistry

14. Formation of a diazonium salt from aniline.

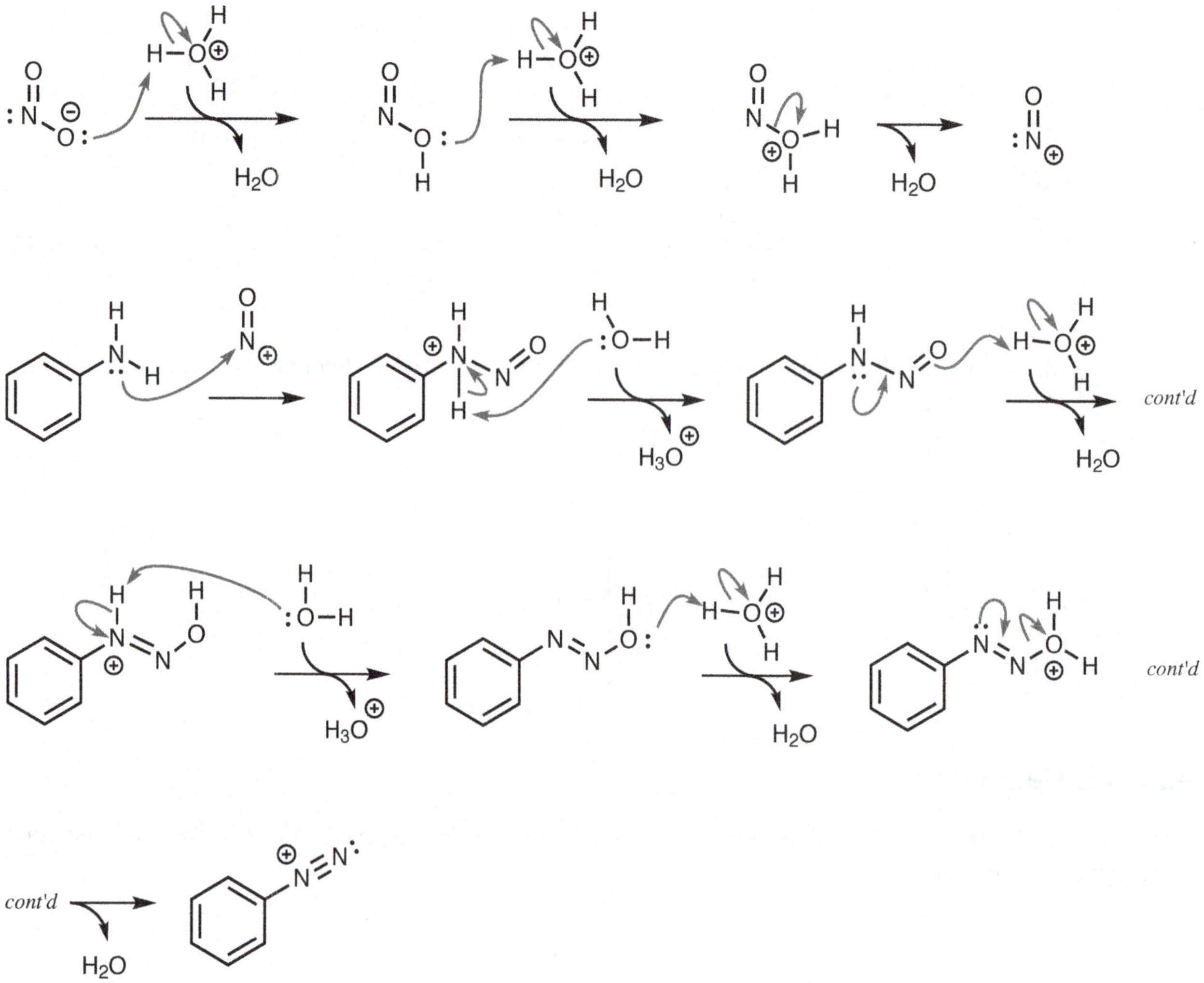

Reaction of diazonium salts.

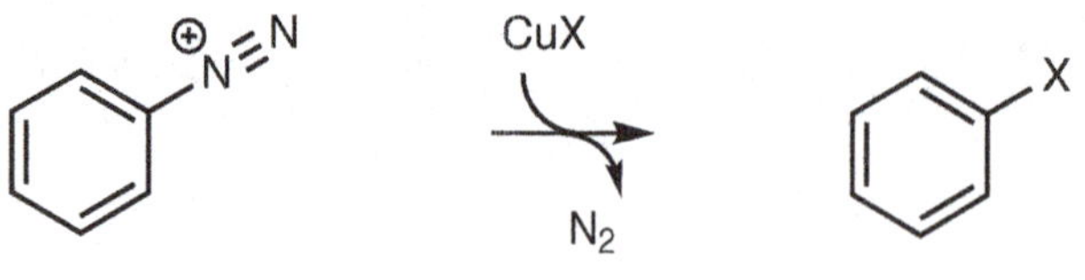

X = Cl, Br, CN, OH

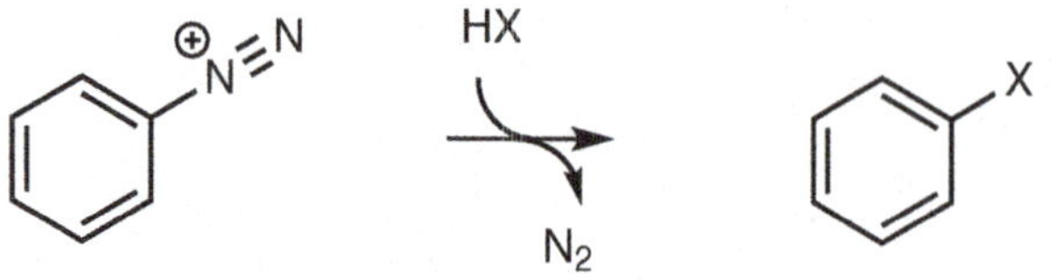

X = I, F, or OH

15 — Carbene and Nitrene Reactions

Carbene Reactions

1. Simmons-Smith carbene addition to cyclohexene to give a bicyclo[4.1.0]heptane. (See *Notes*.)

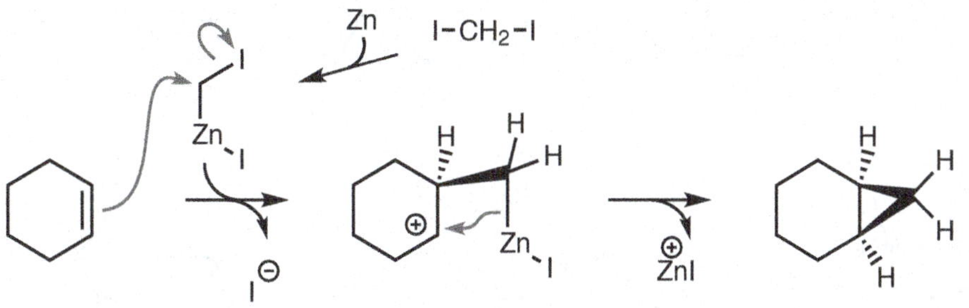

2. Dihalocarbene addition to (*E*)-1-phenylpropene to give a dibromocyclopropane.

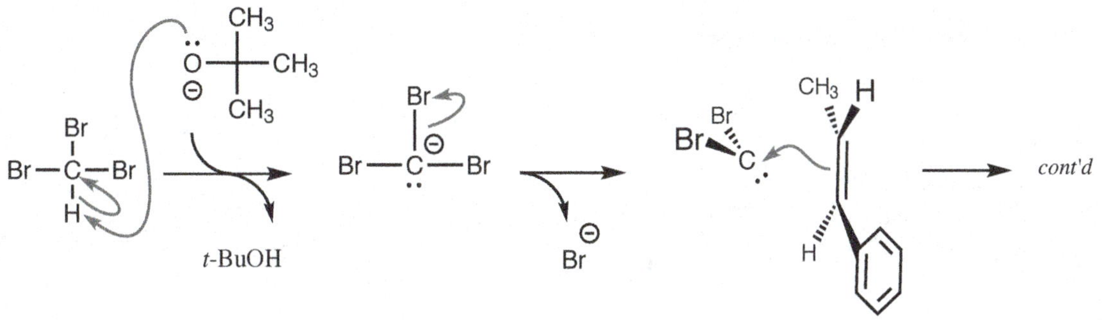

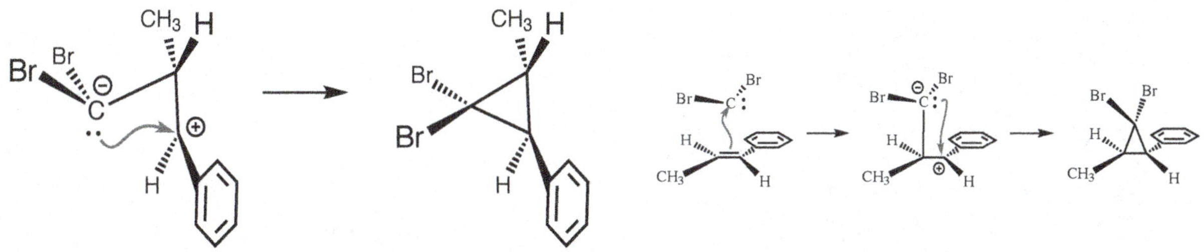

3. Dihalocarbene addition to cyclohexene to give 7,7-dichlorobicyclo[4.1.0]heptane.

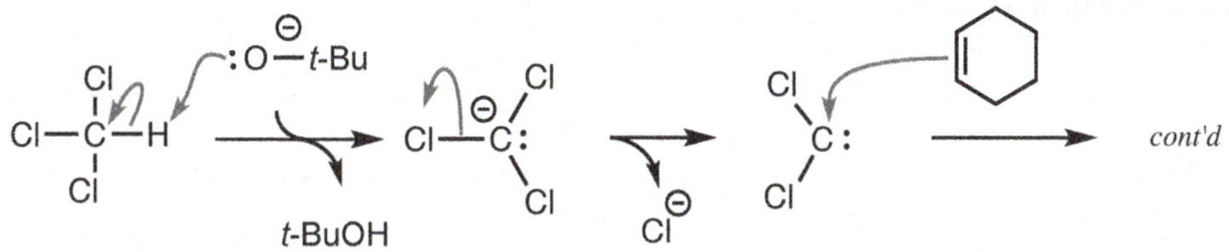

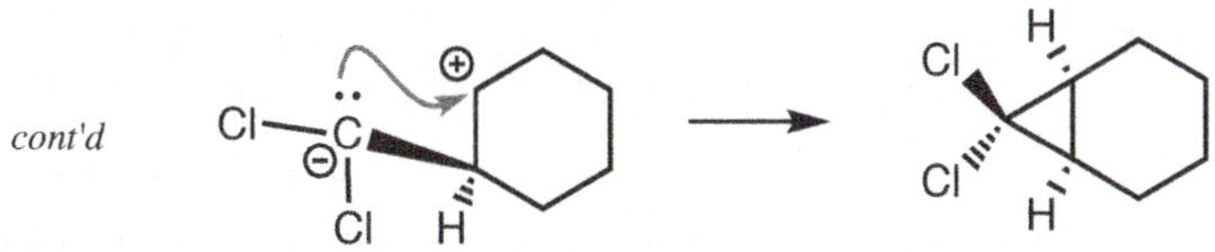

Curtius Rearrangement

4.   Step 1, Curtius rearrangement, reaction of acid chloride with azide and rearrangement via a nitrene intermediate.

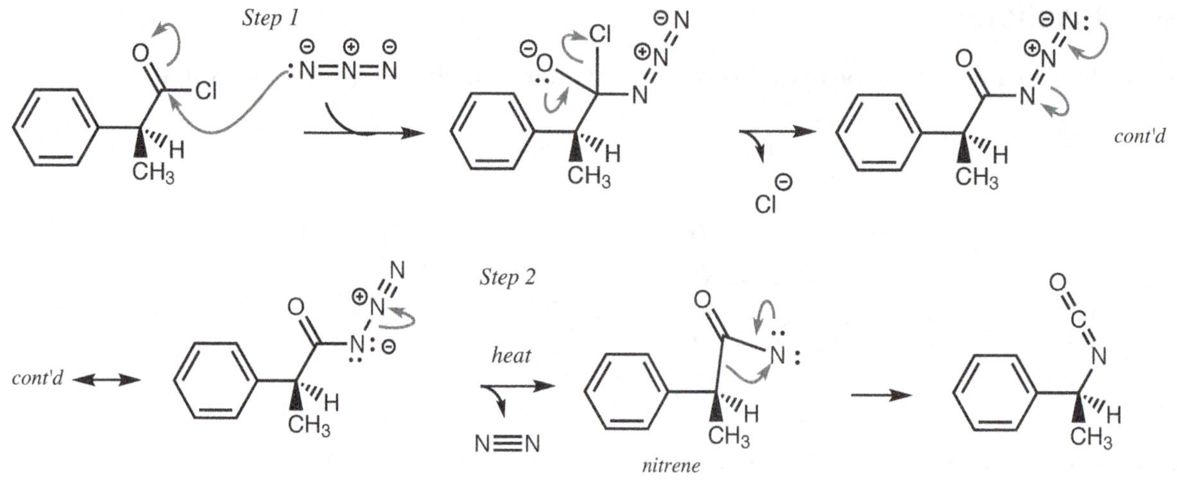

Step 2, hydrolysis of isocyanate to an amine.

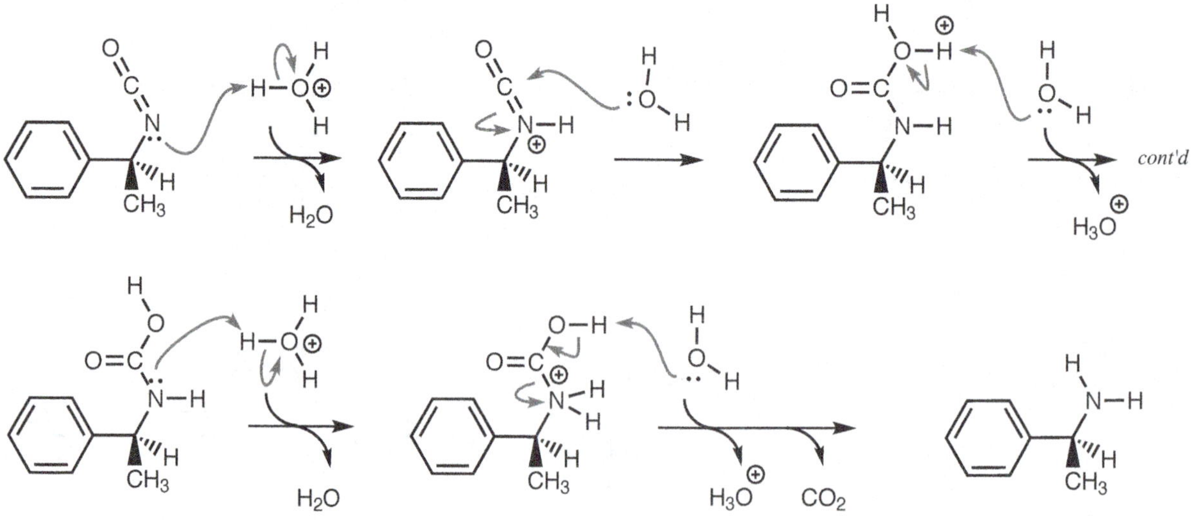

must be isolated from its ammonium salt

Alternate Step 2, hydrolysis of isocyanate to a carbamate. A carbamate is a nitrogen analog of a carbonate ester. The steps are similar to the above reaction with an alcohol replacing the water.

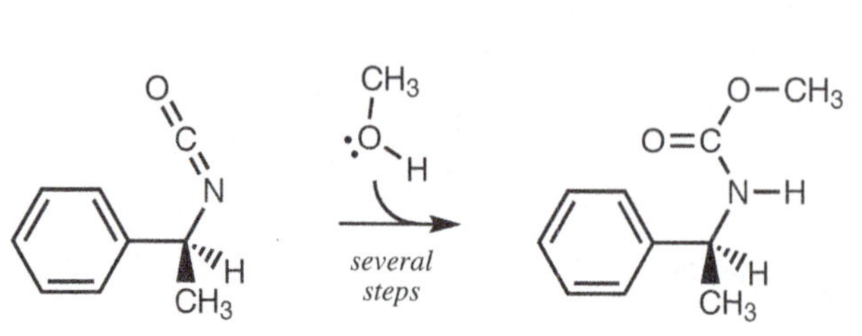

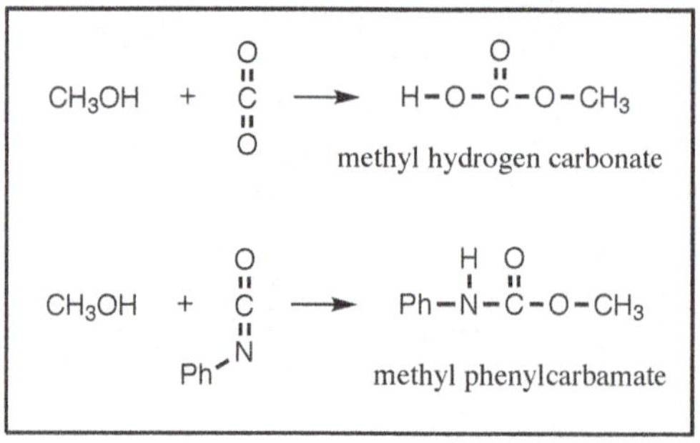

Hoffmann Rearrangement

5. Step 1, Hoffmann rearrangement of a primary amide to give the corresponding amine. The reaction of the amide with sodium hydroxide and bromine.

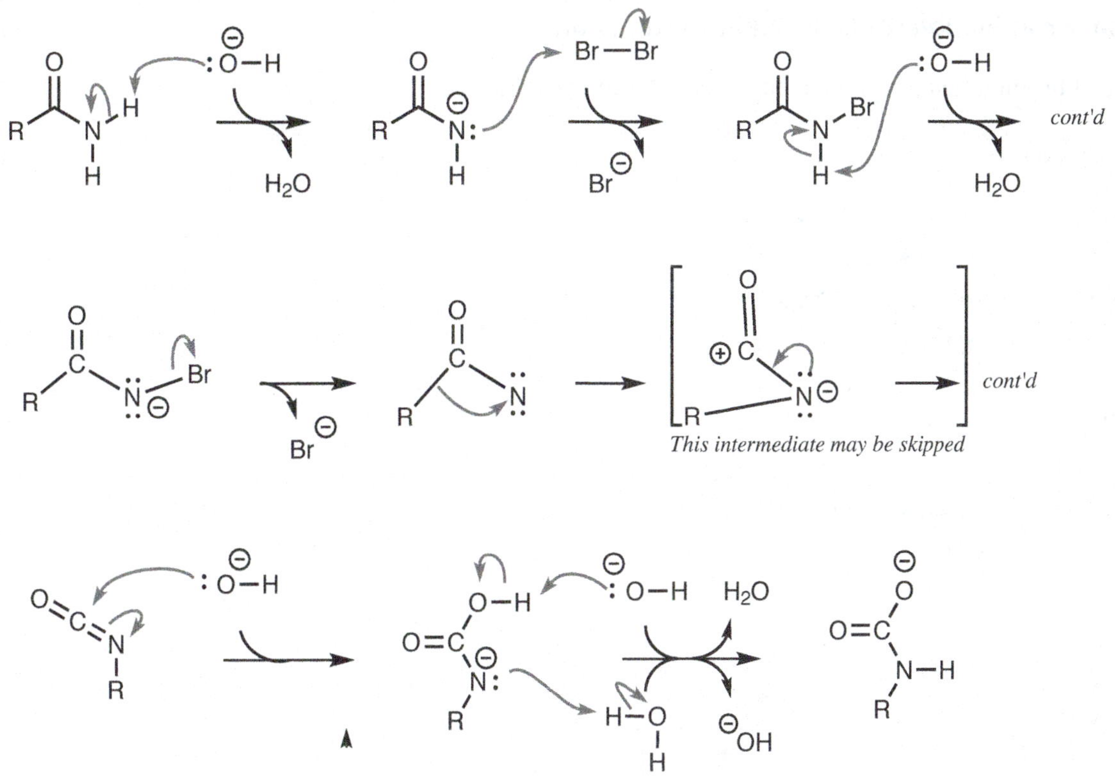
cont'd
cont'd
This intermediate may be skipped

Step 2, Acidification and decarboxylation to give the amine.

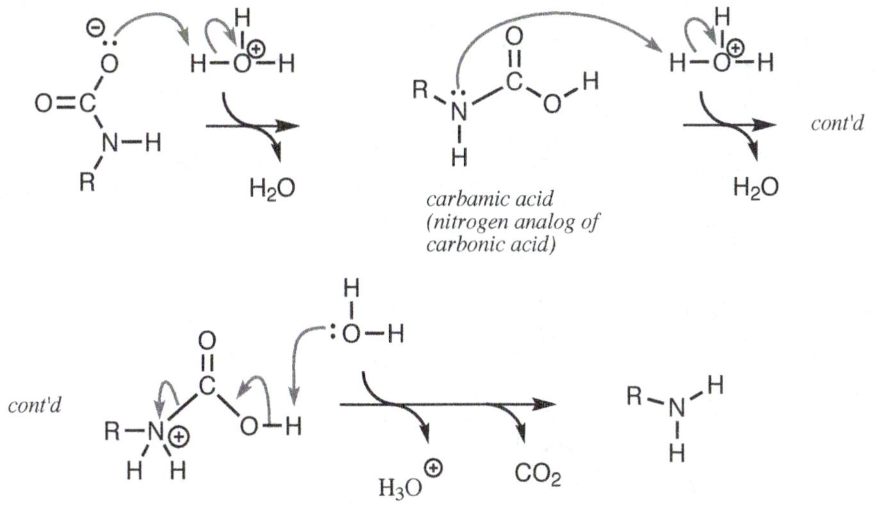
carbamic acid (nitrogen analog of carbonic acid)
cont'd
cont'd

16 — Radical Reactions

For a discussion of radical reactions, see *Notes*.

Free Radical Bromination (or Chlorination) Reaction

1. Free radical bromination of cyclohexane to give bromocyclohexane.

Overall reaction

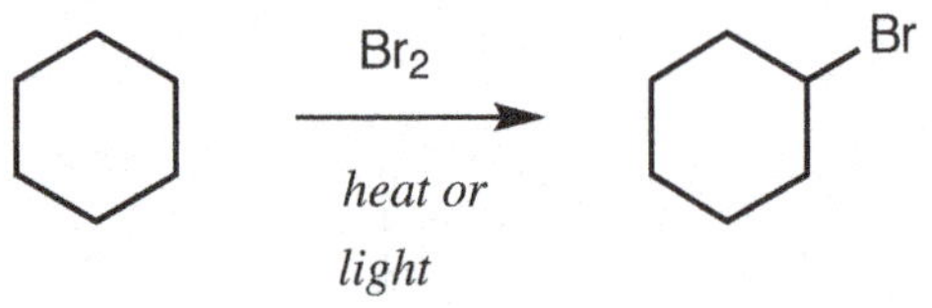

Initiation

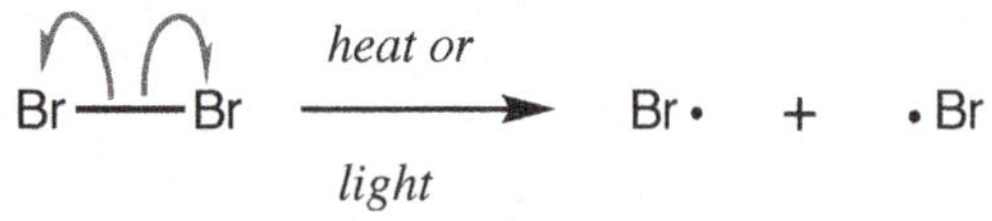

Propagation

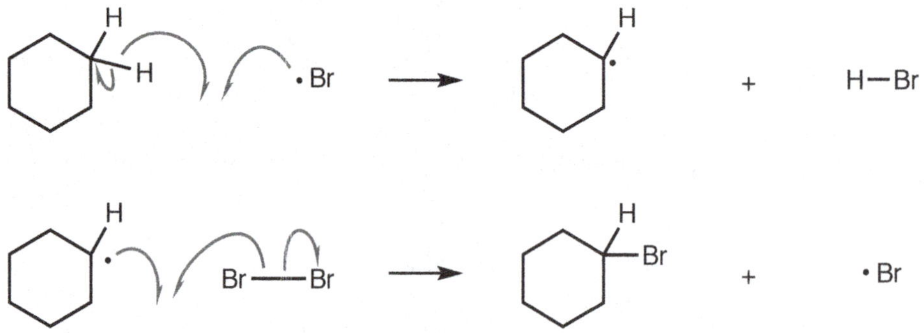

Termination

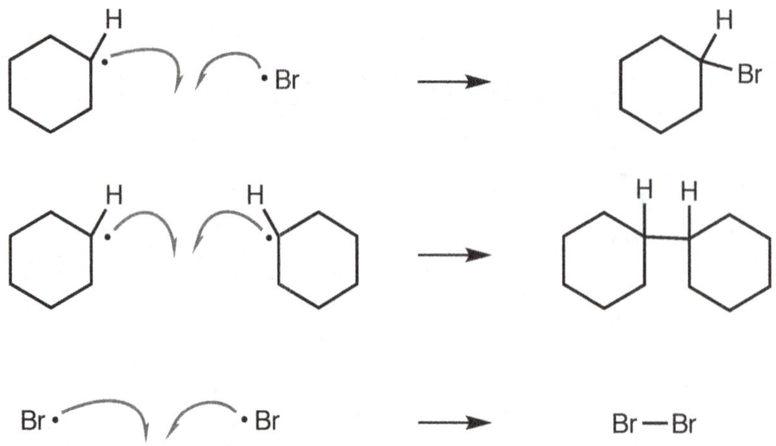

Allylic Bromination with NBS

2. Free radical bromination of cyclohexene with *N*-bromosuccinimide, an allylic bromination.

Overall reaction

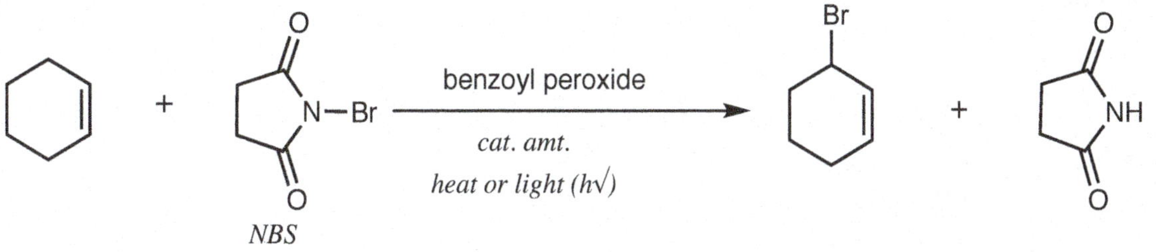

NBS

Initiation (See *Notes*.)

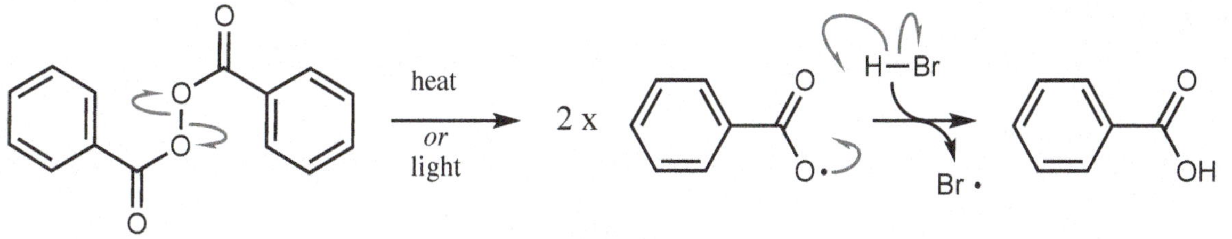

Propagation

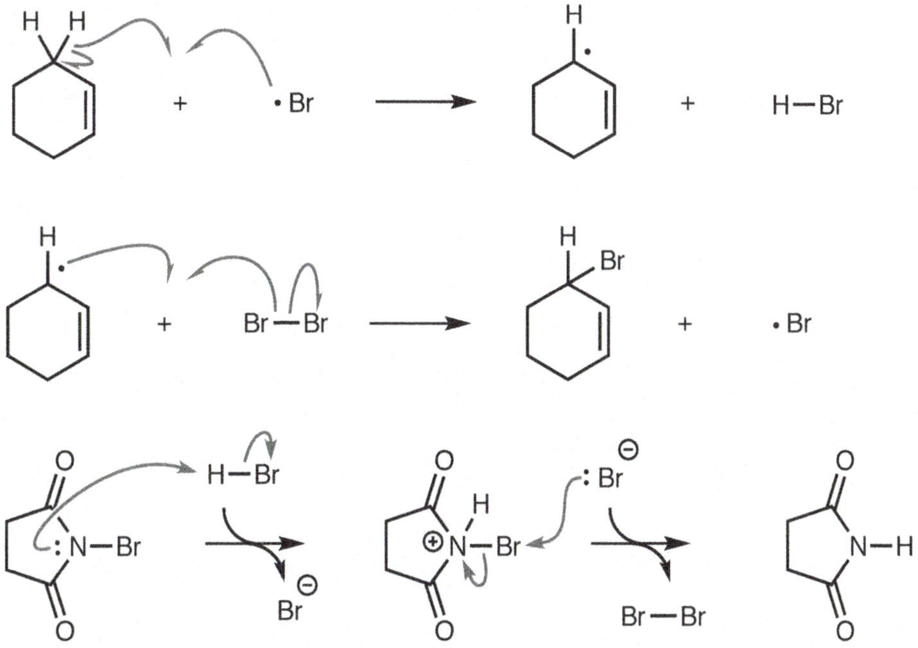

Termination

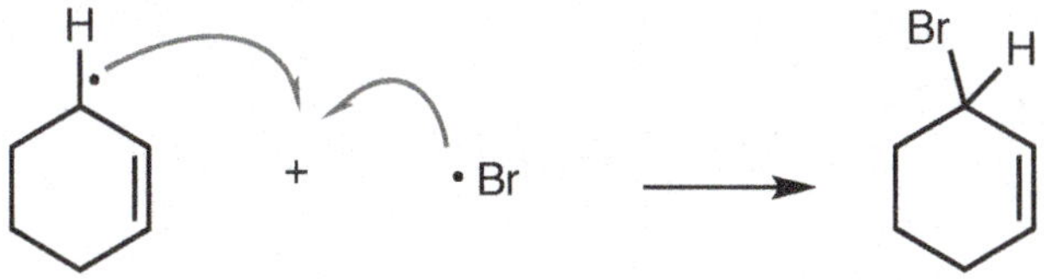

+ *others*

Radical Addition of Hydrogen Bromide

3. Free radical addition of HBr/H_2O_2 to propene to give 1-bromopropane via an *anti*-Markovnikov addition of hydrogen bromide. (See *Notes*.)

Overall reaction

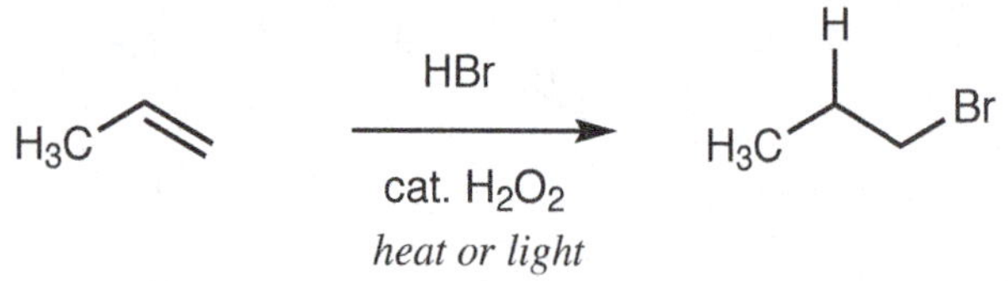

Initiation (See *Notes*.)

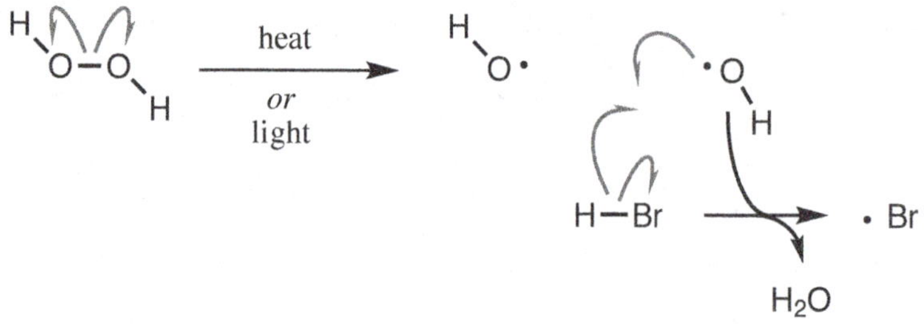

Propagation

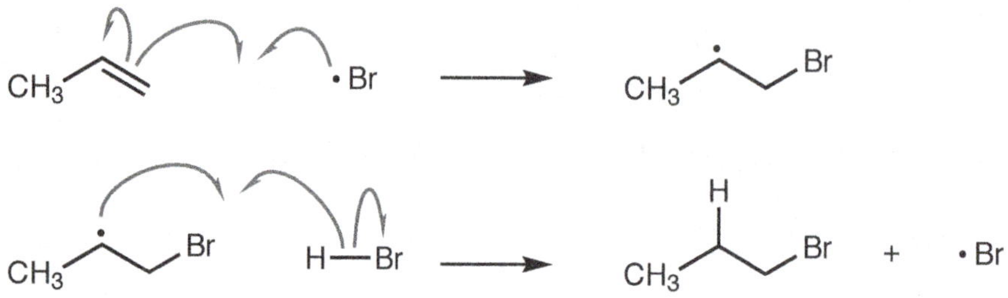

Termination

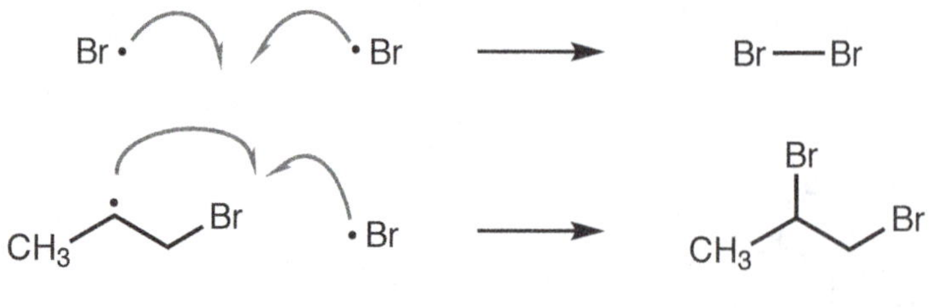

others

Benzylic Bromination with NBS

4. Benzylic bromination of ethylbenzene with NBS to give 1-bromo-1-phenylethane.

Overall reaction

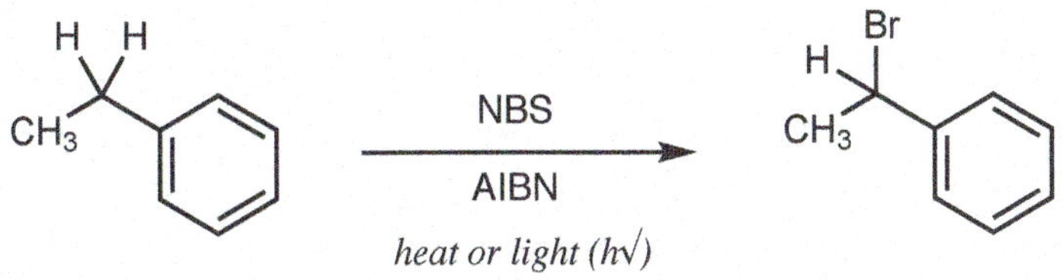

Initiation with azobisisobutyronitrile (AIBN)

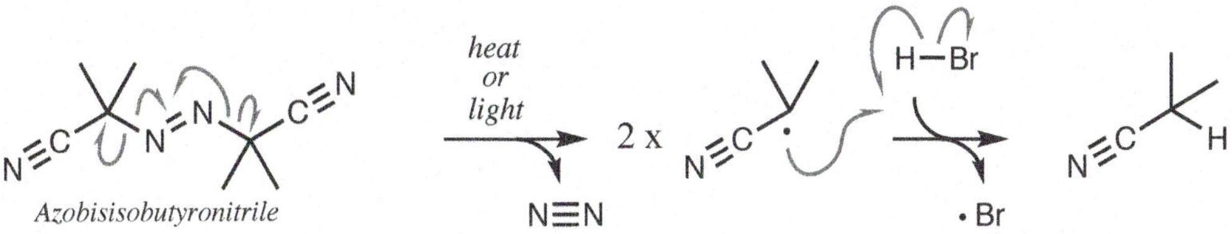

Azobisisobutyronitrile

Propagation

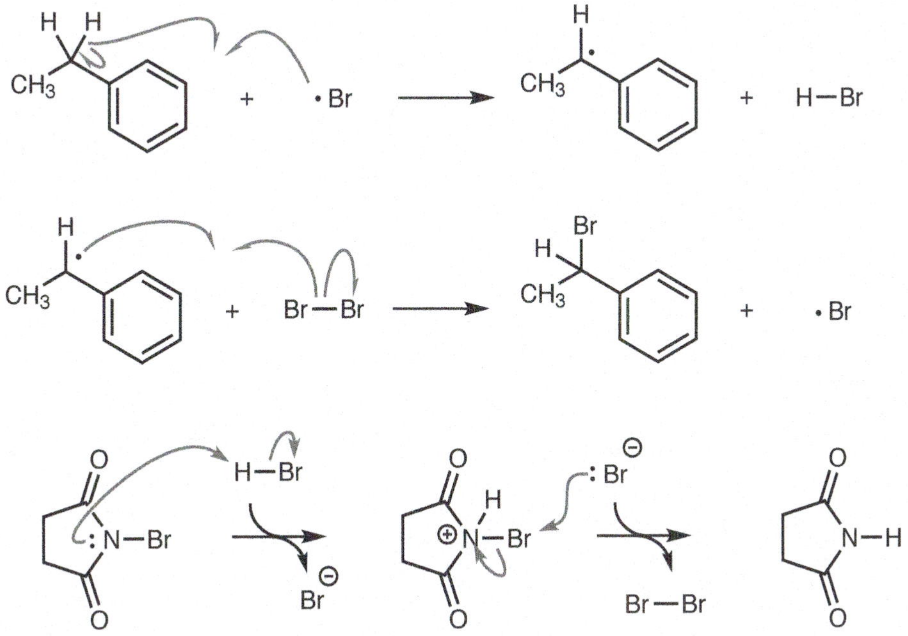

Termination

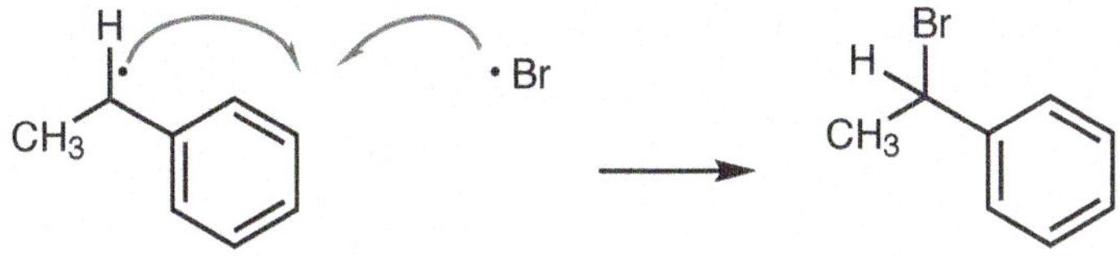

+ *others*

A challenge to maintaining the utility of this book as a guide is to maintain the organization of the different parts. While it would be useful to have notes in different parts, doing so makes it very difficult to maintain the overall organization of the book. It is simpler to keep notes to a minimum within each part and to place notes elsewhere. As a result, additional comments have been added in the following *Notes* section.

Chapter 1   Getting Ready for Reactions

About the Atom

A commonly held view is that, 'There is a strong correlation between the length of a covalent bond (i.e., the distance between the bonding atoms) and the strength of the bond.' This notion is consistent with Coulomb's Law.

An alternate model may be drawn in which bonds are made up of electron pairs that are mutually attracted to pairs of positively charged nuclei. This model is represented in *An Atomic Model for a Hydrogen-Fluorine* $F = \dfrac{kq_1q_2}{r^2}$

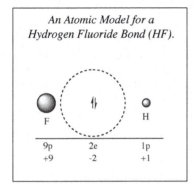

An Atomic Model for a Hydrogen Fluoride Bond (HF).

| 9p | 2e | 1p |
| +9 | -2 | +1 |

Bond on the left. This model is in agreement with the acidities for the atoms C, N, O, and F. It can be used explain why a proton, connected to a triple bond which is shorter, can be more acidic than a proton attached to a double or single bond. Because the C-H bond lengths are the shortest in a terminal triple bond, the electrons are closest to the carbon nucleus and the electron-proton distance is larger. With that model, the key variable for acidity (bond strength) is the proton-electron pair distance. Since the force varies inversely with the square of the distance, a greater distance will result in a weaker bond. For acidity, that longer proton-electron pair distance results in greater acidity.

How Neighboring Atoms Affect Acidity

In the following acid-base reaction, the proton acidity is affected by the groups that are attached to the neighboring atom. The substituent effects are noted in the table. Thus, the electron withdrawing properties can be passed through several atoms, though with decreasing magnitude by distance.

pK$_a$ 50 **A**	*No electron withdrawing effects are present. High pK$_a$*
pK$_a$ 44 **B**	*Two hydrogens have been replaced by an sp^2-carbon. It will withdraw electrons and increase proton acidity.*
pK$_a$ 17 **C**	*Carbon with two hydrogens have been replaced by an sp^2-oxygen. Because oxygen is more electron withdrawing, it markedly increases the acidity.*
pK$_a$ 22 **D**	*The neighboring hydrogen has been replaced by a more electron donating CH$_3$-group and the acidity decreases.*

Although we haven't discussed resonance structures to this point, you will find that the principle of resonance has a great effect on atom acidity. The examples **B**, **C**, and **D** can also be discussed as resonance effects, *see Resonance Structures*. The effect of resonance on a structure is that the electrons can be shared with a neighboring atom.

Resonance Structures

With resonance structures, a neighboring atom can donate electrons or pull electrons away. Examples 14 and 15 (page 4), illustrate electron donation and Examples 16, 17, and 26 (pages 4 & 6) illustrate electrons being pulled to the more electron withdrawing atom. When the neighboring atom donates electrons, it will release the electrons on the atom to which they are being donated. As a result, the electrons of the receiver atom will be held less tightly and can react more easily, see Example 17, Step 2, page 60. Conversely, if the electrons are pulled away, the electrons of the donor atom will now be held more tightly.

Ambident Anions

If the resonance structures have a negative charge that can be distributed over two different atoms, such as a carbon and an oxygen atom, the resonance structure with the negative charge on the more electron withdrawing atom will be the greater resonance contributor, **D**. **D** and **D'** are referred to as ambident anions. If this ambident were to react with bromomethane, we should ask which electrons might extend further from the nucleus, those of the carbon (in **D'**) or the oxygen (in **D**)? Although **D** may contribute more to the stability of this ambident anion, the pair of electrons are shared between a carbon and oxygen atom, the carbon-electron pair distance will be greater than the oxygen-electron pair distance. Therefore, a reaction of the (ambident) anions **D** and **D'** will be more likely to occur on the carbon atom in **D'** because those electrons extend further from the nucleus. If a pair of electrons can exist on two different atoms*, the less electron withdrawing atom (C>N>O) will be the more likely site of reaction. Therefore, the product of this reaction will be **F**.

Amide Regiochemistry

That analysis works for most cases. Consider reactions of a neutral amide and an amide anion. The electrons of a neutral amide react on the oxygen atom. The resonance structures are consistent with this because an oxygen atom depletes the non-bonded electrons of the nitrogen causing them to be held more tightly. This electronic effect may be justified as well by the effects seen in the IR and NMR spectra. If an amide anion is reacted, the oxygen is unable to deplete the non-bonded electrons of the nitrogen and that remains the preferred reactive site. However, this analysis does not work if the atoms are in different rows, for example see Chapter 3, Example 2.

Alkene Stability

Finally, electron donation and electronegativity can contribute to stability. We can anticipate that 2-butene should be more stable than 1-butene because 2-butene has two electron donating carbons attached to two electron withdrawing sp^2 carbons. However, 1-butene has one electron donating carbon and three electron withdrawing hydrogen atoms attached to to the sp^2 carbons. Thus, the increased stability of alkenes with increased substitution is understandable.

1-butene   2-butene

Atomic Charge

Formal charges are very effective in maintaining electronic bookkeeping, i.e. balancing the charges in starting materials and products. However, there is a tendency to think a charge confers a property on an atom containing it that may not be present. In the table in Part C, page 1, my objective was to show that structures with the same numbers of protons and electrons could have different properties. The properties of each structure was affected by the numbers of external and nuclear protons. This principle was used to show how the protons could effect the acidity.

- For example, dissolving hydrogen chloride in water gives hydrochloric acid, a solution of hydronium ions and chloride ions. Despite the opposite charges, this reaction with water shifts to the right and shows the electrons of water can form a stronger bond with a proton/hydrogen atom than a chloride ion can. The negative charge on the chloride ion tells us the electrons surrounding the chloride are in excess. However, because the electrons are in excess, we should not think that that they should be more basic than those of water. We should only expect the electrons of chloride ion are not held as tightly as the electrons of hydrogen chloride.

---

* This pattern seems to work if all atoms of the ambident anion are in the same row (carbon, nitrogen and oxygen). However, if an oxygen were replaced with the more electronegative sulfur, the balance now favors reaction on sulfur. I presume that even though sulfur can donate electrons to another atom, because of the larger number of protons in its nucleus, it holds those electrons more tightly and reduces the electron density (and distance) on the other resonance atoms.

We should also address how a formal charge applies to a structure. I have drawn the Lewis structures for hydronium ion, water, and hydroxide ion below. These structures all contain ten electrons and the oxygen nucleus has eight protons. The formal charges of the oxygen is positive, none, and negative while the true charge of the nucleus is unchanged for all three forms. The formal charge can tell us something about how the electrons might be affected. I have emphasized that protonation (and other nuclei) can change the distance that electrons can be found from a nucleus, but the charge of an electron and proton are unchanged. Therefore, the effect of protonation is to decrease the distance of the non-bonded electrons to the nucleus and thus their availability.

hydronium ion   water   hydroxide ion

- Let's look at hydronium ion. When hydrogen chloride gas is dissolved in water, the name hydrochloric acid tells us more about how the acid was prepared than the physical structure of the resulting solution. Should it be represented by an HCl molecule or a hydronium ion? We know that hydrogen bonding is important property of water. I have drawn a hydronium molecule next to two water molecules. We should expect that if hydrogen bonding prevails in water, it will also be present for hydronium ions. This example shows some of the difficulty chemists have in representing molecules in a consistent and compact form. Hydrogen bonding of hydronium ion shows the positive charge can be distributed over several atoms.

- Another common error is to create a new bond to a nitrogen or oxygen atom that has a positive charge. Despite the formal charge on an oxygen or nitrogen atom, both contain completed octets and therefore cannot accept more electrons. This should be an example of an acid-base reaction.

Summary about Atoms

A lesson I learned from how atoms make chemical bonds is '*the more tightly held a pair of electrons becomes, the weaker will its bond to another nucleus become*'. A corollary to that is '*a loosely held pair of electrons can form a stronger bond or the easier it becomes to form a bond*'. The analogy for an electron pair's ability to attack another nucleus is, '*chemical reactivity is like a boxer. A boxer with a longer reach would have an easier time to hit the nucleus of a another atom*'.

Drawing Resonance Structures

p. 2   I have simplified the resonance structures. Previously, I used a variety of examples. However, I concluded organized examples may be more helpful to teach everyone how to draw resonance structures. It is important that everyone be successful in drawing the curved arrows. Resonance structures are also important as they are the precursors to writing reaction mechanisms.

In the reaction mechanisms and when resonance structures are encountered, I have generally written the greatest resonance contributor. You may wish to draw another resonance structure to help predict a product.

Chapter 2   Acid-Base Chemistry

Acid-Base Problems

p. 7   The problems have been laid out from Parts A, B, and C to increase in difficulty. I have also included writing sentences to represent the electron movement illustrated by the curved arrows. Understanding curved arrows is as important as understanding acid-base equilibria.

- Many times, students face problems in which they do not know how to proceed. 'Trial and error' is an undervalued and underused strategy to solve chemistry problems. Some problems illustrate how to arrive at the same answer if two possible options exist.

- The chemical properties that can be illustrated by the pK_a of a compound can also be used to predict its reaction. The method laid out here is a model for how new problems may be solved.

- While the underlying principle prefers the weakest base, you might also recognize the electrons of the weakest base will correspond with the least available electrons. Because bromide and iodide hold their electrons so tightly, they are more difficult to form bonds with.

Chapter 3   Substitution Reactions

You should consult with your textbook for a more thorough discussion of S_N1 and S_N2 reactions. Substitution reactions can be described in two extremes. In an S_N1 reaction, the leaving group must pull its electrons from the carbon in the rate-limiting step. No bond formation occurs until a carbocation exists long enough for the leaving group to diffuse away and the group having left does not influence any incoming nucleophile. If the RX carbon had been optically active, then the product of the reaction would be completely racemic.

On the other extreme is an S_N2 reaction in which bond formation precedes bond cleavage. The nucleophile must *push* into the carbon nucleus. If the RX carbon were optically active, the product would retain that optical activity but it will be of the opposite configuration due to an inversion that takes place.

In between these extremes will be the reactions for which you have to predict the products and sometimes there are ambiguities or exceptions. Therefore, I find that predicting the products can be challenging. However, there are some simple principles,

- An S_N1 reaction requires conditions that favor carbocation formation, namely a good leaving group preferably on a tertiary carbon, a polar solvent, <u>usually water</u>, and a weak nucleophiles (often uncharged). Tertiary halides can undergo S_N1 reactions while primary halides (unless allylic or benzylic) do not.

- S_N2 reactions require conditions in which a pair of electrons are able to attack the carbon attached to the leaving group. That attack must precede bond cleavage. In order to extend the reach of the electrons, anions and especially anions of weak acids are common. However, nucleophiles whose conjugate acid is greater than 11 increases the amount of elimination products. High nucleophile concentrations and polar aprotic solvents (DMF, DMSO) are common.

- The ease of nucleophilic substitution is: methyl halide>primary halide>secondary halide. Tertiary halides fail to react via S_N2 mechanisms. Secondary halides are more difficult to displace by an S_N2 reaction and the conditions of the reaction are more important in predicting the products.

- Any factors that interfere with S_N2 reactions will increase the amount of elimination. In Chapter 4, Examples 2, 3, and 5 result in elimination reactions. Elimination increases as steric hindrance to an S_N2 reaction increases.

- The solvent effect for S_N2 reactions is rationalized by the formation of reactivity reducing hydrogen bonds (or their absence). Polar aprotic solvents cannot hydrogen bond and therefore facilitate S_N2 reactions. This can explain why fluoride, the most basic halide, can react faster in a polar aprotic solvent, and iodide can react faster in protic solvents. It is presumed that the electrons of fluoride would contain hydrogen bonds and thus limit their availability. Iodide would be more immune to that effect.

For a given set of problems, students will not experience many difficulties in substitution reactions. Just focus on bond making and breaking, and the orientation of the nucleophiles and substrates. However, some nucleophiles may have two different sites to react (ambident anions). When resonance structures exist, only the greatest resonance contributor is usually shown. However, bonds may form to a charged atom of an unshown resonance isomer. It may be useful to write the resonance structure and consider why a reaction might occur on one atom versus another.

Chap. 3.1   This reaction is usually run in acetone because it is aided by the insolubility of NaCl. Sodium chloride precipitates and suppresses a reverse reaction.

Chap. 3.2   The displacement of a primary halide is subject to steric hindrance. As the nucleophile becomes larger, an increasing amount of elimination will accompany the substitution reaction, See Chapter 4.3.

Chap. 3.3   The example shows a variant of the principle discussed on *Ambident Anions*, p. 2-*Notes*. I rationalize this exception as sulfur, with it larger number of protons, as being able to deplete the higher electron density on nitrogen in the second resonance structure.

Chap. 3.4   If hydroxide or an alkoxide is used as the nucleophile, elimination would be the major product. See Chapter 4. Because the bromide is secondary, some elimination product is still found.

Chap. 3.5   This example shows S_N2 reactions are not limited to basic conditions. If HCl is used, the reaction is not as efficient, therefore $ZnCl_2$ is used as a catalyst to accelerate the loss of water. This comprises the Lucas test for primary, secondary, and tertiary alcohols. The mechanism will also change in going to a tertiary alcohol, see Example 22.

Chap. 3.6   The electrons of the cyanide anion in Examples 6, 8, and 9 could react with the those of either the nitrogen or carbon. I argue that it isn't the negative charge that we write on the carbon that determines its reactivity, but rather the distance the electrons extend. As it has fewer nuclear protons, the reaction occurs on carbon. Also note that while the triflate can be displaced, the bromide cannot. The bromide is attached to an sp^2 carbon. *Direct substitution reactions cannot take place on sp^2 carbons.* Later we will see the mechanism for how a halide attached to sp^2 carbons can be displaced via an addition-elimination reaction.

Chap. 3.8   Note, one equivalent signifies the mole to mole ratio. While a di-nitrile could be produced with excess cyanide, in this example, the axial tosylate reacts faster and leads to the major product. Axial substitution being faster than equatorial is a general trend.

Chap. 3.9   Note, one equivalent signifies the mole to mole ratio. A bromide can react faster than a chloride and leads to the major product.

Chap. 3.10   In Example 10, the reaction of an acetylide anion is often the first anion used in carbon-carbon bond formation. It was illustrated with methyllithium, a commonly used commercially available base for this reaction. However, depending on your class and textbook, you may find other bases are used. Note the same reaction with sodium amide (left) or lithium diisopropylamide (LDA, right) for the same reaction. Sodium amide (sodamide) is an older base that is less frequently used in the modern organic chemistry laboratory than LDA.

Chap. 3.11   The enolate of methyl phenylacetate is an example of an ambident anion, p. 2-*Notes*. Draw the resonance structure and answer, "Which electrons of the anion extend further from the nucleus, carbon or oxygen?" Also, note the order of the atoms change. All atoms must be legitimate Lewis structures.

Chap. 3.12   You may find potassium hydroxide reported as base for a Gabriel amine alkylation reaction. Because the reaction product is very sensitive to cleavage with hydroxide, I prefer potassium carbonate or sodium hydride as an alternate. Either can be used in situ. The greatest resonance contributor is shown, but you should apply the principle reported in ambident anions, p. 2-*Notes*.

              This is the first step (alkylation) of a Gabriel amine synthesis. An hydrolysis can be carried out with base, Chapter 8. However, one of the best methods is to hydrolyze one of the imide bonds with base and then use an acidic hydrolysis of the resulting amide, see Chapter 8.

Chap. 3.13   In this Williamson ether alkylation reaction, a carbon alkylation could take place as argued in ambident anions, *p.*2-Notes, however, doing so would lose the energy gained from aromatization. The lower energy path preserves aromaticity by *O*-alkylation.

Chap. 3.15   The S_N2 reaction only cleaves the CH_3-O bond but not the aryl C-O bond. The aryl C-O bond is sp² hybridized and thus cannot undergo an S_N2 displacement reaction.

Chap. 3.17   In Example 17, opening of the epoxide ring occurs at the least hindered carbon. This is a push reaction with bond formation preceding bond cleavage. There is relatively little bond polarization of a carbon-oxygen bond and oxygen is not a good leaving group. The nucleophile must begin bond formation in order to succeed in breaking the carbon-oxygen bond. The reaction conditions (strong nucleophile) are consistent with an S_N2 reaction and sensitive to steric hindrance.

Chap. 3.18   Example 18 is a good example of how an epoxide can react with a very weak nucleophile, an uncharged methanol. Epoxides aren't as reactive as primary halides so prior acidification will pull the electrons toward the oxygen to make it a better leaving group. This reaction blurs the line between S_N1 and S_N2 reactions. The reaction does not occur on the least substituted carbon, which would typify an S_N2 reaction, yet the reaction occurs with inversion of the carbon, which is not characteristic of an S_N1 reaction. The intramolecular nature of the oxygen atom leaving group allows the reaction dynamics to change as a result.

              The reaction consumes and regenerates acid, since it isn't consumed only a catalytic amount is required, but the amount can affect the rate of reaction by increasing the concentrations of protonated species.

Chap. 3.19   Example 19 contains another ambident anion, and the epoxide chirality center is preserved in the product.

Chap. 3.20   In Example 20, solvolysis means the solvent causes the bond to break. In this example iodide is an excellent leaving group and gives a *t*-butyl carbocation intermediate. If iodide is replaced with poorer leaving groups, more alkene by-product will accompany the substitution reaction.

Chap. 3.22   Example 22 shows how reaction conditions control product formation. Examples 20 and 22 have tertiary carbocation intermediates yet different products.

Chap. 3.23   In this solvolysis reaction, the additional chirality center results in a (*3R*)-diastereomeric mixture .

Chap. 3.24   The products of S_N1 reactions can typically contain rearranged products. It can be difficult to predict the amount of rearrangement that might occur. In this example, the rearranged intermediate is more stable than the unrearranged intermediate and is energetically favored. However, even a reaction as shown in Example 5, when done on 2-octanol is accompanied by 7% of 3-bromooctane.

              The kinetics of the reaction must be a slow solvolysis of the bromide which is consistent with an S_N1 reaction followed a fast rearrangement. The rearrangement must accompany bond cleavage or very soon thereafter as the reaction with water will also be a fast reaction. If the rearrangement were slow, then more unrearranged product would accompany the rearranged product.

Chap. 3.25   Example 25 begins similar to Example 22, except the tertiary carbocation is also allylic. The bromide can attack at either carbocation of the resonance structures (not shown). The fastest reaction will occur at the more positive carbon. However, that product can reform the carbocation (as in Example 20). The competing slower reaction on the primary carbocation leads to the more stable product, see *Alkene Stability*, p. 2-*Notes*.

Chap. 3.26   This reaction appears to be a paradoxical result to the same intermediate of Example 25. However, where Example 25 were conditions favoring the thermodynamic product, Example 26 favors the kinetic product. Sodium carbonate has been added to the reaction mixture to prevent protonation of the product alcohol and to reform the carbocation. Therefore, the product of this reaction is the kinetic product of the reaction.

If the isomeric bromide were solvolyzed under the same conditions, the same products result. The bromide starting materials would not solvolyze at the same rates however. The tertiary-allylic bromide below would solvolyze faster than the primary bromide, product of Example 25.

Chap. 3.28   If the chloride were to give a symmetrical allylic carbocation, it could react with a nucleophile from either side and from the front or back. Therefore, if the chloride were optically active, a single racemic product would result. In this instance, a mixture of products will result. The wavy bond indicates both stereoisomers are present.

Chapter 4   Elimination Reactions

You should consult your textbook for a more complete discussion of elimination reactions. We can draw three main variants of an elimination reaction, an E2, an E1, and an E1cb elimination. In practice, we may find it difficult to distinguish between the different mechanisms. The reaction kinetics require a base to cause the reaction to take place, yet products suggest a carbocation intermediate that results in the most stable alkene. On the other hand, we may find the products consistent with formation of a carbanion even though such a carbanion does not seem thermodynamically possible.

Concerted reaction. The base and leaving group participate equally in the elimination reaction.

Two step reaction. The rate of the reaction is solely dependent on the breaking of the bond to the leaving group. Possible rearrangment and loss of stereochemical control may occur.

A two step reaction in which the anion does not result in a concerted loss of the leaving group. Proton acidity is an important factor in determining product distribution.

In the examples, the fastest reactions will occur when the electrons are donated opposite to the leaving group. This is consistent with the antiperiplanar effects and *anti*-elimination reactions. However, *cis*-elimination reactions are also common, though generally at a slower reaction rate. Examples 18, 19, and 20 contain *cis*-elimination reactions.

Chap. 4.1   This example shows part of the challenge of predicting substitution and elimination reactions. If the base were ethoxide, the product would be substitution, see Chap. 3.2. Because *t*-butoxide is the base, the elimination reaction is favored because of the steric bulk and/or a decrease in solvent polarity. Note bullet point two of Chapter 3 Notes.

Chap. 4.2   In this example, several elimination factors are now combined. Since the halide is secondary, elimination is the major product. Secondly, the stereochemistry of the elimination is antiperiplanar (the proton and bromine are on opposite sides). Thirdly, according to the Curtin-Hammett principle, we might expect the product distribution to reflect the relative activation energy barriers for formation of the *cis*- and *trans*-2-butenes and not the ground state distributions.

Chap. 4.3   With a tertiary halide, the carbon-bromine bond is weaker, especially in ethanol. As a consequence, a growing positive charge on the tertiary carbon can attract neighboring electrons. The electrons most readily donated are those attached to the most substituted carbon. Loss of the proton associated with the migrating electrons gives the product that is consistent with formation of the most substituted double bond. These conditions lead to a Zaitsev product.

Chap. 4.4-7   These examples adhere to the role that conformation has in the elimination rate and regiochemistry of the product. In Example 4, a rotation must be made to get the hydrogen and bromide antiperiplanar. Elimination of HBr from that conformation leads to the *cis*-stilbene. In Examples 5 and 6, the conformation with an axial halide and hydrogen lead to the elimination product. In Example 6, a mixture of the alkenes can result and the Zaitsev product is the major product. In Example 7, a tertiary butyl group locks the conformation of the halide as axial and equatorial in each structure. The axial halide is eliminated 500x faster than the equatorial halide.

Chap. 4.8-12   These examples show the effect of base and hydrogen acidity on elimination reactions. As seen in Example 3, the preferred elimination gives the most substituted alkene. With *t*-butoxide base, the preferred elimination is from the least substituted carbon (most acidic). In Example 10, the hydrogen *alpha* to the carbonyl group is the most acidic and gives an enolate. The loss of chloride is a slower process than the deprotonation reaction. Examples 11 and 12 are carbon variants of increased *alpha* hydrogen acidity to an sp^2 carbon.

Chap. 4.13-16 These examples are all E1 eliminations. They proceed through a carbocation and the requirement for an antiperiplanar deprotonation step is lost. However, the elimination reaction does require a conformation allowing an overlap of the electrons. In Example 15, this results in a mixture of alkenes as the carbocation can rotate first. In Example 16, a rearrangement reaction takes place first to relieve ring strain from the cyclobutane ring. Deprotonation results in the most substituted alkene.

Chap. 4.17   Example 17 is a Hofmann elimination reaction. What is different compared to the prior examples is the leaving group is a strong base, by comparison. The pK_a of trimethylammonium ion is 9.76 while HCl is -7. I interpret this to mean the acidity of the proton being removed is a greater factor than that of a tertiary bromide, as Example 3. The product of the Hofmann reaction is the least substituted alkene.

Chap. 4.18   A Cope elimination is an example of a *cis*-elimination reaction. Other *cis*- or *syn*-elimination reactions that use a cyclic mechanism are acetate and xanthate pyrolysis reactions.

Chap. 4.19   This elimination is a good example of the use of selenium in organic chemistry. It is readily oxidized and the selenoxide is readily eliminated at moderate temperatures. Sulfur will undergo an analogous reaction, but requires a higher temperature.

Chap. 4.20-21 This sequence converts a double bond into a triple bond. The elimination is subject to the bromide configuration and conformation. The reaction could have started with the dibromide, but because the configuration is frequently the result of a bromination reaction, that is the starting material shown. Go to Chapter 5 for the bromination mechanism.

               An antiperiplanar elimination occurs to give bromoalkene intermediates. From a *trans*-alkene, the final elimination must be a *cis*-elimination. This reaction is less energetically favored and slower. The *cis*-alkene proceeds through a faster *trans*-elimination.

Chap 4.22-23   Examples 22 and 23 show the formation of a terminal acetylene. In Example 22, *t*-butoxide is the base. While it is not as strong of as an amide base, it does not abstract the terminal acetylene hydrogen and does not require an additional stoichiometric amount of base to be used. In Example 23, LDA is used as the base. It is a convenient modern base alternate to $NaNH_2$. In this case, the product is the acetylide anion. By working up the reaction with D_2O, the expected terminal hydrogen is replaced with deuterium.

Chapter 5   Electrophilic Addition to Alkenes and Alkynes

Chap. 5.3   The hydration of an alkene is an example that may be found in some organic chemistry textbooks, I wanted to explain why I do not use it. Although this reaction can be preformed, it is not a general reaction. The problem is the product is more basic than the starting alkene. As a result, the backwards reaction is favored. If water is not present, a sulfate ester can form instead. It is less basic and so blocks reverting to an alkene. However, the sulfate ester must be hydrolyzed and this would be an awkward reaction for common usage. In the laboratory, it would be easier to perform an oxymercuration-reduction (Example 16) or addition of acetic acid (Example 5) and hydrolyze the ester (Example 8.21).

Chap. 5.7,-9   Often, I am uncertain whether a rearrangement should take place. Example 7 is a frequently used example. However, the mechanism is more convincing that a rearrangement should take place than the actual conversion. Example 8 typifies a kind of rearrangement reaction I find in many texts and examples. Relief of ring strain is a good impetus for rearrangement. I usually assume that if a carbocation can form adjacent to a strained ring, then a rearrangement will occur. I always suggest to students to anticipate a rearrangement reaction with relief of ring strain. Finally, the rearrangement in Example 9 probably occurs to the greatest extent. The aromatic ring can donate electrons to stabilize an adjacent carbocation. After rearrangement, you can draw resonance structures utilizing the aromatic ring.

Chap. 5.10&11 Examples 10 and 11 show different products can occur with dienes (and mixtures are common). Dienes are good examples for producing different thermodynamic and kinetic products. In most reactions, the kinetic and thermodynamic products are the same, but with dienes, a kinetic product can be isolated that can react to give a more stable thermodynamic product.

If the activation energy of a reaction is low, the reaction will be fast. However, it may not be the most stable product. With dienes, that product may continue to react under those conditions to give a more stable product. How would you know that a kinetic or thermodynamic product should be result? A signal that a kinetic product might be expected is if the reaction is carried out at a low temperature.

In Examples 10 and 11, a tertiary versus a primary carbocation more strongly favors the formation of a product from the tertiary carbocation. However, that product is the less stable kinetic product. The tertiary and allylic halide can reverse and react to give the more stable thermodynamic product, the most substituted alkene. If the reaction can be conducted at low temperatures, the kinetic products may be isolated.

You shouldn't assume that the addition will occur in a 1,4-manner. In this example, the 1,2-adduct is the kinetic and thermodynamic product. You should assess the carbocation and alkene stability. The more stable the carbocation, the less stable will its product be. The more substituted the alkene, the more stable it will be.

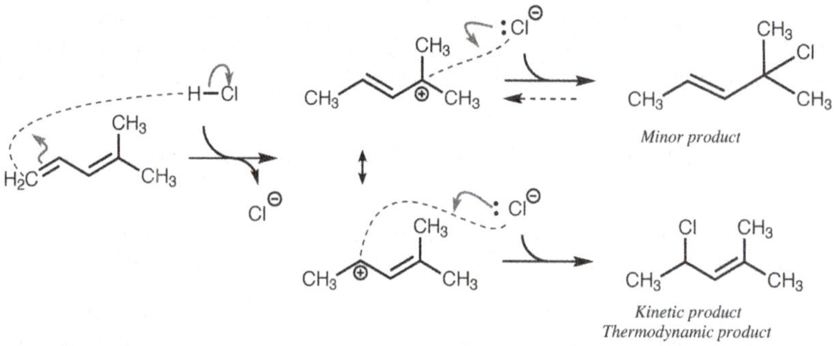

Minor product

Kinetic product
Thermodynamic product

The differential is less with a III° vs a II° carbocation. Therefore more charge will be found on the II° carbocation and the barrier to reaction at that carbon will be less. It also leads to the most stable alkene. Therefore, 1,2-addition leads to the kinetic and thermodynamic product.

Bromination

Chap. 5.12   Many textbooks show bromination as a concerted reaction with three bonds forming and breaking. You may also write a concerted mechanism, but I prefer a stepwise mechanism. In a concerted reaction, the alkene gains and looses electrons equally and the bromine must act as a nucleophile in that it donates more electrons than it accepts in becoming positively charged. We know electron-rich alkenes react faster than electron poor ones. This reactivity is more easily understood in a stepwise mechanism in which the alkene acts as the nucleophile and bromine the electrophile.

In the bromination of cyclohexene, it doesn't matter at which carbon the bromide enters to open the bromonium ion, as it is symmetrical. It only matters that the entering bromide is on the opposite side of the bond being broken in an S$_N$2-like reaction. If an anchoring *t*-butyl group were present, then preference for opening the bromonium ion to give the diaxial product is observed. The entering and leaving electrons can be aligned in an S$_N$2 like reaction as shown below. Note how the bromine is close to the entering trajectory while they are quite distant for a diequatorial product.

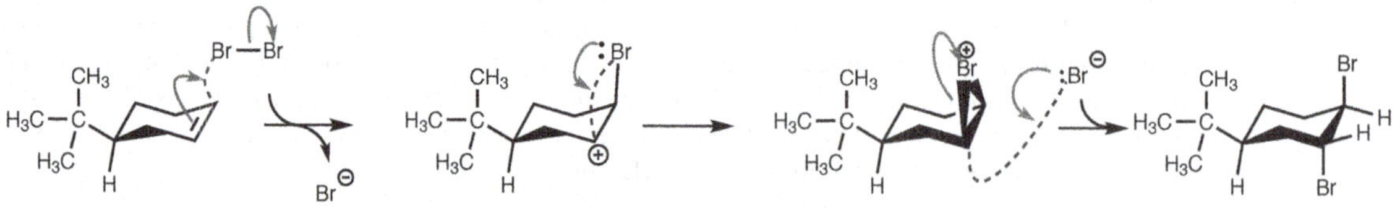

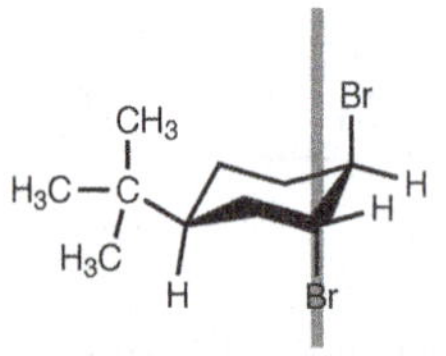

The light line represents the axis of an S$_N$2 reaction having taken place. It is easier to imagine the departing Br being on the axis on the left than on the right. On the left, the bromine electrons are aligned with that axis, but the angle on the right is not.

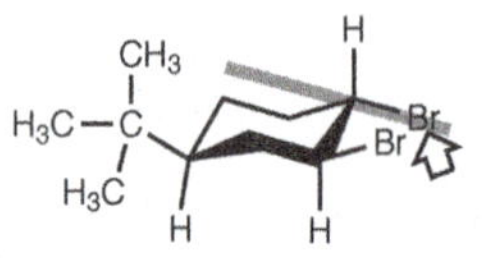

Chap. 5.14   In example 14, the reaction is stereo and regiospecific. In this case, the nucleophile (water) attacks the most substituted carbon and results in inversion. The stereochemical consequences of this reaction are the same as an S$_N$2 reaction. However, as a general rule, S$_N$2 reactions do not take place on tertiary carbons. How is this different? Generally, in an S$_N$2 reaction, the process of bond formation precedes bond cleavage. In this instance, you can think that unlike the S$_N$2 reaction, in which little bond cleavage has occurred to expose the nucleus to attack, the bond may be considered highly polarized and thus the carbon contains considerable positive charge. Therefore, the reaction is more S$_N$1 like than S$_N$2 like. The atom best able to support that charge is now the site of an S$_N$2-like reaction.

Alternately, you may think of it as though a three membered ring were not present (though **incorrect**). The reaction is then like an S$_N$1 reaction in which the neighboring atom and its non-bonded electrons blocks one face of the carbocation and results in attack from the opposite side.

The opening of the bromonium ion will again be *trans*-diaxial, see Example 5.12. The electrons attack the carbon nucleus from the opposite side of the leaving electrons of the bromonium ion. You may draw a Newman projection of a cyclohexane to see the stereochemistry of this reaction.

A bromination reaction in water results in some confusion. Even though bromide has a negative charge, it isn't the strongest base. Note in the equilibrium below, the equilibrium lies to the right. Therefore, water is a stronger base. Since water is the stronger base, it should react faster with the carbocation than bromide io*n*.

pK$_a$ -9   pK$_a$ -1.7

Oxymercuration

Chap. 5.15&16 This reaction is a good analog to the bromination reaction. In the reaction of an electrophilic mercury atom, the alkene will be the nucleophile. We might forget that mercury has ten valence electrons. Therefore, after it reacts with an alkene, mercury has a number of electrons that are available to react with a neighboring positive charge, just as the bromine atom did.

In Example 15, a hydrogen atom is staged for migration to give a tertiary carbocation. If the non-bonded electrons of the mercury did not react with the carbocation, then a rearrangement might take place. The mercury atom, with its electrons, prevents the rearrangement reaction from taking place.

Reductive Demercuration

The mercury is removed by a borohydride reduction followed by loss of metallic mercury. The loss of mercury is reportedly a radical reaction. Two possibilities are written below. However, for expediency, I suggest you write the reaction as indicated in Part D unless directed otherwise by your instructor.

Hydroboration-Oxidation of Alkene

Chap. 5.17   In Example 17, note how similar step one of this reaction is to the prior reactions in this chapter. Boron is the electrophile and reacts with the electrons from the alkene. The second part of this reaction takes place before any atom movement can take place. The electrons from the negatively charged boron are donated to the carbocation. In this case, a proton is attached to the donated electrons.

I formerly wrote to repeat the first reaction 2X. While this is expedient, I found many students didn't understand what was happening with the repeat statement. If in learning the mechanism, when it becomes clear to you that it is the same reaction being repeated three times, you can just write the first hydroboration and then indicate that the hydroboration step is repeated 2X.

In step 2, boron is still an electrophile except the nucleophile is now a hydroperoxy anion. Note, it is the same reaction repeated three times. You can just write the first oxidation step and then indicate that the oxidation step is repeated 2X.

You may also use this alternate mechanism for the hydrolysis. I like the deprotonation to promote the loss of the alkoxide. However, the product of that elimination is an sp^2-hybridized boron with a negative charge. The next step requires that boron attract an additional pair of electrons. The product of that addition is an sp^3-hybridized boron with an alkoxide oxygen neighbor. While I did not use this mechanism, if your textbook or your instructor uses it, then you may use it also.

Stepwise versus Concerted Reactions

Many textbooks write the hydroboration step as a concerted reaction. I have not done so for two reasons. One is that this appears to be a symmetry forbidden 2 + 2 reaction. The symmetry rules can be used to explain why alkenes and HBr do not thermally add to an alkene in a concerted reaction. Because the addition of borane to an alkene is a syn addition, many books write it as a concerted reaction. This will lead to the reaction product in a single step. If a reaction were concerted, then charges should be minimized and differences in reaction rates due to charged intermediates should be small or absent. However, the reaction rates differ in predictable ways, e.g., electron rich alkenes will react faster than electron poor ones. By one definition of a concerted reaction, *any electron movements that occur faster than bond rotation are concerted*. Therefore a reaction may still be concerted and have additional steps provided they occurs faster than atom movements. If a reactant, when drawn as an electrophile or nucleophile, explains the reactivity, it often easier to realize that effect from a stepwise mechanism.

Addition to an Internal Acetylene

Chap. 5.18   The three membered ring immediate in Example 18 contains a kind of misrepresentation. The addition of HCl to the acetylene is an example of a three-centered two-electron bond. As written in Example 18, it implies two bonds to hydrogen. I prefer the representation shown below. If addition of a proton was on a pair of electrons, an intermediate such as on the left would result. The charge of the nuclei would be unchanged, but overall an imbalance would exist. Because the acetylene carbons are sp-hybridized, one less bond is available to donate electrons and stabilize a carbocation. As a result, you may think the electrons are still attached to the acetylene carbons. When the nucleophile approaches one of the carbons, it will only be as the addition occurs that bond cleavage results. I like this model to explain why addition of HCl gives a high amount of *trans*-product.

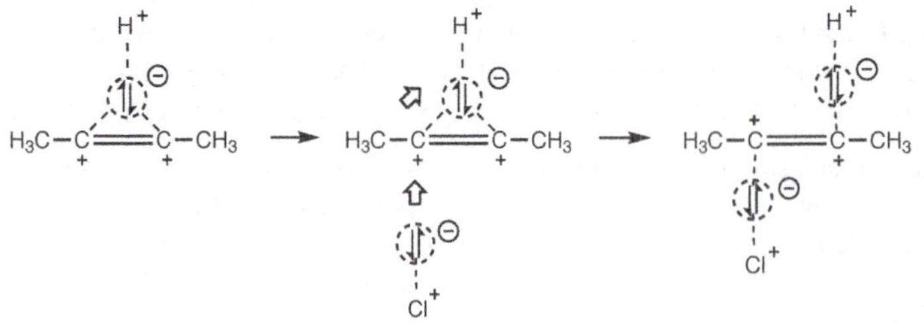

If we compare the above with addition of HCl to propene, the electron density about the carbocation is sufficient to allow the connection with the pair of electrons to break. This gives the most stable carbocation and is consistent with a Markovnikov addition of HCl. It can also be used to explain why addition to alkenes can also give high degrees of anti-addition.

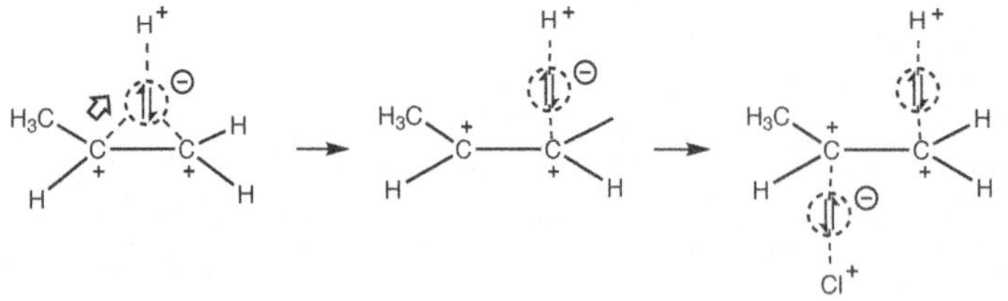

Chap. 5.19   In Example 19, a similar intermediate to Example 18 forms, although in this case, it doesn't matter if it opens *trans* as the stereochemistry is eventually lost. Sulfuric acid is a good acid to use because its conjugate base, bisulfate ion, is a weak nucleophile, its negative charge is distributed over several atoms, and does not compete for product formation. This reaction requires an additional carbon protonation step to give a ketone as the final product. In the last deprotonation step, which proton is more acidic, the one on the oxygen (to give the ketone) or the one on the carbon (to regenerate the enol)?

Addition to a Terminal Acetylene

Chap. 5.20   When two carbons are attached to an acetylene, they donate electrons to the acetylene, and protonation can occur directly . With a terminal acetylene, mercury is needed to convert it into a ketone. The oxymercuration of a terminal alkyne parallels the alkene reaction. However, the product would be an alkenylalcohol which can tautomerize. During the tautomerization, mercury gives up its electrons to reform mercuric ion and a ketone.

Disiamylborane Hydroboration–Oxidation of an Acetylene

Chap. 5.21   A hindered borane is necessary to avoid over reduction of the acetylene. The intermediate product is an alkenylalcohol (enol) which tautomerizes to an aldehyde.

Chapter 6   Rearrangement Reactions

Baeyer-Villiger Oxidation

Chap. 6-1-7   I have written two different types of rearrangement reactions, an acid catalyzed and a base catalyzed. Typically, peracetic acid is more acidic in nature as it is stabilized with sulfuric acid. On the other hand, many rearrangement reactions take place with a buffer to avoid acidic conditions. Many of those examples use trifluoroperacetic acid.

Chap. 6.2   I had previously written the rearrangement reaction of the phenyl group occurring concomitantly with loss of the electrons from the oxygen atom. Then I saw an example in which the phenyl migration utilized the electrons of the aromatic ring. I liked this version as it better shows why an electron-donating group would migrate because it requires a positive charge be placed upon the ring. In a concerted migration, the electron pushing from the oxygen does not make any demands of the phenyl ring.

The key to predicting the Baeyer-Villiger oxidation products is being able to predict which group of the tetrahedral intermediate (I) will migrate, here phenyl or methyl. The order of migration is hydrogen > tertiary alkyl > secondary alkyl > phenyl > primary alkyl > methyl. If there are two aromatic rings, the more electron rich ring migrates most quickly, but there are many exceptions based on the peracid used, reaction conditions, and stereochemistry. The migrating group can be described as the one best able to stabilize a carbocation. The electron movement is similar to elimination reactions with the moving electrons preferring an anti arrangement.

Concerted mechanism

Pinacol Rearrangement

Chap. 6.8   A pinacol rearrangement mechanism with one less intermediate can also be written. The protonated alcohol can be rearranged to the product. The non-bonded electrons of the adjacent oxygen facilitate migration of the methyl group and loss of water.

Chapter 7   Electrocyclic Reactions

Diels-Alder Reactions

Chap 7.1-15   The Diels-Alder reaction is simple, but errors are common. I suggest you label the atoms of each reactant, complete the bonds, label the product, and make sure the bonds correspond with the labels. For bicyclic products, I suggest breaking this into two operations. First, make the correct bonds for the product. Then, convert the two-dimensional structure into a correct three-dimensional structure. I can make an error if trying to do both at the same time.

Chap 7.12-15   The Diels Alder reaction is generally regarded as an electrocyclic reaction. However, this example could also be written as a completely ionic reaction. Doing so would achieve the reaction regiochemistry. It would require the bond forming steps to occur faster then atom rotation to avoid loss of stereochemistry. This example blurs the lines for concerted reactions but is not a stepwise ionic reaction.

Other Electrocyclic Reactions

Chap. 7.16-18   These reactions are often excluded from introductory texts. I have included them as they fit the general model for electrocyclic reactions and extend the model.

Chap. 7.17   In Example 17, it is an exaggeration that hydronium ion is necessary for the tautomerization to take place. The product phenol is a more likely acid that catalyzes the tautomerization. Drawing the phenol as the acid was awkward.

Chapter 8   Carbonyl Addition and Addition-Elimination Reactions

Addition to Carbonyl Groups

Through this chapter (and others), we must consider which intermediates might be present in an addition to a carbonyl group.

We must consider how the reaction conditions determine which intermediates might be present. If a reaction is acid catalyzed, then Route A will predominate, see Example 7. Alternately, if the reaction is basic, then a mechanism such as Route B will predominate, see Example 1. However, many reactions will be more difficult to determine which mechanism is operating. We must also consider Route C in which two neutral compounds react together to give ionic intermediates and which proton shifts might be occurring to complete the reaction. Finally, Route D should also be considered. This example is completely concerted with low charge gradients but high entropic demands.

I cannot predict which intermediate (or analog) might be present in any reaction. You also should not be surprised to find any of Routes A-D are suggested. What you should note is that these mechanisms are schemes for proton transfers. You should ask yourself which scheme is the most compatible with the reaction conditions. If a difference exists, a different intermediate, it will not affect the outcome of the reaction. A mechanism is an attempt to rationalize how a reaction might take place. The objective of that rationalization is to make the chemistry logical.

Example 13 is written consistent with Route C. Since HCl is released during the reaction, could the mechanism become acid catalyzed as in Route A? I similarly thought that since aniline is a weaker base than pyridine, it should be deprotonated first as this would increase the rate of elimination of pyridine in Example 17.

Route A

Route B

Route C

Route C1

Route C2

Route D

Grignard Reaction

Chap. 8.1   Example 1 is a Grignard reaction. Grignard reagents are formed from magnesium metal and an alkyl or aryl halide. The ease of their formation is I>Br>Cl. Magnesium is thought to donate electrons in a single electron transfer process. The product of this reaction is methyl magnesium iodide or methyl Grignard. Again, this reagent can react as though it were an anion. Grignard reagents are very reactive and must be protected from water. Anhydrous ether is the most common solvent.

Chap. 8.2   You may find organic compounds are written left to right as shown in Example 2. When they react, the atom to atom connectivity must be maintained. This is a good example where numbering, lettering or numbering and lettering the reactants and products is useful.

Alkyllithium Addition to a Carbonyl Group (Formation and Properties)

Chap. 8.3   Organolithium reagents are similar to Grignard reagents in their formation and reactions. Lithium has an electron that it donates to form a radical. Radicals are very reactive and thus reacts with a second equivalent of lithium metal to form the organolithium reagent.

Wittig Reaction and Horner-Wadsworth-Emmons Reaction

Chap. 8.5   The Wittig and Horner-Wadsworth-Emmons variation of the Wittig reaction are complementary reactions having different advantages. The details of the Wittig reaction are useful in understanding the stereochemistry of the product. After formation of the reagent (Step 1), the addition (of the ylide) to an aldehyde gives an intermediate in which the more stable conformation predominates. Note the conformation shown below. This has been drawn conventionally and as the corresponding Newman projection (see box). This conformation was not included in the reaction scheme. This conformation must undergo a bond rotation in order to bring the phosphorus and oxygen atoms together. That rotation gives the conformation that results in a *cis*-alkene. The Wittig reaction stereochemistry is determined by the fast addition to the carbonyl group.

Chap. 8.6   For the Horner-Wadsworth-Emmons variation of the Wittig reaction, a very similar set of structures can be written. The difference is the pK$_a$ of the Wittig reagent is lower and thus the addition reaction itself is reversible (see *Michael Addition Reaction*, p. 16-Notes). Because the reaction can reverse, the formation of a *cis*-alkene from the higher energy eclipsed conformation (the marked structure) is unfavorable. When the reaction reverses, the slower forming addition product forms which can adopt a lower energy eclipsed conformation necessary for the cyclization-elimination step to form the *trans*-alkene.

Ketal Formation and Hydrolysis

Chap. 8.7&8   The formation of a ketal and its hydrolysis have been written with single forward reaction arrows. However, these reactions are reversible. The formation of products are controlled by their equilibria (Le Chatelier's Principle). If water is added, the equilibrium favors hydrolysis, and if water is removed, the equilibrium favors ketal formation.

Oxime Formation

Chap. 8.9   Example 9 is a common mechanism with other primary amines. The OH group can be replaced with an alkyl group, NH_2, or NRR' to produce a variety of C=N products. The mechanistic steps are the same or very similar. The acidiy of the nitrogen atoms can affect the order and acids present in proton transfer reactions. For an oxime, the optimal pH is approximately 4, while the imine formation of a Wolff-Kischner reaction occurs under strongly basic conditions.

Chap. 8.10&11 These examples show how reaction conditions can control the products of a reaction. While Le Chatelier's Principle controls ketal formation, pK_a controls the addition of HCN to a ketone or nitrile. Also, the rate of addition is strongly affected by the electron donating properties of the substituents. Aromatic aldehydes are less reactive than alkyl aldehydes and aromatic ketones less reactive still.

Chap. 8.13-19 It is difficult to predict which intermediates are present in an addition reaction to a carbonyl group. Examples 13-15 face a mechanistic quandary. What charges can be present in the intermediates? The variations are detailed on pages 97-98, *Addition to Carbonyl Groups*. However, that scheme only shows addition reactions. We must consider addition and elimination reactions together.

Example 13 is mechanistically simple. It is a better choice in this example because benzoyl chloride is only moderately reactive.

At other times, you must decide which mechanism might be operating. If the base or acid acceptor is sufficiently strong, an alkoxide intermediate may be present. If the reaction is sufficiently acidic, then protonation of the acid chloride might be considered. Anhydrides are less reactive and therefore typically react under acid or base catalyzed conditions. Example 15 uses an acid catalyst. You must decide whether the reaction conditions are compatible with the intermediates found for a given reaction.

Chap. 8.14   Example 14 was particularly difficult to predict the mechanism for. Textbooks gave no guidance. I questioned whether a zwitterionic intermediate might form under the increasingly acidic conditions. Indeed, an acid catalyzed mechanism might be used. However, initially, the reaction must occur before an appreciable amount of acid can be present, therefore it seemed plausible to form the zwitterion. I rationalized its formation by imagining the effect that an oxonium ion and a chloride would have on the electrons of the alkoxide. It seemed reasonable that these substitutents would indeed stabilize the alkoxide. From this intermediate, chloride is the weakest base and should be the first group to leave. The next intermediate is a protonated ester, a stronger acid than the preceding intermediate, and readily loses its proton to the alcohol solvent.

Chap. 8.15   If you have thought about the intermediates in Examples 13 and 14, then in Example 15, you should be familair with the factor to be considered. Since pyridine is the stronger base, it can add in the first step. Then, chloride, as a weaker base then pyridine, therefore it is the best leaving group from the tetrahedral intermediate. Similarly, in the next intermediate, ethanol is a weaker base than pyridine, therefore it must be deprotonated to make pyridine the weakest base. Otherwise the addition of ethanol will reverse to the prior intermediate. The loss of pyridine from the tetrahedral intermediate give the final product. This exemplifies how pyridine can serve as a catalyst in acylation reactions.

Chap. 8.16   Because anhydrides are less reactive than acid chlorides, catalysts are commonly used to accelerate the reactions. If a base were used, a reaction mechanism similar to Example 13 would prevail. In Example 16, an acid is used to accelerate the addition of alcohol to form the tetrahedral intermediate. An intramolecular proton transfer is suggested as intramolecular proton transfers are entropically favored.

Chap 8.17-19 Example 14 is a the oxygen analog for the intermediates present in Examples 17, 18, and 19. You may need to look up the pK_a of aniline and pyridine to make the proper predictions. In these examples, the transferred acyl group is moderately reactive. Consequently, these reactions can be performed in protic solvents like ethanol or water. While these solvents can react with the acid chloride or anhydride, they do not react as fast as an amine nucleophile. Example 17 could be carried out in water with NaOH as acid acceptor under Schotten-Baumen condition.

Chap. 8.20-22 Examples 20 and 21 are the carboxylic acid analog of the ketalization reaction in Examples 7 and 8. The products are controlled by Le Chatelier's Principle. Example 22 is a convenient alternative method to convert an ester to a carboxylic acid. Because the by-products are volatile, work-up is simple as well. Note the change in mechanism.

Chap. 8.23&24 These reactions are similar. In Example 23, it is relatively easy to form the alkoxide. Its pK_a is similar to hydroxide. In Example 24, nitrogen is much more basic. Therefore, to expel a nitrogen atoms under basic conditions, intermediate **1** must be converted to the dianion **2**. Then amide anion becomes the weakest base. The alternates, methyl anion or an oxygen dianion, are not weaker bases in this case. (The pK_a's of CH_4, $BuNH_2$(or NH_3), and H_2O are 50, 36, and 15.7, respectively. H_2O is the weakest base. Therefore HO^- is the weakest base from intermediate **1**. The weakest base from intermediate **2** is $BuNH^-$, the conjugate base of $BuNH_2$).

The products of Examples 23 and 24 are a carboxylic acid and a neutral compound. If you wish to isolate the neutral compound, it should be extracted before acidification. Similarly, if the alcohol is not volatile or water soluble, failure to remove it before acidification will result in its presence with the carboxylic acid.

Chap. 8.25   Example 25 should be compared to Example 24. In Example 24, a dianion had to be formed. Its concentration is likely to be low unless higher concentrations of base are present. In Example 25, the nitrogen atom will make protonation more facile. Then, in the intervening neutral intermediate, the nitrogen atom will be more basic and protonation will facilitate its removal.

Reactions of Esters

Chap 8.26   Mechanistically, Example 26 is simple. However, there are two carbonyl compounds of differing reactivity present during this reaction. What is important to notes is that the starting ester is less reactive than the intermediate methyl ketone. Therefore if less than 1 equivalent of Grignard reagent is used, then unreacted ester will be found as a by-product.

Chap. 8.27   Esters are slow to react with an amine. Electron withdrawing groups attached to the ester, such as trifluoacetates, increase the rate of reaction. In Example 27, a *p*-nitrophenyl ester is used to make the oxygen a weaker base and a better leaving group. The mechanism should be similar for other amine-ester combinations.

Reactions of Nitriles

Chap. 8.28-30   Hydrolysis of nitriles produce amide intermediates which may react further depending on reaction conditions. In Example 28, it is difficult to isolate an amide from the hydrolysis of a nitrile under acidic conditions because the pK_a of the conjugate acid of an amide is greater than that of a nitrile. Therefore, any amide that forms, is protonated more easily than the starting nitrile. As a result, the nitrile will be converted completely to the carboxylic acid. In this respect, the reaction is similar to a Grignard reagent adding to an ester where the intermediate is more reactive (basic) than the starting material. It is possible to stop an acid catalyzed nitrile hydrolysis at the amide stage by limiting the amount of water.

Under basic conditions, the hydrolysis can stop at the amide stage, see Example 29. Because a dianion intermediate is required for an anion of ammonia to form as a leaving group, the reaction conditions must be more vigorous. For the mechanism of an amide hydrolysis to a carboxylic acid, see Example 24.

The reaction of a nitrile with a Grignard reagent results in a ketone because a Grignard reagent can only add one equivalent, Example 30. The intermediate will resist increasing the negative charge on the nitrogen that would result from addition of the second equivalent.

Miscellaneous

Chap. 8.31   This reaction is an alkylation of a carboxylate anion and should be an unusual alkylation method. Because it is often included with carbonyl chemistry, I have placed it here.

Chapter 9   Reactions of Enols and Enolates

Aldol Reaction

Chap. 9.1   Example 1 is a self-condensation reaction. This leads to some confusion as we may not think of what species might be present. The reactants are in equilibrium, therefore an enolate and aldehyde are both present and are able to react with one another.

The dehydration step is frequently accomplished by heat, strong base or strong acid. I infer from heat or excess base to expect a dehydration reaction to occur. Similarly, low temperature or a very low base concentration are often a signal that the dehydration will not occur, see Example 2. The dehydration is an E1cb reaction, see Chapter 4.

Chap 9.2   There are two steps in the directed aldol condensation reaction. In the first, is a kinetic deprotonation of 2-methylcyclohexanone. You may choose either of these arguments. The kinetic deprotonation removes the less sterically hindered hydrogen. Alternately, since I have argued carbon is a better electron donor (p. 1B), then deprotonation occurs on the more acidic hydrogen. However, the kinetic product is not the most stable enolate. The greater the degree of substitution, the more stable it becomes. If an excess of ketone is present, then the thermodynamic enolate results.

Upon addition of the aldehyde, the product has three chirality centers. While there is a major product that results, I have not indicated the stereochemistry.

Chap. 9.4   In the Mannich reaction, two different acids are represented in the mechanism. In Step 1, formation of the iminium salt, a protonation of an alkoxide is shown. The acid for that reaction is an ammonium salt. Later, the intermediate alcohol is protonated with hydronium ion. The Mannich reaction is conducted under mildly acidic conditions. I interpret this to mean that nearly all of the nitrogen atoms will be protonated as ammonium ions. Since they are greatest in concentration, this will be the most prevalent acid. In order to protonate the alcohol, a small amount of hydronium ion may be necessary.

The next problem is the enolization of the ketone. The pK_a of acetone and its conjugate acid are widely spaced (20 and -8).

The Mannich reaction is generally conducted with the amine hydrochloride salt plus a catalytic amount of additional acid. Therefore, I am presuming the catalytic amount of acid is sufficient to protonate some ketone and thus form some enol. The acid concentration cannot be so large as to prevent deprotonation of the ammonium salt necessary for the addition to formaldehyde in formation of the iminium salt.

Claisen Condensation

Chap 9.5   Example 5 is another self-condensation reaction. It requires an excess of sodium ethoxide.

Look up the pK_a of ethyl acetate and ethanol. In which direction would the following equilibrium lie? What effect will that have on the amount of enolate you would find in the reaction mixture?

You will need to determine the equilibrium of the following reaction as well. How will this equilibrium affect the stoichiometry of the reaction? The intermediate is more acidic than the starting materials. Unlike the aldol condensation, the Claisen condensation requires an excess of base because the intermediate consumes base and the concentration of enolate will be low because the equilibrium favors the starting material.

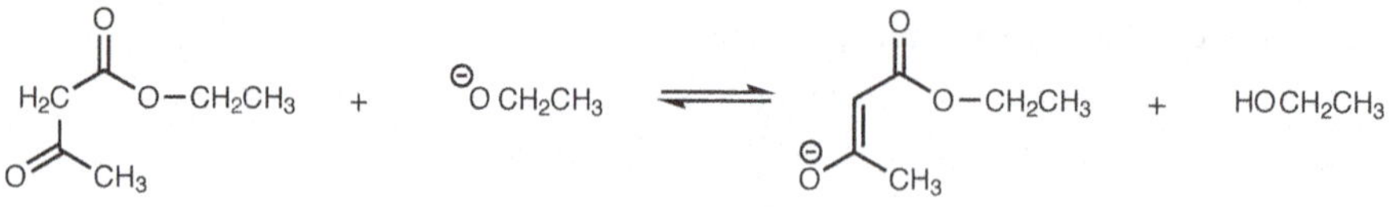

Crossed-Claisen Condensation

Chap. 9.6   Example 6 is a crossed-Claisen condensation. Note that ethyl formate does not have any enolizable hydrogens, therefore it cannot form a nucleophile. With the ketone, methylcyclohexanone, two possible condensation products can and do form. However, the hindered product cannot enolize and therefore the reaction reverses to reform the enolate. Example 11 is also a reverse Claisen reaction.

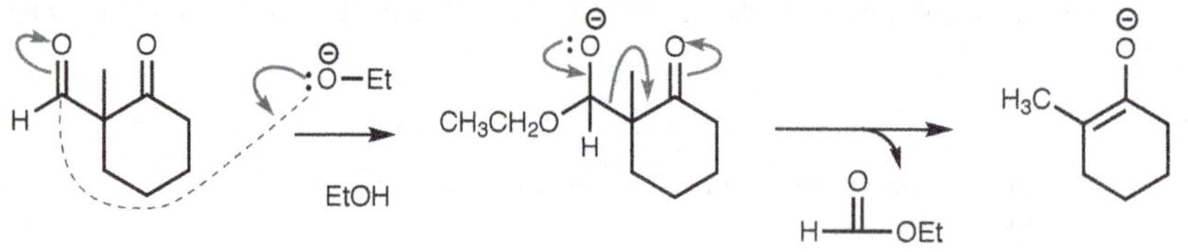

Finally, the neutral product prefers the enol form of the aldehyde. Aldehydes are easily enolized and the result is conjugated to the ketone.

Chap. 9.9-12   Ester enolization reactions are extended to simple alkylation reactions, dianion alkylation, and a nitrile alkylation reaction. In Example 9, a simple alkylation reaction occurs and in Example 10, a dianion is alkylated. The first alkylation reaction occurs on the more basic carbon. That anion is stabilized by one carbonyl oxygen. The second reaction forms the selenide used for the cis-elimination in Chapter 4. Example 11 also has an enolization reaction, but it is formed because the tetrahedral intermediate can give an enolate. The ester (enolate) is less acidic than ethanol, but if formed, the β-ketoester does not reform.

Halogenation of Carbonyl Compounds

Chap. 9.13   In Example 13, you must know what effect the bromine atoms have on the acidity of the ketone hydrogens. Which is more acidic, the starting ketone or the monobrominated (or dibrominated) ketone? Because the product is more acidic than the starting material, it is more rapidly deprotonated than the starting material. As a result, tribromomethyl forms faster than monobromomethyl. Since a tribromomethyl anion is more stable than a methyl anion (see p. 1B), it can form from cleavage of the carbonyl-hydroxide tetrahedral intermediate.

Chap. 9.14   In Example 14, the acid catalyzed bromination, where does HBr come from? If bromine reacted with acetic acid, the products HBr and CH₃COOBr. Also, what are the by-products of the bromination reaction? Answer, HBr. The rate-limiting step of the acid catalyzed bromination is the enolization of the ketone. Furthermore, since bromine removes electron density from the carbonyl group, the starting material is more basic than the brominated product. Therefore, monobromination is more easily done under acidic conditions.

Michael Addition Reaction

Chap. 9.15&16   Examples 15 and 16 are conjugate addition reactions. I rationalize these reactions as two competing modes, a fast addition at the carbonyl (one pair of electrons) and a slower one at the double bond (two pairs of electrons). If tetrahedral adducts **3** or **4** were to form and the nucleophile was a weak base, then either could reverse to reform the starting materials.

If a 1,4-addition reaction took place to form **5**, it too could reverse. However, if a protonation reaction also occurred to give **6**, it is less acidic than **4** and less able to reform starting materials. If the nucleophile were a weak base, it would not be basic enough to regenerate the enolate to reverse the reaction. If the nucleophile is more basic than an alkoxide, then the tetrahedral intermediate **3** will be the most stable product because it will be the weakest base. The best nucleophiles for Michael addition or conjugation addition reactions are less basic than an alkoxide. If we considered the enone (MIchael acceptor), if an alkyl groups is attached to the alkene, because they donate electrons, will also slow conjugate addition reactions.

In Example 16, only a catalytic amount of ethoxide is required for the reaction. Therefore, only a portion of the final product would exist as the indicated anion and as a result, the acidification step could be skipped with little effect on the reaction yield.

Enamine Formation

Chap. 9.17   The mechanism of this reaction parallels other nucleophilic additions to carbonyl groups. A nitrogen atom is better able to add to a carbonyl group and thus does not need a catalyst. Catalysis is likely involved in the dehydration steps. It is similar to the optimal pH for oxime formation. If the reaction pH were too low, the amine would be complexed by the acid. If the pH were high, the oxygen must be lost as hydroxide. In between, the oxygen can be lost as water. You should also note that because only a catalytic amount of acid is present, the predominant acid in the reaction will be an ammonium salt. Even though the equilibrium will favor the ammonium salt over the oxonium salt, this acid must protonate the oxygen to a small extent and allow the reaction to proceed. After water is lost, the product is an iminium salt, the nitrogen isostere of a carbonyl group. If R is H, then deprotonation occurs to give an imine and if R is not H, then the alkyl hydrogen is lost and the enamine results.

Enamine Alkylation and Acylation

Why does an enamine react on carbon rather than the nitrogen atom? A key is to draw the resonance structure of the enamine. The resonance structure has a negative charge on the carbon. On which atom will the electrons extend the furthest? In which form would you predict the electrons are best able to react? However, many reactions may involve some direct reaction on the nitrogen atom. If it is an acylation reaction, that product may be reversible and leads to the same final product.

Chapter 10   Dehydration Agents

This chapter will deal with reagents that result in a net removal of water in which the oxygen atom becomes absorbed by the reagents. For example, one can perform an S_N2 nucleophilic substitution of an alcohol in the presence of HBr to give an alkyl bromide and water, see Chapter 3. In this chapter, we will see similar reactions, but water will not be formed as the leaving group.

Chap. 10.1   In Example 1, the dehydration of an amide with acetic anhydride is one of many reagents for this conversion. The mechanisms are fundamentally the same. They start with a complexation of the oxygen to make it a better leaving group. I like using acetic anhydride because the reaction mixture is simply heated and the by-products can be distilled off.

The site of reaction is the oxygen atom rather than the nitrogen, see *Amide Regiochemistry*, p. 2-*Notes*.

Chap. 10.2   This is one of two possible mechanisms for thionyl chloride reactions with alcohols. Pyridine is shown as a base in the reaction, but it also increases the chloride ion concentration and thus the S_N2 substitution reaction rate. In the absence of pyridine, then the chloride may react from the same side, an S_Ni reaction. Substitution from the same side is thought to be an unfavorable reaction and as an alternate, it is thought that an ion pair mechanism may operate. With a chiral alcohol and pyridine, a chiral chloride of the opposite configuration results. Without pyridine, a chiral chloride of the same configuration results.

Chap. 10.3   Example 3 is similar to the thionyl chloride reaction, Example 2. With toluenesulfonyl chloride, the product is a sulfonate ester rather than a chloride. Why doesn't the chloride ion by-product displace the tosylate ester? The chloride reacts more slowly and if the reaction time is extended, chlorobutane will accompany the tosylate.

Reaction of a Carboxylic Acid with Thionyl Chloride

Chap 10.4&5   Example 4 is the classic reaction of thionyl chloride with a carboxylic acid. Example 5 is one of the best ways to make an acid chloride. It only requires a catalytic amount of DMF.

Halide from an Alcohol with a Phosphorus Reagent

Chap. 10.6   Example 6 is probably too simple. Below is a more likely mechanism(s) for this reaction. You should note the boxed structures. A similar structure is present in the Arbusov reaction with a trialkyl phosphite (Chapter 8, Example 6, Step 1). Here it reacts with HBr to give an alkyl bromide and phosphorus acid. Therefore, formation of a phosphorus salt is a plausible mechanism.

Chap. 10.7   Example 7 is a convenient and effective chlorination reaction. Carbon tetrachloride can be substituted with chlorine or NCS for chlorination reactions and with carbon tetrabromide, bromine, NBS and probably others for bromination reactions.

Chap. 10.8   The Mitsunobu reaction extends the phosphorus chemistry in an interesting way. Two protons are absorbed by the azo-compound and the oxygen becomes attached to the phosphorus for a net loss of water. The displacement is an S_N2 substitution reaction resulting in inversion of the alcohol.

Chapter 11   Reduction Reactions

Chap. 11.1   In borohydride reductions, I illustrated the solvent, methanol, as part of the reaction cycle. Each reduction generates a borane product which acts as a Lewis acid. Reaction with the solvent regenerates a borohydride reductant in which the hydrides are replaced with methoxide or an alkoxide. Thus, all of the hydride hydrogens become replaced.

Lithium Aluminum Hydride Reductions

Chap. 11.4   In lithium aluminum hydride reductions, a cycle similar to borohydride reduction occurs except that since an aprotic solvent is used, the trivalent aluminum must react with any of the Lewis base species present in a reaction cycle. In Example 4, the R-group could be an ethoxy or butoxy group as both are present in the reaction mixture.

Chap. 11.5&6   In the reduction of a carboxamide, Examples 5 and 6, an R-group can form from any of the alkoxy (or amide, Example 6) groups. Also, an oxygen atom must be removed to get to the final amine product. The preferred form in which it can be expelled is shown with the example. I think this might be the most difficult step in the reaction and reductions which involve loss of a complexed oxygen are more sluggish than other LAH reductions. For example, references often show the use of an excess of $LiAlH_4$ to give a successful reduction of an amide. Example 7 begins with formation of hydrogen. I needed an example to remind students of this reaction with an amide and carboxylic acid.

Chap. 11.7   Notice the first step of the lithium aluminum hydride reduction in Examples 6 and 7 is abstraction of a hydrogen. As lithium aluminum hydride reacts at the start of a reaction, it reacts similarly with water (or alcohol) and therefore the entire reduction cycle must be carried out in the absence of moisture. Reductions are carried out in aprotic solvents such as ether, tetrahydrofuran, or dimethoxyethane. The release of hydrogen is also evident from unreacted aluminum hydrides during work-up. Lithium aluminum hydride will react violently with water or low molecular weight alcohols.

Reductive-Amination

Chap. 11.8   The reductive-amination reaction is written as two separate reactions, which it is. However, both reactions can be carried out at the same time. The reactions are compatible with one another and are frequently carried out in a single flask. The imine, as it forms, can be reduced.

Triacetoxyborohydride is one of several reductants that one can use. Others are sodium cyanoborohydride ($NaBH_3CN$) or hydrogen with a palladium or nickel catalyst.

In step 1, the aldehyde is converted into an imine. I've written this as a stepwise process, however others have suggested an intermolecular protonation-deprotonation.

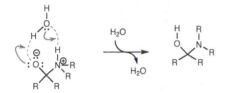

Chap. 11.11   You must use a different arrow for single electron transfers than normal. While most reactions involve reactions with two electrons, this reaction has two single electron transfer steps. They are noted with the single barbed arrow.

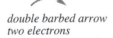

double barbed arrow
two electrons

single barbed arrow
single electron

Chapter 12   Oxidation Reactions

General form for oxidations

The key step in an oxidation reaction is to break an oxygen-oxidant bond (O-X) such that the X-group accepts the electrons, Equation 1. The group X can be a halogen, oxygen, or a metal, frequently Cr, Mn, Ag, and others.

(1)

The first step to an O-X bond formation for monovalent oxidants (Br, Cl, Ag), is a simple displacement reaction, Equation 2. Those reactions are simple and generally don't need further discussion.

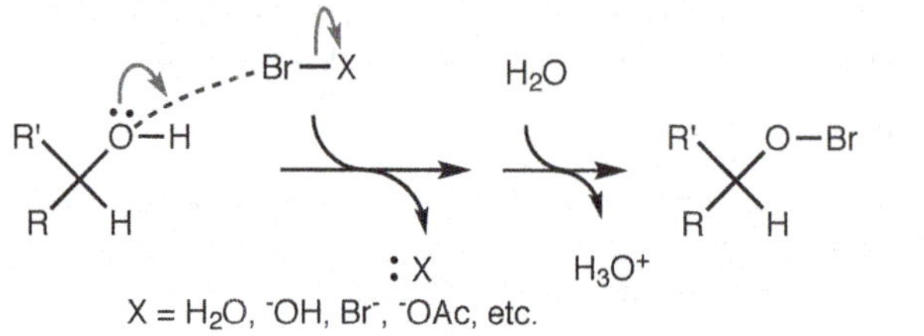

$$X = H_2O, \ ^-OH, \ Br^-, \ ^-OAc, \ etc. \tag{2}$$

For polyvalent oxidants, formation of an O-X bond is lengthier. It is usually a series of proton transfer reactions. However, these proton transfer reactions often distract from the oxidation mechanism itself. Therefore, I am going to display the principles of those reactions, some variances to consider, and an example or two. I will leave it to you to devise plausible steps to complete the reaction.

In the scheme below, four general routes to form an oxidant-ester are shown. The routes differ by the timing of proton transfer reactions, concurrent, zwitterionic, early protonation, or early deprotonation. Route **A** is electronically neutral, but places a high entropic requirement on the reaction. Route B in an ionic route and one must determine whether these intermediates are consistent with the pH requirements of the reaction. Routes **C** and **D** are low and high pH mechanisms. Because these reactions precede the oxidation step, they are fast and not rate limiting, their mechanisms are not known. Little literature data exists to guide us in determining their route. Therefore, the steps showing the formation of the esters of the oxidants are not described.

Route A

Route B

Route C

Route D

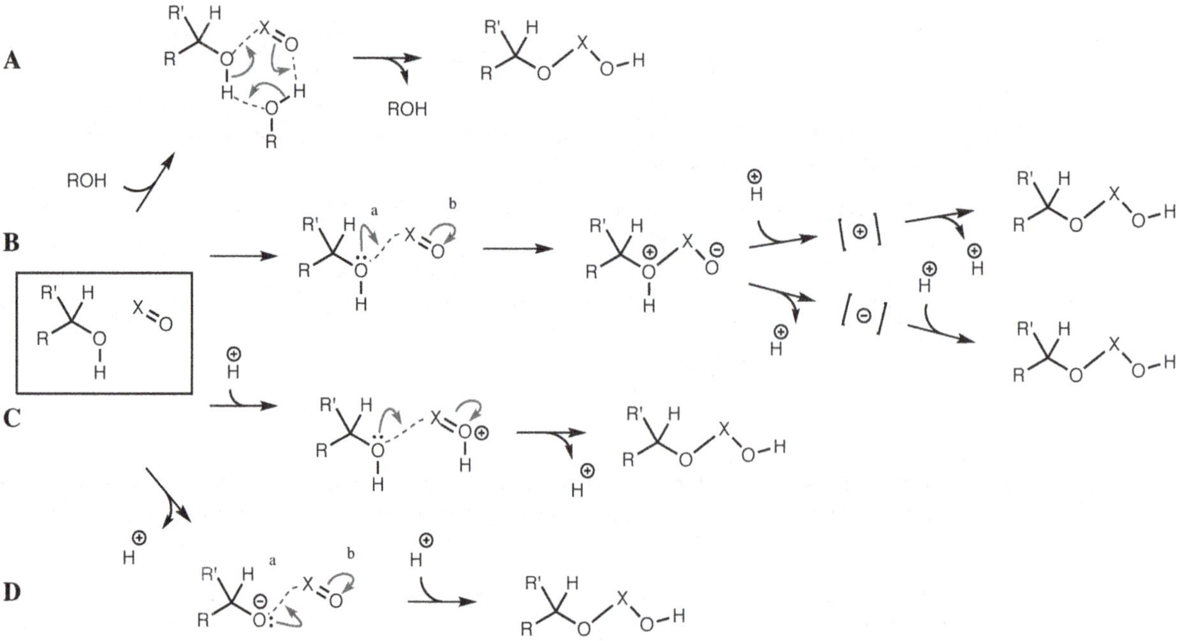

Overall, any reactions that can be written as shown below or contains similar structures may form by a series of proton transfer steps. These reactions take place in protic solvents and are therefore capable of following any of the mechanisms described earlier.

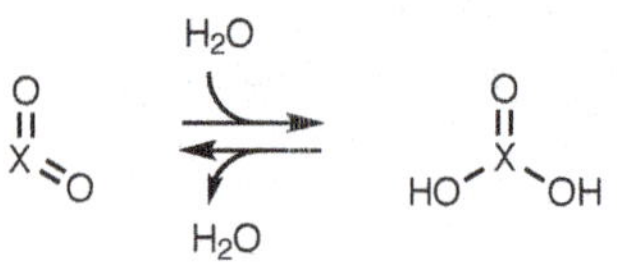

Now that we have discussed the proton transfer reactions that take place in an oxidation reaction, let's move to the oxidatin step, the second step of the reaction. In that step, a pair of electrons is transferred to the oxidant. For chromic acid oxidations, that step is rate limiting. The base could be water, but other bases are possible or it may be an intramolecular reaction.

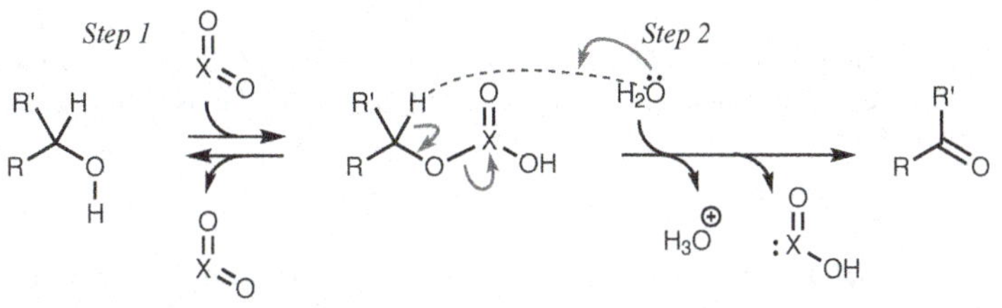

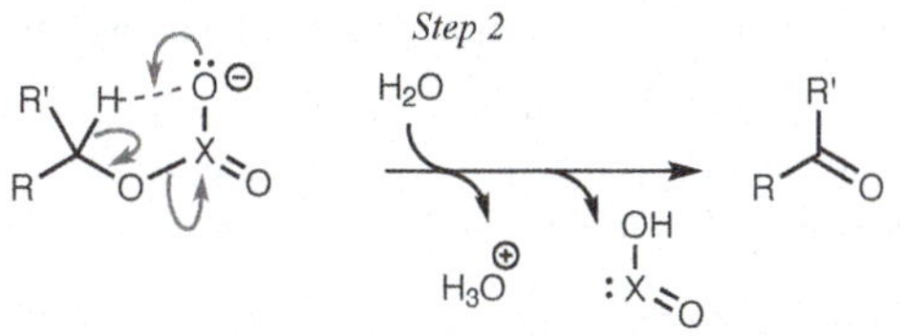

In order to simplify oxidations, the oxidant-esters formation step will not be shown. We will concentrate on the oxidation step. You should suggest a plausible route for the formation of the oxidant-ester. You may consult your textbook, your instructor, or contact me for additional help.

Because these oxidation reactions form metal esters by addition to the metal oxide, I have written the structures of several common oxidants and their hydrated forms. The hydrated form is the product of water addition and any oxidation may be thought to involve either of those forms (unless there is experimental evidence indicating otherwise).

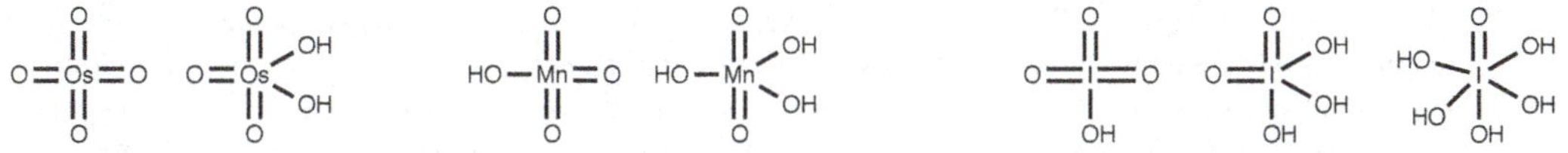

osmic acid   permanganic acid   periodic acid

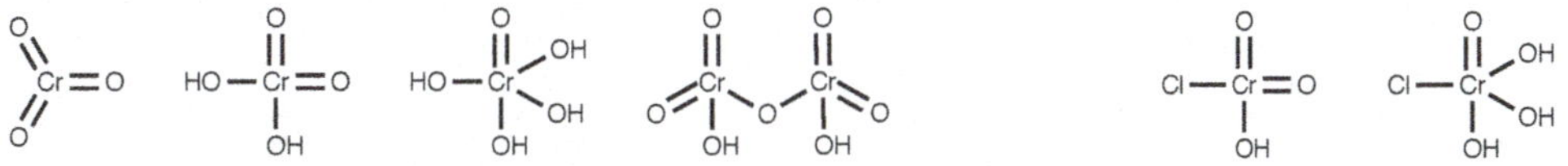

chromic acid   chlorochromic acid

Chap. 12.1   In Example 1, the proton transfer steps have been combined. In addition, the chromate ester is written in the hydrate form. This reaction is a compromise to make the steps appear logical, to highlight the actual oxidation step itself, and to not result in a lengthy series of proton transfer steps. I thought the hydrate form was in keeping with other oxidations in this chapter, though it may differ from your textbook. If you wish, you may write additional proton transfer steps to effect a loss of water to form an anhydro-ester.

A variation that I think is also plausible, is shown below.

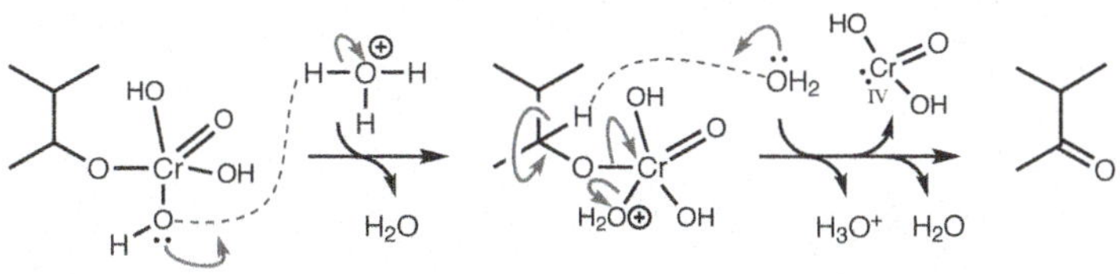

Chap. 12.2   In Example 2, the oxidation step is exactly the same as Example 1. However, the first product, the aldehyde, can add water to the carbonyl group and form a hydrate. (You will need to complete the proton transfer steps.) The hydrate is another alcohol with a geminal hydrogen than can be removed. An oxidation step is repeated to give a carboxylic acid.

In order to isolate an aldehyde, an oxidation that occurs in the absence of water must be performed. The PCC and Swern oxidations, Examples 3 and 6, can be carried out in anhydrous conditions and therefore result in aldehydes from primary alcohols.

Chap. 12.6   The challenge of writing mechanisms in discreet steps is imprudent with the epoxidation reaction. Thus, you see four pairs of electrons moving concurrently in this mechanism.

Chap. 12.7&8 A Swern oxidation, Example 7, represents a common laboratory method to perform oxidations. For a Swern oxidation, oxalyl chloride is used to form the chlorosulfonium salt. This is one of several methods to form an equivalent sulfonium salt. Example 8 represents another method.

After the reaction of the sulfonium salt with an alcohol, the oxidation steps converge. A fraction of the deprotonation/oxidation step may also occur directly with the triethylamine base. Another alkyl group may be used to avoid the malodorous dimethyl sulfide.

Ozone Oxidation

Chap. 12.9   An ozonide is actually a peroxy diacetal. For reductive Step 2, two goals are accomplished, 1) reduction of the O-O bond and 2) conversion of the acetals or ketals into two carbonyl groups. In alternate Step 2, the O-O bond is used to convert an acetal to a carboxylic acid. If a dicarboxylic acid could be formed, then an additional mole of hydrogen peroxide would be required.

Since the oxidation of an ozonide is related to the Baeyer-Villiger oxidation (Chapter 6.2 and 6.3), Baeyer-Villiger products are frequent by-products. It is sometimes advantageous to separate the second oxidation step to avoid such by-products, that is, first do a reductive work-up, isolate the aldehydes and ketones, and then oxidize the aldehyde in a separate step.

Osmium Tetroxide, Potassium Permanganate, and Periodate Oxidations

Chap. 12.10-12   These reagents are similar in structure and function. They differ in that osmium tetroxide does not cleave the C-C bond, periodate only cleaves the C-C bond, and permanganate can do both. Permanganate can be intercepted by hydrolyzing the intermediate cyclic ester. Excess hydroxide and low temperatures are signals for that hydrolysis and isolation of the cis-hydroxylation product.

Chap. 12.11   The hydrolysis of the manganese ester in the first oxidation will generate a dialkoxide. Protonation with water will regenerate the two hydroxide ions and balance the charges.

Permanganate can also oxidize an alcohol or a hemi-acetal (the hydrate of an aldehyde). In the cleavage of a double bond, if the double bond is not tetra-substituted, then the oxidation will continue to a carboxylic acid. The addition of water to the aldehyde, proton transfer steps, and manganese ester formation steps are not shown, and only the oxidation step is shown. Hydroxide will form the salt of the carboxylic acid, therefore, the final step of a permanganate oxidation, if heated, must be neutralization of the potassium salt of the acid.

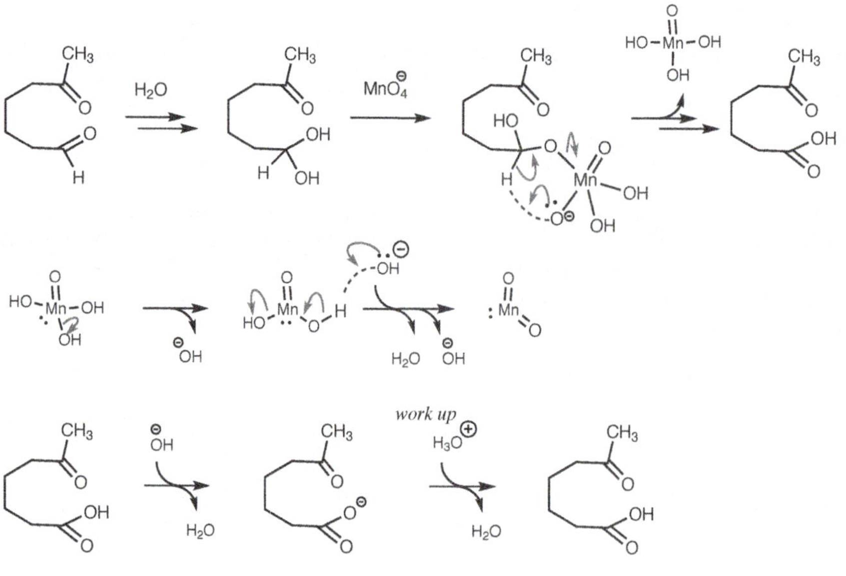

The final disposition of manganese (and other metals) is more complex than shown here. Potassium permanganate can interact with manganese intermediates as well as auto-decompose when heated.

Chapter 13   Organometallic Reactions

Organometallic chemistry often involves reactions in which the mechanism is not understood, incompletely understood, or complex. In some reactions, I show how it *might* take place.

Chap. 13.1   The steps in the Heck reaction in Example 1 are fairly well understood. The Heck reaction is a very useful reaction because it only requires a catalytic amount of palladium, the reaction conditions are very mild, and a wide variety of functional groups are compatible with the reaction. We will study two similar examples, reactions of acyclic and cyclic alkenes. Before the coupling reaction takes place, zero valent palladium metal, the active catalyst, must be formed in a sub-reaction. Palladium acetate reacts with an alkene (much like mercuric acetate, but in an *anti*-Markovnikov orientation). The intermediate is unstable and loses hydridopalladium acetate, which in the presence of base forms an ammonium acetate and zero valent palladium (Pd⁰). Zero valent palladium is commercially available and when used, this sub-reaction is skipped.

The first step of the coupling sub-reaction is insertion into a carbon-halogen bond, in this instance a carbon-bromine bond. This reaction is similar to the formation of a Grignard reagent. The rest of the process is similar to the palladium reduction sub-reaction. The organopalladium compound inserts into an alkene double bond. In order for an elimination to occur, the bonds must rotate. Elimination gives the coupled product and hydridopalladium bromide. The hydridopalladium bromide can lose hydrogen bromide and regenerate palladium zero.

Also important to the reaction are the ligands that donate electrons to palladium. Palladium will form an 18-electron intermediate, therefore the ligands, although not shown, are vital to the success of the reaction. The electrons of an alkene presumably replace the electrons of one of the ligands that coordinate with the palladium.

Catalytic Reduction of an Alkene

Chap. 13.3   This mechanism is hypothetical. I extrapolated from the Heck reaction to similar mechanisms and other metals. This mechanism can broadly explain formation of the products and by-products of catalytic reduction reactions. It also explains how *trans*-fatty acids can be produced.

Gilman Reagents

Chap. 13.4   The mechanism in Example 4 is hypothetical. The mechanism of the Gilman coupling is not known. I've written it to broadly agree with how these reagents *might* react. The groups couple without losing the stereochemistry. While I don't know how that might occur, I anticipated that just as zero-valent palladium could insert into a carbon-halogen bond, so too might a copper insert into a carbon-bromine bond. If it did, then the intermediate on the right would occur. The decomposition of this intermediate parallels that of the palladium (Heck) and mercury (reduction step of oxymercuration) reactions.

If you consider the properties of reagents that break an alkyl, alkenyl, or aryl bromide bond, they are all able to donate electrons. In a Gilman reagent, one can presume that replacing an iodide of copper iodide with a methyl group would increase the electron density on a copper atom. If another methyl group were also added to the copper, then it is again plausible that the electrons on copper would be even more available. Now, without explaining the mechanism of the reaction, each of these reagents, Gilman, magnesium, lithium, and palladium, donate electrons to the carbon-halogen bond in breaking it and leading to a net insertion.

If you look back to the de-mercuration step of the oxymercuration reaction (page 25) this is reported in the literature to be a radical reaction. The decomposition of the alkylcuprate can also be written as a radical reaction. However, I like to write this as a two-electron process as it is similar to other reactions in this book, e.g., oxidation of boranes, Baeyer-Villiger oxidation, loss of PdH₂, and others. I like the two-electron mechanism as this nicely parallels these other reactions. By using a parallel, they are more plausible, easier to understand, remember, and use (which is the objective of this book).

Chapter 14   Aromatic Substitution Reactions

Chap. 14.2   Friedel-Crafts alkylation reactions with primary halides occur with extensive rearrangement.

Electrophilic substitution of substituted aromatic compounds

Chap. 14.5-10   The resonance structures of the reactants can be used to predict whether the atoms of a benzene ring will be **more** or **less** electron rich in an electrophilic aromatic substitution reactions. First, draw the resonance structures for the substituted benzenes, shown below. Examine the resonance structures and answer; "*Will the compound react faster or more slowly than benzene in an electrophilic substitution reaction (such as bromination or nitration)? Where on the molecule will the highest electron density be found? At the ortho-para or meta position?*"

I like this method of examining the resonance structures of the reactants to predict the regiochemistry of the reaction. I believe the reaction profile is more consistent with the slow step being addition to an aromatic ring (and disrupting aromaticity). I am also more satisfied by how the resonance structures predict reactivity because those which add electron density to the ring will be more reactive and those which remove electron density are less reactive. However, halogens do not follow that pattern in that they contribute electrons by resonance, but are less reactive. However, this is consistent with halogens having non-bonded electrons to donate, but because of their electron withdrawing properties, are reluctant to do so.

- Electron donating groups (usually, X = N, O, halogen), *ortho-para* directors

- Electron withdrawing groups (usually Y = C, N, P, or S; Z = N, O, or S), *meta* directors

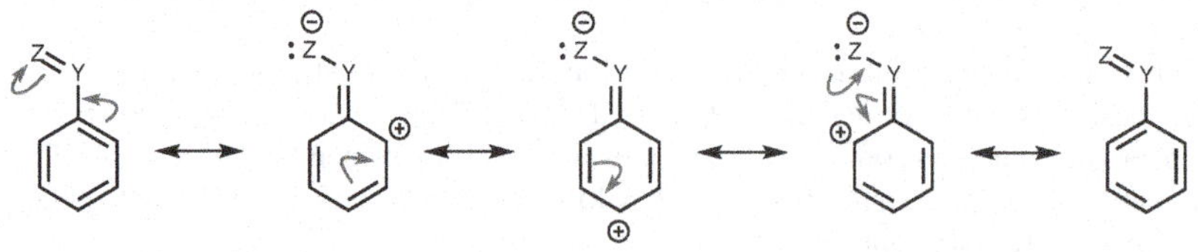

- Anisole, *ortho-para* director

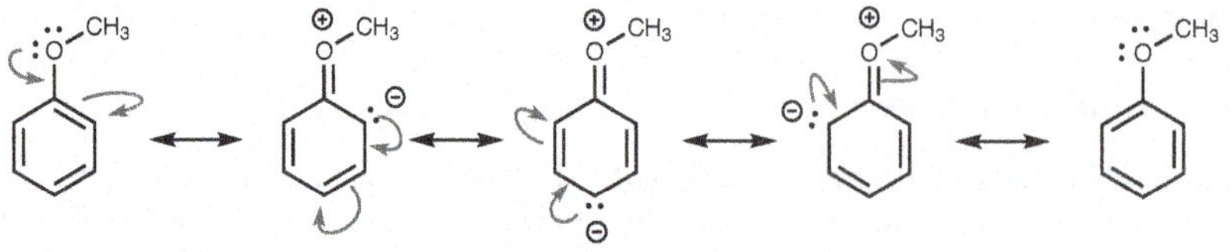

- Acetophenone, *meta* director

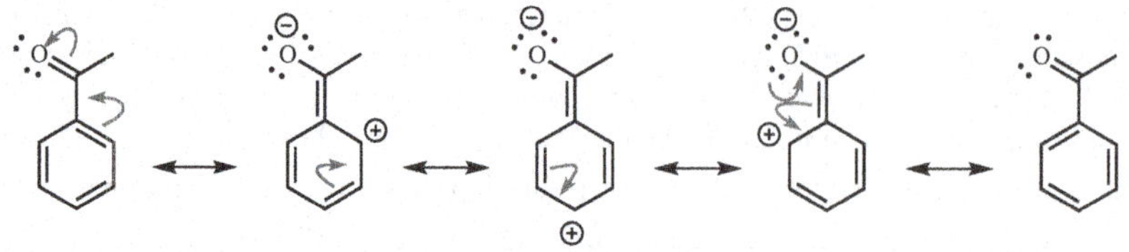

- Toluene, weak *ortho-para* director

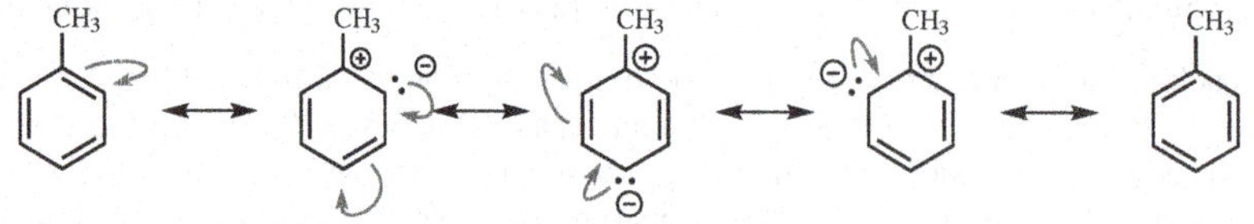

- Benzonitrile, *meta* director

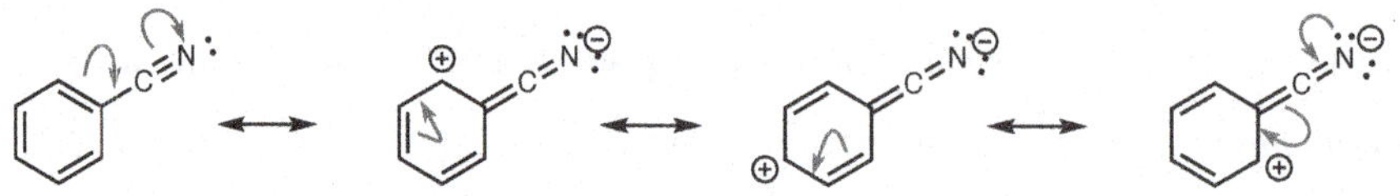

Aromatic Substitution on Multi-substituted Aromatic Rings

If a multi-substituted benzene reacts, the **activating substituents have the greatest effect on determining the regiochemistry of the products**. In the methoxybenzoic acid shown below, the methoxy group controls the substitution pattern whether the carboxy group is *ortho*, *meta*, or *para* to it.

- *p*-Methoxybenzoic acid, *ortho-para* director to CH₃O and *meta* director to COOH

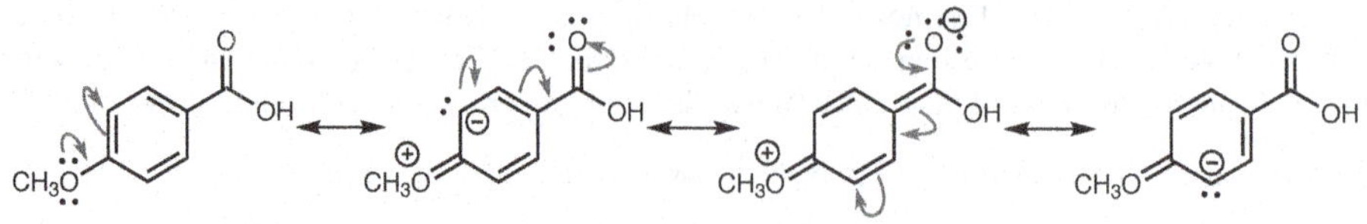

- *m*-Nitroanisole, *ortho-para* director to CH$_3$O and *meta* director to NO$_2$

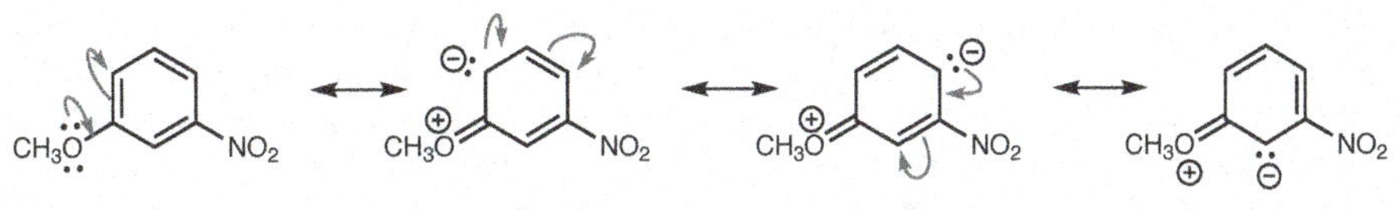

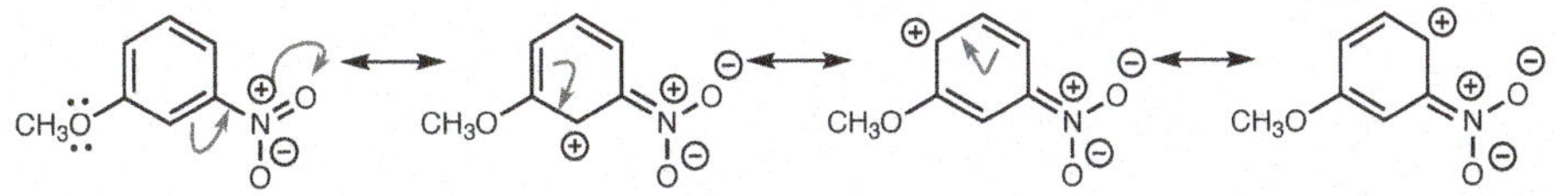

In *m*-nitroanisole, the CH$_3$O-group is an activating and an *ortho-para* director but because of the nitro group, it is of reduced reactivity and regioselectivity. The upper series of resonance structures show where electron density will be increased and direct substitution to those positions. The lower series of resonance structures shows the effect of the nitro group. It will <u>not</u> control the regiochemistry, but it will decrease the electron density (and thus reactivity) created by the methoxy group.

Chap. 14.10   In Example 10, the reaction of the second equivalent of chloromethane and aluminum chloride, no reaction occurs in the *p*-nitrobenzamide ring. Since Friedel-Crafts reactions are the least reactive of electrophilic reagents, no reaction occurs in the ring with the two electron withdrawing groups.

Nucleophilic Aromatic Substitution

Chap. 14.11&12   If you react a substituted aromatic compound with a nucleophile rather than an electrophile, a different reaction mechanism ensues. The nucleophile may add to the aromatic ring provided the aromatic ring contains enough electron withdrawing groups. The electron withdrawing groups must be able to absorb the negative charge of the nucleophile. The reaction is an addition-elimination reaction in a manner similar to an acid chloride. The most effective electron-withdrawing group is a nitro group, but other groups will work at slower rates.

Benzyne Reaction

Chap. 14.13   No reaction will take place if you react a nucleophile with a halobenzene unless there is a strong electron withdrawing group present. Then an addition-elimination reaction can take place. However, if the nucleophile is a sufficiently strong base, a deprotonation may take place that can result in an elimination reaction to form a benzyne intermediate. The benzyne, because the triple bond is in a six-membered ring, is very reactive. Thus, a nucleophile can add to either side of the triple bond and the regiochemistry of the original halobenzene becomes lost.

Diazonium Chemistry

Chap. 14.14   The diazotization reaction in Example 14 is not as complicated as it may look. It has six proton transfer reactions (acid-base chemistry), two dehydration reactions, and a condensation of NO$^+$ with aniline. One of the steps is like an acid catalyzed enolization except the carbon atoms are replaced by two nitrogen atoms. However, the most frequent error that I have noted has been due to an incorrect Lewis structure for sodium nitrite.

Chapter 15   Carbene and Nitrene Reactions

Carbene Reactions

In a simplified model of carbene chemistry, there are two types of carbenes, singlet and triplet, and they have different rates of reactivity and stereochemical outcomes. In general, singlet carbenes are more reactive and experience less loss of stereochemistry than an equivalent triplet carbene.

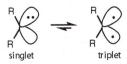
singlet   triplet

The two forms can interconvert. The singlet is closer to a carbocation in its characteristics. It has a vacant orbital which it fills upon reaction. A triplet has two half-filled orbitals. When it reacts with an alkene, it reacts as a diradical and can lose its stereochemistry. Often, the triplet is the more stable form.

Chap. 15.1   The Simmons-Smith reagent has great variety in it reactions. Your textbook may use a concerted addition of the carbene or carbenoid group to an alkene. I prefer this alternate mechanism as it does not violate the symmetry rules for a 2+2 reaction, and it follows the pattern of bromine, mercury, or borane additions to an alkene. I also like it because it can explain the reaction of a Simmons-Smith reagent as a nucleophile and then as electrophile (with loss of iodide in the last step) in other reactions (not shown),

Chapter 16   Radical Reactions

I placed radical reactions in the last chapter. While radical reactions should not be thought of as unusual, you should also note that the driving force for their reactivity is to form paired electrons. While the remainder of the book has been about reactions of paired electrons, that is electron pairs being attracted to nuclei, I wanted radical reactions to be in a chapter of their own. Electrons being attracted to electrons to form an electron pair is a different kind of attraction and reaction. Also, single electron movements are designated differently from electron pair movements, they use a single barbed arrow. You should consult your textbook for information on rates and selectivity of radical reactions.

double barbed arrow
two electrons

single barbed arrow
single electron

Chlorination and bromination reactions are mechanistically the same, though they have different selectivity.

Allylic Bromination with NBS

Chap. 16.2   Several catalysts are capable of initiating radical reactions. Benzoyl peroxide and azobisisobutyronitrile (AIBN) are commonly used. While the reaction generates an oxygen radical to start the reaction, the unpaired electron becomes transferred to a bromine radical. The following is a possible path to generate a bromine radical from an oxygen radical.

Radical Addition of Hydrogen Bromide

Chap. 16.3   Why does NBS lead to an allylic bromination while a radical addition of HBr leads to addition? Both are reactions of a bromine radical with an alkene. If addition of a bromine radical is the expected product, how can one get allylic bromination (eq 2) to occur if the addition of a radical to a terminal position (eq 1) is a faster reaction? If the addition of the bromine radical to a terminal alkene is a fast but reversible reaction, then its success would be aided by quenching the intermediate radical with HBr. If the concentration of HBr is low or absent, then the reaction may reverse. Then a competing abstraction of the allylic hydrogen in a bromination reaction might occur. The use of *N*-bromosuccinimide is effective as the HBr generated by the allylic bromination is consumed by NBS to give succinimide.

(1)

(2)

CPSIA information can be obtained
at www.ICGtesting.com
Printed in the USA
LVOW03s1253060216

474013LV00006B/228/P